T0214309

Lecture Notes in Artificial Intelligence 12179

Subseries of Lecture Notes in Computer Science

More information about this series at http://www.springer.com/series/1244

Rafael Bello · Duoqian Miao ·
Rafael Falcon · Michinori Nakata ·
Alejandro Rosete · Davide Ciucci (Eds.)

Rough Sets

International Joint Conference, IJCRS 2020
Havana, Cuba, June 29 – July 3, 2020
Proceedings

Springer

Editors
Rafael Bello ⓘ
Department of Computer Science
Universidad Central de Las Villas
Santa Clara, Cuba

Rafael Falcon ⓘ
School of Electrical Engineering
and Computer Science
University of Ottawa
Ottawa, ON, Canada

Alejandro Rosete ⓘ
Departamento de Inteligencia Artificial e
Infraestructura de Sistemas Informáticos
Universidad Tecnológica de La Habana
"José Antonio Echeverría" (CUJAE)
Havana, Cuba

Duoqian Miao ⓘ
Department of Computer Science
and Technology
Tongji University
Shanghai, China

Michinori Nakata ⓘ
Department of Management
and Information Science
Josai International University
Togane, Chiba, Japan

Davide Ciucci ⓘ
Department of Informatics, Systems
and Communication
Università degli Studi di Milano-Bicocca
Milan, Italy

ISSN 0302-9743   ISSN 1611-3349   (electronic)
Lecture Notes in Artificial Intelligence
ISBN 978-3-030-52704-4   ISBN 978-3-030-52705-1   (eBook)
https://doi.org/10.1007/978-3-030-52705-1

LNCS Sublibrary: SL7 – Artificial Intelligence

This Springer imprint is published by the registered company Springer Nature Switzerland AG
The registered company address is: Gewerbestrasse 11, 6330 Cham, Switzerland

Preface

Rough Set Theory (RST) is a prominent methodology within the umbrella of Computational Intelligence and Granular Computing (GrC) to handle uncertainty in inconsistent and ambiguous environments. RST has enjoyed widespread success in a plethora of real-world application domains and remains at the forefront of numerous theoretical studies to consolidate and augment its well-established properties.

The International Joint Conference on Rough Sets (IJCRS) is the flagship conference of the International Rough Set Society (IRSS). Held annually, the IJCRS conference series aim at bringing academic researchers and industry practitioners together to discuss and deliberate on fundamental issues around rough sets and unveil successful applications of this vibrant theory in multiple domains. IJCRS provides an excellent opportunity for researchers to present their ideas before the rough set community, or for those who would like to learn about rough sets and find out whether this approach could be useful for their problems.

IJCRS 2020 was originally planned to take place during June 29 – July 3, 2020, at the Meliá Habana Hotel in Havana, Cuba. Due to the COVID-19 pandemic, however, the conference was turned into a virtual forum (https://virtualijcrs2020.uclv.edu.cu/) in order to facilitate the exchange among the conference participants while ensuring their physical safety and well-being.

IJCRS 2020's submission topics revolved around three major groups:

- *Fundamental Rough Set Models and Methods*: e.g. covering rough set models, decision-theoretic rough set methods, dominance-based rough set methods, rough clustering, rough computing, rough mereology, partial rough set models or game-theoretic rough set methods
- *Related Methods and Hybridization*: e.g. artificial intelligence, machine learning, pattern recognition, decision support systems, fuzzy sets and near sets, uncertain and approximate reasoning, information granulation, formal concept analysis, Petri nets or nature-inspired computation models
- *Application Areas*: e.g. medicine and health, bioinformatics, business intelligence, smart cities, semantic web, computer vision and image processing, cybernetics and robotics, knowledge discovery, etc.

This volume is a compilation of the IJCRS 2020 conference proceedings and contains all papers accepted by the Program Committee (except short papers) after a rigorous peer-review process.

IJCRS 2020 received 50 submissions from 119 authors in 18 countries.[1] Every submission was reviewed by at least two Program Committee members. On average,

---

[1] While this number of submissions is lower than previous years, it is also very encouraging that over 100 authors chose to submit their work despite the lingering shadow of COVID-19 and the increasing number of postponed/cancelled conferences because of this pandemic.

each submission received 2.84 reviews. Finally, the Program Committee chairs selected 37 regular submissions based on their originality, significance, correctness, relevance, and clarity of presentation to be included in the IJCRS 2020 proceedings. The Program Committee chairs also accepted two tutorials to be part of the proceedings and two short papers that are available at the virtual forum. We would like to thank all authors for submitting their papers and the Program Committee members for their valuable contribution to the conference through their anonymous, detailed review reports. We also wish to congratulate those authors whose papers were selected for presentation and/or publication in the proceedings.

IJCRS 2020's success was possible thanks to the dedication and support of many individuals and organizations. First and foremost, we want to thank IRSS for kindly accepting Havana, Cuba, as the venue of the 2020 IJCRS edition, which signals the increasing interest RST is amassing across Latin America and the Caribbean. We wish to express our gratitude to our honorary chairs (Andrzej Skowron and Yiyu Yao), the Steering Committee members (Tamás Mihálydeák, Victor Marek, and Sushmita Mitra), and the 70 Program Committee members for their invaluable suggestions, support, and excellent work throughout the organization process.

We are very grateful to our plenary speakers (Witold Pedrycz and Dominik Slezak), tutorial speakers (Mani A. and Oliver Urs Lenz), and our special session organizers for accepting our invitations to share their cutting-edge research work on rough sets and granular computing.

Special thanks go to Lázaro Pérez Lugo (Universidad Central de Las Villas, Cuba) whose relentless efforts building, testing, and deploying the virtual forum ensured the success of this conference in the new COVID-19 circumstances.

Last but certainly not least, we acknowledge the excellent Springer support provided by Aliaksandr Birukou and Anna Kramer. Their diligent work was greatly appreciated as they navigated us in a very professional and smooth manner during the compilation and editing of these proceedings.

Happy reading! We hope that this volume helps spark further interest in RST and other related methodologies.

June 2020

Rafael Bello
Duoqian Miao
Rafael Falcon
Michinori Nakata
Alejandro Rosete
Davide Ciucci

Organization

Honorary Chairs

Andrzej Skowron   University of Warsaw, Poland
Yiyu Yao   University of Regina, Canada

Steering Committee

Davide Ciucci   Università di Milano-Bicocca, Italy
Tamás Mihálydeák   University of Debrecen, Hungary
Victor Marek   University of Kentucky, USA
Sushmita Mitra   Indian Statistical Institute, Kolkata, India

Conference Chairs

Rafael Bello   Universidad Central de Las Villas, Cuba
Duoqian Miao   Tongji University, China

Program Committee Chairs

Rafael Falcon   University of Ottawa, Canada
Michinori Nakata   Josai International University, Japan

Tutorial Chairs

Chris Cornelis   Ghent University, Belgium
Hong Yu   Chongqing University of Posts
       and Telecommunications, China

Local Arrangement Chair

Alejandro Rosete   Universidad Tecnológica de La Habana "José Antonio
       Echevería" (CUJAE), Cuba

Publicity Chair

Mauricio Restrepo   Universidad Militar Nueva Granada, Colombia

Webmaster

Lázaro Pérez Lugo   Universidad Central "Marta Abreu" de las Villas, Cuba

Program Committee

Amedeo Napoli	Université de Lorraine, France
Andrei Paun	University of Bucharest, Romania
Andrzej Szałas	University of Warsaw, Poland
Andrzej Skowron	University of Warsaw, Poland
Anna Gomolinska	University of Białystok, Poland
Bay Vo	Ho Chi Minh City University of Technology, Vietnam
Beata Zielosko	University of Sielsia, Poland
Bing Zhou	Sam Houston State University, USA
Caihui Liu	Gannan Normal University, China
Chien-Chung Chan	University of Akron, USA
Christopher Hinde	Loughborough University, UK
Churn-Jung Liau	Academia Sinica, Taiwan
Claudio Meneses	Universidad Católica del Norte, Chile
Costin-Gabriel Chiru	Politehnica University of Bucharest, Romania
Davide Ciucci	Università di Milano-Bicocca, Italy
Dayong Deng	Zhejiang Normal University, China
Dmitry Ignatov	National Research University Higher School of Economics, Russia
Dongyi Ye	Fuzhou University, China
Georg Peters	Munich University of Applied Sciences and Australian Catholic University, Germany/Australia
Guilong Liu	Beijing Language and Culture University, China
Guoyin Wang	Chongqing University of Posts and Telecommunications, China
Hiroshi Sakai	Kyushu Institute of Technology, Japan
Hung Son Nguyen	University of Warsaw, Poland
Ivo Düntsch	Fujian Normal University, China
Jaroslaw Stepaniuk	Białystok University of Technology, Poland
Jaume Baixeries	Universitat Politècnica de Catalunya, Spain
Jesús Medina	University of Cádiz, Spain
Jingtao Yao	University of Regina, Canada
Jiye Liang	Shanxi University, China
Jouni Jarvinen	University of Turku, Finland
Krzysztof Pancerz	University of Rzeszów, Poland
Loan T. T. Nguyen	University of Warsaw, Poland
Mani A.	Kolkata, India
Marcin Szczuka	University of Warsaw, Poland
Marcin Michalak	Silesian University of Technology, Poland
Marek Sikora	Silesian University of Technology, Poland
Marzena Kryszkiewicz	Warsaw University of Technology, Poland
Masahiro Inuiguchi	Osaka University, Japan
Md. Aquil Khan	Indian Institute of Technology Indore, India
Michal Kepski	University of Rzeszów, Poland
Michinori Nakata	Josai International University, Japan

Mohua Banerjee   Indian Institute of Technology Kanpur, India
Mu-Chen Chen   National Chiao Tung University, Taiwan
Murat Diker   Hacettepe University, Turkey
Nguyen Long Giang   Vien Công nghe thông tin, Vietnam
Nizar Bouguila   Concordia University, Canada
Piotr Artiemjew   University of Warmia and Mazury, Poland
Pradipta Maji   Indian Statistical Institute, India
Rafal Gruszczynski   Nicolaus Copernicus University, Poland
Richard Jensen   Aberystwyth University, UK
Ryszard Janicki   McMaster University, Canada
Ryszard Tadeusiewicz   AGH University of Science and Technology, Poland
Sándor Radeleczki   University of Miskolc, Hungary
Sheela Ramanna   University of Winnipeg, Canada
Soma Dutta   Vistula University, Poland
Tamás Mihálydeák   University of Debrecen, Hungary
Thierry Denoeux   Université de Technologie de Compiègne, France
Tianrui Li   Southwest Jiaotong University, China
Vilem Novak   University of Ostrava, Czech Republic
Vladimir Parkhomenko   Peter the Great St. Petersburg Polytechnic University,
                               Russia
Wojciech Ziarko   University of Regina, Canada
Xiuyi Jia   Nanjing University of Science and Technology, China
Yan Yang   Southwest Jiaotong University, China
Yiyu Yao   University of Regina, Canada
Yoo-Sung Kim   Inha University, South Korea
Zbigniew Suraj   University of Rzeszów, Poland
Zbigniew Ras   University of North Carolina at Charlotte, USA
Zied Elouedi   Institut Supéur de Gestion de Tunis, Tunisia
Zoltán Ernó Csajbók   University of Debrecen, Hungary

Contents

Data Summarization

Community Detection

General Rough Sets

Rough Forgetting

Patrick Doherty[1,2] and Andrzej Szałas[1,3(✉)]

[1] Department of Computer and Information Science, Linköping University,
581 83 Linköping, Sweden
{patrick.doherty,andrzej.szalas}@liu.se
[2] School of Intelligent Systems and Engineering, Jinan University (Zhuhai Campus),
Zhuhai, China
[3] Institute of Informatics, University of Warsaw, Banacha 2, 02-097 Warsaw, Poland
andrzej.szalas@mimuw.edu.pl

Abstract. Recent work in the area of Knowledge Representation and Reasoning has focused on modification and optimization of knowledge bases (KB) through the use of *forgetting operators* of the form $forget(KB, \bar{R})$, where $\bar{R}$ is a set of relations in the language signature used to specify the KB. The result of this operation is a new KB where the relations in $\bar{R}$ are removed from the KB in a principled manner resulting in a more efficient representation of the KB for different purposes. The forgetting operator is also reflected semantically in terms of the relation between the original models of the KB and the models for the revised KB after forgetting. In this paper, we first develop a rough reasoning framework where our KB's consist of rough formulas with a semantics based on a generalization of Kleene algebras. Using intuitions from the classical case, we then define a forgetting operator that can be applied to rough KBs removing rough relations. A constructive basis for generating a new KB as the result of applying the forgetting operator to a rough KB is specified using second-order quantifier elimination techniques. We show the application of this technique with some practical examples.

1 Introduction and Motivations

In Artificial Intelligence, the field of Knowledge Representation and Reasoning (KRR) deals with the use of logical languages to represent knowledge or beliefs and the use of inference in some logic to derive additional knowledge or belief implicit in a base theory represented as a set of logical formulas. The explicit base theory is often called a *Knowledge Base* (KB). Consequences A of the KB are derived through a consequence relation, KB $\models A$. A signature Σ (vocabulary) is

The first author has been supported by the ELLIIT Network Organization for Information and Communication Technology, Sweden; the Swedish Foundation for Strategic Research SSF (SymbiKBot Project); and a guest professor grant from Jinan University (Zhuhai Campus). The second author has been supported by grant 2017/27/B/ST6/02018 of the National Science Centre Poland.

© Springer Nature Switzerland AG 2020
R. Bello et al. (Eds.): IJCRS 2020, LNAI 12179, pp. 3–18, 2020.
https://doi.org/10.1007/978-3-030-52705-1_1

associated with the logical language and is used for specifying the legal relations, functions, constants, etc., used in the syntax of formulas.

In recent years, there has been much interest in the topic of *forgetting* operations in KRR [12]. Intuitions for such operators are based loosely on the fact that humans often forget what they know or believe for reasons of efficiency in reasoning. Mapping this loose intuition over to KRR results in some very powerful and useful techniques for dealing with redundant information in KB's, optimizing query retrieval in relation to KB's [10,17], progressing databases [18], forgetting with description logics [6,31] and rule based languages [26,27], dealing with missing information and dataset reduction [14], forgetting sets of literals in first-order logic [28], in addition to other techniques. In general, one major type of forgetting aims at removing information from a KB in a controlled manner where the syntactic elimination has a principled semantic correlation characterized in model theory.

Given a KB and a signature Σ, a common type of forgetting, $forget(KB, \Sigma')$, can be formulated where $\Sigma' \subsetneq \Sigma$ and KB' is the result of forgetting the components in Σ' in KB. One interesting question is the relation between the models and consequences of KB and the consequences of KB' after the forgetting operation is applied to KB. Initial intuitions for this type of forgetting can be traced all the way back to Boole [2] and his use of variable elimination. Assume a propositional language with signature $\Sigma = \{p, q, r\}$. Given a propositional formula A and a signature $\Sigma' = \{p\}$, the result of forgetting p in A, $forget(A, \Sigma')$, is $A_p^+ \vee A_p^-$, where A_p^+ is the result of replacing all occurrences of p in A with 'true' and A_p^- is the result of replacing all occurrences of p in A with 'false' and simplifying the result.

In KRR application areas such as robotics, the knowledge or beliefs robots have about different aspects of the world, is often incomplete and/or uncertain. Consequently, one wants to find a concise way to model this. Rough set theory [7,8,22,23] has been used to model different types of incompleteness using indiscernibility and approximations. The general idea is to begin with a universe of individuals and define an indiscernibility relation over these individuals. In the classical case, this generates an equivalence relation over individuals. A rough set is defined by specifying a lower and upper approximation, each consisting of a number of equivalence classes generated by the indiscernibility relation. All individuals in equivalence classes included in the lower approximation are in the rough set, all equivalence classes in the upper approximation intersect with the rough set, and the individuals in the remaining equivalence classes lie outside the set. This brings to mind a division of individuals into a tripartite division reminiscent of three-valued logics. Later in the paper, this intuition will be formalized more precisely. In the context of KRR, there has been interest in generalizations of logical languages and inference to include rough logical languages and inference using rough theories [8]. This generalization will be used as a vehicle for specifying rough forgetting operators applied to rough relations in such logics.

Another application area for rough sets and logics is with big data applications. According to [14], "most of the attribute values relating to the timing of big data [...] are missing due to noise and incompleteness. Furthermore, the number of missing links between data points in social networks is approximately 80% to 90% and the number of missing attribute values within patient reports transcribed from doctor diagnoses are more than 90%."

Rough sets are discussed in [14] as one of remedies to deal with missing data. In this context, the combination of rough sets with the use of forgetting operations might prove to be very useful. In cases where important information is missing, it might be useful to forget the relation or find a relation's explicit definition and – using the definition – complete parts of the missing content. In fact, the second-order quantifier elimination techniques which we describe in this paper and use as a tool for forgetting, provide us with definitions of eliminated (forgotten) relations as a *side effect*.

This paper is primarily about developing a first-order logical framework for rough theories that can be used to construct rough KB's, with a formal semantics based on rough relational structures. Given such a logic, we then define a forgetting operator that can be applied to rough theories and we provide the semantics for such an operator. The forgetting operator is based on second-order quantifier elimination techniques developed for rough theories. In previous work, we have shown how second-order quantifier elimination techniques can be automated for well-behaved fragments of second-order logic. We expand on these results in the context of rough theories.

The paper is structured as follows. In Sect. 2 we discuss the rough reasoning framework used throughout the paper. In Sect. 3, we recall definitions for forgetting used with classical logic and then generalize these and introduce rough forgetting. Next, in Sect. 4, we provide second-order quantifier elimination theorems with proofs which can serve as foundations for algorithmic techniques for rough forgetting. Section 5 provides a number of examples showing how the proposed techniques work in practice. Finally, Sect. 6 concludes the paper.

2   Rough Reasoning Framework

Rough sets [21,22] have been defined in many ways (see, e.g., [4,5,7,8,16,23, 25,29,30] and numerous references there). Three- and many-valued approaches have been intensively studied in the context of rough sets [3,4,15,16]. In the current paper we will follow the presentation of [16].

Definition 1 (Approximation space). *Let $\mathcal{U}$ be a set of objects and E be an equivalence relation on $\mathcal{U}$, Then $\mathcal{A} = \langle \mathcal{U}, E \rangle$ is called an approximation space. By the* lower approximation *(s^+) and* upper approximation *($s^\oplus$) of a set $s \subseteq \mathcal{U}$ we mean:*

$$s^+ \stackrel{\text{def}}{=} \{x \in \mathcal{U} \mid \forall y \big(E(x,y) \to y \in s \big)\}; \quad s^\oplus \stackrel{\text{def}}{=} \{x \in \mathcal{U} \mid \exists y \big(E(x,y) \wedge y \in s \big)\}. \tag{1}$$

A set $s \subseteq \mathcal{U}$ is definable *in $\mathcal{A}$ iff s is a union of equivalence classes of E.* ∎

In rough sets, E represents an indiscernibility relation. Approximations are interpreted as follows, where $s \subseteq \mathcal{U}$ is a set:

– the lower approximation s^+ represents objects certainly belonging to s;
– the upper approximation $s^\oplus$ represents objects possibly belonging to s.

Definition 2 (Rough sets). *For an approximation space $\mathcal{A}$, the ordered pair $\langle s_l, s_u \rangle$, where $s_l \subseteq s_u$ and s_l, s_u are definable sets, is called a* rough set *(wrt $\mathcal{A}$).*[1]   ■

Remark 1. In the literature, the equivalence relation used to define rough approximations has been argued to be too strong for many application areas [8,24,25]. In fact, seriality of E (i.e., the property that $\forall x \exists y(E(x, y))$) has been proposed as the weakest well-behaved requirement on E. This ensures that the lower approximation is included in the upper approximation of a rough set [11,29].

Note also that, according to [19, Section 19.3], every reflexive similarity relation can be refined to an equivalence relation in a natural way. So reflexivity can be used as a basic requirement on indiscernibility relations.[2]   ■

As shown in [16], there is a close correspondence between rough sets and Kleene algebras defined below.

Definition 3 (Kleene algebra). *An algebra $\mathcal{K} = \langle K, \cup, \cap, -, \bot, \top \rangle$ is called a* Kleene algebra *if the following hold.*

1. *$\mathcal{K}$ is a De Morgan algebra, i.e., $\langle K, \cup, \cap, \bot, \top \rangle$ is a distributive lattice with the greatest element $\top$ and the least element $\bot$, and for all $s, t \in K$,*
   (a) $-(s \cap t) = -s \cup -t$ (De Morgan property),
   (b) $--s = s$ (involution).
2. *$s \cap -s \leq t \cup -t$, for all $s, t \in K$ (Kleene property).*   ■

Note that in Definition 3 we refer to "greatest" and "least" elements. As usual in lattice theory, we mean the ordering:

$$s \leq t \stackrel{\text{def}}{\equiv} (s = s \cap t), \text{ (equivalently: } t = s \cup t). \tag{2}$$

For rough sets, a subclass of Kleene algebras, rough Kleene algebras, will have the role of Boolean algebras for classical sets.

Definition 4 (Rough Kleene algebra). *Let $\mathcal{U}$ be a set of objects. A Kleene algebra $\mathcal{K} = \langle K, \cup, \cap, -, \bot, \top \rangle$ is called a* rough Kleene algebra *over $\mathcal{U}$ iff:*

---

[1] The set s_l serves as the *lower approximation* and s_u – as the upper approximation of a set.
[2] Note that reflexivity implies seriality.

– K consists of pairs of sets $\langle s_l, s_u \rangle$ such that $s_l \subseteq s_u \subseteq \mathcal{U}$;

– $\bot \stackrel{\text{def}}{=} \langle \emptyset, \emptyset \rangle, \top \stackrel{\text{def}}{=} \langle \mathcal{U}, \mathcal{U} \rangle$;

– $-\langle s_l, s_u \rangle \stackrel{\text{def}}{=} \langle -s_u, -s_l \rangle$.

By a generalized rough set we mean any element of $\mathcal{K}$. ∎

As the logical counterpart of rough Kleene algebras we will use the three valued logic of Kleene, K_3, with truth values T (true), F (false) and U (unknown), ordered by:

$$F < U < T, \tag{3}$$

with connectives $\vee, \wedge, \neg$. The semantics of connectives is defined by:

$$\tau_1 \vee \tau_2 \stackrel{\text{def}}{=} \max\{\tau_1, \tau_2\}; \quad \tau_1 \wedge \tau_2 \stackrel{\text{def}}{=} \min\{\tau_1, \tau_2\}; \tag{4}$$

$$\neg F \stackrel{\text{def}}{=} T; \quad \neg U \stackrel{\text{def}}{=} U; \quad \neg T \stackrel{\text{def}}{=} F, \tag{5}$$

where $\tau_1, \tau_2 \in \{F, U, T\}$ and $\max, \min$ are the maximum and minimum wrt (3).

Let us now define the syntax of rough formulas used in this paper. In addition to connectives $\neg, \wedge, \vee$ and quantifiers $\forall, \exists$ of Kleene logic K_3, we add two connectives: $\in$ and $\subseteq$. Their intended meaning is rough set membership and rough set inclusion, respectively.

Definition 5 (Syntax of rough formulas). *Let V be a set of first-order variables (representing domain elements), C be a set of constants and R be a set of relation symbols. Then:*

– Kleene formulas, *KF*, *are defined by the grammar:*

$$\langle KF \rangle ::= \langle R \rangle \mid \neg \langle KF \rangle \mid \langle KF \rangle \vee \langle KF \rangle \mid \langle KF \rangle \wedge \langle KF \rangle \mid$$
$$\exists \langle V \rangle \langle KF \rangle \mid \forall \langle V \rangle \langle KF \rangle;$$

– rough formulas, *RF*, *are defined by the grammar, where $\overline{C \cup V}$ denotes tuples consisting of constants and/or variables:*

$$\langle RF \rangle ::= \langle KF \rangle \mid \langle \overline{C \cup V} \rangle \in \langle KF \rangle \mid \langle KF \rangle \subseteq \langle KF \rangle \mid$$
$$\neg \langle RF \rangle \mid \langle RF \rangle \vee \langle RF \rangle \mid \langle RF \rangle \wedge \langle RF \rangle \mid$$
$$\exists \langle V \rangle \langle RF \rangle \mid \forall \langle V \rangle \langle RF \rangle.$$

An occurrence of a variable is called bound *in a formula if it appears inside the scope of a quantifier. It is called* free *when it is not bound.* ∎

Rough theories (rough knowledge bases) are defined below.

Definition 6 (Rough theories, rough knowledge bases). *Finite sets of rough formulas are called* rough theories *(or rough knowledge bases). A finite set of formulas T is understood as a single formula being the conjunction of formulas in T: $\bigwedge_{A \in T} A$.* ∎

Remark 2. In the rest of the paper we will often use the traditional syntax for relations. For example, rather than writing $\forall x \exists y\big((x,y) \in r\big)$, we will write $\forall x \exists y\big(r(x,y)\big)$. ∎

Definition 7 (Rough literals and facts). *By a* rough literal *we mean an expression of the form* $\pm r(\bar{e})$, *where* $\pm$ *is the empty symbol or* $\neg$, *r is a relation symbol and $\bar{e}$ is a tuple of constants and/or variables. By a* rough fact *we mean a rough literal not containing variables.* ∎

The following important property, justifying the use of K_3 in the context of rough forgetting, is an immediate consequence of Theorems 8, 11, 15, proved in [16]. Below:

- $A_{\mathcal{K}}$ is the class of Kleene algebras;
- RS is the class of rough Kleene algebras;
- $A \models_{t,f} B$ iff for every assignment $w : RF \longrightarrow \{F, U, T\}$,
  - $w(A) = T$ implies $w(B) = T$, and
  - $w(B) = F$ implies $w(A) = F$.

Corollary 1. *For any rough formulas* $A, B \in RF$:

$$A \models_{A_{\mathcal{K}}} B \text{ iff } A \models_{t,f} B \text{ iff } A \models_{RS} B, \tag{6}$$

where $\models_{A_{\mathcal{K}}}$ *and* $A \models_{RS}$ *are semantic consequence relations for* $A_{\mathcal{K}}$ *and* RS, *respectively.* ∎

To define the semantics of rough formulas, we first need a generalization of relational structures to their rough version.

Definition 8 (Rough relational structures). *Let* $\mathcal{U}$ *be a set of objects,* $\mathcal{K}$ *be a rough Kleene algebra over* $\mathcal{U}$ *and* $n \geq 1$ *be a natural number. By an n-argument* rough relation *over* $\mathcal{U}$ *we mean any generalized rough set consisting of tuples of the Cartesian product* $\mathcal{U}^n$. *By a* rough relational structure *we mean* $\langle \mathcal{U}, r_1, \ldots, r_k \rangle$ *where for* $1 \leq i \leq k$, r_i *is an n_i-argument rough relation over* $\mathcal{U}$. *One-argument rough relations are called* rough concepts *and two-argument ones are called* rough roles. ∎

The semantics of rough formulas is defined below, where $A(x \leftarrow a)$ denotes the formula obtained from A by substituting all free occurrences of variable x in A by constant a.

Definition 9 (Semantics of rough formulas). *Let* $\mathcal{U}$ *be a set,* $\mathcal{K} = \langle K, \cup, \cap, -, \bot, \top \rangle$ *be a rough Kleene algebra over* $\mathcal{U}$ *and* $\mathcal{R} = \langle \mathcal{U}, r_1, \ldots, r_k \rangle$ *be a rough relational structure,*

1. *The* value *of a rough formula,* $vs_{\mathcal{R}} : KF \longrightarrow K$, *is inductively defined by:*
   - *for a relation symbol* r, $vs_{\mathcal{R}}(r) \overset{\text{def}}{=} r^{\mathcal{R}}$ *where* $r^{\mathcal{R}}$ *is the relation* r *in* $\mathcal{R}$;[3]
   - $vs_{\mathcal{R}}(\neg A) \overset{\text{def}}{=} -vs_{\mathcal{R}}(A)$;
   - $vs_{\mathcal{R}}(A \vee B) \overset{\text{def}}{=} vs_{\mathcal{R}}(A) \cup vs_{\mathcal{R}}(B)$;
   - $vs_{\mathcal{R}}(A \wedge B) \overset{\text{def}}{=} vs_{\mathcal{R}}(A) \cap vs_{\mathcal{R}}(B)$;
   - $v_{\mathcal{R}}(\exists x(A(x))) \overset{\text{def}}{=} \bigcup\limits_{a \in \mathcal{U}} vs_{\mathcal{R}}(A(x \leftarrow a))$;
   - $v_{\mathcal{R}}(\forall x(A(x))) \overset{\text{def}}{=} \bigcap\limits_{a \in \mathcal{U}} vs_{\mathcal{R}}(A(x \leftarrow a))$.

2. *The* truth value *of a rough formula,* $v_{\mathcal{R}} : RF \longrightarrow \{F, U, T\}$, *is defined inductively:*
   - *for a Kleene formula* A *with* k *free variables,* $\bar{a} \in \mathcal{U}^k$, *and* $vs_{\mathcal{R}}(A) = \langle r_l, r_u \rangle$,
   $$v_{\mathcal{R}}(\bar{a} \in A) \overset{\text{def}}{=} \begin{cases} T \text{ when } a \in r_l; \\ U \text{ when } a \in r_u \setminus r_l; \\ F \text{ when } a \in U \setminus r_u. \end{cases}$$
   - $v_{\mathcal{R}}(A \subseteq B) \overset{\text{def}}{=} \begin{cases} T \text{ when for all } \bar{a} \in \mathcal{U}^k, v_{\mathcal{R}}(\bar{a} \in A) \leq v_{\mathcal{R}}(\bar{a} \in B); \\ F \text{ otherwise,} \end{cases}$
   where A, B *are Kleene formulas with* k *free variables, and* $\leq$ *is the reflexive closure of* (3);
   - $v_{\mathcal{R}}(\neg A) \overset{\text{def}}{=} \neg v_{\mathcal{R}}(A)$, *for* $\circ \in \{\vee, \wedge\}, v_{\mathcal{R}}(A \circ B) \overset{\text{def}}{=} v_{\mathcal{R}}(A) \circ v_{\mathcal{R}}(B)$,
   where the semantics of $\neg, \vee, \wedge$ *on truth values is defined by* (4)–(5);
   - $v_{\mathcal{R}}(\exists x(A(x))) \overset{\text{def}}{=} \max\limits_{a \in \mathcal{U}} \{v_{\mathcal{R}}(A(x \leftarrow a))\}$, *where* $\max$ *is the maximum wrt* (3);
   - $v_{\mathcal{R}}(\forall x(A(x))) \overset{\text{def}}{=} \min\limits_{a \in \mathcal{U}} \{v_{\mathcal{R}}(A(x \leftarrow a))\}$, *where* $\min$ *is the minimum wrt* (3).

   We write $\mathcal{R} \models A$ *to indicate that* $v_{\mathcal{R}}(A) = T$. *We say that formulas* A *and* B *are equivalent, iff for every* R, $v_R(A) = v_R(B)$. ∎

3   Forgetting and Rough Forgetting

In the rest of the paper, we assume that knowledge bases are given in the form of finitely axiomatizable theories. As indicated in Definition 6, each theory consisting of a finite set of axioms is understood as a single formula, being the conjunction of the axioms.

---

[3] To simplify notation, we use the same notation for relation symbols and corresponding rough relations. Similarly, objects in $\mathcal{U}$ are identified with constants denoting them.

3.1 Forgetting

The following definition, theorem and example have been formulated in [18].

Definition 10 (Forgetting). *Let r be a relation symbol and $\mathcal{M}_1$, $\mathcal{M}_2$ be relational structures. Then $\mathcal{M}_1 \sim_r \mathcal{M}_2$ denotes the fact that $\mathcal{M}_1$ differs from $\mathcal{M}_2$ at most in the interpretation of r.*

Let T be a theory. A theory T' is a result of forgetting r in T iff for any relational structure $\mathcal{M}'$, $\mathcal{M}' \models T'$ iff there is a relational structure $\mathcal{M}$ such that $\mathcal{M} \models T$ and $\mathcal{M} \sim_r \mathcal{M}'$. By $forget(T; r)$ we denote the result of forgetting r in T. ∎

In the rest of the paper $T(r \leftarrow X)$ denotes the formula resulting from $T(r)$ by replacing every occurrence of r in T by X.

Theorem 1. *Let r be a relation symbol and X be a second-order variable with the same number of arguments as r. Then $forget(T; r) \equiv \exists X \big(T(r \leftarrow X)\big)$.* ∎

Example 1. Let $T \equiv \big((student(joe) \vee student(john)) \wedge teacher(john)\big)$. Note that:

$$\big((student(joe) \vee student(john)) \wedge teacher(john)\big)\big(student \leftarrow X\big) = \tag{7}$$
$$(X(joe) \vee X(john)) \wedge teacher(john).$$

Using Theorem 1 and (7) we have:

$$forget(T; student) = \exists X \big((X(joe) \vee X(john)) \wedge teacher(john)\big). \tag{8}$$

It can be easily shown that the formula $\exists X \big((X(joe) \vee X(john)) \wedge teacher(john)\big)$, thus $forget(T; student)$ too, is equivalent to $teacher(john)$. ∎

Theorem 1 shows that the problem of computing $forget(T; r)$ can be reduced to second-order quantifier elimination. For this purpose, in the current paper we will adapt the techniques of [1, 20] to rough theories.[4]

3.2 Rough Forgetting

Rough forgetting is defined by analogy with Definition 10.

Definition 11 (Rough forgetting). *Let r be a relation symbol and $\mathcal{R}_1$, $\mathcal{R}_2$ be rough relational structures. Then $\mathcal{R}_1 \approx_r \mathcal{R}_2$ denotes the fact that $\mathcal{R}_1$ differs from $\mathcal{R}_2$ at most in the interpretation of r.*

Let $\mathcal{T}$ be a rough theory. A theory $\mathcal{T}'$ is a result of rough forgetting r in $\mathcal{T}$ iff for any rough relational structure $\mathcal{R}'$, $\mathcal{R}' \models \mathcal{T}'$ iff there is a rough relational structure $\mathcal{R}$ such that $\mathcal{R} \models \mathcal{T}$ and $\mathcal{R} \approx_r \mathcal{R}'$. By $rforget(\mathcal{T}; r)$ we denote the formula being the result of rough forgetting of r in $\mathcal{T}$. ∎

---

[4] For a broad discussion of related second-order quantifier elimination techniques see [13].

As in the case of classical forgetting, we have the following theorem analogous to Theorem 1, where we use a second-order quantifier, whose semantics is defined by:[5]

$$v_{\mathcal{R}}\big(\exists X\big(\mathcal{T}(X)\big)\big) \overset{\text{def}}{=} \max_{s\in K}\{v_{\mathcal{R}}\big(\mathcal{T}(X\!\leftarrow\! s)\big)\}, \qquad (9)$$

where max, min are the maximum and minimum wrt (3).

Theorem 2. *Let r be a rough relation symbol and X be a second-order variable with the same number of arguments as r. Then for every rough relational structure $\mathcal{R}$:*

$$v_{\mathcal{R}}\big(rforget(\mathcal{T};r)\big) = v_{\mathcal{R}}\big(\exists X\big(\mathcal{T}(r\!\leftarrow\! X)\big)\big). \qquad ■$$

Comparing to classical forgetting, in rough forgetting we deal with rough relations rather than with the classical relations. Thus, $\exists X$ in Theorem 2 is a second-order quantification over rough sets rather than over the classical ones.

4   Eliminating Second-Order Quantifiers from Rough Formulas

To formalize second-order quantifier elimination methods and related concepts, we need a notation $A(X\!\leftarrow\! B[\bar{z}])$ defined as follows. Let A, B be rough formulas such that A contains an n-argument second-order variable X and $\bar{z}$ is a tuple of n first-order variables with free occurrences in formula B. Then:

$$A(X\!\leftarrow\! B[\bar{z}])$$

denotes the result of substituting all occurrences of the second-order variable X by $B(\bar{z})$, where $\bar{z}$ in B is respectively substituted by actual parameters of X (possibly different in different occurrences of X). For example,

$$\big(\underbrace{X(a) \vee X(b)}_{A(X)}\big)\big(X\!\leftarrow\!\underbrace{r(z,y)}_{B(z,y)}[z]\big) \text{ is } (r(a,y) \vee r(b,y)).$$

The quantifier elimination techniques we develop are based on a monotonicity property, defined as follows.

Definition 12 (Monotonicity). *Let X be a second-order variable representing n-argument relations and let $\bar{z}$ be a tuple consisting of n (first-order) variables. We say that a rough formula $A(X)$ is monotone in X iff for every rough relational structure $\mathcal{R}$ and rough formulas B, C not containing X and with $\bar{z}$ being all variables with free occurrences, one of the following properties holds:*

$$v_{\mathcal{R}}(B(\bar{z})) \leq v_{\mathcal{R}}(C(\bar{z})) \text{ implies } v_{\mathcal{R}}\big(A(X\!\leftarrow\! B[\bar{z}])\big) \leq v_{\mathcal{R}}\big(A(X\!\leftarrow\! C[\bar{z}])\big); \quad (10)$$

$$v_{\mathcal{R}}(B(\bar{z})) \leq v_{\mathcal{R}}(C(\bar{z})) \text{ implies } v_{\mathcal{R}}\big(A(X\!\leftarrow\! C[\bar{z}])\big) \leq v_{\mathcal{R}}\big(A(X\!\leftarrow\! B[\bar{z}])\big). \quad (11)$$

---

[5] Recall that K is the universe of a rough Kleene algebra $\mathcal{K}$, fixed earlier.

Properties (10) *and* (11) *are called* up-monotonicity *and* down-monotonicity *of A, respectively.* ∎

The following theorem adapts Ackermann's Lemma [1,9,13] to rough theories.

Theorem 3. *Let X be an n-argument second-order variable. Let $\bar{z}$ be an n-tuple of variables, $A(\bar{z})$ be a rough formula containing no occurrences of X, with variables $\bar{z}$ occurring free, and let $B(X)$ be a rough formula with X as a free variable.*

1. *If $B(X)$ is down-monotone in X then for every rough relational structure $\mathcal{R}$,*

$$v_{\mathcal{R}}\Big(\exists X\big(\forall\bar{z}(A(\bar{z}) \subseteq X(\bar{z})) \wedge B(X)\big)\Big) = v_{\mathcal{R}}\Big(B\big(X{\leftarrow}A[\bar{z}]\big)\Big). \qquad (12)$$

2. *If $B(X)$ is up-monotone in X then for every rough relational structure $\mathcal{R}$,*

$$v_{\mathcal{R}}\Big(\exists X\big(\forall\bar{z}(X(\bar{z}) \subseteq A(\bar{z})) \wedge B(X)\big)\Big) = v_{\mathcal{R}}\Big(B\big(X{\leftarrow}A[\bar{z}]\big)\Big). \qquad (13)$$

Proof. Let us prove (12).[6] Let $\mathcal{R}$ be an arbitrary rough relational structure. We have to prove three equivalences $v_{\mathcal{R}}(lhs) = \tau$ iff $v_{\mathcal{R}}(rhs) = \tau$ for $\tau \in \{\mathrm{T},\mathrm{U},\mathrm{F}\}$, where lhs and rhs are respectively the letfthand and the righthand side of Equation (12):

1. ($\rightarrow$) Assume that $v_{\mathcal{R}}\Big(\exists X\big(\forall\bar{z}(A(\bar{z}) \subseteq X(\bar{z})) \wedge B(X)\big)\Big) = \mathrm{T}$. In this case, there is X such that $v_{\mathcal{R}}\big(\forall\bar{z}(A(\bar{z}) \subseteq X(\bar{z}))\big) = \mathrm{T}$ and $v_{\mathcal{R}}\big(B(X)\big) = \mathrm{T}$. Thus, by Definition 9, for every $\bar{z}$, $v_{\mathcal{R}}\big(A(\bar{z})\big) \leq v_{\mathcal{R}}\big(X(\bar{z})\big)$. By down-monotonicity of $B(X)$ in X we conclude that $v_{\mathcal{R}}\Big(B\big(X{\leftarrow}A[\bar{z}]\big)\Big) = \mathrm{T}$.

   ($\leftarrow$) Assume that $v_{\mathcal{R}}\Big(B\big(X{\leftarrow}A[\bar{z}]\big)\Big) = \mathrm{T}$. To show that there is X satisfying $v_{\mathcal{R}}\big(\forall\bar{z}(A(\bar{z}) \subseteq X(\bar{z})) \wedge B(X)\big) = \mathrm{T}$ it suffices to set $\forall\bar{z}\big(X(\bar{z}) \overset{\text{def}}{=} A(\bar{z})\big)$.

2. ($\rightarrow$) Assume that $v_{\mathcal{R}}\Big(\exists X\big(\forall\bar{z}(A(\bar{z}) \subseteq X(\bar{z})) \wedge B(X)\big)\Big) = \mathrm{U}$. In this case,[7] there is X such that $v_{\mathcal{R}}\big(\forall\bar{z}(A(\bar{z}) \subseteq X(\bar{z}))\big) = \mathrm{T}$ and $B(X) = \mathrm{U}$. Thus, by Definition 9, for every $\bar{z}$, $v_{\mathcal{R}}\big(A(\bar{z})\big) \leq v_{\mathcal{R}}\big(X(\bar{z})\big)$. By down-monotonicity of $B(X)$ in X we conclude that $v_{\mathcal{R}}\Big(B\big(X{\leftarrow}A[\bar{z}]\big)\Big) \geq \mathrm{U}$. Suppose that:

$$v_{\mathcal{R}}\Big(B\big(X{\leftarrow}A[\bar{z}]\big)\Big) = \mathrm{T}. \qquad (14)$$

   However, by 1.($\leftarrow$), (14) implies $v_{\mathcal{R}}\Big(\exists X\big(\forall\bar{z}(A(\bar{z}) \subseteq X(\bar{z})) \wedge B(X)\big)\Big) = \mathrm{T}$, contradicting the assumption. Therefore, $v_{\mathcal{R}}\Big(B\big(X{\leftarrow}A(\bar{z})[\bar{z}]\big)\Big) = \mathrm{U}$.

   ($\leftarrow$) Here, like in the previous point, it suffices to set $\forall\bar{z}\big(X(\bar{z}) \overset{\text{def}}{=} A(\bar{z})\big)$.

---

[6] The proof of (13) is analogous, so we skip it here.

[7] Note that $\subseteq$ is two-valued, i.e., its truth value can only be T or F.

3. ($\rightarrow$) Assume that $v_{\mathcal{R}}\Big(\exists X\big(\forall \bar{z}(A(\bar{z}) \subseteq X(\bar{z})) \wedge B(X)\big)\Big) = \mathrm{F}$. By points 1.($\leftarrow$) and 2.($\leftarrow$), the value $v_{\mathcal{R}}\Big(B\big(X{\leftarrow}A[\bar{z}]\big)\Big)$ can neither be T nor U (since, as before, this would contradict the assumption). Therefore we can only conclude that $v_{\mathcal{R}}\Big(B\big(X{\leftarrow}A[\bar{z}])\big)\Big) = \mathrm{F}$.

4. ($\leftarrow$) Here, like in the previous points, it suffices to set $\forall \bar{z}(X(\bar{z}) \stackrel{\mathrm{def}}{=} A(\bar{z}))$. ∎

The following theorem adapts the fixpoint theorem proved in [20] to rough theories, where $\mathrm{L{\small FP}}\, X\big[A(X)\big]$ and $\mathrm{G{\small FP}}\, X\big[A(X)\big]$ stand for the least and the greatest fixpoint of $A(X)$ wrt X. Note that we deal with complete lattices and will always make sure that $A(X)$ is up-monotone in X, such fixpoints exist by Knaster and Tarski fixpoint theorem.

Theorem 4. *Let X be an n-argument second-order variable. Let $\bar{z}$ be an n-tuple of variables, $A(X, \bar{z})$ be a rough formula in which variables X and $\bar{z}$ are free. Let $A(X, \bar{z})$ be up-monotone in X and let $B(X)$ be a rough formula with X being a free variable.*

1. If $B(X)$ is down-monotone in X then for every rough relational structure $\mathcal{R}$,

$$v_{\mathcal{R}}\Big(\exists X\big(\forall \bar{z}(A(X, \bar{z}) \subseteq X(\bar{z})) \wedge B(X)\big)\Big) = \\ v_{\mathcal{R}}\Big(B\big(X{\leftarrow}\mathrm{L{\small FP}}\, X\big[A(X, \bar{z})[\bar{z}]\big]\big)\Big). \tag{15}$$

2. If $B(X)$ is up-monotone in X then for every rough relational structure $\mathcal{R}$,

$$v_{\mathcal{R}}\Big(\exists X\big(\forall \bar{z}(X(\bar{z}) \subseteq A(X, \bar{z})) \wedge B(X)\big)\Big) = \\ v_{\mathcal{R}}\Big(B\big(X{\leftarrow}\mathrm{G{\small FP}}\, X\big[A(X, \bar{z})[\bar{z}]\big]\big)\Big). \tag{16}$$

Proof. (Sketch) The proof is similar to the proof of Theorem 3. In the case of (15) it suffices to notice that the least X satisfying the lefthand side of the equality is defined by the least fixpoint of $A(X)$. In the case of (16) the suitable X is defined by the greatest fixpoint of $A(X)$. ∎

Remark 3. Theorems 3 and 4 provide us with definitions of the least and the greatest rough relations interpreting eliminated relation symbols:

- if the lefthand side of (12) is true then the least relation X satisfying the formula $\forall \bar{z}(A(\bar{z}) \subseteq X(\bar{z})) \wedge B(X)$ is defined by $\forall \bar{z}(X(\bar{z}) \stackrel{\mathrm{def}}{=} A(\bar{z}))$;
- if the lefthand side of (13) is true then the greatest relation X satisfying the formula $\forall \bar{z}(X(\bar{z}) \subseteq A(\bar{z})) \wedge B(X)$ is defined by $\forall \bar{z}(X(\bar{z}) \stackrel{\mathrm{def}}{=} A(\bar{z}))$;
- if the lefthand side of (15) is true then the least relation X satisfying the formula $\forall \bar{z}(A(X, \bar{z}) \subseteq X(\bar{z})) \wedge B(X)$ is defined by $\forall \bar{z}(X(\bar{z}) \stackrel{\mathrm{def}}{=} \mathrm{L{\small FP}}\, X\big[A(X, \bar{z})\big])$;

– if the lefthand side of (16) is true then the greatest relation X satisfying the formula $\forall \bar{z}(A(X, \bar{z}) \subseteq X(\bar{z})) \wedge B(X)$ is defined by $\forall \bar{z}(X(\bar{z}) \overset{\text{def}}{=} \text{GFP } X[A(X, \bar{z})])$.

These definitions can be used for computing lower and upper approximations of the eliminated relations.   ∎

The following lemma shows monotonicity properties of connectives, useful in second-order quantifier elimination. It directly follows from Definition 9.

Lemma 1.

1. $\bar{z} \in X$ is up-monotone in X;
2. $X \subseteq Y$ is down-monotone in X and up-monotone in Y;
3. for $\circ \in \{\vee, \wedge\}$, $X \circ Y$ is up-monotone in X and in Y;
4. $\neg X$ is down-monotone in X;
5. for $Q \in \{\forall, \exists\}$, $Qx(X(x))$ is up-monotone in X.   ∎

5   Applications and Examples

5.1   The Scenario

Below we will use the following notation:

– x, y are variables denoting places and $p_1, \ldots, p_n$ are constants denoting places;
– $ice(x)$ stands for "x being covered by ice", $rain(x)$ – for "rain in x", $freezing(x)$ – for "temperature in x being close to $0^{\circ}C$", $safe(x)$ – for "x being safe" and $base(x)$ indicating that "there is a base in place x";
– $connected(x, y)$ stands for "places x, y being (directly) connected", $slippery(x, y)$ – for "connection from x to y being slippery", and $sconnected(x, y)$ – for "x, y being safely connected" (perhaps indirectly, via a chain of connections $connected()$).

Let us consider a scenario formalized by the following theory T:

$$\forall x \forall y ((ice(x) \vee ice(y)) \subseteq slippery(x, y)) \wedge \tag{17}$$

$$\forall x \forall y ((x = y) \vee connected(y, x)) \subseteq connected(x, y)) \wedge \tag{18}$$

$$\forall x \forall y (((connected(x, y) \wedge \neg slippery(x, y)) \vee$$
$$\exists z(sconnected(x, z) \wedge sconnected(z, y))) \subseteq sconnected(x, y)) \wedge \tag{19}$$

$$\forall x (base(x) \subseteq safe(x)) \wedge \tag{20}$$

$$\forall x (safe(x) \subseteq (base(x) \vee \exists y(sconnected(x, y) \wedge base(y)))) \wedge \tag{21}$$

$$\forall x ((rain(x) \wedge freezing(x)) \subseteq ice(x)). \tag{22}$$

Note that the relations used in (17)–(22) are rough relations which can be specified as a part of the considered theory. For example, given that there are n places, and:

- *connected*()'s lower approximation is $\{\langle p_1, p_{14}\rangle, \ldots, \langle p_{20}, p_n\rangle\}$ and its upper approximation is the complement of $\{\langle p_7, p_9\rangle, \ldots, \langle p_{20}, p_{30}\rangle, \ldots\}$;
- *rain*()'s lower approximation is $\{p_3, \ldots, p_{48}\}$ and its upper approximation is the complement of $\{p_1, p_2\}$;
- *ice*()'s lower and upper approximation is $\{p_1, \ldots, p_{17}\}$;
- *freezing*()'s lower and upper approximation is $\{p_1, \ldots, p_n\}$,

one can add to the theory the following conjunction of rough facts:

$$connected(p_1, p_{14}) \wedge \ldots \wedge connected(p_{20}, p_n) \wedge \tag{23}$$

$$\neg connected(p_7, p_9) \wedge \ldots \wedge \neg connected(p_{20}, p_{30}) \wedge \ldots \wedge$$

$$rain(p_3) \wedge \ldots \wedge rain(p_{48}) \wedge \neg rain(p_1) \wedge \neg rain(p_2) \wedge \tag{24}$$

$$ice(p_1) \wedge \ldots \wedge ice(p_{17}) \wedge \neg ice(p_{18}) \wedge \ldots \wedge \neg ice(p_n) \wedge \tag{25}$$

$$freezing(p_1) \wedge \ldots \wedge freezing(p_n). \tag{26}$$

Remark 4. It is important to note that the conjunction of rough facts, as specified by (23)–(26), does not affect the applicability of the second-order quantifier elimination techniques provided by Theorems 3 and 4. ∎

5.2   Forgetting Rough Concepts

In the first example, let us forget *ice*() in the scenario theory above. That is, we consider *rforget*$(T; ice())$ and, according to Theorem 2, we eliminate $\exists X$ from formula:

$$\exists X \Big(\underbrace{\forall x \big((rain(x) \wedge freezing(x)) \subseteq X(x) \big)}_{\text{corresponding to (22)}} \wedge$$

$$\underbrace{\forall x \forall y \big((X(x) \vee X(y)) \subseteq slippery(x, y) \big)}_{\text{corresponding to (17)}} \wedge B \Big), \tag{27}$$

where $B \overset{\text{def}}{=} \big((18) \wedge (19) \wedge (20) \wedge (21) \big)$. According to Lemma 1, the part of (27) corresponding to (17) is down-monotone in X thus, using equality (12) of Theorem 3, we obtain the following formula equivalent to (27):

$$\forall x \forall y \big(((rain(x) \wedge freezing(x)) \vee (rain(y) \wedge freezing(y))) \subseteq slippery(x, y) \big) \wedge B.$$

In the second example, let us forget *base*() in the scenario theory above. We consider *rforget*$(T; base())$ and apply Theorem 2 to eliminate $\exists X$ from:

$$\exists X \Big(\underbrace{\forall x \big(X(x) \subseteq safe(x) \big)}_{\text{corresponding to (20)}} \wedge$$

$$\underbrace{\forall x \big(safe(x) \subseteq (X(x) \vee \exists y (sconnected(x, y) \wedge X(y))) \big)}_{\text{corresponding to (21)}} \wedge C \Big), \tag{28}$$

where $C \overset{\text{def}}{=} ((17) \wedge (18) \wedge (19) \wedge (22))$. According to Lemma 1, the part of (28) corresponding to (21) is up-monotone in X thus, using equality (13) of Theorem 3, the equivalent of (28) is $\forall x (safe(x) \subseteq (safe(x) \vee \exists y (sconnected(x,y) \wedge safe(y))))$.

Observe that the resulting formula is equivalent to T, so $rforget(T; base())$ is equivalent to C. Indeed, when $base()$ is forgotten, the theory no longer provides useful information about $safe()$, too.

5.3 Forgetting Many-Argument Relations

Forgetting rough relations with more than one argument is very similar to forgetting rough concepts. To illustrate the use of Theorem 4, let us forget $connected()$. That is, consider $rforget(T; connected())$ and, according to Theorem 2, we eliminate $\exists X$ from:

$$\exists X \Big(\underbrace{\forall x \forall y ((x = y \vee X(y,x)) \subseteq X(x,y))}_{\text{corresponding to (18)}} \wedge$$
$$\underbrace{\forall x \forall y (((X(x,y) \wedge \neg slippery(x,y)) \vee}_{\text{corresponding to (19), line 1}} \tag{29}$$
$$\underbrace{\exists z (sconnected(x,z) \wedge sconnected(z,y))) \subseteq sconnected(x,y))}_{(19), \text{ line 2}} \wedge D \Big),$$

where $D \overset{\text{def}}{=} ((17) \wedge (20) \wedge (21) \wedge (22))$.

According to Lemma 1, the part of (29) corresponding to (19) is down-monotone in X thus, using equality (15) of Theorem 4, we obtain the following equivalent of (29):

$$\forall x \forall y \big(((\text{LFP } X(x,y)[x = y \vee X(y,x)](x,y) \wedge \neg slippery(x,y)) \vee \\ \exists z (sconnected(x,z) \wedge sconnected(z,y))) \subseteq sconnected(x,y) \big) \wedge D. \tag{30}$$

Note that $\text{LFP } [\dots]$ in (30) is equivalent to $x = y$, so (30) can further be simplified to:

$$\forall x \forall y \big(((x = y \wedge \neg slippery(x,y)) \vee \\ \exists z (sconnected(x,z) \wedge sconnected(z,y))) \subseteq sconnected(x,y) \big) \wedge D.$$

6 Conclusions

In this paper, we provided basic foundations for the specification and application of a forgetting operator for rough theories. To do this, we defined a logical language for rough theories consisting of rough formulas and a semantics for such formulas containing rough relations, in terms of rough Kleene algebras. Using intuitions from work with forgetting operators in classical logic, we then specified a rough forgetting operator in the context of rough relational theories. We then showed how the constructive generation of the result of applying a forgetting operator to a rough theory could be achieved by using second-order quantifier elimination techniques. These foundations open up opportunities for the use of these rough logics for KRR applications and the study of additional types of forgetting operators in this context, in particular of forgetting in rule languages that use a Kleene logic-based semantics. Also, algorithmic techniques based on insights using second-order quantifier elimination techniques, are worth investigating as a basis for forgetting operators used with rough relational theories.

References

1. Ackermann, W.: Untersuchungen über das eliminationsproblem der mathematischen logik. Mathematische Annalen **110**, 390–413 (1935)
2. Boole, G.: An Investigation of The Laws of Thought on Which are Founded the Mathematical Theories of Logic and Probabilities. Macmillan, London, UK (1854)
3. Ciucci, D.: Orthopairs: a simple and widely used way to model uncertainty. Fundam. Inform. **108**(3–4), 287–304 (2011)
4. Ciucci, D., Dubois, D.: Three-valued logics, uncertainty management and rough sets. In: Peters, J.F., Skowron, A. (eds.) Transactions on Rough Sets XVII. LNCS, vol. 8375, pp. 1–32. Springer, Heidelberg (2014). https://doi.org/10.1007/978-3-642-54756-0_1
5. D'eer, L., Cornelis, C.: A comprehensive study of fuzzy covering-based rough set models: definitions, properties and interrelationships. Fuzzy Sets Syst. **336**, 1–26 (2018)
6. Del-Pinto, W., Schmidt, R.: ABox abduction via forgetting in ALC. In: The 33rd AAAI Conference on Artificial Intelligence, pp. 2768–2775. AAAI Press (2019)
7. Demri, S., Orłowska, E.: Incomplete Information: Structure, Inference, Complexity. EATCS Monographs. Springer, Heidelberg (2002). https://doi.org/10.1007/978-3-662-04997-6
8. Doherty, P., Łukaszewicz, W., Skowron, A., Szałas, A.: Knowledge Representation Techniques. A Rough Set Approach. Studies in Fuzziness and Soft Computing, vol. 202. Springer, Heidelberg (2006). https://doi.org/10.1007/3-540-33519-6
9. Doherty, P., Łukaszewicz, W., Szałas, A.: Computing circumscription revisited. J. Autom. Reason. **18**(3), 297–336 (1997)
10. Doherty, P., Łukaszewicz, W., Szałas, A.: Computing strongest necessary and weakest sufficient conditions of first-order formulas. In: 17th IJCAI, pp. 145–151 (2001)
11. Doherty, P., Szałas, A.: On the correspondence between approximations and similarity. In: Tsumoto, S., Słowiński, R., Komorowski, J., Grzymała-Busse, J.W. (eds.) RSCTC 2004. LNCS (LNAI), vol. 3066, pp. 143–152. Springer, Heidelberg (2004). https://doi.org/10.1007/978-3-540-25929-9_16
12. Eiter, T., Kern-Isberner, G.: A brief survey on forgetting from a knowledge representation and reasoning perspective. KI **33**(1), 9–33 (2019)
13. Gabbay, D., Schmidt, R., Szałas, A.: Second-Order Quantifier Elimination. Foundations, Computational Aspects and Applications, Studies in Logic, vol. 12. College Pub. (2008)
14. Hariri, R., Fredericks, E., Bowers, K.: Uncertainty in big data analytics: survey, opportunities, and challenges. J. Big Data **6**, 44 (2019)
15. Konikowska, B., Avron, A.: Reasoning about covering-based rough sets using three truth values. J. Appl. Log. IfCoLoG J. Log. Appl. **6**(2), 361–382 (2019)

16. Kumar, A., Banerjee, M.: Kleene algebras and logic: Boolean and rough set representations, 3-valued, rough set and Perp semantics. Studia Logica **105**(3), 439–469 (2017)

17. Lin, F.: On strongest necessary and weakest sufficient conditions. In: Cohn, A., Giunchiglia, F., Selman, B. (eds.) Proceedings of the 7th International Conference, KR2000, pp. 167–175. Morgan Kaufmann (2000)

18. Lin, F., Reiter, R.: Forget it! In: Proceedings of the AAAI Fall Symposium on Relevance, pp. 154–159 (1994)

19. Nguyen, L.A., Szałas, A.: Logic-based roughification. In: Skowron, A., Suraj, Z. (eds.) Rough Sets and Intelligent Systems - Professor Zdzisław Pawlak in Memoriam. Intelligent Systems Reference Library, vol. 42, pp. 517–543. Springer, Berlin (2013). https://doi.org/10.1007/978-3-642-30344-9_19

20. Nonnengart, A., Szałas, A.: A fixpoint approach to second-order quantifier elimination with applications to correspondence theory. In: Orłowska, E. (ed.) Logic at Work: Essays Dedicated to the Memory of Helena Rasiowa. Studies in Fuzziness & Soft Computing, vol. 24, pp. 307–328. Springer, Heidelberg (1998)

21. Pawlak, Z.: Information systems - theoretical foundations. Inf. Syst. **6**, 205–218 (1981)

22. Pawlak, Z.: Rough Sets. Theoretical Aspects of Reasoning about Data. Kluwer Academic Publishers, Dordrecht (1991)

23. Polkowski, L.: Rough Sets. Mathematical Foundations, Advances in Intelligent and SoftRough Sets. Mathematical Foundations, Advances in Intelligent and Soft, vol. 15. Physica-Verlag, Heidelberg (2002). https://doi.org/10.1007/978-3-7908-1776-8

24. Skowron, A., Stepaniuk, J.: Tolerance approximation spaces. Fundamenta Informaticae **27**, 245–253 (1996)

25. Słowiński, R., Vanderpooten, D.: A generalized definition of rough approximations based on similarity. IEEE Trans. Knowl. Data Eng. **12**(2), 331–336 (2000)

26. Wang, Y., Wang, K., Zhang, M.: Forgetting for answer set programs revisited. In: Rossi, F. (ed.) Proceedings of the IJCAI 2013, pp. 1162–1168. IJCAI/AAAI (2013)

27. Wang, Z., Wang, K., Zhang, X.: Forgetting and unfolding for existential rules. In: McIlraith, S., Weinberger, K. (eds.) Proceedings of the 32nd AAAI Conference on Artificial Intelligence, pp. 2013–2020. AAAI Press (2018)

28. Wernhard, C.: Literal projection for first-order logic. In: Hölldobler, S., Lutz, C., Wansing, H. (eds.) JELIA 2008. LNCS (LNAI), vol. 5293, pp. 389–402. Springer, Heidelberg (2008). https://doi.org/10.1007/978-3-540-87803-2_32

29. Yao, Y.Y., Wong, S.K.M., Lin, T.Y.: A review of rough set models. In: Lin, T.Y., Cercone, N. (eds.) Rough Sets and Data Mining, pp. 47–75. Springer, Boston (1997). https://doi.org/10.1007/978-1-4613-1461-5_3

30. Zhang, Q., Xie, Q., Wang, G.: A survey on rough set theory and its applications. CAAI Trans. Intell. Technol. **1**(4), 323–333 (2016)

31. Zhao, Y., Schmidt, R.A.: FAME(Q): an automated tool for forgetting in description logics with qualified number restrictions. In: Fontaine, P. (ed.) CADE 2019. LNCS (LNAI), vol. 11716, pp. 568–579. Springer, Cham (2019). https://doi.org/10.1007/978-3-030-29436-6_34

Functional Extensions of Knowledge Representation in General Rough Sets

A. Mani[1,2(✉)] (ID)

[1] HBCSE, Tata Institute of Fundamental Research, Mumbai 400088, India
a.mani.cms@gmail.com, mani@hbcse.tifr.res.in
[2] Indian Statistical Institute, Kolkata 700108, India
https://www.logicamani.in

Abstract. A number of low and high-level models of general rough sets can be used to represent knowledge. Often binary relations between attributes or collections thereof have deeper properties related to decisions, inference or vision that can be expressed in ternary functional relationships (or groupoid operations) – this is investigated from a minimalist perspective in this research by the present author. General approximation spaces and reflexive up-directed versions thereof are used by her as the basic frameworks. Related semantic models are invented and an interpretation is proposed in this research. Further granular operator spaces and variants are shown to be representable as partial algebras through the method. An analogous representation for all covering spaces does not necessarily hold. Applications to education research contexts that possibly presume a distributed cognition perspective are also outlined.

Keywords: General approximation spaces · Up-directedness · Rough objects · Mereology · Groupoidal semantics · Parthood · Knowledge · AI · Higher granular operator spaces · Contamination problem · Education research

1 Introduction

In relational approach to general rough sets various granular, pointwise or abstract approximations are defined, and rough objects of various kinds are studied [1–6]. These approximations may be derived from information tables or may be abstracted from data relating to human (or machine) reasoning. A *general approximation space* is a pair of the form $S = \langle \underline{S}, R \rangle$ with $\underline{S}$ being a set and R being a binary relation (S and $\underline{S}$ *will be used interchangeably throughout this paper*). Approximations of subsets of $\underline{S}$ may be generated from these and studied at different levels of abstraction in theoretical approaches to rough sets. Because approximations and related semantics are of interest here, the relational system is much more than a general frame. *Often it happens that S is interpreted as a set of attributes and that any two elements of S may be associated with a third element through a mechanism of reasoning, by preference, or via decision-making*

© Springer Nature Switzerland AG 2020
R. Bello et al. (Eds.): IJCRS 2020, LNAI 12179, pp. 19–34, 2020.
https://doi.org/10.1007/978-3-030-52705-1_2

guided by an external mechanism. The purpose of this research is to study these situations from a minimalist perspective. This approach also directly adds to the concept of knowledge in classical rough sets [7] and in general rough sets [3, 8–10] and therefore the study is referred to as an extension of the same.

Mereology, the study of parts and wholes, has been studied from philosophical, logical, algebraic, topological and applied perspectives. In the literature on mereology [9,11,12], it is argued that most ideas of binary *part of* relations in human reasoning are at least antisymmetric and reflexive. *A major reason for not requiring transitivity of the parthood relation is because of the functional reasons that lead to its failure* (see [11]), and to accommodate *apparent parthood* [12]. The study of mereology in the context of rough sets can be approached in at least two essentially different ways. In the approach aimed at reducing contamination by the present author [1,2,8,10], the primary motivation is to avoid intrusion into the data by way of additional assumptions about the data relative to the semantic domain in question. In numeric function based approaches [13], the strategy is to base definitions of parthood on the degree of rough inclusion or membership – this differs substantially from the former approach. Rough Y-systems and granular operator spaces, introduced and studied extensively by the present author [1,2,8,10,12], are essentially higher order abstract approaches in general rough sets in which the primitives are ideas of approximations, parthood, and granularity. Part-of relations can also be the subject of considerations mentioned in the first paragraph, and the relation R in a general approximation space can be a parthood. Specific versions of parthood spaces have been investigated in a forthcoming joint work by the present author. Relative to that work new results on parthood spaces are proved, up-directedness is studied in classical approximation spaces, and the formalism on granular operator spaces and variants are improved in this research. Applications to education research contexts are also outlined.

1.1 Background

An *information table* $\mathcal{I}$, is a tuple of the form

$$\mathcal{I} = \langle S, \, \mathbb{A}, \, \{V_a : a \in \mathbb{A}\}, \, \{f_a : a \in \mathbb{A}\}\rangle$$

with S, $\mathbb{A}$ and V_a being sets of *objects*, *attributes* and *values* respectively. Information tables generate various types of relational or relator spaces which in turn relate to approximations of different types and form a substantial part of the problems encountered in general rough sets.

In classical rough sets [7], equivalence relations of the form R are derived by the condition $x, y \in S$ and $B \subseteq \mathbb{A}$, let $(x, y) \in R$ if and only if $(\forall a \in B)\nu(a, x) = \nu(a, y)$. $\langle S, R \rangle$ is then an *approximation space*. On the power set $\wp(S)$, lower and upper approximations of a subset $A \in \wp(S)$ operators, (apart from the usual Boolean operations), are defined as per: $A^l = \bigcup_{[x] \subseteq A}[x]$, $A^u = \bigcup_{[x] \cap A \neq \emptyset}[x]$, with $[x]$ being the equivalence class generated by $x \in S$. If $A, B \in \wp(S)$, then A is said to be *roughly included* in B ($A \sqsubseteq B$) if and only if

$A^l \subseteq B^l$ and $A^u \subseteq B^u$. A is roughly equal to B ($A \approx B$) if and only if $A \sqsubseteq B$ and $B \sqsubseteq A$ (the classes of $\approx$ are rough objects).

The rough domain corresponds to rough objects of specific type, while the classical and hybrid one correspond to all and mixed types of objects respectively [2]. Boolean algebra with approximation operators forms a classical rough semantics. This fails to deal with the behavior of rough objects alone. The scenario remains true even when R in the approximation space is replaced by arbitrary binary relations. In general, $\wp(S)$ can be replaced by a set with a parthood relation and some approximation operators defined on it as in [2]. The associated semantic domain is the classical one for general Rough sets. The domain of discourse associated with roughly equivalent sets is a *rough semantic domain*. Hybrid domains can also be generated and have been used in the literature [1].

The problem of reducing confusion among concepts from one semantic domain in another is referred to as the contamination problem. Use of numeric functions like rough membership and inclusion maps based on cardinalities of subsets are also sources of contamination. The rationale can also be seen in the definition of operations like $\sqcup$ in pre-rough algebra (for example) that seek to define interaction between rough objects but use classical concepts that do not have any interpretation in the rough semantic domain. Details can be found in [14]. In machine learning practice, whenever inherent shortcomings in algorithmic framework being used are the source of noise then the frameworks may be said to be contaminated.

Key concepts used in the context of general rough sets (and also high granular operator spaces [1, 10]) are mentioned next.

- A *crisp object* is one that has been designated as *crisp* or is an approximation of some other object.
- A *vague object* is one whose approximations do not coincide with itself or that which has been designated as a *vague* object.
- An object that is explicitly available for computations in a rough semantic domain (in a contamination avoidance perspective) is a *discernible object*.
- Many definitions and representations are associated with the idea of *rough objects*. From the representation point of view these are usually functions of definite or crisp or approximations of objects. Objects that are invariant relative to an approximation process are said to be *definite objects*. In rough perspectives of knowledge [7, 8], algebraic combinations of definite objects (in some sense) or granules are assumed to correspond to crisp concepts, and knowledge to specific collections of crisp concepts. *It should be mentioned that non algebraic definitions are excluded in the present author's axiomatic approach* [1, 2, 10].

Definition 1. *A partial algebra (see [15]) P is a tuple of the form*

$$\langle \underline{P}, f_1, f_2, \ldots, f_n, (r_1, \ldots, r_n) \rangle$$

with $\underline{P}$ being a set, f_i's being partial function symbols of arity r_i. The interpretation of f_i on the set $\underline{P}$ should be denoted by $f_i^{\underline{P}}$, but the superscript will be *dropped in this paper as the application contexts are simple enough. If predicate symbols enter into the signature, then P is termed a* partial algebraic system.

In this paragraph the terms are not interpreted. For two terms s, t, $s \stackrel{\omega}{=} t$ shall mean, if both sides are defined then the two terms are equal (the quantification is implicit). $\stackrel{\omega}{=}$ is the same as the existence equality (also written as $\stackrel{e}{=}$) in the present paper. $s \stackrel{\omega^*}{=} t$ shall mean if either side is defined, then the other is and the two sides are equal (the quantification is implicit). $\stackrel{\omega^*}{=}$ is written as $\stackrel{s}{=}$ in [18]). Note that the latter equality can be defined in terms of the former as

$$(s \stackrel{\omega}{=} s \longrightarrow s \stackrel{\omega}{=} t) \,\&\, (t \stackrel{\omega}{=} t \longrightarrow s \stackrel{\omega}{=} t)$$

2   Relations and Groupoids

Under certain conditions, partial or total groupoid operations can correspond to binary relations on a set.

Definition 2. *In a general approximation space $S = \langle \underline{S}, R \rangle$ consider the following conditions:*

$$(\forall a, b)(\exists c) Rac \,\&\, Rbc \qquad \text{(up-dir)}$$
$$(\forall a) Raa \qquad \text{(reflexivity)}$$
$$(\forall a, b)(Rab \,\&\, Rba \longrightarrow a = b) \qquad \text{(anti-sym)}$$

If S satisfies up-dir, then it shall said to be a up-directed approximation space. If it satisfies the last two then it shall said to be a parthood space and a up-directed parthood space when it satisfies all three.

The condition up-dir is equivalent to the set $U_R(a, b) = \{x : Rax \,\&\, Rbx\}$ being nonempty for every $a, b \in S$ and is also referred to as *directed* in the literature. It is avoided because it may cause confusion.

The problem of rewriting the semantic content of binary relations of different kinds using total or partial operations has been of much interest in algebra (for example [16,17]). Results on using partial operations for the purpose are of more recent origin [18,19].

Definition 3. *If R is a binary relation on S, then a* type-1 partial groupoid operation *(1PGO) determined by R is defined as follows:*

$$(\forall a, b) \, a \circ b = \begin{cases} b & \text{if } Rab \\ c & c \in U_R(a, b) \,\&\, \neg Rab \\ \text{undefined} & \text{otherwise} \end{cases}$$

If R is up-directed, then the operation is total. In this case, the collection of groupoids satisfying the condition will be denoted by $\mathfrak{B}(S)$ and an arbitrary element of it will be denoted by $\mathsf{B}(S)$. If R is not up-directed, then the collection of partial groupoids associated will be denoted by $\mathfrak{B}_p(S)$. The term 'a $\circ$ b' will be written as 'ab' for convenience.

Theorem 1. *The partial operation $\circ$ corresponds to a binary relation R if and only if*

$$(\forall a, b)(\exists z)(ab \neq b \,\&\, az = bz = z \rightarrow a(ab) = b(ab) = ab)$$
$$(\forall a, b, c)(ab = c \rightarrow c = b \text{ or } (\exists z)az = bz = z)$$

The following results have been proved for relational systems in [18, 19].

Theorem 2. *For a groupoid A, the following are equivalent*

- *A reflexive up-directed approximation space S corresponds to A*
- *A satisfies the equations*

$$aa = a \,\&\, a(ab) = b(ab) = ab$$

Definition 4. *If A is a groupoid, then two general approximation spaces corresponding to it are $\Re(A) = \langle \underline{A}, R_A \rangle$ and $\Re^*(A) = \langle \underline{A}, R_A^* \rangle$ with*

$$R_A = \{(a, b) : ab = b\}$$
$$R_A^* = \bigcup\{(a, ab), (b, ab)\}$$

Theorem 3. – *If A is a groupoid then $\Re^*(A)$ is up-directed.*
- *If a groupoid $A \models a(ab) = b(ab) = ab$ then $\Re(A) = \Re^*(A)$.*
- *If S is an up-directed approximation space then $\Re((B)(S)) = S$.*

Theorem 4. *If $S = \langle \underline{S}, R \rangle$ is a up-directed approximation space, then*

- *R is reflexive $\Leftrightarrow \mathsf{B}(S) \models aa = a$.*
- *R is symmetric $\Leftrightarrow \mathsf{B}(S) \models (ab)a = a$.*
- *R is transitive $\Leftrightarrow \mathsf{B}(S) \models a((ab)c) = (ab)c$.*
- *If $\mathsf{B}(S) \models ab = ba$ then R is antisymmetric.*
- *If $\mathsf{B}(S) \models (ab)a = ab$ then R is antisymmetric.*
- *If $\mathsf{B}(S) \models (ab)c = a(bc)$ then R is transitive.*

Morphisms between up-directed approximation spaces are preserved by corresponding groupoids in a nice way. This is an additional reason for investigating the algebraic perspective.

3   Up-Directed General Approximation Spaces

In general, partial/quasi orders, and equivalences need not satisfy up-dir. When they do satisfy the condition, then the corresponding general approximation spaces will be referred to as *up-directed general approximation spaces*.

For any element $a \in S$, the neighborhood granule $[a]$ and inverse neighborhood $[a]_i$ associated with it in a general approximation space shall be given by $[a] = \{x : Rxa\}$ and $[a]_i = \{x : Rax\}$ respectively.

Definition 5. *For any subset $A \subseteq S$, the following approximations can be defined:*

$$A^l = \bigcup\{[a] : [a] \subseteq A\} \qquad \text{(lower)}$$

$$(1)$$

$$A^u = \bigcup\{[a] : \exists z \in [a] \cap A\} \qquad \text{(upper)}$$

If inverse neighborhoods are used instead, then the corresponding approximations will be denoted by l_i and u_i respectively.

3.1 Classical Approximation Spaces

If an approximation space is up-directed, then it is essentially redundant with respect to the relation. Proof of the following theorem is not hard and can be found in a forthcoming paper due to the present author.

Theorem 5. *Let S be an approximation space, then all of the following hold:*

- *If R is up-directed, then $S^2 = R$.*
- *If R is not up-directed, then the groupoid operation of Definition 3 is partial and it satisfies*

$$(\forall a, b, c)(ab = c \longrightarrow b = c)$$

- *For each $x \in S$, $[x]$ is closed under $\circ$ and so every equivalence class is a total groupoid that satisfies:*

$$(\forall a, b, c)\, aa = a\ \&\ (ab)a = a\ \&\ a((ab)c) = (ab)c$$

Definition 6. *On the power set $\wp(S)$, the partial operation $\circ$ induces a total operation as in Eq. 2.*

$$(\forall A, B \in \wp(S))\, A \odot B = \{x : (\exists a \in A)(\exists b \in B)\, ab = x\} \qquad (2)$$

Proposition 1. *If S is an approximation space then $\left\langle \wp(S), \cup, \cap, {}^c, l, u, \circ, \bot, \top \right\rangle$ is a Boolean algebra with operators enhanced by a groupoid operation that satisfies all of the following (apart from the well known conditions):*

$$(\forall a, b)\, aa \cap a = aa\ \&\ ab \cap b = ab \qquad \text{(pre-refl)}$$

$$(\forall a, b, c)\, ((a \cup b)c) \cap ((ac) \cup (bc)) = (ac) \cup (bc) \qquad \text{(pre-mo)}$$

$$(\forall a, b, c)\, (a \cup b = b \longrightarrow (ac) \cup (bc) = bc) \qquad \text{(mo)}$$

$$(\forall a, b)\, (ab)^l \cup b^l = b^l \qquad \text{(l-mo)}$$

$$(\forall a, b)\, (ab)^u \cap b^u = (ab)^u \qquad \text{(u-mo)}$$

Proof. – Note that by the definition of the partial groupoid operation, for any two sets $a, b \in \wp(S)$ ab must be a subset of b. So the pre-refl property holds.

– pre-mo is again a consequence of pre-refl.

- If a is a subset of b, then ac must again be a subset of bc which in turn would be a subset of c. This can be verified by a purely set-theoretic argument.
- ab must be a subset of b. So $(ab)^l$ must be subset of b^l. It follows that their union must be the latter.

Because classes are closed under the groupoid operation, it follows that

Theorem 6. *On the set of definite elements $\delta(S)$ of an approximation space S, the induced operations from the algebra in Proposition 1 again forms a Boolean subalgebra with groupoid operations that satisfies reflexivity $(\forall a)\, aa = a$.*

It should be noted that up-directedness is not essential for a relation to be represented by groupoid operations. The following construction that differs in part from the above strategy can be used for partially ordered sets as well, and has been used by the present author in [20] in the context of knowledge generated by approximation spaces. The method relates to earlier algebraic results including [21, 22]. The groupoidal perspective can be extended for quasi ordered sets.

If $S = \langle \underline{S}, R \rangle$ is an approximation space, then define (for any $a, b \in S$)

$$a \circledast b = \begin{cases} a \text{ if } Rab \\ b \text{ if } \neg Rab \end{cases} \tag{3}$$

Relative to this operation, the following theorem (see [21]) holds:

Theorem 7. $\langle S, \circledast \rangle$ *is a groupoid that satisfies the following axioms (braces are omitted under the assumption that the binding is to the left, e.g. 'abc' is the same as '(ab)c'):*

$$xx = x \tag{E1}$$
$$x(az) = (xa)(xz) \tag{E2}$$
$$xax = x \tag{E3}$$
$$azxauz = auz \tag{E4}$$
$$u(azxa)z = uaz \tag{E5}$$

3.2   Parthood Spaces

Definition 7. *Let S be a parthood space, then let $S_{lu} = \{x : x = a^l \text{ or } x = a^u \,\&\, a \in S\}$. On S_{lu}, the following operations can be defined (apart from l and u by restriction):*

$$a \sqcap b = (a \cap b)^l \tag{Cap}$$
$$a \sqcup b = (a \cup b)^u \tag{Cup}$$
$$\bot = \emptyset; \ \top = S^u \tag{iu34}$$

The resulting algebra $S_{lu} = \langle \underline{S_{lu}}, \sqcap, \sqcup, \cup, l, u, \bot, \top \rangle$ will be called the algebra of approximations in a up-directed space *(UA algebra). If R is a up-directed* *parthood relation or a reflexive up-directed relation respectively, then it shall said to be a up-directed parthood algebra of approximations (AP algebra) or a reflexive up-directed algebra of upper approximations (AR algebra) respectively.*

Theorem 8. *A AP algebra S_{lu} satisfies all of the following (universal quantifiers have been omitted):*

$$a \barwedge a = a \ \& \ (a \Cup a) \barwedge a = a \tag{idp3}$$

$$a \Cup a = a^u \tag{qidp4}$$

$$a \barwedge b = b \barwedge a \ \& \ a \Cup b = b \Cup a \tag{com12}$$

$$a \barwedge (b \Cup a) = a \tag{habs}$$

$$a \Cup (b \Cup c) = (a \Cup b^u) \Cup c^u \tag{qas1}$$

$$(a \Cup (b \Cup c)) \Cup ((a \Cup b) \Cup c) =$$
$$((a \Cup a) \Cup (b \Cup b)) \Cup (c \Cup c \Cup c) \tag{qas0}$$

Proof. **idp3** $a \barwedge a = (a \cap a)^l = a^l = a$ and $a \Cup a = a^u$ and $a^u \cap a = a$
qidp4 $a \Cup a = (a \cup a)^u = a^u$.
com12 This follows from definition.
habs $a \barwedge (b \Cup a) = (a \cap (b \cup a)^u)^l = ((a \cap a^u) \cup (a \cap b^u))^l$ which is equal to $(a \cup (a \cap b^u))^l = a^l = a$
qas1 $a \Cup (b \Cup c) = (a \cup (b \cup c)^u)^u = (a^u \cup b^{uu} \cup c^{uu})$ and this is $(a \cup b^u))^u \cup c^{uu} = (a \cup b^u) \Cup c^u$
qas0 This can be proved by writing all terms in terms of $\cup$. In fact $(a \Cup (b \Cup c)) \Cup ((a \Cup b) \Cup c) = a^{uuu} \cup b^{uuu} \cup c^{uuu}$. The expression on the right can be rewritten in terms of $\Cup$ by qidp4.

The above two theorems in conjunction with the properties of approximations on the power set, suggest that it would be useful to enhance UA-, AP-, and AR-algebras with partial operations for defining an abstract semantics.

Definition 8. *A partial algebra of the form*

$$S_{lu}^* = \langle \underline{S_{lu}}, \barwedge, \Cup, \cup, \sqcap, {}^\kappa, l, u, \bot, \top \rangle$$

will be called the algebra of approximations in a up-directed space *(UA partial algebra) whenever $S_{lu} = \langle \underline{S_{lu}}, \barwedge, \Cup, \cup, l, u, \bot, \top \rangle$ is a UA algebra and $\sqcap$ and ${}^\kappa$ are defined as follows ($\cap$ and ^{c} being the intersection and complementation operations on $\wp(S)$):*

$$(\forall a, b \in S_{lu}) \, a \sqcap b = \begin{cases} a \cap b & if \ a \cap b \in S_{lu} \\ undefined & otherwise \end{cases} \tag{4}$$

$$(\forall a \in S_{lu}) \, a^\kappa = \begin{cases} a^c & if \ a^c \in S_{lu} \\ undefined & otherwise \end{cases} \tag{5}$$

If R is an up-directed parthood relation or a reflexive up-directed relation respectively, then it shall said to be a up-directed parthood partial algebra of approximations (AP partial algebra) or a reflexive algebra of upper approximations (AR partial algebra) respectively.

Theorem 9. *If S is an up-directed approximation space, then its associated enhanced up-directed parthood partial algebra $S_{lu}^* = \langle \underline{S_{lu}}, \sqcap, \uplus, \cup, \sqcap, \circ, ^\kappa, l, u, \bot, \top \rangle$ satisfies all of the following:*

$$\langle \underline{S_{lu}}, \sqcap, \uplus, \cup, l, u, \bot, \top \rangle \text{ is a AP algebra} \tag{app1}$$

$$a \sqcap a = a \,\&\, a \sqcap \bot = \bot \,\&\, a \sqcap \top = a \tag{app2}$$

$$a \sqcap b \overset{\omega}{=} b \sqcap a \,\&\, a \sqcap (b \sqcap c) \overset{\omega}{=} (a \sqcap b) \sqcap c \tag{app3}$$

$$a \sqcap a^u = a = a \sqcap a^l \,\&\, a^{\kappa\kappa} \overset{\omega}{=} a \tag{app4}$$

$$a \sqcap (b \cup c) \overset{\omega}{=} (a \sqcap b) \cup (a \sqcap c) \tag{app5.0}$$

$$a \cup (b \sqcap c) \overset{\omega}{=} (a \cup b) \sqcap (a \cup c) \tag{app5.1}$$

$$(a \sqcap b)^\kappa \overset{\omega}{=} a^\kappa \cup b^\kappa \,\&\, (a \cup b)^\kappa \overset{\omega}{=} a^\kappa \sqcap b^\kappa \tag{app6}$$

$\circ$ *is the partial groupoid operation induced from its power set.*

Proof. The theorem follows from the previous theorems in this section.

If the parthood relation is both up-directed and also transitive, then it is possible to have an induced groupoid operation on the set of definite elements $(\delta_{l_i u_i}(S) = \{x : x_i^l = x_i^u \,\&\, x \in \wp(S)\}$. If $A, B \in \delta_{lu}(S)$, then let $A \cdot B = \{ab \, a \in A, \,\&\, b \in B\}$. $\delta_{lu}(S)$ is closed under set union and intersection, and the pseudo-complementation $^+$ is defined from $[x]_i^+ = \bigcup \{A : A \in \delta_{lu(S)} \,\&\, A \cap [x]_i = \emptyset$ for any $x \in S$.

Theorem 10. *If S is an up-directed parthood space in which R is transitive, then $\langle \delta_{l_i u_i}(S), \cdot, \cap, \cup, ^+, 0, 1 \rangle$ is a Heyting algebra with an extra groupoid operation induced by the groupoid operation on S.*

Proof. The proof that $\langle \delta_{l_i u_i}(Q), \cap, \cup, ^+, \bot, \top \rangle$ is a Heyting algebra is analogous to the proof in [23].

If $a, b \in [x]_i$, for an inverse neighborhood granule, then there exists a c such that Rac and Rbc, but by the definition of $[x]$, $c \in [x]$ follows. Therefore $[x]_i$ is a subgroupoid of S for each $x \in S$.

Suppose $A, B \in \delta_{l_i u_i}(S)$, then (as any element in these sets must be in some granules) for any $a \in [x]_i \subseteq A$ and $b \in [z]_i \subseteq B$, ab is in the order filter generated by $[x] \cup [z]$. So AB must be an element of $\delta_{l_i u_i}(S)$. This essentially proves the theorem.

3.3   Examples, Meaning and Interpretation

Abstract examples are easy to construct for the situations covered and many are available in other papers [11,12,18] by the present author and others. So an application strategy to student-centric learning (a constructive teaching method in which students learn by explorative open-ended activities) is proposed. It should be noted that education researchers adhere to various ideas of distributed cognition (that the environment has a key role in cognitive process that are inherently personal [24]) and so the basic assumptions of formal concept analysis may be limiting [25]. Suppose a student has access to a set K of concepts and is likely to arrive at a another set of potentially vague concepts H. Teachers typically play the role of facilitators, are not required to be the sole source of knowledge, and would need to direct the activity to an improved set of concepts H^+. In the construction of these sets, groupoidal operations can play a crucial role. Equations of the form $ab = c$ can be read in terms of concepts – c can be a better relevant concept for the activity in comparison to the a and b. Note that no additional order structure on the set of concepts is presumed. *This is important also because concepts may not be structured as in lattice-theoretic perspective of formal concept analysis or classical rough sets.*

In classical rough sets, definite concepts correspond to approximations (definite objects). From the present study, it can be seen that the induced total groupoid operation on the set of definite objects is the part of a concept b that can be read from another concept a. This interpretation is primarily due to the relation R being symmetric and reflexive. When the approximation space is up-directed, then it happens that every object is indiscernible from every other object. So the property of up-directedness is not of much interest in the classical context. The ⊛ operation concerns choice between two things and so is relevant for pairwise comparisons [26].

In parthood and other up-directed general approximation spaces, a groupoid operation typically corresponds to answering the question *which attribute or object is preferable to two given attributes or objects?* Therefore collections of all possible definable groupoid operations correspond to all answers. Ideas of vision then must be about choices of subsets or subclasses among possible definable operations. Formally,

Definition 9. *A* vision *for an up-directed approximation space, S is a subset $\mathcal{V}(S)$ of $\mathfrak{B}(S)$.*

4   Formalism of Higher Granular Operator Spaces

Granular operator spaces and variants [1,8,10,27] are abstract frameworks for extending granularity and parthood in the context of general rough sets, and are also variants of rough Y-systems studied by the present author [2]. In this section, it will be shown that all types of granular operator spaces and variants can be transformed into partial algebras that satisfy additional conditions. This is also nontrivial because all covering approximation spaces cannot be transformed in the same way.

Definition 10. *A* High General Granular Operator Space *(GGS)* $\mathbb{S}$ *shall be a partial algebraic system of the form* $\mathbb{S} = \langle \underline{\mathbb{S}}, \gamma, l, u, \mathbf{P}, \leq, \vee, \wedge, \bot, \top \rangle$ *with* $\underline{\mathbb{S}}$ *being a set, γ being a unary predicate that determines $\mathcal{G}$ (by the condition γx if and only if $x \in \mathcal{G}$) an* admissible granulation*(defined below) for* $\mathbb{S}$ *and l, u being operators :* $\underline{\mathbb{S}} \longmapsto \underline{\mathbb{S}}$ *satisfying the following ($\underline{\mathbb{S}}$ is replaced with $\mathbb{S}$ if clear from the* *context. $\vee$ and $\wedge$ are idempotent partial operations and $\mathbf{P}$ is a binary predicate. Further γx will be replaced by $x \in \mathcal{G}$ for convenience.):*

$$(\forall x)\mathbf{P}xx \tag{PT1}$$

$$(\forall x, b)(\mathbf{P}xb \ \& \ \mathbf{P}bx \longrightarrow x = b) \tag{PT2}$$

$$(\forall a, b)a \vee b \overset{\omega}{=} b \vee a \ ; \ (\forall a, b)a \wedge b \overset{\omega}{=} b \wedge a \tag{G1}$$

$$(\forall a, b)(a \vee b) \wedge a \overset{\omega}{=} a \ ; \ (\forall a, b)(a \wedge b) \vee a \overset{\omega}{=} a \tag{G2}$$

$$(\forall a, b, c)(a \wedge b) \vee c \overset{\omega}{=} (a \vee c) \wedge (b \vee c) \tag{G3}$$

$$(\forall a, b, c)(a \vee b) \wedge c \overset{\omega}{=} (a \wedge c) \vee (b \wedge c) \tag{G4}$$

$$(\forall a, b)(a \leq b \leftrightarrow a \vee b = b \ \leftrightarrow \ a \wedge b = a) \tag{G5}$$

$$(\forall a \in \mathbb{S}) \, \mathbf{P}a^l a \ \& \ a^{ll} = a^l \ \& \ \mathbf{P}a^u a^{uu} \tag{UL1}$$

$$(\forall a, b \in \mathbb{S})(\mathbf{P}ab \longrightarrow \mathbf{P}a^l b^l \ \& \ \mathbf{P}a^u b^u) \tag{UL2}$$

$$\perp^l = \perp \ \& \ \perp^u = \perp \ \& \ \mathbf{P}\top^l\top \ \& \ \mathbf{P}\top^u\top \tag{UL3}$$

$$(\forall a \in \mathbb{S}) \, \mathbf{P}\perp a \ \& \ \mathbf{P}a\top \tag{TB}$$

Let $\mathbb{P}$ stand for proper parthood, defined via $\mathbb{P}ab$ if and only if $\mathbf{P}ab \ \& \ \neg\mathbf{P}ba$). A granulation is said to be admissible if there exists a term operation t formed from the weak lattice operations such that the following three conditions hold:

$$(\forall x \exists x_1, \dots x_r \in \mathcal{G}) \, t(x_1, x_2, \dots x_r) = x^l$$

$$\text{and } (\forall x) \, (\exists x_1, \dots x_r \in \mathcal{G}) \, t(x_1, x_2, \dots x_r) = x^u, \qquad \text{(Weak RA, WRA)}$$

$$(\forall a \in \mathcal{G})(\forall x \in \mathbb{S}) \, (\mathbf{P}ax \longrightarrow \mathbf{P}ax^l), \qquad \text{(Lower Stability, LS)}$$

$$(\forall x, a \in \mathcal{G})(\exists z \in \mathbb{S}) \, \mathbb{P}xz, \ \& \ \mathbb{P}az \ \& \ z^l = z^u = z, \qquad \text{(Full Underlap, FU)}$$

The conditions defining admissible granulations mean that every approximation is somehow representable by granules in a algebraic way, that every granule coincides with its lower approximation (granules are lower definite), and that all pairs of distinct granules are part of definite objects (those that coincide with their own lower and upper approximations). Special cases of the above are defined next.

Definition 11. – *In a GGS, if the parthood is defined by $\mathbf{P}ab$ if and only if $a \leq b$ then the GGS is said to be a* high granular operator space GS.
– *A* higher granular operator space *(HGOS) $\mathbb{S}$ is a GS in which the lattice operations are total.*
– *In a higher granular operator space, if the lattice operations are set theoretic union and intersection, then the HGOS will be said to be a set HGOS.*

Theorem 11. *In the context of Definition 10, the binary predicates $\mathbf{P}$ can be replaced by partial two-place operations 1PGO $\odot$ and γ is replaceable by a total unary operation h defined as follows:*

$$hx = \begin{cases} x & \text{if } \gamma x \\ \perp & \text{if } \neg\gamma x \end{cases} \tag{6}$$

Consequently $\mathbb{S}^+ = \langle \underline{\mathbb{S}}, h, l, u, \odot, \vee, \wedge, \perp, \top \rangle$ *is a partial algebra that is seman-tically (and also in a category-theoretic sense) equivalent to the original GGS* $\mathbb{S}$.

Proof. Because of the restriction UL3 on $\perp$ and the redundancy of $\leq$ (because of G5), the result follows.

Definition 12. *The partial algebra formed in the above theorem will be referred to a* high granular operator partial algebra *(GGSp).*

Problem 1. All covering approximation spaces considered in the rough set liter-ature actually assume partial Boolean or partial lattice theoretical operations. Some authors (especially in modal logic perspectives) [3,5,28] presume that all Boolean operations are admissible – this view can be argued against. A natural question is *Are the modal logic semantics themselves only a possible interpreta-tion of the actuality?* All this suggests the problem of finding minimal operations involved in the context.

Because all covering approximation spaces do not use granular approxima-tions in the sense mentioned above, it follows that they do not form GGSo always. In the next example, the applicability of the above to activity based mathematics teaching is considered.

Example 1. In constructivist activity based learning, teachers almost always set learning goals ahead of initiating activities. Therefore knowledge constructed in such contexts are constrained by concept maps (typically directed) accessible to the teacher in question [29–31]. As a consequence desired concept granules (and explicit or implicit ontology) can be specified by teachers. But students and teachers are likely to make use of a number of additional vague or exact concepts in any specific activity. In addition, general ideas of parthood as specified in the definition of GGS can be interpreted over the collection of vague and exact concepts. It may even make sense to define additional groupoid operations apart from the ones induced by the relations. [30,31] do not make room for vague concepts and presume a transitive parthood that operates over the teacher's goals.

For example, in [32], the goal of the game activity is to understand and apply Pythagoras theorem in few situations. The board game (see Fig. 1) involves students throwing a pair of die, form the square root of the sum of the squares of the values obtained and round off the result to a whole number and advance that many squares on the board. The goal of the game (for students) is to reach the finish block. It can be seen that concepts such as sample space, events, floor and ceiling functions, and vague variants thereof, incorrect concepts of biased dies are all part of the potential learning space. All these can be approximated (irrespective of consequence) relative to the teacher's specification of granules. Moreover they may improperly specify the relation between concepts that are of lesser interest to the lesson plan.

It is not hard to see that the generalized scenario described in the last two paragraphs can be modeled by a GGSp.

An expanded version of the last example will appear separately.

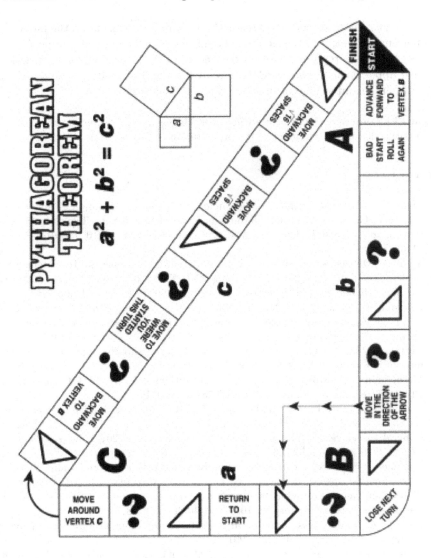

Fig. 1. Board game

5   Further Directions and Remarks

In this research methods of representing important ideas of decisions or preferences inherent in information tables (related to data including those relating to human reasoning) have been invented and the semantics considered in two types of rough sets. A representation theorem is proved for transitive parthood spaces. Further the formalism of higher granular operator spaces and variants are shown to be representable as partial algebras. Examples illustrating key aspects of the research in education research contexts have also been constructed.

Among the many directions of research motivated by this paper, the following are more important: a finer algebraic classification of the derived groupoids and partial groupoids, representation of derived partial algebras as quasi varieties, detailed application to education research contexts (especially in the direction indicated in Example 1), and in self-organizing systems.

Acknowledgement. The present author would like to thank the reviewers for useful remarks that led to improvement of the presentation of this research.

References

1. Mani, A.: Algebraic methods for granular rough sets. In: Mani, A., Düntsch, I., Cattaneo, G. (eds.) Algebraic Methods in General Rough Sets. Trends in Mathematics, pp. 157–336. Birkhauser, Basel (2018). https://doi.org/10.1007/978-3-030-01162-8_3

2. Mani, A.: Dialectics of counting and the mathematics of vagueness. In: Peters, J.F., Skowron, A. (eds.) Transactions on Rough Sets XV. LNCS, vol. 7255, pp. 122–180. Springer, Heidelberg (2012). https://doi.org/10.1007/978-3-642-31903-7_4

3. Pagliani, P., Chakraborty, M.: A Geometry of Approximation: Rough Set Theory: Logic Algebra and Topology of Conceptual Patterns. Trends in Mathematics. Springer, Dordrecht (2008). https://doi.org/10.1007/978-1-4020-8622-9

4. Cattaneo, G.: Algebraic methods for rough approximation spaces by lattice interior-closure operations. In: Mani, A., Düntsch, I., Cattaneo, G. (eds.) Algebraic Methods in General Rough Sets. Trends in Mathematics, pp. 13–156. Birkhäuser, Cham (2018). https://doi.org/10.1007/978-3-030-01162-8_2

5. Pagliani, P.: Three lessons on the topological and algebraic hidden core of rough set theory. In: Mani, A., Cattaneo, G., Düntsch, I. (eds.) Algebraic Methods in General Rough Sets. TM, pp. 337–415. Springer, Cham (2018). https://doi.org/10.1007/978-3-030-01162-8_4

6. Cattaneo, G., Ciucci, D.: Algebraic methods for orthopairs and induced rough approximation spaces. In: Mani, A., Düntsch, I., Cattaneo, G. (eds.) Algebraic Methods in General Rough Sets. Trends in Mathematics, pp. 553–640. Birkhäuser, Basel (2018). https://doi.org/10.1007/978-3-030-01162-8_7

7. Pawlak, Z.: Rough Sets: Theoretical Aspects of Reasoning About Data. Kluwer Academic Publishers, Dodrecht (1991)

8. Mani, A.: Knowledge and consequence in AC semantics for general rough sets. In: Wang, G., Skowron, A., Yao, Y., Ślęzak, D., Polkowski, L. (eds.) Thriving Rough Sets. SCI, vol. 708, pp. 237–268. Springer, Cham (2017). https://doi.org/10.1007/978-3-319-54966-8_12

9. Mani, A.: Algebraic semantics of proto-transitive rough sets. In: Peters, J.F., Skowron, A. (eds.) Transactions on Rough Sets XX. LNCS, vol. 10020, pp. 51–108. Springer, Heidelberg (2016). https://doi.org/10.1007/978-3-662-53611-7_3

10. Mani, A.: High Granular Operator Spaces and Less-Contaminated General Rough Mereologies, pp. 1–77 (2019, forthcoming)

11. Seibt, J.: Transitivity. In Burkhardt, H., Seibt, J., Imaguire, G., Gerogiorgakis,S. (eds.) Handbook of Mereology, pp. 570–579. Philosophia Verlag, Germany (2017)

12. Mani, A.: Dialectical rough sets, parthood and figures of opposition-I. In: Peters, J.F., Skowron, A. (eds.) Transactions on Rough Sets XXI. LNCS, vol. 10810, pp. 96–141. Springer, Heidelberg (2019). https://doi.org/10.1007/978-3-662-58768-3_4

13. Polkowski, L.: Approximate Reasoning by Parts. Springer, Heidelberg (2011). https://doi.org/10.1007/978-3-642-22279-5
14. Mani, A.: Contamination-free measures and algebraic operations. In: 2013 IEEE International Conference on Fuzzy Systems (FUZZ), pp. 1–8. IEEE (2013)
15. Ljapin, E.S.: Partial Algebras and Their Applications. Academic, Kluwer (1996)
16. Poschel, R.: Graph algebras and graph varieties. Algebra Universalis **27**, 559–577 (1990)
17. Chajda, I., Langer, H.: Directoids: An Algebraic Approach to Ordered Sets. Heldermann, Lemgo (2011)
18. Chajda, I., Langer, H., Sevcik, P.: An algebraic approach to binary relations. Asian Eur. J. Math. **8**(2), 1–13 (2015)
19. Chajda, I., Langer, H.: Groupoids assigned to relational systems. Math. Bohemica **138**, 15–23 (2013)
20. Mani, A.: Towards logics of some rough perspectives of knowledge. In: Suraj, Z., Skowron, A. (eds.) Intelligent Systems Reference Library dedicated to the memory of Prof. Pawlak ISRL. Intelligent Systems Reference Library, vol. 43, pp. 419–444. Springer, Heidelberg (2013). https://doi.org/10.1007/978-3-642-30341-8_22
21. Jezek, J., Mcenzie, R.: Variety of equivalence algebras. Algebra Universalis **45**, 211–219 (2001)
22. Freese, R., Jezek, J., Jipsen, J., Markovic, P., Maroti, M., Mckenzie, R.: The variety generated by order algebras. Algebra Universalis **47**, 103–138 (2002)
23. Kumar, A., Banerjee, M.: Algebras of definable and rough sets in quasi order-based approximation spaces. Fundamenta Informaticae **141**(1), 37–55 (2015)
24. Werner, K.: Enactment and construction of the cognitive niche: toward an ontology of the mind-world connection. Synthese **197**, 1313–1341 (2020). https://doi.org/10.1007/s11229-018-1756-1
25. Wille, R.: FCA as mathematical theory of concepts and concept hierarchies. In: Ganter, B., et al. (eds.) Formal Concept Analysis. LNAI, vol. 3626, pp. 1–33. Springer (2005)
26. Greco, S., Matarazzo, B., Słowiński, R.: Dominance-based rough set approach on pairwise comparison tables to decision involving multiple decision makers. In: Yao, J.T., Ramanna, S., Wang, G., Suraj, Z. (eds.) RSKT 2011. LNCS (LNAI), vol. 6954, pp. 126–135. Springer, Heidelberg (2011). https://doi.org/10.1007/978-3-642-24425-4_19
27. Mani, A.: Antichain based semantics for rough sets. In: Ciucci, D., Wang, G., Mitra, S., Wu, W.-Z. (eds.) RSKT 2015. LNCS (LNAI), vol. 9436, pp. 319–330. Springer, Cham (2015). https://doi.org/10.1007/978-3-319-25754-9_30
28. Samanta, P., Chakraborty, M.K.: Interface of rough set systems and modal logics: a survey. In: Peters, J.F., Skowron, A., Ślęzak, D., Nguyen, H.S., Bazan, J.G. (eds.) Transactions on Rough Sets XIX. LNCS, vol. 8988, pp. 114–137. Springer, Heidelberg (2015). https://doi.org/10.1007/978-3-662-47815-8_8
29. White, S.: Conceptual structures for STEM data. In: Pfeiffer, H.D., Ignatov, D.I., Poelmans, J., Gadiraju, N. (eds.) ICCS-ConceptStruct 2013. LNCS (LNAI), vol. 7735, pp. 1–21. Springer, Heidelberg (2013). https://doi.org/10.1007/978-3-642-35786-2_1
30. Kharatmal, M.: Concept mapping for eliciting students' understanding of science. Indian Educ. Rev. **45**(2), 31–43 (2009)

31. Kharatmal, M., Nagarjuna, G.: An analysis of growth of knowledge based on concepts and predicates - a preliminary study. In Chunawala, S., Kharatmal, M., et al. (eds.) Proceedings of epiSTEME, vol. 4, pp. 144–149. Macmillan (2011)
32. Erickson, D., Stasiuk, J., Frank, M.: Bringing pythagoras to life. Math. Teach. **744–747**, (1995)

Similarity Based Granules

Dávid Nagy[(⊠)], Tamás Mihálydeák, and Tamás Kádek

Department of Computer Science, Faculty of Informatics, University of Debrecen,
Egyetem tér 1, Debrecen 4010, Hungary
{nagy.david,kadek.tamas}@inf.unideb.hu,
mihalydeak.tamas@unideb.hu

Abstract. In the authors' previous research, a possible usage of the correlation clustering in rough set theory was investigated. Correlation clustering is based on a tolerance relation and its output is a partition. The system of granules can be derived from the partition and as a result, a new approximation space appears. This space focuses on the similarity (represented by a tolerance relation) itself and it is different from the covering type approximation space relying on a tolerance relation. In real-world applications, the number of objects is very high. So it can be effective only if a portion of the data points is used. Previously we provided a method that chooses the necessary number of objects that represent the data set. These members are called representatives and it can be useful to apply them in the approximation of an arbitrary set. A new approximation pair can be defined based on the representatives. In this paper, some very important properties are checked for this approximation pair and the system of granules.

Keywords: Rough set theory · Correlation clustering · Set approximation · Representatives · Granules

1 Introduction

Nowadays a huge amount of data is stored in databases. The stored data is usually represented by objects with (maybe different) properties. Properties are handled in two steps: attributes and the corresponding attribute values. Generally, a finite number of attributes and a finite number of the corresponding attribute values are used. Usually, there are more objects than attribute values. Therefore, more than one object may have the same attribute values (not considering the IDs), so they are indiscernible based on the available knowledge. Naturally, indiscernible objects have to be treated in the same way. Pawlak's original system of rough sets shows the consequences of indiscernibility. In many practical cases, not only indiscernible objects have to be treated in the same way, but objects with the same attribute values of some (and not all) attributes. This is one of the theoretical bases of the generalizations of Pawlak's original theory. Some objects have to be treated in the same way. In rough set theory the objects, that are treated in the same way, belong to a base set. In our previous study, we

© Springer Nature Switzerland AG 2020
R. Bello et al. (Eds.): IJCRS 2020, LNAI 12179, pp. 35–47, 2020.
https://doi.org/10.1007/978-3-030-52705-1_3

examined whether the partition, generated by correlation clustering, can be considered as the system of base sets in an application. Correlation clustering is a clustering method in data mining which creates a partition of the input data set based on a tolerance relation (representing similarity). The clusters gained this way contain similar objects. In our previous paper [11,12] we showed that it is worth to generate the system of base sets from the partition. This way, the base sets contain objects that are typically similar to each other and the generated approximation space (similarity based rough sets) possesses several very useful properties. Informally, in granular computing a granule contains objects which have to be treated in the same way. Granules play—as the most fundamental concept—a crucial role in granular computing. It means that granules (and not objects belonging to them) are in the focus of investigations. The clusters generated by the correlation clustering can be considered as granules. In order to use granules, one has to give their names. In order to preserve the connection between a granule and its objects, the name of the granule can be an object belonging to the granule. This object can represent the given granule. In a very general case to choose representatives is not a trivial problem. In the case of a system relying on an indiscernible relation any object of a granule can be its name, can represent the corresponding granule. When similarity (represented by a tolerance relation) is used to get granules, then the method of correlation clustering gives a possibility to define representatives [5,10]. In [10] a new approximation pair is proposed that is completely based on the representatives. Professor Mihir Chakraborty proposed some very important properties of granules (presented at the International Workshop on Modern and Unconventional Approaches to Reasoning and Computing in 2017). In this paper, we examined these properties along with some other ones for our introduced granules. We also show that the clusters gained from the correlation clustering satisfy all the minimal properties of the granules. Therefore, the clusters can be really treated as granules. The structure of the paper is the following: we begin by introducing the theoretical background of rough set theory. In Sect. 3 correlation clustering is defined. In Sect. 4 we present our previously introduced approximation space. In Sect. 5 we show the definition of the approximation pairs that are based on the representatives. After this, we show which of the defined properties hold for the proposed approximation pair. Finally, we conclude the results.

2   Theoretical Background

From the granular point of view a Pawlakian approximation space [13–15] is an ordered 5-tuple $\langle U, \mathfrak{G}, \mathfrak{D}, \mathsf{l}, \mathsf{u} \rangle$ generated by an equivalence relation $\mathcal{R}$ (which represents indiscernibility), where:

- $U \neq \emptyset$ is the universe of objects
- $\mathfrak{G}$ is the set of granules for which the following properties hold:
  - $\mathfrak{G} \neq \emptyset$
  - if $G \in \mathfrak{G}$ then $G \subseteq U$ and $G \neq \emptyset$

- $\mathfrak{G} = \{G \mid G \subseteq U,$ and $x, y \in G$ if $x\mathcal{R}y\}$
- $\mathfrak{D}$ is the set of definable sets which can be given by the following inductive definition:
  1. $\mathfrak{G} \subseteq \mathfrak{D}$;
  2. $\emptyset \in \mathfrak{D}$;
  3. if $D_1, D_2 \in \mathfrak{D}$, then $D_1 \cup D_2 \in \mathfrak{D}$.
- The functions l, u form a Pawlakian approximation pair $\langle \mathsf{l}, \mathsf{u} \rangle$ if the followings are true for an arbitrary set $S \subseteq U$:
  1. $Dom(\mathsf{l}) = Dom(\mathsf{u}) = 2^U$
  2. $\mathsf{l}(S) = \bigcup\{G \mid G \in \mathfrak{G}$ and $G \subseteq S\}$;
  3. $\mathsf{u}(S) = \bigcup\{G \mid G \in \mathfrak{G}$ and $G \cap S \neq \emptyset\}$.

3 Correlation Clustering

Cluster analysis is an unsupervised learning method in data mining. The goal is to group the objects so that the objects in the same group are more similar to each other than to those which are in other groups. In many cases, the similarity is based on the attribute values of the objects. Although there are some cases when the properties of objects can be difficult to be quantified, but something about their similarity or dissimilarity can still be said. For example, let's consider the humans. We cannot describe someone's looks using only a number, but we can make simple statements on whether two people are similar or dissimilar. These opinions are dependent on the person making the statements. Someone can say that two people are similar while others treat them as dissimilar. If we want to formulate the similarity and dissimilarity using mathematics, we need a tolerance relation (i.e. a reflexive and symmetric relation). If this relation holds for two objects, we can say that they are similar. If this relation does not hold, then they are dissimilar. This relation is reflexive because every object is similar to itself. It is also symmetric because if some object is similar to another one, then the similarity is equivalent the other way round. However transitivity does not necessarily hold. If we take a human and a mouse, then due to their inner structure they are considered similar. This is the reason mice are used in many drug experiments. A human and a mannequin are also similar, this time according to their shape. This is why these dolls are used in display windows. However, a mouse and a mannequin are dissimilar (except that both are similar to the same object). Correlation clustering is a clustering technique based on a tolerance relation [6,7,17].

The task is to find an $R \subseteq U \times U$ equivalence relation which is *closest* to the tolerance relation. A (partial) tolerance relation $\mathcal{R}$ [8,16] can be represented by a matrix M. Let matrix $M = (m_{ij})$ be the matrix of the partial relation $\mathcal{R}$ of similarity: $m_{ij} = 1$ if objects i and j are similar, $m_{ij} = -1$ if objects i and j are dissimilar, and $m_{ij} = 0$ otherwise.

A relation is called partial if there exist two elements (i, j) such that $m_{ij} = 0$. It means that if we have an arbitrary relation $R \subseteq U \times U$ we have two sets of pairs. Let R_{true} be the set of those pairs of elements for which R holds and

R_{false} be the one for which R does not hold. If R is partial, then $R_{true} \cup R_{false}$ is a proper subset of $U \times U$. If R is total, then $R_{true} \cup R_{false} = U \times U$.

A partition of a set S is a function $p : S \to \mathbb{N}$. Objects $x, y \in S$ are in the same cluster at partitioning p, if $p(x) = p(y)$. For a conflict one of the following two cases holds:

– Two dissimilar objects end up in the same cluster
– Two similar objects end up in different clusters

The cost function is the number of these disagreements. The formal definition can be seen in [11]. For a relation, the partition with the minimal cost function value is called *optimal*. Solving a correlation clustering problem is equivalent to minimising its cost function for the fixed relation. If the cost function's value is 0, the partition is called *perfect*. Given the $\mathcal{R}$ and R we call the value f the distance of the two relations. With this definition, the partition generates an equivalence relation. This relation can be considered to be the closest to the tolerance relation.

It is easy to check that we cannot necessarily find a perfect partition for an arbitrary similarity relation. Consider the simplest such case, given three objects A, B and C, and A is similar to both B and C, but B and C are dissimilar. In this situation, the following 5 partitions can be given:

$$\{\{A, B, C\}, \{\{A, B\}, \{C\}\}, \{\{A, C\}, \{B\}\}, \{\{B, C\}, \{A\}\}, \{\{A\}, \{B\}, \{C\}\}\}.$$

It is easy to see that in every of one them there is at least 1 conflict. The number of partitions can be given by the Bell number [1], which grows exponentially. So the optimal partition cannot be determined in reasonable time. In a practical case a quasi optimal partition can be sufficient, so a search algorithm can be used.

The main advantage of the correlation clustering is that the number of clusters does not need to be specified in advance like in many clustering algorithms, and this number is optimal based on the similarity. However, as the number of partitions grows exponentially it is an NP-hard problem.

4   Similarity Based Granules

The system of granules is based on the background knowledge embedded in an information system. The granules represent the background knowledge (or its limit). In the Pawlakian systems, two objects are treated as indiscernible if all of their known attribute values are the same. The indiscernibility property can be represented by an equivalence relation. In practical applications not only the indiscernible objects must be handled in the same way but also those that are similar to each other based on some property. (Irrelevant differences for the purpose of the given applications should not be taken into account.) Some covering approximation spaces use tolerance relations, which represent similarity, instead of equivalence relations, but the usage of these relations is very special.

It emphasizes the similarity to a given object and not the similarity of objects 'in general'. This means that a granule contains objects which are similar to a distinguished object. In these systems, each object generates a granule. With correlation clustering, a quasi-optimal partition of the universe can be obtained [2–4]. The members of a partition are called clusters. They contain objects that are typically similar to each other and not just to a distinguished member. In our previous research, we investigated if the partition can be understood as a system of granules [9,11,12]. According to our results, it is worth to generate a partition with correlation clustering. Singleton clusters represent very little information (its member is only similar to itself). Without increasing the number of conflicts its member cannot be considered similar to any objects. So, they always require an individual decision. By deleting the singletons, a partial system of granules can be defined. The formal definition of the proposed approximation space (similarity based rough sets) can be seen in the following definition.

Definition 1. *Similarity based rough set approximation space can be represented by an ordered 6-tuple $\langle U, \mathfrak{G}, \mathfrak{D}, \mathsf{I}, \mathsf{u}, \mathfrak{S} \rangle$ based on a tolerance relation (representing similarity) $\mathcal{R}$. Let p be the partition gained from the correlation clustering (based on $\mathcal{R}$).*

- *the definition of $U, \mathfrak{D}, \mathsf{I}$ and u are the same as in the Pawlakian space.*
- *$\mathfrak{S}$ denotes the set of the singleton members.*
- *$\mathfrak{G} = \{G \mid G \subseteq U \setminus \mathfrak{S}, \text{ and } x, y \in G \text{ if } p(x) = p(y)\}$*

The introduced approximation space has some useful features:

- the similarity of objects relying on their properties (and not the similarity to a distinguished object) plays an important role in the definition of granules;
- the system of granules consists of disjoint sets, so the lower and upper approximations are closed in the following sense: Let S be a set and $x \in U$. If $x \in \mathsf{I}(S)$, then we can say, that every object $y \in U$ which is in the same cluster as x is in $\mathsf{I}(S)$. If $x \in \mathsf{u}(S)$, then we can say, that every object $y \in U$ which is in the same cluster as x is in $\mathsf{u}(S)$.
- the number of clusters is not predefined because the algorithm finds the optimal number. This way, only the necessary number of granules appear (in applications we have to use an acceptable number of granules);
- the size of the granules is not too small, nor too big.

The amount of daily produced data is unbelievable. There are around 2.5 quintillion bytes of data created each day at our current pace and it is only accelerating with the growth of the Internet of Things (IoT). In data sciences, it is extremely important that certain methods can be used for a large amount of data. Due to the exploding volume and speed of data growth, the resource need and execution time of the algorithms show an increasing trend. In data mining to mitigate this problem, it is common to use samples. There are numerous ways to choose a part of the input dataset which can be treated as a sample. In every method, it is crucial that the chosen objects must represent the entire population. In this case, representativeness means that the specific properties are as similar in the sample as in the entire set. Without this property, important information might be disregarded. Imagine that a product is needed to be sold, for example, a toy to a group of children. In almost every group of youngsters, there is at least one child whose decision has the most influence on the group's life. In this case, one child is enough to be found and convinced to buy the toy. The rest of the group will follow them. This child can be treated as the representative of the group. It means that in the computations only this child should be considered instead of the whole group. In a pawlakian system, any object can be the representative of a certain granule. In the covering systems (based on a tolerance relation) the representatives are obvious in each granule. In the similarity based rough set approximation space, the situation is not that simple. In each granule, we need to choose an object that is the most similar in the set. Naturally, it can happen that the entire granule cannot be represented by only one member. In [5] we proposed an algorithm that produces the necessary number of representatives for each granule. The algorithm assigns a rank value to each object. This value shows how much the given object represents the granule.

Definition 2. *The object with the highest rank value is called primary representative. If there is more than one object with the same rank, then the primary representative is chosen randomly.*

Generally speaking, we can say that a granule represents a property. A represented property can be characterized by attributes and the corresponding attribute values. For example, the property 'being red apple' can be characterized by color and fruit type as attributes and by red and apple as corresponding attribute values. If P is a property, then P can be an intension of a granule G. The granule itself is a set of objects that possess the property described by its intension. In our system, a granule contains objects that are typically similar to each other. Every granule has a primary representative which represents the entire granule the most. In an information system, every object has attributes and attributes values. The list of these attribute values describes a certain property.

Definition 3. *The intension of a granule is the property described by its primary representative.*

5   Approximation Based on Representatives

In the classical sense, the lower approximation of a set S is the union of those granules that are subsets of S. In order to get these granules, every object in each granule must be considered. It can be a time-consuming task if the number of points is high. The effectiveness of the representatives lies in situations when the number of objects is very large. It can be practical to use the strength of representatives in the approximation process. For each granule, let us consider only its representatives. Let $G \in \mathfrak{G}$ be a granule, and $REP(G)$ be the set of its representatives such that $REP(G) \subseteq G$ and $REP(G) \neq \emptyset$ for all $G \in \mathfrak{G}$ (and so $\emptyset \notin \mathfrak{G}$). The approximation pair are defined as the following:

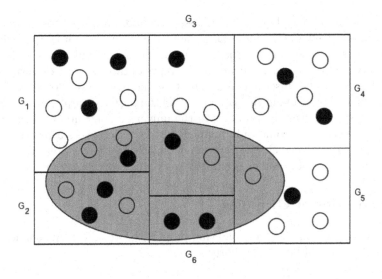

Fig. 1. Approximation based on representatives

- $\mathsf{l}_r(S) = \bigcup\{G \mid G \in \mathfrak{G} \text{ and } REP(G) \subseteq S\}$ (and so $\mathsf{l}_r(S) \in \mathfrak{D}$);
- $\mathsf{u}_r(S) = \bigcup\{G \mid G \in \mathfrak{G} \text{ and } REP(G) \cap S \neq \emptyset\}$ (and so $\mathsf{u}_r(S) \in \mathfrak{D}$).

This way, the lower approximation of a set S becomes the union of those granules for which every representative is a member of S. A granule belongs to the upper approximation if at least one of its representatives is in the set S. Naturally, the certainty of the lower approximation might be lost, but as the number of points is increasing, it can be very useful.

In Fig. 1 a simple example is provided for the method. The granules are denoted by solid-line rectangles, and the set we wish to approximate (S) is denoted by a grey ellipse. For each granule, the black circles symbolise the representatives.

The approximation of the set S is the following based on the representatives:

- $\mathsf{l}_r(S) = G_2 \cup G_6$
- $\mathsf{u}_r(S) = G_1 \cup G_2 \cup G_3 \cup G_6$

The approximation of the set S is the following based on the classical approximation pair:

- $\mathsf{l}(S) = G_2 \cup G_6$
- $\mathsf{u}(S) = G_1 \cup G_2 \cup G_3 \cup G_5 \cup G_6$

The lower approximation is the same in both cases. The upper approximation differs in one granule (G_5). When there is a huge number of points and there are several sets to be approximated, we recommend approximation using representatives. In this case, the method can reduce the run-time of the approximation significantly. Determining the approximation with the classical functions 32 objects needed to be considered. Using the proposed method, only 13 of them had to be tested, so almost 60% of the original points were discarded. Of course, with 32 to 13 points is not a significant change, but in the case of millions of objects, it can be very useful. Working with only the representatives, we can always save time and resources because we can be sure that the number of representatives is less than that of U. Proving this is very straightforward. Naturally, there cannot be more representatives than objects in the universe. Their numbers cannot be equal either because it could only happen if every object were a representative which implies that every cluster were singleton. Using these system is pointless because the system of granules is empty (every singleton cluster is discarded).

6   Properties of Granules

In this section, we examine the following properties (we call them as axioms) of granules (by Prof. Mihir Chakraborty):

   I $\forall G \in \mathfrak{G} : G \neq \emptyset$
   II $\forall G \in \mathfrak{G} : \exists a \in U$ such that G may be associated with a. Notation: G_a
   III if $b \in G_a$ then $a \in G_b$
   IV $\forall G \in \mathfrak{G} : \mathsf{l_r}(G) = G$
   V $\forall G \in \mathfrak{G} : \mathsf{u_r}(G) = G$
   VI $\forall G \in \mathfrak{G} : \mathsf{l_r}(\mathsf{l_r}(G)) = \mathsf{l_r}(G)$
   VII $\forall G \in \mathfrak{G} : \mathsf{u_r}(\mathsf{u_r}(G)) = \mathsf{u_r}(G)$
   VIII $\forall G \in \mathfrak{G} : \mathsf{u_r}(\mathsf{l_r}(G)) = \mathsf{l_r}(G)$
   IX $\forall G \in \mathfrak{G} : \mathsf{l_r}(\mathsf{u_r}(G)) = \mathsf{u_r}(G)$
   X $\mathsf{l_r}(G)$ and $\mathsf{u_r}(G)$ are duals

Theorem 1. *In $\langle U, \mathfrak{G}, \mathfrak{D}, \mathsf{l}, \mathsf{u} \rangle$ (classical Pawlakian approximation space), all of the aforementioned axioms hold.*

Theorem 2. *All the existing granules admit Axiom I, II and IV.*

Theorem 3. *In $\langle U, \mathfrak{G}, \mathfrak{D}, \mathsf{l_r}, \mathsf{u_r}, \mathfrak{S} \rangle$ (similarity based rough sets approximation space based on the representatives) all of the aforementioned axioms hold except for the duality property.*

Proof (Axiom I). This axiom trivially holds because in the similarity based rough sets approximation space every granule contains at least 2 objects.

Proof (Axiom II). The axiom holds as every granule has at least one representative. We can associate the granule with one of the representatives of the granule.

Proof (Axiom III). If representative b is in the granule of representative a, then it could only happen if $G_a = G_b$. Let us suppose that $G_a \neq G_b$. From Axiom II we know that $b \in G_b$. So if representative b is in G_a, then $G_a \cap G_b = \{b\}$ which means that G_a and G_b are not disjoint. This is a contradiction, therefore G_a and G_b must be the same set.

Proof (Axiom IV). $\mathsf{l_r}(G) = \bigcup\{G' \mid G' \in \mathfrak{G} \text{ and } \forall x \in REP(G') : x \in G\}$. The granules are pairwise disjoint, so there is no granule whose representatives is a member of G (other than G itself). Naturally, every representative of G is the member of G. Therefore, the set $\{G' \mid G' \in \mathfrak{G} \text{ and } \forall x \in REP(G') : x \in G\}$ contains only G from which $\mathsf{l_r}(G) = G$ follows.

Proof (Axiom V). The proof of the fifth axiom is very similar to the proof of the fourth axiom. $\mathsf{u_r}(G) = \bigcup\{G' \mid G' \in \mathfrak{G} \text{ and } \exists x \in REP(G') : x \in G\}$. The granules are pairwise disjoint, so there is no granule whose representatives is a member of G (other than G itself). If $\forall x \in REP(G) : x \in G$ is true, then $\exists x \in REP(G) : x \in G$ will be also true. Therefore, the set $\{G' \mid G' \in \mathfrak{G} \text{ and } \forall x \in REP(G') : x \in G\}$ contains only G from which $\mathsf{u_r}(G) = G$ follows.

Proof (Axiom VI–IX). If Axiom 4 and 5 hold, then Axiom VI–IX follow.

Proof (Axiom X). The duality property holds if the following two equalities hold for any granule G (C denotes the complement operator):

1. $\mathsf{l_r}(G) = \mathsf{u_r}(G^{\mathsf{C}})^{\mathsf{C}}$
2. $\mathsf{u_r}(G) = \mathsf{l_r}(G^{\mathsf{C}})^{\mathsf{C}}$

Let $U = \{a, b, c, d, e\}$, $\mathfrak{G} = \{G_1, G_2\}$, $G_1 = \{a, b\}$, $G_2 = \{c, d\}$, $REP(G_1) = \{a\}$, $REP(G_2) = \{c\}$. In this example, $\mathsf{l_r}(G_1) = \{a, b\}$ and $G_1^{\mathsf{C}} = \{c, d, e\}$. From this $\mathsf{u_r}(G_1^{\mathsf{C}}) = \{c, d\}$ follows. However, $\mathsf{u_r}(G_1^{\mathsf{C}})^{\mathsf{C}} = \{a, b, e\} \neq \{a, b\}$. Therefore the duality property does not hold.

6.1   Properties of Approximation Pairs

In the previous section, the axioms only focused on the granules. In this section, we examine some additional properties of the proposed approximation pair. Here, the properties to be checked are based on definable and arbitrary sets not only granules. The most essential features of approximation pairs are specified as follows.

Monotonicity
l and u are said to be monotone if $S \subseteq S'$ then $\mathsf{l}(S) \subseteq \mathsf{l}(S')$ and $\mathsf{u}(S) \subseteq \mathsf{u}(S')$
Weak approximation property
$\forall S \in 2^U : \mathsf{l}(S) \subseteq \mathsf{u}(S)$
Strong approximation property
$\forall S \in 2^U : \mathsf{l}(S) \subseteq S \subseteq \mathsf{u}(S)$
Normality of l
$\mathsf{l}(\emptyset) = \emptyset$
Normality of u
$\mathsf{u}(\emptyset) = \emptyset$

Theorem 4. *In* $\langle U, \mathfrak{G}, \mathfrak{D}, \mathsf{l_r}, \mathsf{u_r}, \mathfrak{S} \rangle$ *(similarity based rough sets approximation space based on the representatives), the monotonicity, the weak approximation property and the normality of* $\mathsf{l_r}$ *and* $\mathsf{u_r}$ *hold and the strong approximation property does not hold.*

Proof (Monotonicity). Let S and S' be two arbitrary set such that $S \subset S'$ which means that there is an object x which is a member of S' but not a member of S. The following cases can be true for x:

1. $x \in \mathfrak{G}$, then $\mathsf{l_r}(S) = \mathsf{l_r}(S')$ and $\mathsf{u_r}(S) = \mathsf{u_r}(S')$
2. x is a non-representative, then $\mathsf{l_r}(S) = \mathsf{l_r}(S')$ and $\mathsf{u_r}(S) = \mathsf{u_r}(S')$
3. x is a representative of a granule G, then the following cases can happen:
   (a) if $\neg \exists y (y \in REP(G) \wedge x \neq y \wedge y \in S)$, then $\mathsf{l_r}(S) = \mathsf{l_r}(S')$ and $\mathsf{u_r}(S) \subset \mathsf{u_r}(S')$
   (b) if $\exists y (y \in REP(G) \wedge x \neq y \wedge y \in S)$, then $\mathsf{l_r}(S) = \mathsf{l_r}(S')$ and $\mathsf{u_r}(S) = \mathsf{u_r}(S')$
   (c) if $\forall y (y \in REP(G) \wedge x \neq y \rightarrow y \notin S)$, then $\mathsf{l_r}(S) \subset \mathsf{l_r}(S')$ and $\mathsf{u_r}(S) \subset \mathsf{u_r}(S')$

In every case, we found that $\mathsf{l_r}(S) \subseteq \mathsf{l_r}(S')$ and $\mathsf{u_r}(S) \subseteq \mathsf{u_r}(S')$, therefore the monotonicity holds.

Proof (Weak approximation property). Let S be an arbitrary set and let us assume that there is a granule G such that $G \subseteq \mathsf{l_r}(S)$ but $G \not\subseteq \mathsf{u_r}(S)$. Due to the definition of the lower approximation, we know that $\forall x \in REP(G) : x \in S$ is true, so $\exists x \in REP(G) : x \in S$ is also true. This implies that $G \subseteq \mathsf{u_r}(S)$. We reached a contradiction, therefore the weak approximation property holds.

Proof (Strong approximation property). Let $U = \{a, b, c\}$ be the universe, $G = \{a, b, c\}$ a granule, $\mathfrak{G} = \{G\}$ be the system of granules, $S = \{a, b\}$ be the set to be approximated and $REP(G) = \{b\}$ be the representatives of G. In this case $\mathsf{l_r}(S) = G = \{a, b, c\}$ which means that $\mathsf{l_r}(S) \not\subseteq S$. So the strong approximation property does not hold.

Proof (Normality of $\mathsf{l_r}$ and $\mathsf{u_r}$). The empty set does not have a representative. Therefore the condition in the definition of the lower and upper approximation is false for every granule. This implies that $\mathsf{l_r}(\emptyset) = \mathsf{u_r}(\emptyset) = \emptyset$.

Theorem 5. *Let $G \in \mathfrak{G}$ and $D \in \mathfrak{D}$. If $a \in G$ and $a \in D$ then $G \subseteq D$.*

Proof. If $a \in D$ then there exists a $G' \in \mathfrak{G}$ such that $a \in G'$ and $G' \subseteq D$. The members of $\mathfrak{G}$ are pairwise disjoint, so it is true for all $G_1, G_2 \in \mathfrak{G}$ that $G_1 \cap G_2 \neq \emptyset$ only if $G_1 = G_2$. Therefore $G = G'$ hence $a \in G$ and $a \in G'$. Earlier we have found that $G' \subseteq D$ and so $G \subseteq D$.

Theorem 6. *$\mathsf{l_r}(D) \subseteq D$ for all $D \in \mathfrak{D}$.*

Proof. We indirectly suppose, that there exists a $D \in \mathfrak{D}$ so that $\mathsf{l_r}(D) \not\subseteq D$. Therefore there exists an $a \in \mathsf{l_r}(D)$ so that $a \notin D$. If $a \in \mathsf{l_r}(D)$ then there exists a $G \in \mathfrak{G}$ where $REP(G) \subseteq D$ such that $a \in G$. $REP(G) \neq \emptyset$ so there exists a $b \in REP(G)$ and so $b \in D$. Because $REP(G) \subseteq G$ it is also true that $b \in G$. Based on Theorem 5, if $b \in G$ and $b \in D$ then $G \subseteq D$. Because of $a \in G$ the $a \in D$ contradiction appears.

Theorem 7. $u_r(D) \subseteq D$ *for all* $D \in \mathfrak{D}$.

Proof. We indirectly suppose, that there exists a $D \in \mathfrak{D}$ so that $u_r(D) \not\subseteq D$. Therefore there exists an $a \in u_r(D)$ so that $a \notin D$. If $a \in u_r(D)$ then there exists a $G \in \mathfrak{G}$ where $REP(G) \cap D \neq \emptyset$ such that $a \in G$. So there exists a $b \in REP(G) \cap D$ so obviously $b \in REP(G)$ and $b \in D$. Because $REP(G) \subseteq G$ it is also true that $b \in G$. Based on Theorem 5, if $b \in G$ and $b \in D$ then $G \subseteq D$. Because of $a \in G$ the $a \in D$ contradiction appears.

Definition 4 (Weak approximation pair). *An approximation pair* $\langle l, u \rangle$ *is a weak approximation pair on* U *if:*

- l *and* u *are monotone (monotonicity)*
- $u(\emptyset) = \emptyset$ *(normality of* u*)*
- *if* $D \in \mathfrak{D}$, *then* $l(D) = D$ *(granularity of* $\mathfrak{D}$*)*
- *if* $\forall S \in 2^U : l(S) \subseteq u(S)$ *(weak approximation property)*

Theorem 8. $\langle l_r, u_r \rangle$ *is a weak approximation pair.*

Proof. Previously we proved that l_r and u_r are monotone and the normality of u_r and the weak approximation property hold. We need to prove that the granularity of $\mathfrak{D}$ also holds. From Theorem 6 we know that $l_r(D) \subseteq D$ for any definable set. We just need to prove that $D \subseteq l_r(D)$ for any definable set. Let's indirectly suppose that $D \not\subseteq l_r(D)$. It means that there is a granule G' such that $G' \subseteq D$ but $G' \not\subseteq l_r(D)$. Therefore, there must be a representative member r of G' such that $r \notin D$. By definition $r \in G'$. If $G' \subseteq D$, then every member of G' is a member of D. However $r \in G'$ but $r \notin D$, therefore G' cannot be a subset of D. This contradicts our original assumption. So $D \subseteq l_r(D)$.

7   Conclusion

In [11,12] the authors introduced a partial approximation space relying on the tolerance relation (representing similarity). The genuine novelty of this new approximation space is the way in which the system of base sets is defined: it is the result of correlation clustering, and so the similarity is taken into consideration generally. In granular computing, a granule is a collection of objects that are treated in the same way. In correlation clustering, a cluster contains entities that are typically similar to each other. In this case, the objects that are in the same cluster are treated in the same way. Therefore, we can treat the clusters and so the base sets as granules. In data sciences, it is very common to use only a subset of the original dataset instead of the entire collection. The members of this subset can be called as representatives. A very important criterion is that these objects must have the same properties as the whole data set. In [5,10] we provided a possible way to choose the necessary number of representatives of a set. We also introduced a new approximation pair which is based on the representatives. In this paper, we examined some essential properties of granules (proposed by Prof. Mihir Chakraborty). We showed that the system of granules generated by the correlation clustering satisfies all the minimal properties of the granules. Therefore, the clusters can be really treated as granules. We also proved that the introduced approximation pair is a weak approximation pair.

Acknowledgement. This work was supported by the construction EFOP-3.6.3-VEKOP-16-2017-00002. The project was supported by the European Union, co-financed by the European Social Fund.

References

1. Aigner, M.: Enumeration via ballot numbers. Discret. Math. **308**(12), 2544–2563 (2008). https://doi.org/10.1016/j.disc.2007.06.012. http://www.science direct.com/science/article/pii/S0012365X07004542
2. Aszalós, L., Mihálydeák, T.: Rough clustering generated by correlation clustering. In: Ciucci, D., Inuiguchi, M., Yao, Y., Ślęzak, D., Wang, G. (eds.) RSFDGrC 2013. LNCS (LNAI), vol. 8170, pp. 315–324. Springer, Heidelberg (2013). https://doi.org/10.1007/978-3-642-41218-9_34
3. Aszalós, L., Mihálydeák, T.: Rough classification based on correlation clustering. In: Miao, D., Pedrycz, W., Ślęzak, D., Peters, G., Hu, Q., Wang, R. (eds.) RSKT 2014. LNCS (LNAI), vol. 8818, pp. 399–410. Springer, Cham (2014). https://doi.org/10.1007/978-3-319-11740-9_37
4. Aszalós, L., Mihálydeák, T.: Correlation clustering by contraction. In: 2015 Federated Conference on Computer Science and Information Systems (FedCSIS), pp. 425–434. IEEE (2015)
5. Aszalós, L., Nagy, D.: Iterative set approximations based on tolerance relation. In: Mihálydeák, T., et al. (eds.) IJCRS 2019. LNCS (LNAI), vol. 11499, pp. 78–90. Springer, Cham (2019). https://doi.org/10.1007/978-3-030-22815-6_7
6. Bansal, N., Blum, A., Chawla, S.: Correlation clustering. Mach. Learn. **56**(1–3), 89–113 (2004)
7. Becker, H.: A survey of correlation clustering. In: Advanced Topics in Computational Learning Theory, pp. 1–10 (2005)
8. Mani, A.: Choice inclusive general rough semantics. Inf. Sci. **181**(6), 1097–1115 (2011)
9. Mihálydeák, T.: Logic on similarity based rough sets. In: Nguyen, H.S., Ha, Q.-T., Li, T., Przybyła-Kasperek, M. (eds.) IJCRS 2018. LNCS (LNAI), vol. 11103, pp. 270–283. Springer, Cham (2018). https://doi.org/10.1007/978-3-319-99368-3_21
10. Nagy, D., Aszalós, L.: Approximation based on representatives. In: Mihálydeák, T., et al. (eds.) IJCRS 2019. LNCS (LNAI), vol. 11499, pp. 91–101. Springer, Cham (2019). https://doi.org/10.1007/978-3-030-22815-6_8
11. Nagy, D., Mihálydeák, T., Aszalós, L.: Similarity based rough sets. In: Polkowski, L., Yao, Y., Artiemjew, P., Ciucci, D., Liu, D., Ślęzak, D., Zielosko, B. (eds.) IJCRS 2017. LNCS (LNAI), vol. 10314, pp. 94–107. Springer, Cham (2017). https://doi.org/10.1007/978-3-319-60840-2_7
12. Nagy, D., Mihálydeák, T., Aszalós, L.: Similarity based rough sets with annotation. In: Nguyen, H.S., Ha, Q.-T., Li, T., Przybyła-Kasperek, M. (eds.) IJCRS 2018. LNCS (LNAI), vol. 11103, pp. 88–100. Springer, Cham (2018). https://doi.org/10.1007/978-3-319-99368-3_7

13. Pawlak, Z.: Rough sets. Int. J. Parallel Prog. **11**(5), 341–356 (1982)
14. Pawlak, Z., Skowron, A.: Rudiments of rough sets. Inf. Sci. **177**(1), 3–27 (2007)
15. Pawlak, Z., et al.: Rough sets: theoretical aspects of reasoning about data. In: System Theory, Knowledge Engineering and Problem Solving, vol. 9. Kluwer Academic Publishers, Dordrecht (1991)
16. Skowron, A., Stepaniuk, J.: Tolerance approximation spaces. Fundamenta Informaticae **27**(2), 245–253 (1996)
17. Zimek, A.: Correlation clustering. ACM SIGKDD Explor. Newslett. **11**(1), 53–54 (2009)

Approximate Reaction Systems Based on Rough Set Theory

Andrea Campagner[1(✉)], Davide Ciucci[1], and Valentina Dorigatti[2]

[1] Dipartimento di Informatica, Sistemistica e Comunicazione,
University of Milano–Bicocca, Viale Sarca 336/14, 20126 Milan, Italy
`a.campagner@campus.unimib.it`
[2] Dipartimento di Scienze Teoriche e Applicate, University of Insubria,
Via J.H. Dunant 3, 21100 Varese, Italy

Abstract. In this work we investigate how Rough Set Theory could be employed to model uncertainty and information incompleteness about a Reaction System. The approach that we propose is inspired by the idea of an abstract scientific experiment: we define the notion of *test*, which defines an approximation space on the states of a Reaction System, and *observation*, to represent the interactive process of knowledge building that is typical of complex systems. We then define appropriate notions of reducts and study their characterization in terms of both computational complexity and relationships with standard definitions of reducts in terms of Information Tables.

Keywords: Complex systems · Reaction Systems · Rough sets

1 Introduction

Complex systems, that are characterized by the mutual interaction of basic components, represent currently one of the topics of major interest in many disciplines. This interest has been fostered both by the potential impact that these systems have in the real world and also by the difficulty that they pose with respect to the modeling and formalization point of view. Indeed, as interaction represents one of the main features of complex systems, there has been increasing attention towards developing mathematical and formal models that are explicitly based on the notion of interaction: some prominent examples are cellular automata [5], membrane computing [14], formalisms to describe concurrent processes [4,16], reaction systems [8]. This latter class of models has recently been proposed as a simple and abstract formalization of biochemical processes involving *substances* and *reactions*, by which the states (i.e., collections of substances) are transformed. While interesting from a computational or purely mathematical point of view, one of the major limitations of this framework (and, more in general, of abstract *idealized* models of complex systems), as recently acknowledged in [6], relates to the fact that these models *ignore* the realistic aspects that are *intrinsic* in complexity, in particular with respect to the fact that information

© Springer Nature Switzerland AG 2020
R. Bello et al. (Eds.): IJCRS 2020, LNAI 12179, pp. 48–60, 2020.
https://doi.org/10.1007/978-3-030-52705-1_4

available about these systems is usually only *partial, uncertain* and *incomplete* and acquired through *interaction* with the system.

Rough Set theory [15] has originally been proposed to explicitly deal with this type of information: both with respect to the representation of uncertain and potentially incomplete information [13] (through the notion of lower and upper approximations) and also with respect to knowledge acquisition [3, 10, 22] (through the notion of reducts and rule extraction). Indeed, the relationship between these two mathematical frameworks have been investigated, under the perspective of Interactive Granular Computing [19], in [6, 18] where Rough Set Theory is integrated with Reaction Systems in order to be able to account for uncertainty and incomplete knowledge in the latter formalism.

In this work, we also discuss how to relate these two modeling frameworks, though under a different perspective. Indeed, the main purpose of this article is to investigate how Rough Set Theory can be used to study Reaction Systems, both from the modeling point of view and from the uncertainty representation and management one. More specifically, we will consider the case where states of a Reaction System are not directly perceived as is, but only through the observation of the results of some *experiments* or *tests* that have been performed on those states, as would be the case in a realistic scientific experiment. As such, the reaction system in intrinsically built on information that can be affected by different forms of uncertainty. Notably, while we will focus on the specific case of Reaction Systems, the methodology that we propose mainly considers the *graph of the dynamics* that underlies the model and thus, at least in principle, should be easily generalizable to any class of discrete dynamical systems.

The rest of this paper will be structured as follows: in Sect. 2 we recall the necessary background concerning both Reactions Systems, Rough Sets and their linking; while in Sect. 3 we present the mathematical framework that we propose. Finally, in Sect. 4 we discuss the obtained results and possible future research directions.

2   Mathematical Background

In this section, the basic notions on both reaction systems and rough sets are given.

2.1   Introduction to Reaction Systems

Reaction Systems are a model of computation inspired by biochemical reactions involving reactants, inhibitors and products from a finite background set.

Definition 1. *A* Reaction System *is an ordered pair (S, A) such that S is a finite set of* substances *or* entities, *and A is a set of reactions in S.*

A reaction can be formally defined as follows.

Definition 2. *A reaction, in a reaction system* (S, A)*, is a triplet* $a = (R_a, I_a, P_a)$ *where* $R_a \subseteq S$ *is the set of reactants,* $I_a \subseteq S$ *is the set of inhibitors, and* $P_a \subseteq S$ *is the set of products.*

The result of applying reaction a to a set $X \subseteq S$, denoted by $res_a(X)$, is conditional: if R_a is included in X and I_a is disjoint with X, then a is enabled on X, otherwise a is not enabled on X. If a is enabled on X, then a transforms the set of reactants into the product set. Thus, formally:

$$res_a(X) = \begin{cases} P_a & R_a \subseteq X \text{ and } I_a \cap X = \emptyset \\ \emptyset & \text{otherwise} \end{cases} \tag{1}$$

For a reaction system (S, A), the result function of A is $res_A : 2^S \to 2^S$, and for each $T \subseteq S$ it is defined as:

$$res_A(T) = \bigcup_{a \in en_A(T)} P_T \tag{2}$$

where $en_A(T)$ is the set of reactions of A enabled in T.

Given a RS $R = (S, A)$ the associated *graph of the dynamics* is the graph $G[R] = (V, E)$ where $V = 2^S$ and $(v_1, v_2) \in E$ if $res_A(v_1) = v_2$. An example of a Reaction System is illustrated in Example 1.

Example 1. *Let* $R = (S, A)$ *be a Reaction System where:*

- $S = \{A, B, C\}$;
- $A = \{(\emptyset, ABC, BC), (A, C, AB), (B, C, AB), (C, AB, AC), (AB, \emptyset, ABC)\}$.

The graph of the dynamics of R *is shown in Fig. 1.*

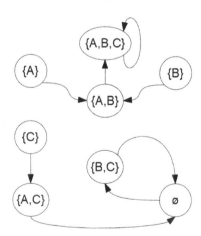

Fig. 1. Graph of the dynamics for the Reaction System described in Example 1.

We refer the reader to [7] for a recent overview and tutorial on Reaction Systems.

2.2 Introduction to Rough Sets

Rough sets are an approach to imperfect knowledge proposed by Zdziław Pawlak to model uncertain and incomplete knowledge [15]. For recent overviews on Rough Set Theory and applications we refer the reader to [1,23]. The basic notion of Rough Set Theory is that of an information table.

Definition 3. *An* Information Table *is an ordered pair* $IT = (U, Att)$ *such that* U *is a finite non-empty set of* objects *and Att is a finite non-empty set of* attributes, *where each* $a \in Att$ *is a function* $a : U \mapsto V_a$ *and* V_a *is the set of possible values of* a.

Given an IT, we say that two objects u, u' are *indiscernible* w.r.t. $B \subset Att$ if $\forall b \in B, b(u) = b(u')$. Indiscernibility defines an equivalence relation where the equivalence class of an object u is denoted as $[u]_B$.

Given an Information Table $IT = (U, Att)$ and $B \subseteq Att$, we can define for $X \subseteq U$ its *rough approximation* (or, *rough set*) as $B(X) = \langle l_B(X), u_B(X) \rangle$, where $l_B(X) = \bigcup_{[u]_B \subseteq X} [u]_B$ is the *lower approximation* of X and $u_B(X) = \bigcup_{[u]_B \cap X \neq \emptyset} [u]_B$ is the respective *upper approximation*. We denote with $\mathcal{R}_B(U)$ the set of rough sets on U determined by $B \subseteq Att$.

The lower approximation of a set consists of all the elements that *surely* belong to that set, while the upper approximation of a set is made of all the element that *possibly* belong to the set. The boundary region can be defined as $Bnd(X) = u(X) \setminus l(X)$ and can be understood as the collection of elements whose belonging to the set is not certain.

Given an Information Table $IT = (U, Att)$, a *super-reduct* [21] is a subset of attributes $R \subseteq Att$ such that $\forall x, [x]_R = [x]_{Att}$. A super-reduct R is a *reduct* if no subset of R is also a super-reduct. We denote by $RED(IT)$ the set of reducts of IT, the *core* of an IT is defined as $Co(IT) = \bigcap_{R \in RED(IT)} R$.

Finally, we notice that sometimes the starting point for defining rough sets is a so-called approximation space (U, R), with U a set of instances and R an equivalence relation (or, equivalently, a partition of U). Thus, any Information Table induces an approximation space, which is a more general notion. The lower and upper approximations are, then, defined exactly as above.

2.3 Related Work on Linking Rough Sets and Reaction Systems

The importance of linking Rough Set Theory and Reaction Systems, with the goal of augmenting the formalism of Reaction Systems with notions of partial information and incompleteness, has been recognized in [6,18]. Intuitively, in these studies, the basic concept is that of a situation that could be understood as a state of the system under observation. Situations can only be perceived through attributes (that could represent physical experiments or other properties) and for the observed situations (which represents the objects in an Information Table) we are able to precisely tell whether a given substance was present or not in that situation. However, we can give a lower and an upper approximation of the present substances.

Formally, in this framework, the authors start from the substances s of a Reaction System $R = (S, A)$ and, for each such substance, they define a Decision Table $DT(s) = (U, Att_s, d_s)$, that is an Information Table (U, Att_s) plus a decision $d_s : U \mapsto \{0, 1\}$, where U is a set of physical situations, Att_s are attributes through which the physical situation is perceived and $d_s(u) = 1$ iff substance s is present in situation u. Then, the set of situations in which s is present is represented by the decision class $D(s) = \{u | d_s(u) = 1\}$. Since it can happen that the attributes Att_s do not carry enough information to take a clear decision, the decision class can be approximated via the information given by the attributes Att_s using the standard Rough Set notions of lower and upper approximations, thus defining, $L(D(s))$ and $U(D(s))$. Then, the authors define how a state $\hat{S}$ could be represented by aggregation of the decision systems $DT(s)$ for $s \in \hat{S}$.

The approach that we take in the following is similar in spirit, in that we also take states as the basic notion of our framework and we assume that, in general, these states are not completely recognizable but only perceived via tests that affirm whether some substances are present or not in the current situation. A fundamental difference, however, relates to the fact that the decision attribute in the framework of [6,18] can be seen as an a-priori notion that is independent of the attributes, in that it is already represented in the decision system. As we will see in the following sections, in the approach that we propose the decision w.r.t. a substance being present or not in a situation is only an a-posteriori notion that is entirely defined by the values of the attributes or, as we will call them, *tests*. Indeed, the result of the tests is the only information that we have about a state and we are able to state that a given substance s is present in a given situation only inasmuch the result of the tests is able to do so.

The notion of test that we will introduce resembles the notion of a *sensor* in complex dynamical systems [11]: both represent available information about the state of a complex system and, in both cases, one of the most interesting problem is related to finding a minimal and sufficient set of tests (resp. sensors) that are able to accurately *describe* the dynamics of the whole, partially unobservable system. The main differences between these two notions relate to the fact that: sensors are defined in the context of *classical* (i.e. based on dynamical systems theory), typically continuous, complex systems while the notion of test that we will introduce is based on *discrete* dynamical systems; furthermore, the underlying theory for minimal set of sensors are based on ideas from statistical mechanics, control theory and related disciplines, while the theory that we develop for tests is based on Rough Sets and graph theory.

3   Methods

As argued in Sect. 1, one of the main features of real complex systems which is lacking in the formalism of Reaction Systems is the ability to model partial or uncertain information about the states of the system. Further, a Reaction System is fully specified in terms of the reactions, while in reality the model is usually construed via gradual observation of the behavior of the system. In this section we will formalize both concepts through application of Rough Set Theory to Reaction Systems. We assume that the dynamics of the complex system that we observe is fully described by an underlying Reaction System which, however, may be unknown. The goal is then to understand, given a certain set of experimental tests that we may perform, whether these tests are sufficient to accurately describe the dynamics of the system. In order to do so, in Sect. 3.1 we will formalize the notion of partial observability of a Reaction System through the notion of *Approximate Reaction System* and *tests*. Further we will consider the issue of dynamic acquisition of knowledge about a Reaction System, formalized via *observations*, that is states of partial knowledge about the graph of the dynamics of a Reaction System. In Sect. 3.2 we will describe reducts for Approximate Reaction Systems, their existence conditions and characterization.

3.1 Approximate Reaction Systems

Definition 4. *An* Approximate Reaction System *(ARS) is a triple* $R = (S, A, T)$, *where* S *is the set of substances,* A *is the set of reactions and* T *is the set of tests. A test* $t \in T$ *is a function* $t : S \mapsto \{\perp, \top\}$, *we denote with* $supp(t) = \{s \in S | t(s) = \top\}$ *the support of* t. *The result of test* t *on state*

$$X \subseteq S \text{ is } r_t(X) = \bigvee_{s \in X} t(s) = \begin{cases} \top & supp(t) \cap X \neq \emptyset \\ \perp & otherwise \end{cases} \tag{3}$$

Definition 5. *We say that a test* t identifies *a substance* $s \in S$ *if* $supp(t) = \{s\}$. *As all tests* t *that identify a given substance* s *are isomorphic, we will denote any such test as* t_s.

Intuitively, a test represents a piece of information about the state of a Reaction System that tells an observer whether some given substances are present, or not, in the state. In particular a test is given a *disjunctive interpretation*, it is only able to tell us whether at least one (but not necessarily all) of the substances it tests for are present in the given state. The intuition for this definition derives from the concept of a *chemical test*, that is a qualitative or quantitative procedure designed to identify, quantify, or characterise a chemical compound or chemical group: so, a test that identifies a substance represents a chemical test that is able to precisely detect a single chemical compound (e.g. a test for blood sugar), while chemical test for recognizing chemical groups, e.g. acids, can be represented by a general test. Then, an Approximate Reaction System represents the uncertain and partial knowledge that we have on the behaviour of a real underlying reaction system given that we are only able to observe its states through the tests specified by T.

We observe that a set of tests T defines an indiscernibility partition of the states:

$$X \sim_T Y \text{ iff } \forall t \in T, t(X) = t(Y) \tag{4}$$

We denote by $[X]_T$ the equivalence class of state $X \subseteq 2^S$ determined by the set of tests T. Thus, it follows that the set of test determines an approximation space $(2^S, \sim_T)$. The rough approximations of the states are formally defined as follows: let $X \subseteq 2^S$ be a state, then, its *rough approximation* determined by T is given by $r(X) = \langle l(X), u(X) \rangle$ where

$$l(X) = \bigcap_{Y \in [X]_T} Y \tag{5}$$

$$u(X) = \bigcup_{Y \in [X]_T} Y \tag{6}$$

Given an ARS $R = (S, A, T)$, the associated graph of the dynamics is the graph $G[R] = (V_T, E_T)$ where:

- $V_T = \mathcal{R}_T(2^S)$;
- $(v_1 = \langle l(X), u(X) \rangle, v_2 = \langle l(Y), u(Y) \rangle) \in E_T$ iff $\exists l(X) \subseteq W \subseteq u(X), l(Y) \subseteq Z \subseteq u(Y)$ s.t. (W, Z) is an edge in the graph of the non-approximated reaction system.

An example of an ARS and its associated graph of the dynamics is shown in Example 2.

Example 2. *Let R be a ARS $R = (S, A, T)$: where S and A are as detailed in Example 1, while $T = \{t_1, t_2, t_3\}$ where $supp(t_1) = \{B\}$, $supp(t_2) = \{A, C\}$ and $supp(t_3) = \{B, C\}$. Figure 2 illustrates the related graph of the dynamics.*

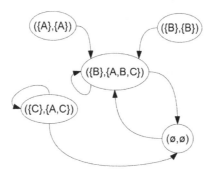

Fig. 2. Graph of the dynamics for the ARS in Example 2.

We notice that this graph features a form of non-determinism as, for example, there are multiple outgoing arcs from the node labeled $\langle B, ABC \rangle$. We notice, furthermore, that the state where only substance B is distinct from the state $\langle B, ABC \rangle$ even though the lower approximation of the latter one is exactly B.

As illustrated in Example 2, one can observe that, in general, the incompleteness and uncertainty determined by the fact that the result of tests is the only information available about the states, the resulting graph of the dynamics could feature a form of *non-determinism*, while the graph of the dynamics of a standard Reaction System is necessarily deterministic. This suggests that, having fixed a set of tests, if we observe the evolution of a system and we derive that the resulting graph of the dynamics is non-deterministic then, the employed test are not sufficient to properly describe the system (at least, if we assume that the underlying phenomenon *could* be modeled as a Reaction System).

While tests formalize the notion of partial observability in terms of the substances, they do not provide a formalization of the idea that, in general, knowledge about a complex system is acquired iteratively by repeatedly observing its evolution over time from an initial state. We formalize this other notion via *observations*:

Definition 6. *Given the graph* $G[R] = (V_T, E_T)$ *of an ARS R, we denote as* $G[R]_x = (V_x, E_x)$ *the set of all maximal paths starting from x.*

We say that an observation *of an ARS is a collection* $\mathcal{O}(R) = \{G[R]_{x_1}, ..., G[R]_{x_n}\}$ *for* $x_1, ..., x_n \in V_T$. *We denote with* $V_{\mathcal{O}(R)}, E_{\mathcal{O}(R_T)}$, *respectively, the set of nodes and edges in* $\mathcal{O}(R)$.

Given an observation of an ARS we can define the respective Information Table as:

Definition 7. *An ARS Information Table* $I[\mathcal{O}(R)]$ *for an observation* $\mathcal{O}(R)$ *of an ARS* $R = (S, A, T)$ *is an ordered pair* (U, T), *where* $U = \bigcup_{G[R]_x \in \mathcal{O}(R)} V_x$.

Thus an ARS Information Table represents two different types of partial, incomplete information about an underlying Reaction System: first, the incompleteness of information w.r.t. the global dynamics of the Reaction System as only the dynamics involving the states under observation is known; second, the incompleteness of information w.r.t. the states, as these are only observed through the set of tests that are performed.

We notice that the definition of *identifies* that we previously defined applies only to single tests. In order to generalize this notion we would need to consider set of tests. Intuitively a set of tests F identifies s if we know with certainty that, given a state X, if $\exists t \in F, t(X) = \top$ then $s \in X$.

Formally,

Definition 8. *A set of tests* $F \subseteq T$ identifies $s \in S$ *if*

$$\forall X \subseteq 2^S, \exists t \in F \text{ s.t. } s \in supp(t) \text{ and } t(X) = \top \wedge$$
$$\forall s' \in supp(t) \setminus \{s\}, \exists t' \neq t \in F \text{ s.t. } s' \in supp(t') \wedge t'(X) = \bot. \quad (7)$$

This notion allows to define an alternative formulation of lower and upper approximations that is not explicitly based on the equivalence relation on the states:

Definition 9. *The lower and upper approximation defined by the relation T identifies s are, respectively:*

$$l'(X) = \{s \in S | \forall t \in T \text{ s.t. } s \in supp(t), t(X) = \top \wedge T \text{ identifies } s\} \quad (8)$$
$$u'(X) = \{s \in S | \forall t \in T \text{ s.t. } s \in supp(t), t(X) = \top\} \quad (9)$$

Then, we can prove the following result, that states that the two alternative formulations are equivalent:

Lemma 1. $l(X) = l'(X)$ *and* $u(X) = u'(X)$.

Proof. Consider first the upper approximation $u(X)$: by Definition 6 a substance s is in $u(X)$ iff $\exists Y \in [X]_T.s \in Y$. Thus, $\forall t \in T.s \in supp(t)$, $t(Y) = \top$; but, by definition of $\sim_T$ this also means that $t(X) = \top$. Since s was arbitrary we can see that Definition 6 implies Definition 8. For the converse we can consider two cases:

1. $s \in X$, then obviously Definition 6 holds;
2. $s \notin X$ but $\forall t \in T$ s.t. $s \in supp(t).t(X) = \top$. Let $Y = X \cup \{s\}$, then evidently $Y \in [X]_T$ but this means that Definition 6 follows.

As regards the lower approximation, we showed that the first part of the Definition characterizes the substances that are in the upper approximation, then we must show that the condition that T identifies s is necessary and sufficient for saying that s is also in the lower approximation. Let us assume that T identifies s, and t be the test that satisfies the condition for state X. Similarly for each substance s' let $t^{s'}$ be the test s.t. $s' \in supp(t^{s'})$ and $t^{s'}(X) = \bot$. Then if $supp(t) = \{s\}$ the implication obviously follows, so let us focus on the case where $\{s\} \subset supp(t)$. Consider the equivalence class $[X]_T$, then evidently $\forall Y \in [X]_T, t(Y) = \top$ and $t^{s'}(Y) = \bot$ which means that $s' \notin Y$ and since this holds $\forall s' \neq s \in supp(t)$ is must hold that $s \in Y$, so Definition 5 follows.

For the converse, notice that if Definition 5 holds then if $s \in l(X)$, then $\exists t \in T.s \in supp(t) \wedge t(Y) = \top$, otherwise there would be a state $Z = Y \setminus \{s\}$ with both $Z, Y \in [X]_T$. If $\{s\} = supp(t)$ then Definition 8 follows. On the contrary, consider $s' \neq s \in supp(t)$ such that $\exists t' \neq t$ with $s' \in supp(t')$. If $\exists Y$ s.t. $t'(Y) = \bot$ then we are done. Otherwise we can notice that such a couple t, t' must exist otherwise it must exists $Z, Y \in [X]_T$ s.t. $Z = (Y \setminus \{s\}) \cup \{s'\}$ but this is an absurd as we assumed that $s \in l(X)$. Thus Definition 8 really is a characterization of lower approximations.

In the following section we will define the concept of reduct for an ARS.

3.2   Reducts

Given an $ARS = (S, A, T)$, we may ask whether the given set of tests is sufficient to describe the dynamics that we could observe, if we had been able to fully observe the states of the Reaction System. More in general, one may assume that the provided set of tests is all we can have (e.g., for a certain situation the provided tests are the most precise and powerful that are known) to describe the Reaction System: in this case we can ask whether all the tests available are necessary or there exists some test that is redundant.

Both these two concepts correspond to the idea of a reduct in Rough Set Theory.

Definition 10. *Given an ARS $R_T = (S, A, T)$ and an observation $\mathcal{O}(R_T)$, for a given $F \subseteq T$, we define $R_F = (S, A, F)$. Then, we say that F is a:*

- complete super-reduct *if $G[R_F] = G[(S, A)]$;*
- relative super-reduct *if $G[R_F] = G[R_T]$;*
- weak super-reduct *if $G[R_F]_{|\mathcal{O}(R_T)} = G[R_T]_{|\mathcal{O}(R_T)}$*

where $G[R_F]_{|\mathcal{O}(R_T)}$ is the restriction of $G[R_F]$ to $\mathcal{O}(R_T)$. We say that F is a complete (resp. relative, weak) reduct if it is a complete (resp. relative, weak) super-reduct and it is minimal w.r.t. this property.

The following result characterizes (complete, relative, weak) reducts in terms of Information Tables:

Proposition 1. *Let $R_T = (S, A, T)$ be an ARS and $\mathcal{O}(R_T)$ an observation. Then $F \subseteq T$:*

- *is a weak reduct iff it is a reduct for $I[\mathcal{O}(R_T)]$;*
- *is a relative reduct iff it is a reduct for $I[R_T] = (2^S, T)$;*
- *is a complete reduct iff it is a reduct for $I^*[R_T] = (2^S, T \cup \{t_s | s \in S\})$, where t_s is a test that identifies s.*

Proof. The case of weak reducts follows directly from Definition 10.

The condition for relative reducts is equivalent to saying that $\mathcal{R}_T(2^S) = \mathcal{R}_F(2^S)$, that is the set of rough sets of states are the same when considering the full set of tests or the reduct F. We can notice that an equivalent condition for F being a relative reduct would be being a reduct for the ARS information system $I[R_T] = (\{[X]_T : X \subseteq 2^S\}, T)$ in which the equivalence classes determined by T are made explicit.

On the other hand, the condition for complete reducts states that F must be able to identify all the substances $s \in S$.   □

Corollary 1. *The smallest complete reduct RED_{min} of an ARS where $\forall s \in S$ $\exists t_s \in T$ has $|RED_{min}| = |S|$.*

We notice that while the definition of (complete, relative, weak) reducts suggests an algorithm for checking whether $F \subseteq T$ is a reduct (e.g. by constructing the discernibility matrix for the corresponding Information Table), the time complexity of this algorithm is linear in the size of the ARS Information Table but, in general, exponential in $|S|$. We can see from Corollary 1 that, at least for complete reducts, a simple algorithm for finding reducts (and hence for testing them) when we restrict to the case where $\forall s \in S, t_s \in T$ and that operates in time linear in $|T|$ can be given.

A different, but equivalent, characterization of reducts can be formulated in terms of the *identifies* relation defined in Sect. 3.1.

Theorem 1. *Let $ARS = (S, A, T)$ be an ARS and let $S_T = \{s \in S | T \text{ identifies } s\}$. Then $F \subseteq T$:*

- is a complete super-reduct iff $S = S_F$;
- is a relative super-reduct iff $S_T = S_F$.

Proof. This follows from the fact that if the condition holds then $\forall X$ the lower and upper approximations remain equal.

Notice that while Theorem 1 and Proposition 1 are equivalent characterizations, the former result suggests an algorithm for testing reducts whose runtime is $O(|S|^2|T|^2)$. Algorithm 1 describes the algorithm for the case of complete reducts, the case for relative reducts is equivalent. The consequence of this result is that the problem of finding complete and relative reducts is in NP not only when considering the graph of the dynamics as the size of the problem, but also when considering the size of the Reaction System. As finding reducts in general Information Tables is NP-complete [20], we conjecture that the problem of finding (complete, relative) reducts lies in the same complexity class.

Algorithm 1. A polynomial-time algorithm for the verification of complete reducts.

---

    procedure CHECK-COMPLETE-REDUCT
Require: $R = (S, A)$ Reaction System , F a reduct
       $check \leftarrow \top$
      for all $s \in S$ **do**
         for all $f \in F : s \in supp(f)$ **do**
            $temp \leftarrow \top$
            for all $s' \neq s \in supp(f)$ **do**
               $temp \leftarrow temp \wedge \exists f' \neq f.s' \in supp(f') \wedge s \notin supp(f')$
            end for
            if $temp = \top$ **then**
               $check \leftarrow \top$
               Break
            else
               $check \leftarrow \bot$
            end if
         end for
      end for
      Return check
    end procedure

Notice that while a similar characterization could be given also for checking weak reducts, in that case the complexity would still be polynomial w.r.t. the number of states in the *observation*, hence, in the worst case, exponential in the number of substances.

4 Conclusion

In this paper we considered the study of mathematical methods to model complex systems, focusing on the formalism of Reaction Systems, in particular with respect to their ability to model incomplete and partial information. As these characteristics are commonly represented through Rough Set Theory, and also acknowledging a recent research direction towards the linking of Reaction Systems and Rough Sets, we developed a mathematical framework, based on core Rough Set theoretic concept to study these issues. We introduced the notion of partial observability of the states of a Reaction System, through the notion of tests, and after observing that this induces an approximation space we applied ideas from Rough Set Theory to define lower and upper approximations; reducts that could be used to automatically model Reaction Systems based on (potentially uncertain and incomplete) observations. In order to further the applications of Rough Set Theory to the study of complex systems, we think that the following open problems may be of interest:

- We provided a characterization of complete and relative reducts based on tests and their ability to identify the substances. This characterization suggests that the problem of finding (complete, relative) reducts is in NP not only w.r.t. the size of the graph of the dynamics (which is in general exponential in the number of substances) but also w.r.t. the size of the Reaction System. Similar characterizations for weak reducts would be interesting;
- We considered reducts as sets of tests that are able to represent, without loss of information, the graph of the dynamics of the Reaction System (or Approximate Reaction System). It is not hard, however, to observe that this definition may be too restrictive: indeed, if one's interest only concerns the general dynamics of a system, then an approximated graph may be tolerable as long as it has the same properties of the original graph (e.g., w.r.t. the reachability of states). It would then be interesting to give a definition of reducts that characterizes this property of invariance w.r.t. the satisfaction of properties expressed in a given logic [2];
- While in this work we considered approximations and reducts, Rough Set Theory also encompasses methods for *rule induction* [9,17] in order to *explain* a Decision Table via sets of rules. Applying these approaches in the context of Approximate Reaction Systems and observations (and, more in general, complex systems) could enable the interactive and iterative learning and updating of Reaction System models [12] based on observed dynamics;
- Finally, while the present work applies to Reaction Systems, we argued that, as the proposed methods mainly use the graph of the dynamics, these notions could also be extended to other discrete complex systems formalisms: in order to do so, appropriate definitions of tests should be considered.

References

1. Akama, S.: Topics in Rough Set Theory: Current Applications to Granular Computing. Intelligent Systems Reference Library, vol. 168. Springer, Cham (2020). https://doi.org/10.1007/978-3-030-29566-0
2. Azimi, S., Gratie, C., Ivanov, S., Manzoni, L., Petre, I., Porreca, A.E.: Complexity of model checking for reaction systems. Theoret. Comput. Sci. **623**, 103–113 (2016)

3. Bello, R., Falcon, R.: Rough sets in machine learning: a review. In: Wang, G., Skowron, A., Yao, Y., Ślęzak, D., Polkowski, L. (eds.) Thriving Rough Sets. SCI, vol. 708, pp. 87–118. Springer, Cham (2017). https://doi.org/10.1007/978-3-319-54966-8_5
4. Bergstra, J.A., Ponse, A., Smolka, S.A.: Handbook of Process Algebra. Elsevier, Amsterdam (2001)
5. Chopard, B., Droz, M.: Cellular Automata. Springer, Dordrecht (1998). https://doi.org/10.1007/978-94-015-9153-9
6. Dutta, S., Jankowski, A., Rozenberg, G., Skowron, A.: Linking reaction systems with rough sets. Fundamenta Informaticae **165**(3–4), 283–302 (2019)
7. Ehrenfeucht, A., Petre, I., Rozenberg, G.: Reaction systems: a model of computation inspired by the functioning of the living cell. In: Konstantinidis, S., Moreira, N., Reis, R., Shallit, J. (eds.) The Role of Theory in Computer Science: Essays Dedicated to Janusz Brzozowski, pp. 1–32. World Scientific, Singapore (2017)
8. Ehrenfeucht, A., Rozenberg, G.: Reaction systems. Fundamenta informaticae **75**(1–4), 263–280 (2007)
9. Hu, M., Yao, Y.: Structured approximations as a basis for three-way decisions in rough set theory. Knowl. Based Syst. **165**, 92–109 (2019)
10. Lin, T.Y., Cercone, N.: Rough Sets and Data Mining: Analysis of Imprecise Data. Springer, Boston (2012). https://doi.org/10.1007/978-1-4613-1461-5
11. Liu, Y.Y., Slotine, J.J., Barabási, A.L.: Observability of complex systems. Proc. Natl. Acad. Sci. **110**(7), 2460–2465 (2013)
12. Męski, A., Koutny, M., Penczek, W.: Reaction mining for reaction systems. In: Stepney, S., Verlan, S. (eds.) UCNC 2018. LNCS, vol. 10867, pp. 131–144. Springer, Cham (2018). https://doi.org/10.1007/978-3-319-92435-9_10
13. Orlowska, E.: Incomplete Information: Rough Set Analysis. Studies in Fuzziness and Soft Computing, vol. 13. Physica, Heidelberg (2013). https://doi.org/10.1007/978-3-7908-1888-8
14. Păun, G., Rozenberg, G.: A guide to membrane computing. Theoret. Comput. Sci. **287**(1), 73–100 (2002)
15. Pawlak, Z.: Rough sets. Int. J. Comput. Inf. Sci. **11**(5), 341–356 (1982)
16. Reisig, W.: Petri Nets: An Introduction, vol. 4. Springer, Heidelberg (2012)
17. Sakai, H., Nakata, M.: Rough set-based rule generation and apriori-based rule generation from table data sets: a survey and a combination. CAAI Trans. Intell. Technol. **4**(4), 203–213 (2019)
18. Skowron, A., Dutta, S.: Rough sets: past, present, and future. Nat. Comput. **17**(4), 855–876 (2018). https://doi.org/10.1007/s11047-018-9700-3
19. Skowron, A., Jankowski, A., Dutta, S.: Interactive granular computing. Granul. Comput. **1**(2), 95–113 (2016)
20. Skowron, A., Rauszer, C.: The discernibility matrices and functions in information systems. In: Słowiński, R. (ed.) Intelligent Decision Support, pp. 331–362. Springer, Dordrecht (1992). https://doi.org/10.1007/978-94-015-7975-9_21
21. Ślęzak, D., Dutta, S.: Dynamic and discernibility characteristics of different attribute reduction criteria. In: Nguyen, H.S., Ha, Q.-T., Li, T., Przybyła-Kasperek, M. (eds.) IJCRS 2018. LNCS (LNAI), vol. 11103, pp. 628–643. Springer, Cham (2018). https://doi.org/10.1007/978-3-319-99368-3_49
22. Thangavel, K., Pethalakshmi, A.: Dimensionality reduction based on rough set theory: a review. Appl. Soft Comput. **9**(1), 1–12 (2009)
23. Wang, G., Skowron, A., Yao, Y., Ślęzak, D., Polkowski, L. (eds.): Thriving Rough Sets. SCI, vol. 708. Springer, Cham (2017). https://doi.org/10.1007/978-3-319-54966-8

Weighted Generalized Fuzzy Petri Nets and Rough Sets for Knowledge Representation and Reasoning

Zbigniew Suraj[1]([⊠]) [iD], Aboul Ella Hassanien[2] [iD], and Sibasis Bandyopadhyay[3]

[1] Institute of Computer Science, University of Rzeszów, Rzeszów, Poland
zbigniew.suraj@ur.edu.pl
[2] Faculty of Computers and Artificial Intelligence, Cairo University, Giza, Egypt
aboitcairo@gmail.com
[3] Asansol, India

Abstract. In this paper, we consider the decision tables provided by experts in the field. We construct an algorithm for executing a highly parallel program represented by a fuzzy Petri net from a given decision table. The constructed net allows objects to be identified in decision tables to the extent that appropriate decisions can be made. Conditional attribute values given by experts are propagated by the net at maximum speed. This is done by properly organizing the net's work. Our approach is based on rough set theory and weighted generalized fuzzy Petri nets.

Keywords: Decision system · Information system · Rough set · Decision rule · Weighted generalized fuzzy Petri net

1 Introduction

Rough set theory, proposed by Pawlak in 1982 [18], is a mathematical tool for dealing with unclear, imprecise, incoherent and uncertain knowledge. It has been observed for many years that both research and applications of rough set theory are attracting more and more attention of researchers. It can be successfully used in many areas of application alone or in combination with other approaches. Here, we use this theory to support modeling of decision-making systems using weighted generalized fuzzy Petri nets.

In this paper, we assume that a decision table S representing experimental knowledge is given [17]. It consists of a number of rows labeled by elements from a set of objects U, which contain the results of measurements, observations, reviews etc. represented by a value vector of conditional attributes (conditions) from A together with a decision d corresponding to this vector. Values of conditions are provided by experts in the field. In some applications the values of conditional attributes can be interpreted as states of local processes in a complex system and the decision value is related to the global state of that system [13,16,24]. Sometimes it is necessary to transform a given experimental decision

© Springer Nature Switzerland AG 2020
R. Bello et al. (Eds.): IJCRS 2020, LNAI 12179, pp. 61–77, 2020.
https://doi.org/10.1007/978-3-030-52705-1_5

table by taking into account other relevant features (new conditional attributes) instead of the original ones. This step is necessary when the decision algorithm constructed directly from the original decision table yields an inadequate classification of unseen objects or when the complexity of decision algorithm synthesis from the original decision table is too high. In this case some additional time is necessary to compute the values of new features after the original values are given. The input for our algorithm consists of a decision table (if necessary, pre-processed as described above).

We shall construct a fuzzy Petri net allowing to make a decision as soon as a sufficient number of conditional attribute values is known and conclusions drawn from the knowledge encoded in S (cf. [22]). In the paper we formulate this problem and present its solution.

First, we assume that knowledge encoded in S is represented by rules automatically extracted from S. We consider acceptable rules in S, i.e. rules for which the accuracy factor need not necessarily be equal to 1 [23]. We assume that the knowledge encoded in S is complete in the sense that invisible objects have attribute value vectors consistent with rules extracted from S. This assumption may be too restrictive, because the rules for the classification of new objects should be generated only from appropriate features (attributes). The rule is active if the values of all attributes on its left side are given. Our algorithm should propagate information from attributes to other attributes as soon as possible. This is the reason for generating true decision rules corresponding to relative reducts with respect to the decision in S [22]. The last step of our algorithm is the implementation of the set of generated rules using fuzzy Petri nets. Each step of a computation of the constructed fuzzy Petri net consists of two phases. In the first phase, it is checked that all condition values are known, and if so, in the second phase, new information about the values is sent through the net at maximum speed. The whole computation process is carried out by proper organization of the net's work.

In the paper, we use fuzzy Petri nets [3,5,10,12,30] as a model of the target decision-making system. Net properties can be verified using tools for the analysis of Petri nets (see e.g. [28]).

Over the past few decades, there has been a series of modifications to the classic fuzzy Petri nets (FPNs) [12] to deal with complex decision-making systems. Chen [4] introduced weight factors into FPNs and proposed a weighted FPN (WFPN) model. Ha et al. [7] extended his work by adding input and output weight factors into WFPNs. Then the intuitionistic fuzzy sets were integrated into FPNs, and an intuitionistic FPN was presented in [11,26]. Skowron and Suraj [23] developed a parallel algorithm for real-time decision-making based on rough set theory and classic Petri nets. Peters et al. [20] combined the theory of FPNs, rough sets, and colored Petri nets to develop a rough fuzzy Petri net model. Suraj and Fryc [27] introduced time factor to approximate Petri nets, which plays a vital role in developing real-time decision-making systems. Bandyopadhyay et al. [1] proposed to link Petri nets and soft sets and introduced a soft Petri net model. Suraj and Hassanien [29] combined the theory of FPN and sets of fuzzy intervals to avoid the problem of determining the exact membership or truth value.

This paper establishes some relationships between rough set theory and fuzzy Petri nets. Parameter values such as rule certainty coefficients, input and output weights of arcs in the net model are calculated automatically from a given decision table. The empirical example provided here shows the effectiveness of the proposed model.

The rest of this paper is organized in the following way. Section 2 contains some background knowledge regarding rough set theory. In Sect. 3, the weighted generalized fuzzy Petri net formalism is given. Section 4 describes three structural forms of decision rules and a method for transformation of decision tables into weighted generalized fuzzy Petri nets. An example illustrating the approach presented in this paper is provided in Sect. 5. Finally, Sect. 6 suggests some directions for further research related to our approach.

2 Preliminaries of Rough Set Theory

In this section we recall basic notions of rough set theory. Among them are those of information systems, indiscernibility relations, dependencies of attributes, relative reducts, significance of attributes and rules [14,15].

2.1 Information Systems and Decision Systems

An *information system* is a pair $S = (U, A)$, where U is a non-empty finite set of *objects* called the *universe* and A is a non-empty finite set of *attributes* such that $a : U \rightarrow V_a$ for every $a \in A$. The set V_a is called the *value set* of a, and $V = \bigcup_{a \in A} V_a$ is said to be the *domain* of A.

Let $S = (U, A)$ be an information system and let $B \subseteq A$ and $X \subseteq U$. Then there is associated an equivalence relation $\mathrm{ind}(B)$: $\mathrm{ind}(B) = \{(u, u') \in U \times U :$ for every $a \in B$ $a(u) = a(u')\}$. $\mathrm{ind}(B)$ is called the *B-indiscernibility relation*. If $(u, u') \in \mathrm{ind}(B)$, then objects u and u' are indiscernible from each other by attributes from B. The equivalence classes of the B-indiscernibility relation are denoted $[u]_B$.

We can approximate X using only the information contained in B, constructing the *B-lower* and *B-upper approximations* of X, denoted by $\underline{B}X$ and $\overline{B}X$ respectively, where $\underline{B}X = \{u : [u]_B \subseteq X\}$ and $\overline{B}X = \{u : [u]_B \cap X \neq \emptyset\}$. The objects in $\underline{B}X$ can be with certainty classified as members of X on the basis of knowledge in B, while the objects in $\overline{B}X$ can be only classified as possible members of X on the basis of knowledge in B. The set X is *rough* if $\overline{B}X - \underline{B}X \neq \emptyset$.

A *decision system* (a *decision table*) is any information system of the form $S = (U, A \cup \{d\})$, where $d \notin A$ is a distinguished attribute called *decision attribute* (*decision*). The elements of A are called *conditional attributes* (*conditions*).

Let $S = (U, A \cup \{d\})$ be a decision system. The cardinality of the image $d(U) = \{k : d(u) = k$ for some $u \in U\}$ is called the *rank* of d and is denoted by $r(d)$. We assume that the set V_d of values of the decision d is equal to $\{1, ..., r(d)\}$.

Let us observe that the decision d determines a partition $\{X_1, ..., X_{r(d)}\}$ of the universe U, where $X_k = \{u \in U : d(u) = k\}$ for $1 \leq k \leq r(d)$. The set X_i is called the i-th *decision class of S*. If $X_1, ..., X_{r(d)}$ are the decision classes of S, then the set $\underline{B}X_1 \cup ... \cup \underline{B}X_{r(d)}$ is called the B-*positive region of S* and is denoted by $POS_B(d)$.

Any decision system $S = (U, A \cup \{d\})$ can be represented by a data table with the number of rows equal to the cardinality of the universe U and the number of columns equal to the cardinality of the set $A \cup \{d\}$. On the position corresponding to the row u and column a the value $a(u)$ appears.

Example 1. A small decision system is shown in Table 1. We have a set of objects (patients) $U = \{1, 2, 3, 4, 5, 6\}$, a set of conditional attributes (symptoms) $A = \{$H (Headache), M (Muscle-pain), T (Temperature)$\}$. The decision attribute is denoted by F (Flu). The possible values of attributes from $A \cup \{$F$\}$ are equal to no, yes, normal, high, or very high and $r($F$) = 2$. The decision F defines a partition $\{X_1, X_2\}$ of U, where $X_1 = \{1, 2, 3, 6\}$, $X_2 = \{4, 5\}$. Each row of the table can be seen as information about specific patient.

Table 1. An example of a decision system

$U/A \cup \{d\}$	H	M	T	F
1	no	yes	high	yes
2	yes	no	high	yes
3	yes	yes	very high	yes
4	no	yes	normal	no
5	yes	no	high	no
6	no	yes	very high	yes

2.2 Dependency of Attributes

An important issue in data analysis is discovering of dependencies between attributes. Intuitively, a set of attributes C depends totally on a set of attributes B, denoted by $B \Rightarrow C$, if there exists a functional dependency between values of C and B.

Let $S = (U, A)$ be an information system and let $B, C \subseteq A$.

We say that the set C *depends on B in degree k* $(0 \leq k \leq 1)$, denoted by $B \Rightarrow_k C$, if $k = \gamma(B, C) = \frac{|POS_B(C)|}{|U|}$, where $POS_B(C) = \bigcup_{X \in U/C} \underline{B}(X)$ and $|X|$ denotes the cardinality of $X \neq \emptyset$. The set $POS_B(C)$ is called a *positive region* of the partition U/C with respect to B. In fact, it is the set of all elements of U that can be uniquely classified to blocks of the partition U/C by means of B.

Let $B, C \subseteq A$, and $B' \subseteq B$. A set B' is a C-*reduct of B* (or B' *is a relative reduct of B with respect to C*), if B' is a minimal subset of B and $\gamma(B, C) = \gamma(B', C)$.

Example 2. Consider once again the decision system presented in Table 1. For example, for the dependency $\{H, M, T\} \Rightarrow_k \{F\}$ we get $k = 2/3$. However, for the dependency $\{T\} \Rightarrow_k \{F\}$, we get $k = 1/2$. The attribute T offers a worse classification than the entire set of attributes H, M, T. It is worth to noting that neither H nor M can be used to recognize flu, because for both dependencies $\{H\} \Rightarrow_k \{F\}$ and $\{M\} \Rightarrow_k \{F\}$ we have $k = 0$. In Table 1 there are two relative reducts with respect to $\{F\}$, $R_1 = \{H, T\}$ and $R_2 = \{M, T\}$ of the set of conditions $\{H, M, T\}$.

2.3 Significance of Attributes

Significance of an attribute a in a decision system $S = (U, A \cup \{d\})$ can be evaluated by measuring the effect of removing of an attribute $a \in A$ from the attribute set A on the positive region defined by the table S.

Let $B \subseteq A$. Significance of an attribute $a \in A$ is defined as follows:
$$\sigma(B, d, a) = \gamma(B, \{d\}) - \gamma(B - \{a\}, \{d\}) = \frac{|POS_B(\{d\})| - |POS_{B-\{a\}}(\{d\})|}{|U|}, \text{ and}$$
is simply denoted by $\sigma(a)$ when B and $\{d\}$ are understood.

This numerical factor measures the difference between $\gamma(B, \{d\})$ and $\gamma(B - \{a\}, \{d\})$, i.e. it says how the factor $\gamma(B, \{d\})$ changes when an attribute a is removed.

Note that the following relationship is also met: $0 \leq \sigma(B, d, a) \leq 1$.

Example 3. Using the above formula for the decision system from Example 1, we obtain the following results for Table 1:

1. For the set of conditional attributes A: $\sigma(H) = 0$, $\sigma(M) = 0$, $\sigma(T) = 1/2$
2. For the relative reduct R_1: $\sigma(H) = 1/6$, $\sigma(T) = 2/3$
3. For the relative reduct R_2: $\sigma(M) = 0$, $\sigma(T) = 3/4$

2.4 Rules in Decision Systems

Rules express some of the relationships between values of the attributes described in decision tables. In this subsection we recall the definition of rules as well as other related concepts.

Let $S = (U, A \cup \{d\})$ be a decision system, $B \subseteq A \cup \{d\}$, and $V = \bigcup_{a \in A} V_a \cup V_d$.

Atomic formulae over B and V are expressions of the form $a = v$. They are called *descriptors* over B and V, where $a \in B$ and $v \in V_a$. The set $DESC(B, V)$ of formulae over B and V is the least set containing all atomic formulae over B and V and closed with respect to the propositional connectives OR (disjunction), AND (conjunction) and NOT (negation).

Let $\tau \in DESC(B, V)$. $\|\tau_S\|$ denotes the meaning of τ in the decision system S which is the set of all objects in U with the property τ. These sets are defined as follows:

1. if τ is of the form $a = v$ then $\|\tau_S\| = \{u \in U : a(u) = v\}$
2. $\|(\tau \text{ OR } \tau')_S\| = \|\tau_S\| \cup \|\tau'_S\|$; $\|(\tau \text{ AND } \tau')_S\| = \|\tau_S\| \cap \|\tau'_S\|$; $\|\text{NOT } \tau_S\| = U - \|\tau_S\|$.

The set $\text{DESC}(A, V_a)$, $a \in A$, is called the set of *conditional formulae of S*.

A *decision rule* r for S is any expression of the form IF τ THEN $d = v$, where $\tau \in \text{DESC}(A, V_a)$, $v \in V_d$ and $\|\tau_S\| \neq \emptyset$. Formulae τ and $d = v$ are called the *predecessor* and the *successor* of the decision rule r. $\|\tau_S\|$ is the non-empty set of objects *matching* the decision rule and $\|\tau_S\| \cap \|(d = v)_S\|$ is the set of objects *supporting* the rule. With every decision rule r we can associate several numerical factors. The *accuracy factor* of the decision rule r is the number $acc(r) = \frac{\|\|\tau_S\| \cap \|(d=v)_S\|\|}{\|\|\tau_S\|\|}$, while the *strength factor* of the decision rule r is understood as $str(r) = \frac{\|\|\tau_S\| \cap \|(d=v)_S\|\|}{|U|}$. The decision rule r is *true* in S, if $acc(r) = 1$, otherwise it is *acceptable* in S.

It is also easy to see that $0 \leq str(r) \leq acc(r) \leq 1$ for every the decision rule r in S.

Example 4. Let us consider the decision system table S from Example 1 presented in Table 1. Using the method for generating decision rules in S [22], we get the following rules, corresponding to the relative reduct $R_1 = \{\mathsf{H}, \mathsf{T}\}$ along with the numerical factors defined above:

– r_1: IF H=no AND T=very high THEN F=yes; $str(r_1) = 1/6$, $acc(r_1) = 1$
– r_2: IF H=yes AND T=very high THEN F=yes; $str(r_2) = 1/6$, $acc(r_2) = 1$
– r_3: IF H=no AND T=high THEN F=yes; $str(r_3) = 1/6$, $acc(r_3) = 1$
– r_4: IF H=yes AND T=high THEN F=yes; $str(r_4) = 1/6$, $acc(r_4) = 1/2$
– r_5: IF H=yes AND T=high THEN F=no; $str(r_5) = 1/6$, $acc(r_5) = 1/2$
– r_6: IF H=no AND T=normal THEN F=no; $str(r_6) = 1/6$, $acc(r_6) = 1$

Note that the rules r_1, r_2, r_3, r_6 are true in Table 1, while the other rules are acceptable in this table.

For a systematic overview of rule synthesis, see e.g. [9, 15, 21].

3 Weighted Generalized Fuzzy Petri Nets

Fuzzy Petri nets are a modification of classic Petri nets to deal with imprecise, unclear or incomplete information in knowledge-based systems that are widely used to model fuzzy production rules and rule-based reasoning.

In this section, we define weighted generalized fuzzy Petri nets (WGFP-net). The new model is a modification of generalized fuzzy Petri nets, proposed in [25]. The main difference between the current net model and the previous one concerns the weights of arcs. Weights are now added to the input and output arcs. They are any numbers from 0 to 1, automatically calculated from the data table and interpreted in the concepts of rough set theory (see Sect. 4) (cf. [2, 10]).

In this paper WGFP-nets are used as a tool for computing a parallel program from a given decision table. After modeling a decision table by a WGFP-net the states are identified in the net to an extent allowing to take the appropriate decisions.

We also assume that the reader knows the basic concepts of classic Petri nets [6] and triangular norms [8].

Let $[0,1]$ denotes the set of real numbers between 0 and 1.

A *weighted generalized fuzzy Petri net* is a tuple $N = (P, T, I, O, M_0, S, \alpha, \beta, \gamma, Op, \delta)$, where: (1) $P = \{p_1, p_2, \ldots, p_n\}$ is a finite set of places; (2) $T = \{t_1, t_2, \ldots, t_m\}$ is a finite set of transitions; (3) $I: P \times T \to [0,1]$ is the input function that maps directed arcs from places to output transitions of those places. If a directed arc (p,t) exists between a place p and a transition t, then $I(p,t) > 0$, otherwise 0. The values of $I(p,t)$ for $(p,t) \in P \times T$ are called input weights of transitions t and are denoted by iw; (4) $O: T \times P \to [0,1]$ is the output function that maps directed arcs from transitions to output places of those transitions. If a directed arc (t,p) exists between a transition t and a place p, then $O(t,p) > 0$, otherwise 0. The values of $O(t,p)$ for $(t,p) \in T \times P$ are called output weights of transitions t and are denoted by ow; (5) $M_0: P \to [0,1]$ is the initial marking; (6) $S = \{s_1, s_2, \ldots, s_n\}$ is a finite set of statements; (7) $\alpha: P \to S$ is the statement binding function; (8) $\beta: T \to [0,1]$ is the truth degree function; (9) $\gamma: T \to [0,1]$ is the threshold function; (10) Op is a union of t-norms and s-norms called the set of operators, and the sets P, T, S, Op are pairwise disjoint; (11) $\delta: T \to Op \times Op \times Op$ is the operator binding function.

We also accept that if $I(p,t) = 0$ $(O(p,t) = 0)$ then the directed arc from input (output) place p to transition t does not exist in the net drawing. Similarly, if $M_0(p) = 0$ then the token does not exist in the place p. In addition, if $I(p,t) = 1$ $(O(t,p) = 1)$, then the weight of the arc equal to 1 is also disregarded in the net drawing. The numbers $\beta(t)$ and $\gamma(t)$ are placed in a net picture under the transition t. The first number is interpreted as the truth degree of an implication corresponding to a given transition t. The role of the second one is to limit the possibility of transition firings, i.e., if the input operator In value for all values corresponding to input places of the transition t is less than a threshold value $\gamma(t)$ then this transition cannot be fired (activated). The operator binding function δ connects transitions with triples of operators (In, Out_1, Out_2). The first operator in the triple is called the input operator, and two remaining ones are the output operators. The input operator In concerns the way in which all input places are connected with a given transition t (more precisely, statements corresponding to those places). However, the output operators Out_1 and Out_2 concern the way in which the next marking is computed after firing the transition t. In the case of the input operator we assume that it can belong to one of two classes, i.e., t- or s-norm, whereas the second one belongs to the class of t-norms and the third to the class of s-norms.

Let N be a WGFP-net. A marking of N is a function $M: P \to [0,1]$.

The dynamic behavior of the system is represented by the *firing* of the corresponding transition, and the evolution of the system is represented by a *firing sequence* of transitions. We assume that the networks built in the form presented in this paper operate according to the *firing rule* consisting of the following three steps:

1. A transition $t \in T$ is *enabled* (or *ready for firing*) for marking M if the number produced by input operator In for all input places of the transition t by M multiplied by the relevant weights of arcs is positive and greater than, or equal to the number being a value of threshold function γ corresponding to the transition t. Formally, the following condition for $\gamma(t)$ should be satisfied: $In(iw_{i1} \cdot M(p_{i1}), iw_{i2} \cdot M(p_{i2}), ..., iw_{ik} \cdot M(p_{ik})) \geq \gamma(t) > 0$, where In is an input operator of the transition t, iw_{ij} is an input weight of t and $M(p_{ij})$ is a marking of a place p_{ij} for $j = 1, 2, ..., k$.
2. A transition can fire only if it is enabled.
3. If M is a marking of N enabling transition t and M' is the marking derived from M by firing transition t, then for each $p \in P$ a procedure for computing the next marking M' is as follows: (1) Tokens in all output places of t are modified in the following way: at first the value of input operator In for all input places of t is computed, next the value of output operator Out_1 for the value of In and for the value of truth degree function $\beta(t)$ is determined, and finally, a value corresponding to $M'(p)$ for each $p \in O(p)$ is obtained as a result of output operator Out_2 for the value of Out_1 multiplied by the weight ow and the current marking $M(p)$. (2) Tokens in the remaining places of net N are not changed.

Formally, for each $p \in P$

$$M'(p) = \begin{cases} Out_2(ow \cdot Out_1(In(iw_{i1} \cdot M(p_{i1}), iw_{i2} \cdot M(p_{i2}), ..., iw_{ik} \cdot M(p_{ik})), \beta(t)), \\ M(p)) \text{ if } p \in O(t) \\ M(p) \text{ otherwise} \end{cases}$$

We also assume that if several transitions are simultaneously enabled in the same marking (i.e. transitions are · *concurrent*) then they can be fired by an application of the firing rule described above in one and the same step and the resulting marking is computed according to this rule.

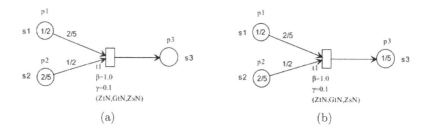

(a)   (b)

Fig. 1. A WGFP-net with the initial marking: (a) before firing t_1, (b) after firing t_1

Example 5. Consider a WGFP-net in Fig. 1. For the net we have: the set of places $P = \{p_1, p_2, p_3\}$, the set of transitions $T = \{t_1\}$, the input function I and the output function O in the form: $I(p_1, t_1) = iw_1 = 2/5$, $I(p_2, t_1) = iw_2 = 1/2$, $I(p_3, t_1) = iw_3 = 0$, $O(t_1, p_1) = ow_1 = 0$, $O(t_1, p_2) = ow_2 = 0$, $O(t_1, p_3) = ow_3 = 1$, and the initial marking $M_0 = (1/2, 2/5, 0)$, the set of statements $S = \{s_1, s_2, s_3\}$, the statement binding function α: $\alpha(p_1) = s_1$, $\alpha(p_2) = s_2$, $\alpha(p_3) = s_3$, the truth degree function β: $\beta(t_1) = 1.0$, the threshold function γ: $\gamma(t_1) = 0.1$, the set of operators $Op = \{ZtN, GtN, ZsN\}$, the operator binding function δ: $\delta(t_1) = (ZtN, GtN, ZsN)$, where $ZtN(a, b) = \min(a, b)$ (minimum, Zadeh t-Norm), $GtN(a, b) = a \cdot b$ (algebraic product, Goguen t-Norm), and $ZsN(a, b) = \max(a, b)$ (maximum, Zadeh s-Norm). The transition t_1 is enabled by the initial marking M_0, since $ZtN(I(p_1, t_1) \cdot M_0(p_1), I(p_2, t_1) \cdot M_0(p_2)) = \min(1/5, 1/5) = 1/5 \geq 0.1 = \gamma(t_1)$. Firing transition t_1 by the marking M_0 transforms M_0 to the resulting marking $M' = (1/2, 2/5, 1/5)$, because $ow_3 \cdot GtN(1/5, \beta(t_1)) = 1 \cdot GtN(1/5, 1.0) = 1/5$ and $ZsN(1/5, M_0(p_3)) = \max(1/5, 0) = 1/5$. Note that in this case the transition t_1 is still enabled by M', but when it is fired at this marking, the result marking is the same as M'. We omit the detailed description of the relevant calculations illustrating the transformation from the marking M' to M' after firing t_1. They run similarly to these above.

4 Transformation of Decision Systems into WGFP-nets

Now we present a method for transforming decision rules representing a given decision system into a WGFP-net.

We assume that a decision system is represented by decision rules of the form IF τ THEN $d = v$.

Let $S = (U, A \cup \{d\})$ be a decision system, and $DESC(A, V_a)$ be the set of the set of conditional formulae of S.

In the paper, we consider three structural forms of decision rules with a list of numerical factors enclosed in square brackets '[' and ']' characterizing these rules (cf. [4, 7, 10]).

Type 1: A simple decision rule

$$r_1 : \text{IF } a = v \text{ THEN } d = v'$$
$$[b; \; \sigma(a), str(r_1); \; acc(r_1)]$$

where $a = v$ and $d = v'$ denote descriptors such that $a = v \in DESC(A, V_a)$ and $v' \in V_d$, b is the truth degree value of $a = v$, $\sigma(a)$ is significance of the attribute a, while $str(r_1)$ and $acc(r_1)$ are the strength factor and the accuracy factor of the rule r_1, respectively.

A WGFP-net structure of the decision rule r_1 is shown in Fig. 2, where iw is the input weight of the transition r_1 and interpreted as $\sigma(a)$, while ow is the output weight of r_1 and interpreted as $str(r_1)$ (see Subsect. 2.3 and 2.4). A larger value of iw or ow means a stronger corresponding connection. However, the value $\beta(r_1) = c$ is interpreted as $acc(r_1)$. Similarly as before, the larger value of β the

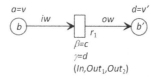

Fig. 2. A WGFP-net representation of the rule of type 1

more credible the rule is. The value of γ represents the threshold value. Larger value b requires greater truth degree of the rule precedence, i.e., $a = v$. The operator In and the operators Out_1, Out_2 represent the input operator and the output operators, respectively. According to Fig. 3 the token value in an output place p' of a transition t corresponding to the decision rule r_1 is calculated as $b' = ow \cdot Out_1(b \cdot iw, c)$, if $b \cdot iw \geq d$, where $d = \gamma(r_1)$ and $\gamma(r_1)$ is the threshold value associated to the transition r_1 and it is given by an expert in the field during the simulation process of the network.

If the predecessor or the successor of a decision rule contains AND or OR (propositional connectives), it is called a *composite decision rule*. Below, two types of composite decision rules are presented together with their WGFP-net representation (see Fig. 3 and Fig. 4).

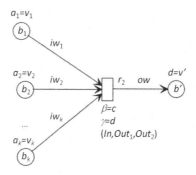

Fig. 3. A WGFP-net representation of the rule of type 2

Type 2: A composite conjunctive decision rule in the predecessor of the rule

$$r_2 : \text{IF } a_1 = v_1 \text{ AND } a_2 = v_2 \cdots \text{ AND } a_k = v_k \text{ THEN } d = v'$$
$$[b_1, b_2, \ldots, b_k; \ \sigma^1(a), \sigma^2(a), \ldots, \sigma^k(a), str(r_2); \ acc(r_2)]$$

where $a_1 = v_1$, $a_2 = v_2$, ..., $a_k = v_k$, $d = v'$ denotes descriptors, and b_1, b_2, ..., b_k, b' their truth degree values, respectively. The meaning of all numerical factors characterizing this rule is similar to the meanings of the relevant factors described for the rule of type 1. The token value b' is calculated in the output place as follows (Fig. 3): $b' = Out_1(In(b_1 \cdot iw_1, b_2 \cdot iw_2, \ldots, b_k \cdot iw_k), c)) \cdot ow)$, if $In(b_1 \cdot iw_1, b_2 \cdot iw_2, \ldots, b_k \cdot iw_k) \geq d$, where $d = \gamma(r_2)$.

Type 3: A composite disjunctive decision rule in the successor of the rule

$$r_3 : \text{ IF } a' = v' \text{ THEN } d = v_1 \text{ OR } d = v_2 \cdots \text{ OR } d = v_n$$

$$[b'; \sigma^1(a'), \sigma^2(a'), \ldots, \sigma^n(a'), str^1(r_3), str^2(r_3), \ldots, str^n(r_3); acc^1(r_3), acc^2(r_3), \ldots, acc^n(r_3)]$$

where $a' = v'$, $d = v_1$, $d = v_2$, $\ldots$, $d = v_n$ denotes descriptors, and b' is the truth degree value of $a' = v'$. The token value for the type 3 is calculated in each output place as follows (Fig. 4): $b_j = ow_j \cdot Out_1(b' \cdot iw, c_j)$, if $b' \cdot iw \geq d_j$, where $d_j = \gamma^j(r_3), j = 1, \ldots, n$.

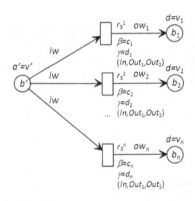

Fig. 4. A WGFP-net representation of the rule of type 3

Remarks:

1. It is easy to see that the rule of type 1 is a particular case of the rule of type 2, as in the case of the rule of type 1, there is only one descriptor in the predecessor. Type 3 can also be easily converted to type 1. Therefore, without losing generality, we can only consider the rules of type 1 and 2.

2. As the rules of type 1 and 3 have only one descriptor in their predecessors, we may omit the input operator In in Fig. 2 and 4. Nevertheless, for better readability of these figures we leave the operator where it is. What's more, the rule of type 3 can be generalized in the case when in the predecessor of the rule instead of one descriptor we have a conjunction of descriptors (as in the rule of type 2). Then the net modeling of such a rule in relation to its predecessor is similar to the one done for the rule of type 2.

3. We assume that the initial markings of output places are equal to 0 in all net models corresponding to the considered rule types. Therefore, in the descriptions of the token values in output places we do not regard the output operator Out_2. In the opposite case, i.e., for non-zero markings of output places, we should take into account this output operator. Thus, in each formula presented above the final token value a' should be computed as follows:

$b' = Out_2(b'', M(p'))$, where b'' denotes the token values computed for suitable rule types by means of formulas presented above, and $M(p')$ is a marking of output place p'. Intuitively, a final token value corresponding to $M'(p')$ for each output place p' of a transition representing a decision rule r is obtained as a result of Out_2 operation for the computed Out_1 operation value and the current marking $M(p')$.

Using the method described above, we can formulate a simple algorithm that constructs a WGFP-net based on a given set of rules extracted from a decision system S. This algorithm transforms the rule into a WGFP-net depending on the form of the transformed rule.

Let $S = (U, A \cup \{d\})$ be a decision system.

Algorithm 1: Construction of WGFP-net using a set of decision rules in S

Input   : A finite set R of decision rules in with a list of parameters
Output: A WGFP-net N_S
$F \leftarrow \emptyset$; (* The empty set. *)
for each $r \in R$
if *r is a rule of type 1* **then**
  $\llcorner$ construct a subnet N_r as shown in Fig. 2;
if *r is a rule of type 2* **then**
  $\llcorner$ construct a subnet N_r as shown in Fig. 3;
if *r is a rule of type 3* **then**
  $\llcorner$ construct a subnet N_r as shown in Fig. 4;
$F \leftarrow F \cup \{N_r\}$;
integrate all subnets from a family F on joint places and create a result net N_S;
return N_S;

---

5   An Example

To illustrate our methodology, let's reconsider the decision rules corresponding to the relative reduct R_1 from Example 4 along with a full list of parameters needed to build a structure of WGFP net model:

- r_1: IF H=no AND T=very high THEN F=yes [σ(H) $= 1/6$, σ(T) $= 2/3$, $str(r_1) = 1/6$; $acc(r_1) = 1$]
- r_2: IF H=yes AND T=very high THEN F=yes [σ(H) $= 1/6$, σ(T) $= 2/3$, $str(r_2) = 1/6$; $acc(r_2) = 1$]
- r_3: IF H=no AND T=high THEN F=yes [σ(H) $= 1/6$, σ(T) $= 2/3$, $str(r_3) = 1/6$; $acc(r_3) = 1$]
- r_4: IF H=yes AND T=high THEN F=yes [σ(H) $= 1/6$, σ(T) $= 2/3$, $str(r_4) = 1/6$; $acc(r_4) = 1/2$]

- r_5: IF H=yes AND T=high THEN F=no [$\sigma(H) = 1/6$, $\sigma(T) = 2/3$, $str(r_5) = 1/6$; $acc(r_5) = 1/2$]
- r_6: IF H=no AND T=normal THEN F=no [$\sigma(H) = 1/6$, $\sigma(T) = 2/3$, $str(r_6) = 1/6$; $acc(r_6) = 1$]

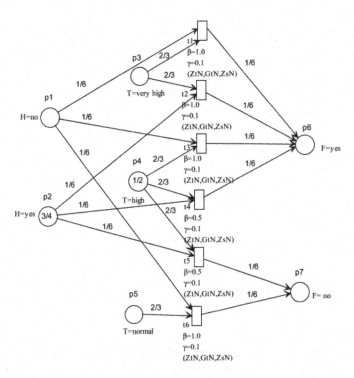

Fig. 5. An example of the WGFP-net model for the diagnosis of flu diseases with the initial marking

Using Algorithm 1 (Sect. 4) for constructing a WGFP-net on the base of a given set of rules, we present the WGFP-net model corresponding to these rules. This net model is shown in Fig. 5. Note that the places p_2 and p_4 include the truth degree values 3/4 and 1/2 corresponding to the descriptors H=yes and T=high, respectively. The remaining places on the net model are empty. In this example, input weights iw attached to arcs belong to the interval [0,1] and are shown in Fig. 5. Moreover, there are: the truth degree function β: $\beta(t_1) = \beta(t_2) = \beta(t_3) = \beta(t_6) = 1.0$ and $\beta(t_4) = \beta(t_5) = 0.5$, the threshold function γ: $\gamma(t_i) = 0.1$ for $i = 1, 2, ..., 6$, the set of operators $Op = \{$ZtN, GtN, ZsN$\}$ and the operator binding function δ defined as follows: $\delta(t_i) = ($ZtN, GtN, LsN$)$ for all transitions in the net.

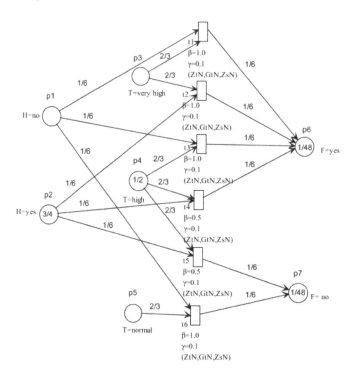

Fig. 6. An example of the WGFP-net model for the diagnosis of flu diseases with the final marking after firing the transitions t_4, t_5

Assessing the statements (descriptors) attached to places p_2 and p_4, we observe that transitions t_4 and t_5 are enabled in the initial marking (see Fig. 5). After firing these transitions in any order we obtain the same values for the decisions F=yes, F=no equal to 1/48 (see Fig. 6). This means that an unambiguous decision does not exist in this case. In the net model with parameters (and this is the model presented in the paper) the problem of ambiguity of decisions is easier to solve than in the model without parameters. In a situation like this, the ambiguity of decisions could be relatively easily resolved if the weights of the output arcs for t_4 and t_5 were different. This situation is possible with a different interpretation of the weights of the input and/or output arcs in this net model. We intend to address this problem in more detail in our future research work.

It is also visible in this figure that in the current marking the transitions t_4 and t_5 are still enabled. Firing these two transitions in the current marking does not change this marking, therefore the simulation of the net operation is already completed. We omit the particular calculation in this case, because it runs similarly as in Example 5 (Sect. 3).

6   Conclusion

Trying to make fuzzy Petri nets more realistic with regard to the perception of physical reality, in this paper we established the relationship between fuzzy Petri nets and rough set theory. This link is of a methodological nature and shows the possible application of rough set methodology to transform the WGFP-net into a more realistic net model. In the proposed model, the weights of arcs and the function β are interpreted using appropriate concepts from the rough set theory, thanks to which their values are calculated from data tables. Decision rules are also automatically generated from these tables, which are the basis for building the net model of the decision algorithm. In addition, the considered net model allows the use of any triangular norms to describe the behavior of the WGFP-nets. The approach developed seems promising and one could try to apply it to problems that can be solved in a similar way.

It is worth noting that the presented net model allows relatively quickly identify the objects specified in a given decision table. However, the algorithm described does not propagate information from attributes to other attributes as soon as possible. If such an algorithm did this, we would achieve even faster decision making in the net model. It is well known that this aspect is extremely important in real-time systems. This is the reason to consider in the next study the rules in minimal form, i.e. with a minimal number of descriptors on its left hand side. Another interesting problem arises when we are unable to determine the exact membership or value of truth, then we should focus our attention on e.g. interval fuzzy sets [19] to indicate their scope instead of exact values. Therefore, it seems useful to examine the WGFP-net in the context of interval t-norms. This should make the model proposed here even more flexible, general and practical. These are just some examples of problems that we would like to examine using the approach presented in the paper.

Acknowledgement. This work was partially supported by the Center for Innovation and Transfer of Natural Sciences and Engineering Knowledge at the University of Rzeszów. We would like to thank the anonymous referees for critical remarks and useful suggestions to improve the quality of the paper.

References

1. Bandyopadhyay, S., Suraj, Z., Nayak, P.K.: Soft Petri net. In: Mihálydeák, T., et al. (eds.) IJCRS 2019. LNCS (LNAI), vol. 11499, pp. 253–264. Springer, Cham (2019). https://doi.org/10.1007/978-3-030-22815-6_20
2. Bandyopadhyay, S., Suraj, Z., Grochowalski, P.: Modified generalized weighted fuzzy Petri net in intuitionistic fuzzy environment. In: Flores, V., et al. (eds.) IJCRS 2016. LNCS (LNAI), vol. 9920, pp. 342–351. Springer, Cham (2016). https://doi.org/10.1007/978-3-319-47160-0_31
3. Cardoso, J., Camargo, H. (eds.): Fuzziness in Petri Nets. Springer, Heidelberg (1999)
4. Chen, S.M.: Weighted fuzzy reasoning using weighted fuzzy Petri nets. IEEE Trans. Knowl. Data Eng. **14**(2), 386–397 (2002)

5. Chen, S.M., Ke, J.S., Chang, J.F.: Knowledge representation using fuzzy Petri nets. IEEE Trans. Knowl. Data Eng. **2**(3), 311–319 (1990)
6. David, R., Alla, H.: Petri Nets and Grafcet: Tools for Modelling Discrete Event Systems. Prentice-Hall, Upper Saddle River (1992)
7. Ha, M.H., Li, Y., Wang, X.F.: Fuzzy knowledge representation and reasoning using a generalized fuzzy Petri net and a similarity measure. Soft Comput. **11**(4), 323–327 (2007)
8. Klement, E.P., Mesiar, R., Pap, E.: Triangular Norms. Kluwer, Dordrecht (2000)
9. Komorowski, J., Pawlak, Z., Polkowski, L., Skowron, A.: Rough sets: a tutorial. In: Pal, S.K., Skowron, A. (eds.): Rough Fuzzy Hybridization: A New Trend in Decision-making, pp. 3–98, Springer, Singapore (1999)
10. Liu, H.-C., You, J.-X., Li, Z.-W., Tian, G.: Fuzzy Petri nets for knowledge representation and reasoning: a literature review. Eng. Appl. Artif. Intell. **60**, 45–56 (2017)
11. Liu, H.-C., You, J.-X., You, X.Y., Su, Q.: Fuzzy Petri nets using intuitionistic fuzzy sets and ordered weighted averaging operators. IEEE Trans. Cybern. **46**(8), 1839–1850 (2016)
12. Looney, C.G.: Fuzzy Petri nets for rule-based decision-making. IEEE Trans. Syst. Man Cybern. **18**(1), 178–183 (1988)
13. Pancerz, K., Suraj, Z.: Rough sets for discovering concurrent systems models from data tables. In: Hassanien, A., Suraj, Z., Ślęzak, D., Lingras, P. (eds.) Rough Computing: Theories, Technologies and Applications, pp. 239–268. IGI Global, Pennsylvania (2008)
14. Pawlak, Z., Skowron, A.: Rudiments of rough sets. Inf. Sci. **177**, 3–27 (2007)
15. Pawlak, Z., Skowron, A.: Rough sets and boolean reasoning. Inf. Sci. **177**, 41–73 (2007)
16. Pawlak, Z.: Concurrent versus sequential the rough sets perspective. Bull. EATCS **48**, 178–190 (1992)
17. Pawlak, Z.: Rough Sets - Theoretical Aspects of Reasoning about Data. Kluwer, Dordrecht (1991)
18. Pawlak, Z.: Rough sets. Int. J. Comput. Inf. Sci. **11**, 341–356 (1982)
19. Pedrycz, W.: Fuzzy Control and Fuzzy Systems, 2nd extended edn. Wiley, Hoboken (1993)
20. Peters, J., Skowron, A., Suraj, Z., Pedrycz, W., Ramanna, S.: Approximate real-time decision making: concepts and rough fuzzy Petri net models. Int. J. Intell. Syst. **14**, 805–839 (1999)
21. Skowron, A.: Synthesis of adaptive decision systems from experimental data. In: Proceedings of the Fifth Scandinavian Conference on Artificial Intelligence, Trondheim, Norway, 29–31 May 1995. IOS Press (1995). Frontiers. Artif. Intell. Appl. 28: 220–238
22. Skowron, A.: A synthesis of decision rules: applications of discernibility matrices. In: Proceedings of a Workshop on Intelligent Information Systems, Practical Aspects of AI II, Augustów, Poland, 7–11 June 1993, pp. 30–46 (1993)
23. Skowron, A., Suraj, Z.: A parallel algorithm for real-time decision making: a rough set approach. J. Intell. Inf. Syst. **7**, 5–28 (1996)
24. Skowron, A., Suraj, Z.: Rough sets and concurrency. Bull. PAS Tech. Sci. **41**(3), 237–254 (1993)
25. Suraj, Z.: A new class of fuzzy Petri nets for knowledge representation and reasoning. Fundam. Inform. **128**(1–2), 193–207 (2013)

26. Suraj, Z., Bandyopadhyay, S.: Generalized weighted fuzzy Petri net in intuitionistic fuzzy environment. In: Proceedings the IEEE World Congress on Computational Intelligence (IEEE WCCI 2016), Vancouver, Canada, 25–29 July 2016, pp. 2385–2392. IEEE (2016)
27. Suraj, Z., Fryc, B.: Timed approximate Petri net. Fundam. Inform. **71**, 83–99 (2006)
28. Suraj, Z., Grochowalski, P.: Petri nets and PNeS in modeling and analysis of concurrent systems. In: Proceedings International Workshop on Concurrency, Specification and Programming (CS& P 2017), Warsaw, Poland, 25–27 September 2017, pp. 1–12 (2017)
29. Suraj, Z., Hassanien, A.E.: Fuzzy Petri nets and interval analysis working together. In: Bello, R., Falcon, R., Verdegay, J.L. (eds.) Uncertainty Management with Fuzzy and Rough Sets. SFSC, vol. 377, pp. 395–413. Springer, Cham (2019). https://doi.org/10.1007/978-3-030-10463-4_20
30. Zhou, K.-Q., Zain, A.M.: Fuzzy Petri nets and industrial applications: a review. Artif. Intell. Rev. **45**(4), 405–446 (2015). https://doi.org/10.1007/s10462-015-9451-9

Rough Sets Meet Statistics - A New View on Rough Set Reasoning About Numerical Data

Marko Palangetić[1]([✉]) [iD], Chris Cornelis[1] [iD], Salvatore Greco[2,3] [iD],
and Roman Słowiński[4,5] [iD]

[1] Ghent University, Ghent, Belgium
{marko.palangetic,chris.cornelis}@ugent.be
[2] University of Catania, Catania, Italy
salgreco@unict.it
[3] University of Portsmouth, Portsmouth, UK
[4] Poznań University of Technology, Poznań, Poland
roman.slowinski@cs.put.poznan.pl
[5] Systems Research Institute, Polish Academy of Sciences, Warsaw, Poland

Abstract. In this paper, we present a new view on how the concept of rough sets may be interpreted in terms of statistics and used for reasoning about numerical data. We show that under specific assumptions, neighborhood based rough approximations may be seen as statistical estimations of certain and possible events. We propose a way of choosing the optimal neighborhood size inspired by statistical theory. We also discuss possible directions for future research on the integration of rough sets and statistics.

Keywords: Rough sets · Statistical learning · Neighborhood based rough sets

1 Introduction

Zdzisław Pawlak introduced rough sets in 1982 to deal with inconsistencies within information tables [15]. His approach is applied to the representation of classes of objects in an information table using two new sets called lower and upper approximation. The lower approximation contains objects which certainly belong to the approximated class, while the objects which are possibly in the approximated class are included in the upper approximation. Formulated in another way, the approach identifies the objects which are certainly consistent with the available knowledge and the objects which are possibly consistent with it. The original method is designed to deal with categorical data or data with a finite domain.

The extension of the model to numerical data faces some difficulties. One possibility to deal with numerical data is to discretize the attributes in the information table and make them categorical [7]. However, such an approach

© Springer Nature Switzerland AG 2020
R. Bello et al. (Eds.): IJCRS 2020, LNAI 12179, pp. 78–92, 2020.
https://doi.org/10.1007/978-3-030-52705-1_6

may lead to a loss of information, since discretization considers a set of values as one single value. The other option are neighborhood based rough sets where the equivalence class from Pawlak's approach is replaced with the neighborhood of an object in a high dimensional Euclidean space [9]. They are related to similarity based rough sets [21], and are part of the more general family of covering based rough sets [26]. The third approach are fuzzy rough sets which use fuzzy generalizations of equivalence relations suitable for application to numerical data [5]. In this paper, we use probability and statistics instead of fuzziness to model uncertainty in data.

From the very beginning, it was acknowledged that Pawlak's approach runs into limitations when it comes to problems which are more probabilistic than deterministic in nature [27]. In general, data consist of true values affected by some noise. Therefore, the first step in data analysis is to remove that noise in order to use the real values to solve the problem of interest. As a robust version of rough sets, the Variable Precision Rough Set (VPRS) approach was proposed by Ziarko [27]. It was also the first attempt to integrate the probabilistic approach and rough sets. Other probabilistic versions of rough sets were presented later, including decision theoretic rough sets [25] and parameterized rough sets [6]. Later on, Ziarko also introduced the assumption that the data are just a sample from an unknown space [28] into rough sets. That is a widely used assumption in statistics and machine learning: data are a realization of a random variable. With this assumption, we seek for a deeper integration of rough sets and statistics. In this paper, we propose a new view on the definition of rough sets, and provide a new definition independent of the type of data. It leads to a natural extension of the initial rough set approach to numerical data. We provide an example how to calculate rough sets for numerical data, elaborate on some of issues we are facing and present some ideas about how to direct the future research on integration of rough sets and statistics.

The paper is organized as follows. In the next section we recall basic concepts of rough set theory. In Sect. 3, statistical learning theory for Pawlak's rough sets is introduced. Section 4 presents rough approximations for numerical data. Section 5 identifies and discusses some potential pitfalls and drawbacks identified in Sect. 4 together with ideas for improvement. Conclusions are provided in Sect. 6.

2 Preliminaries

2.1 Rough Sets

An information table is a 4-tuple $<U, Q \cup \{d\}, X \cup Y, f>$ where $U = \{u_1, \ldots, u_n\}$ is a finite set of objects or alternatives, $Q = \{q_1, \ldots, q_m\}$ is a finite set of condition attributes, d is a decision attribute; $X = \cup_{q \in Q} X_q$, where X_q is the domain of attribute $q \in Q$ while Y is the domain of d. The information function $f : U \times Q \cup \{d\} \rightarrow X \cup Y$ satisfies that $\forall u \in U, \forall q \in Q : f(u, q) \in X_q$ and that $f(u, d) \in Y$. Denote by $X_Q = \prod_{q \in Q} X_q$ the joint domain of condition attributes, while $f(u, Q) \in X_Q$ represents the $|Q|$-tuple of values $f(u, q)$ for $q \in Q$. If X_q is finite, we say that q is categorical, while if $X_q \subseteq \mathbb{R}$ we say that q is numerical.

First we assume that all condition attributes are categorical. We define the equivalence relation $\equiv$ on objects u and v as $u \equiv v \Leftrightarrow \forall q \in Q, f(u, q) = f(v, q)$. This means that two objects are related (indiscernible) if they are equally evaluated on all attributes. Let $[u]_\equiv$ denote the equivalence class of object u, and $A \subseteq U$. We recall Pawlak's lower and upper approximations on U:

$$\underline{apr}_\equiv(A) = \{u \in U | [u]_\equiv \subseteq A\}, \quad \overline{apr}_\equiv(A) = \{u \in U | [u]_\equiv \cap A \neq \emptyset\}.$$

In the lower approximation of A, we include objects u for which all identically evaluated objects are also in A. Therefore, we may conclude that u for sure belongs to A based on available knowledge, since all the instances with the same values are also in A. We include object u in the upper approximation of A if there is an instance in A identically evaluated as u. Hence, we may say that u is possibly in A if some instances, identically evaluated as u, are in A. In this way, we distinguish certain and possible knowledge. Below, we list the important properties of inclusion and duality [15]:

- (inclusion) $\underline{apr}_\equiv(A) \subseteq \overline{apr}_\equiv(A)$,
- (duality) $\underline{apr}_\equiv(A^c) = (\overline{apr}_\equiv(A))^c$, $\overline{apr}_\equiv(A^c) = (\underline{apr}_\equiv(A))^c$.

A question arises: how to apply a similar reasoning when we have numerical data? If we apply the reasoning presented above, the equivalence classes will mostly consist of only one object since it is almost impossible that two objects with numerical characteristics will be identically evaluated on all attributes. This means that all objects from A belong to the lower approximations of A, i.e., all objects from A certainly belong to A. However, in this way we ignore the fact that the noise present in data affects the certainty of objects belonging to a set. The noise is related to imprecision of numerical attributes and, even if the measurement of numerical attributes is precise, to human perception of these precise values.

A way to handle this problem is the neighborhood based rough set approach. Assume now that condition attributes are taking real values and let d be Euclidean distance on $X_Q \subseteq \mathbb{R}^m$. Here, any distance metrics can be used, but Euclidean distance corresponds with the later statistical approach we will use. For object $u \in U$ we define its ϵ-neighborhood $n_\epsilon(u) = \{v \in U; d(f(u, Q), f(v, Q)) < \epsilon\}$. We define the approximations in the following way [9]:

$$\underline{apr}_\epsilon(A) = \{u \in U; n_\epsilon(u) \subseteq A\}, \quad \overline{apr}_\epsilon(A) = \{u \in U; n_\epsilon(u) \cap A \neq \emptyset\}.$$

Here, object u certainly belongs to A if its close neighborhood only contains objects from A. Object u possibly belongs to A if its close neighborhood contains at least one object from A. Equivalent properties of inclusion and duality also hold in this case [9].

From the definition we may see that the approximations heavily depend on the parameter ϵ. The question is, what is the optimal neighborhood size which will identify certain and possible knowledge. Later on we will see that statistical techniques may be useful for this purpose.

2.2 Value-Based Definitions and Inconclusive Regions

Pawlak defines the approximations as sets of objects (SO). The main goal of these definitions is to distinguish possible knowledge from certain knowledge and for this we do not need to refer exactly to the set of objects. We can define the approximations as sets of values (SV), i.e., the sets which will only contain values from the domain of condition attributes. Let $x \in X_Q$. Similarly as in [8] we define sets $[x] = \{u \in U; f(u, Q) = x\}$. The SV approximations are

$$\underline{\text{apr}}^{SV}(A) = \{x; [x] \neq \emptyset \wedge [x] \subseteq A\}, \quad \overline{\text{apr}}^{SV}(A) = \{x; [x] \cap A \neq \emptyset\}.$$

We refer to this definition as SV definition while the original one will be called SO definition. We note that the SV definition keeps the same knowledge as the SO definition. The SO approximations can be obtained from the SV definition by collecting all objects with condition values belonging to the SV approximations (lower or upper). The SV approximations can be obtained from the SO definition as a set of unique condition values $f(u, Q)$ of the objects from the SO approximations. Therefore, in terms of Pawlak's environment of categorical data, SO and SV definitions are equivalent.

We notice that there are values from the domain which cannot be assigned to any approximation. In particular, the condition $\|[x]\| > 0$ is necessary in the definitions. Otherwise a value x for which $\|[x]\| = 0$ would belong to the lower approximations of A and A^c at the same time, i.e., it would certainly belong to two opposite classes. Of course, that is not possible and such values from the domain are called inconclusive. We denote the set $I \subseteq X_Q$ of inconclusive values by

$$I = \{x; x \in X_Q \wedge [x] = \emptyset\}$$

The inclusion property is clearly preserved while duality still holds if the complement operator on X_Q excludes inconclusive values i.e., if it is defined as: $S^c = X_Q - I - S$ for $S \subseteq X_Q$.

On the other hand, for the SV extension in the neighborhood based approximations, neighborhood may be defined for any value from the domain X_Q. If $X_Q \subseteq \mathbb{R}^m$ and $x \in X_Q$ we define $n_\epsilon(x) = \{u \in U; d(x, f(u, Q)) < \epsilon\}$. The SV approximations are:

$$\underline{\text{apr}}_\epsilon^{SV}(A) = \{x; n_\epsilon(x) \neq \emptyset \wedge n_\epsilon(x) \subseteq A\}$$

$$\overline{\text{apr}}_\epsilon^{SV}(A) = \{x; n_\epsilon(x) \cap A \neq \emptyset\}.$$

An arbitrary value $x \in X_Q$ is in the lower approximation of A if its ϵ-neighborhood contains only objects from A while it is in the upper approximation if it contains at least one object from A. Here again we consider the inconclusive areas, i.e., values in which neighborhood there are no objects from

U. As for the SV definitions for Pawlak's rough sets, the inclusion property is preserved while duality holds with exclusion of the inconclusive areas. The SO and SV definitions are not equivalent in this case since SV is more general, and SO can be obtained from it, but not vice versa. For example, there can exist a value $x \in X_Q$ such that its neighborhood contains exactly one object $u \in A$ and no elements from A^c, and such that u is not in the SO lower approximation of A. The latter holds in particular if there exists some $v \in A^c$ such that $d(f(u, Q), f(v, Q)) < \epsilon$. However, x belongs to the SV lower approximation, and such x cannot be reconstructed from the SO lower approximation.

We will use the SV definition to derive a statistical extension of rough sets to numerical data.

3   A Statistical View of Pawlak's Rough Sets

One widely used assumption in statistics and machine learning (ML) is that data are realizations of a joint random variable. Let objects be outcomes of the joint random variable $\mathcal{U} = (\mathcal{X}, \mathcal{Y})$ where $\mathcal{X}$ is a random variable corresponding to the condition attributes, while $\mathcal{Y}$ corresponds to the decision attribute. Since we are dealing with classification problems, we know that $\mathcal{Y}$ is always discrete, while $\mathcal{X}$ is discrete if we work with categorical data, or $\mathcal{X}$ takes values from $\mathbb{R}^m$ if we have numerical data. Those random variables are unknown in practice, so using data as their realizations, we explain the relations between $\mathcal{X}$ and $\mathcal{Y}$.

The idea here is to redefine the approximations in terms of random variables instead of data. The SV approximations were defined on the domain w.r.t. neighborhood operators, while here the approximations are defined on the domain w.r.t. a random variable. In terms of statistics these are the "true" approximations dependent on unknown random variables. The SV approximations on data will play the role of estimators of such approximations.

Since $\mathcal{Y}$ is discrete, assume that its domain is the set $\{0, 1, \ldots, K\}$ for some K. Classification tasks in machine learning often refer to calculation of the conditional probabilities of the particular classes. More formally, for class $k \in \{0, 1, \ldots, K\}$ we want to model the expression $P(\mathcal{Y} = k | \mathcal{X} = x)$ as a function of x for all x from the domain space (either a space of categories or $\mathbb{R}^m$). Assume now that the domain X_Q of $\mathcal{X}$ is finite i.e., $\mathcal{X}$ is discrete. If certainty is modeled in a probabilistic environment, we say that an event is certain if its probability is 1 while an event is possible if its probability is greater than 0. We want to know if value $x \in X_Q$ certainly belongs to class k, i.e., if $P(\mathcal{Y} = k | \mathcal{X} = x) = 1$. In practice, we do not have exact knowledge about the conditional distribution of $\mathcal{Y}$ on $\mathcal{X}$, so we need to estimate it. We recall the set of objects $U = \{u_i = (x_i, y_i) | i = 1 \ldots n\}$ which is now a set of realizations of random variable $\mathcal{U}$, known as a sample. The empirical estimation of the above mentioned conditional probability is

$$\hat{P}(\mathcal{Y} = k | \mathcal{X} = x) = \frac{\sum_{i=1}^n \mathbf{1}_{\{y_i = k, x_i = x\}}}{\mathbf{1}_{\{x_i = x\}}} = \frac{|\{\hat{y} = k\}| \cap |\{\hat{x} = x\}|}{|\{\hat{x} = x\}|},$$

where $\mathbf{1}_A$ is the indicator function, $|\{\hat{y} = k\}|$ is the number of objects y_i equal to k, while $|\{\hat{x} = x\}|$ is the number of objects x_i equal to x. To estimate the set of values x for which $P(\mathcal{Y} = k | \mathcal{X} = x) = 1$, we use the estimated probability instead of the true one. We have that:

$$\frac{|\{\hat{y} = k\}| \cap |\{\hat{x} = x\}|}{|\{\hat{x} = x\}|} = 1 \Leftrightarrow |\{\hat{y} = k\}| \cap |\{\hat{x} = x\}| = |\{\hat{x} = x\}| \wedge |\{\hat{x} = x\}| > 0$$

$$\Leftrightarrow \{\hat{x} = x\} \subseteq \{\hat{y} = k\} \wedge |\{\hat{x} = x\}| > 0.$$

We obtain

$$\{x \in X_Q; \hat{P}(\mathcal{Y} = k | \mathcal{X} = 1)\} = \{x \in X_Q; |\{\hat{x} = x\}| > 0 \wedge \{\hat{x} = x\} \subseteq \{\hat{y} = k\}\}.$$

The right side of the latter equality is identical to the SV definition of Pawlak's rough sets, where $[x]$ is replaced by $\{\hat{x} = x\}$ while A is replaced with $\{\hat{y} = k\}$. Here, it can be noticed that the SV lower approximation may be seen as an estimation of the unknown lower approximation dependent on random variables. A similar procedure may be used for the upper approximation. This leads to the definition of the lower and upper approximations of the class k with respect to random variable $\mathcal{X}$:

$$\underline{\mathrm{apr}}_{\mathcal{X}}^{RV}(\mathcal{Y} = k) = \{x; P(\mathcal{Y} = k | \mathcal{X} = x) = 1\}, \tag{1}$$
$$\overline{\mathrm{apr}}_{\mathcal{X}}^{RV}(\mathcal{Y} = k) = \{x; P(\mathcal{Y} = k | \mathcal{X} = x) > 0\}.$$

We call this the RV definition of rough sets. Such defined "true" approximations do not require any assumptions on $\mathcal{X}$ ($\mathcal{X}$ being discrete or continuous) as long as the conditional probability is defined. This version of the approximations provides a natural extension of rough sets to numerical data (and all other types of data). In practice, approximation estimates for categorical and numerical data are different since the probability estimation is different in the discrete and the continuous case. We have already seen the estimation of the lower approximation for categorical data. Later on it will be shown how to estimate the approximations in the numerical case. The RV rough set definitions can be taken out of the context of classification and they can be extended to arbitrary events. Let A be an event and $\mathcal{X}$ be a random variable. The lower and upper approximations of A w.r.t. $\mathcal{X}$ are defined as:

$$\underline{\mathrm{apr}}_{\mathcal{X}}^{RV}(A) = \{x; P(A | \mathcal{X} = x) = 1\}, \quad \overline{\mathrm{apr}}_{\mathcal{X}}^{RV}(A) = \{x; P(A | \mathcal{X} = x) > 0\}.$$

However, such general definition will not play an important role for our goal, but it may find some other applications in data analysis.

4   Rough Approximations for Numerical Data

In the previous section we have seen how the approximations may be estimated in practice when we deal with categorical data, and that such estimation coincides with Pawlak's approach. Since the approximations do not depend on the type of data, the question is how to estimate them for numerical data. To make things simpler, we assume that classification is binary, i.e., $K = 1$, and we only have two values for the variable $\mathcal{Y}$, 0 and 1. Assume also that the domain of $\mathcal{X}$ is $X_Q \subseteq \mathbb{R}^m$ i.e., $\mathcal{X}$ is a continuous random variable. By $f_{\mathcal{X}}$ we denote the probability density function (PDF) of $\mathcal{X}$, while by $f_{\mathcal{Y}}(k) = P(\mathcal{Y} = k)$ we denote the PDF of the binary random variable $\mathcal{Y}$. The joint PDF of $\mathcal{Y}$ and $\mathcal{X}$ is denoted as $f_{\mathcal{Y},\mathcal{X}}$. From probability theory it holds that $f_{\mathcal{Y}}(0) + f_{\mathcal{Y}}(1) = 1$, $f_{\mathcal{X}}(x) > 0$ for $x \in X_Q$ and $\int_{X_Q} f_{\mathcal{X}}(x)dx = 1$. We calculate the approximations of class 1. Probability theory tells us that:

$$P(\mathcal{Y} = 1 | \mathcal{X} = x) = \frac{f_{\mathcal{Y},\mathcal{X}}(1, x)}{f_{\mathcal{X}}(x)} = 1 - \frac{f_{\mathcal{X}}(x) - f_{\mathcal{Y},\mathcal{X}}(1, x)}{f_{\mathcal{X}}(x)} = 1 - \frac{f_{\mathcal{Y},\mathcal{X}}(0, x)}{f_{\mathcal{X}}(x)}.$$

For the lower approximation we have that

$$P(\mathcal{Y} = 1 | \mathcal{X} = x) = 1 \Leftrightarrow 1 - \frac{f_{\mathcal{Y},\mathcal{X}}(0, x)}{f_{\mathcal{X}}(x)} = 1 \Leftrightarrow \frac{f_{\mathcal{Y},\mathcal{X}}(0, x)}{f_{\mathcal{X}}(x)} = 0 \Leftrightarrow f_{\mathcal{Y},\mathcal{X}}(0, x) = 0.$$

The last equality can be divided by $f_{\mathcal{Y}}(0)$ and we get the condition $f_{\mathcal{X}|\mathcal{Y}=0}(x) = 0$. Here $f_{\mathcal{X}|\mathcal{Y}=0}$ stands for the conditional PDF of $\mathcal{X}$ on event $\{\mathcal{Y} = 0\}$. For the upper approximation we have:

$$P(\mathcal{Y} = 1 | \mathcal{X} = x) > 0 \Leftrightarrow \frac{f_{\mathcal{Y},\mathcal{X}}(1, x)}{f_{\mathcal{X}}(x)} > 0 \Leftrightarrow f_{\mathcal{Y},\mathcal{X}}(1, x) > 0.$$

The last equality can be divided by $f_{\mathcal{Y}}(1)$ and we get the condition $f_{\mathcal{X}|\mathcal{Y}=1}(x) > 0$.

The conclusion we may derive from the calculations is that x certainly belongs to class 1 if the conditional PDF of $\mathcal{X}$ on $\{\mathcal{Y} = 0\}$ evaluated in x is 0. We have that x possibly belongs to class 1 if the conditional PDF of $\mathcal{X}$ on $\{\mathcal{Y} = 0\}$ evaluated in x is greater than 0. These conditions depend on conditional PDFs which are unknown in practice and have to be estimated. More precisely, we need to estimate the so-called level sets, i.e., areas on which the PDF is smaller or greater than some value [2]. In our case, the thresholds we consider for the PDFs are when they are equal to 0 and greater than 0 (lower and upper approximation).

The estimation of level sets is an emerging field in statistics and ML [2,3,20]. Such estimations are essentially different from estimating the PDF itself since we are searching for good estimators for a particular area of the PDF, not for the whole PDF.

Below we present a naive approach of estimating level sets using the estimation of the PDF. Density estimation is a well studied area of statistics [18,19,23]. The main methods are histogram density estimation, kernel density estimation (KDE) and nearest neighbour density estimation. Histograms are known for performing badly in high dimensions [18], while the nearest neighbour methods do not assume that there are areas where the PDF is equal to 0 [14]. For these reasons, KDE appears the most appropriate choice to calculate level sets. We refer the reader to [19] for an overview of density estimation methods.

4.1   Rough Sets and KDE

A kernel $K : \mathbb{R}^m \times \mathbb{R}^m \to \mathbb{R}$ is a positive and symmetric mapping for which it holds that $\forall t \in \mathbb{R}^M, \int_{\mathbb{R}^m} K(t,s)ds = 1$ [24]. It may be seen as a measure of similarity between points from $\mathbb{R}^m$. The kernel density estimator is defined as:

$$\hat{f}^K(t) = \frac{1}{n} \sum_{i=1}^{n} K(t,t_i),$$

where $\{t_1, t_2, \ldots, t_n\}$ is a given sample from the unknown PDF f. The motivation behind this definition is that if x has more points in its proximity, then value $\hat{f}^K(x)$ will be larger, which indicates an area of higher density.

Similarity measures are usually based on distances between points since, intuitively, the closer points are, the more similar they are to each other. Therefore, we use kernels based on Euclidean distance, called radial kernels [12]:

$$K(x,y) = \frac{1}{h} k \left(\frac{\|x - y\|}{h} \right).$$

The notation $\|\cdot\|$ stands for the standard norm on $\mathbb{R}^m$, h is a positive real parameter called bandwidth while k is a univariate positive function. Using radial kernels, the PDF estimator becomes:

$$\hat{f}^{k,h}(x) = \frac{1}{nh^m} \sum_{i=1}^{n} k \left(\frac{\|x - x_i\|}{h} \right). \tag{2}$$

From before we have that the lower approximation can be formulated as:

$$\underline{\mathrm{apr}}_{\mathcal{X}}^{RV} (\mathcal{Y} = 1) = \{x; f_{\mathcal{X}|\mathcal{Y}=0}(x) = 0\}.$$

Therefore, using (2) we get the estimator of the lower approximation:

$$\underline{\mathrm{apr}}_{\hat{\mathcal{X}}}^{RV} (\mathcal{Y} = 1) = \{x; \hat{f}_{\mathcal{X}|\mathcal{Y}=0}^{k,h}(x) = 0\}.$$

Although it is not possible that $f_{\mathcal{X}|\mathcal{Y}=0}(x) = 0$ and $f_{\mathcal{X}|\mathcal{Y}=1}(x) = 0$ at the same time, it may happen that $\hat{f}_{\mathcal{X}|\mathcal{Y}=0}^{k,h}(x) = 0$ and $\hat{f}_{\mathcal{X}|\mathcal{Y}=1}^{k,h}(x) = 0$ for some x. Such values we will denote as inconclusive and we will exclude them from the approximations, as before. Following this, we redefine the estimation of the lower approximation:

$$\underline{\mathrm{apr}}_{\hat{\mathcal{X}}}^{RV} (\mathcal{Y} = 1) = \{x; \hat{f}_{\mathcal{X}|\mathcal{Y}=0}^{k,h}(x) = 0 \wedge \hat{f}_{\mathcal{X}|\mathcal{Y}=1}^{k,h}(x) > 0\}. \tag{3}$$

Henceforth we will focus on the lower approximation. A very similar procedure can be used to estimate the upper approximation.

We have to decide which area satisfies the condition from (3). To estimate $f_{\mathcal{X}|\mathcal{Y}=0}$ we use objects from class 0 and to estimate $f_{\mathcal{X}|\mathcal{Y}=1}$ we use objects from class 1. Recall $U = \{(x_1, y_1), \ldots, (x_n, y_n)\}$ as the set of objects or the sample. Set U is split into two subsets; objects which belong to class 0, and objects which belong to class 1. We denote those sets $U^0 = \{(x_1^0, 0), (x_2^0, 0), \ldots, (x_{n_0}^0, 0)\}$ and $\{U^1 = (x_1^1, 1), (x_2^1, 1), \ldots, (x_{n_1}^1, 1)\}$. To estimate the conditional PDFs $f_{\mathcal{X}|\mathcal{Y}=0}$ and $f_{\mathcal{X}|\mathcal{Y}=1}$ we use the objects from U^0 and U^1 respectively. To estimate the level set $f_{\mathcal{X}|\mathcal{Y}=0}(x) = 0$ we have to find values of x for which $\hat{f}_{\mathcal{X}|\mathcal{Y}=0}^{k,h}(x) = 0$ and to estimate $f_{\mathcal{X}|\mathcal{Y}=1}(x) > 0$ we are searching for x where $\hat{f}_{\mathcal{X}|\mathcal{Y}=1}^{k,h}(x) > 0$. It follows that:

$$\frac{1}{nh} \sum_{i=1}^{n_0} k\left(\frac{\|x - x_i^0\|}{h}\right) = 0 \Leftrightarrow \forall i \in \{1, \ldots n_0\};\ k\left(\frac{\|x - x_i^0\|}{h}\right) = 0.$$

$$\frac{1}{nh} \sum_{i=1}^{n_1} k\left(\frac{\|x - x_i^1\|}{h}\right) > 0 \Leftrightarrow \exists i \in \{1, \ldots n_1\};\ k\left(\frac{\|x - x_i^0\|}{h}\right) > 0.$$

The derivation up to now is general and holds for all functions k and bandwidths h. The question is, which kernel best suits the last condition. The most used kernel in practice is the Gaussian kernel which is also radial: $k(x) = \frac{1}{\sqrt{(2\pi)^m}} e^{-\frac{1}{2}x^2}$. Its main drawback is that it is nowhere equal to 0. It is used under the assumption that there are no impossible or certain events which is not the case here. Therefore, a better choice would be a kernel with different assumptions. In particular, we require a kernel for which k is bigger than 0 on a bounded set i.e., a kernel with bounded support (Fig. 1).

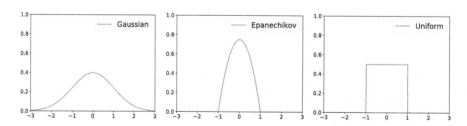

Fig. 1. Kernel examples in univariate case

The theory developed in [13] states that the smallest estimation error under certain conditions is achieved for the Epanechikov kernel. The Epanechikov kernel is radial with

$$k(x) = \max\left\{0, \frac{m+2}{2c_m}(1 - x^2)\right\},$$

where c_m is the volume of the m-dimensional unit ball. According to the definition, its support is the unit hypersphere, which implies that it is bounded. Another kernel with bounded support is the spherical uniform kernel, i.e., the constant radial kernel for which

$$k(x) = \begin{cases} \frac{1}{c_m} & \text{if } x \in (0,1) \\ 0 & \text{otherwise.} \end{cases}$$

Let h_e and h_u be the bandwidths corresponding to the Epanechikov kernel and spherical uniform kernel, respectively. For the Epanechikov kernel, we have that:

$$k\left(\frac{\|x - x_i^0\|}{h_e}\right) = 0 \Leftrightarrow \frac{m+2}{2c_m}\left(1 - \frac{\|x - x_i^0\|^2}{h_e^2}\right) \leq 0 \Leftrightarrow \|x - x_i^0\| \geq h_e,$$

$$k\left(\frac{\|x - x_i^1\|}{h_e}\right) > 0 \Leftrightarrow \frac{m+2}{2c_m}\left(1 - \frac{\|x - x_i^1\|^2}{h_e^2}\right) > 0 \Leftrightarrow \|x - x_i^1\| < h_e,$$

while for the spherical uniform kernel it holds that:

$$k\left(\frac{\|x - x_i^0\|}{h_u}\right) = 0 \Leftrightarrow \|x - x_i^0\| \geq h_u, \quad k\left(\frac{\|x - x_i^1\|}{h_u}\right) > 0 \Leftrightarrow \|x - x_i^1\| < h_u,$$

In both cases, value x certainly belongs to class 1 if in the neighborhood there are no objects from the opposite class and there are some objects from the same class. Hence, by using kernels with bounded support, we obtain simple conditions for estimating the lower approximations.

4.2   Relationship to Neighborhood Based Rough Sets

We summarize the results obtained so far: we defined the lower approximation of class $\{\mathcal{Y} = 1\}$ as : $\underline{\text{apr}}_{\mathcal{X}}^{RV}(\mathcal{Y} = 1) = \{x; f_{\mathcal{X}|\mathcal{Y}=0}(x) = 0\}$ for continuous random variable $\mathcal{X}$. We estimated the approximation by estimating the PDF from the expression using kernel density estimators as:

$$\underline{\text{apr}}_{\hat{\mathcal{X}}}^{RV}(\mathcal{Y} = 1) = \{x; \hat{f}_{\mathcal{X}|\mathcal{Y}=0}^K(x) = 0 \wedge \hat{f}_{\mathcal{X}|\mathcal{Y}=1}^K(x) > 0\}.$$

We have shown that the estimators for certain radial kernels with bounded support lead to the expression:

$$\underline{\text{apr}}_{\hat{\mathcal{X}}}^{RV}(\mathcal{Y} = 1) = \{x; \forall i : \|x - x_i^0\| \geq h \wedge \exists i : \|x - x_i^1\| < h\},$$

for some h. Let us write the neighborhood definition replacing ϵ with h: $n_h(x) = \{x_i \in U; d(x, x_i) < h\}$, where d is the Euclidean distance. Condition $\exists i : \|x - x_i^1\| < h$ means that there is at least one object from U^1 in $n_h(x)$, i.e., $n_h(x) \neq \emptyset$, while $\forall i : \|x - x_i^0\| \geq h$ means that there are no objects from U^0 in $n_h(x)$, i.e., $n_h(x) \subseteq U^1$. It follows that the approximation estimator can be written as:

$$\underline{\text{apr}}_{\hat{\mathcal{X}}}^{RV}(\mathcal{Y} = 1) = \{x; n_h(x) \neq \emptyset \wedge n_h(x) \subseteq U^1\}.$$

The latter expression is exactly the SV (set of values) definition of the neighborhood based rough sets. We can conclude that the estimators of the RV approximations coincide with the SV definition of the neighborhood based rough sets.

The advantage of this representation of the neighborhood based rough sets is that we have proper mathematical tools to calculate the neighborhood size in order to get better results. We are now able to use statistical methods to obtain a proper bandwidth which plays the role of the neighborhood size.

In the following subsection, we will outline a procedure to select the bandwidths in theory, that is: we provide some insights on how the bandwidths can be calculated independently from data, using only the chosen kernel and the original PDF.

4.3   Bandwidth Selection - An Example

This subsection relies on the work presented in [19]. Using the KDE theory, we are able to construct the proper bandwidths for different kernels in order to obtain the best possible estimator of PDFs (or at least close to the best). The bandwidths are chosen to minimize the error of the PDF estimation. A widely used error function is Mean Integrated Square Error (MISE):

$$MISE(\hat{f}^{k,h}) = \int_{X_Q} E((\hat{f}^{k,h}(x) - f(x))^2) dx$$

where E stands for the expected value. When n is significantly larger than the number of attributes m, the MISE of radial kernels can be approximated as:

$$MISE(\hat{f}^{k,h}) \approx C_1 h^4 + \frac{C_2}{nh^m}.$$

The latter expression is also called AMISE or Asymptotic MISE. By minimizing the expression above, we get the optimal bandwidth:

$$h^{opt} = C_3 n^{-\frac{1}{m+4}}.$$

Constants C_1, C_2 and C_3 are dependent on the kernel and on the actual probability density function f. Assuming that our data are normally distributed (or something close to normal with bounded support), we are able to calculate the optimal bandwidths. Under normality assumption, the optimal bandwidths for the Epanechikov and spherical uniform kernels are:

$$h_e^{opt} = [8(d+4)c_m^{-1}(2\sqrt{\pi})^d n^{-1}]^{\frac{1}{m+4}}, \quad h_u^{opt} = [4(d+2)c_m^{-1}(2\sqrt{\pi})^d n^{-1}]^{\frac{1}{m+4}}.$$

From the AMISE expression, we may see that the rate of convergence is not dependent on constant C_3. Therefore, in order to avoid the assumptions and to achieve better results one can try to tune constant C_3 using data. Under h^{opt} for some kernel we also ensure that:

$$\lim_{n \to \infty} MISE(\hat{f}^{k,h^{opt}}) = 0.$$

That ensures that for a sufficiently large sample size n, the inconclusive areas will become negligible. That is also intuitive since with more data we acquire more knowledge which leaves less space for uncertainty.

5   Discussion

We have presented a new way to calculate the neighborhood size in neighborhood based rough sets. A question arises: does it provide satisfactory results in practice?

It is well known that rough sets are widely used in attribute selection [4,10]. The attribute selection in rough sets focuses on preservation of certain knowledge; we delete attributes as long as the lower approximations of all classes remain unchanged.

We have run a series of experiments applying the attribute selection using neighborhood based rough sets together with the calculated bandwidths. Unfortunately, the results were not satisfactory. First, we simulated data with normal distribution to fulfill the assumption from the previous subsection. We have noticed that for lower dimensions, both h_e^{opt}-neighborhood and h_u^{opt}-neighborhood are too wide, meaning that they cover a large amount of data. Consequently, the lower approximations obtained with them consist of a low percentage of data which is unrealistic. With higher dimensions, we observed the opposite problem; the neighborhoods are too narrow which leads to the lower approximation containing almost all data, which is also unrealistic. We can conclude that the naive approach of estimating PDF and searching for the optimal bandwidth is not the best idea. The reason for the failure, even under the normality assumption, may lie in the fact that the optimal bandwidths are mainly useful in the following cases.

– The number of objects in the sample is significantly larger than the number of attributes since the bandwidth optimality is asymptotic.
– The MISE error is calculated using l_2 norm (the integral of the squared difference). Our interest is to get the optimal bandwidth for the level set where PDF is equal to 0. The l_2 convergence does not guarantee that the estimator also uniformly converges to the actual PDF [17]. Thus, we may have that h^{opt} is suitable for the higher density regions where the PDF is significantly larger than 0 and that it may have poor performance for the regions where the PDF is close to 0.

We have also applied the procedure on real data for which the normality assumption does not hold. As soon as the assumption is not fulfilled, the results are getting worse. For example, we considered binary classification in mammographic data from UCI [1] for which $n = 830$ and $m = 5$. In all cases, the lower approximations contained less than 7 % of data, meaning that only 7 % of data can be certainly classified. Keeping in mind that the classification accuracy we obtained with SVM on this dataset is around 85%, 7 % of certainty is unrealistic.

To overcome the limitations of the theoretical bandwidth selection, we identify the following options for future integration of rough sets, KDE and statistics in general.

– **Data driven estimation.** The calculation of bandwidths may be data driven. There is also a statistical theory on how to calculate bandwidths based on data (again [19]). Data driven bandwidths will help us to overcome any a priori assumptions on the distribution of data.

– **Robust approaches.** Having 0 probability regions is a strong assumption which usually does not coincide with reality. Mostly, numerical data exhibit rare events, which may occur in the training data and/or during the prediction process. Having the assumption that data lie in a bounded region may be misleading in many cases and it can produce bad results. The 0 probability regions can be eliminated by applying robust approaches similar to Variable Precision Rough Sets (VPRS).

– **Direct level set estimation.** The bandwidth calculation needs to be more adjusted to the problem of the level set estimation, rather than to the PDF estimation. After we identify the regions of interest, we have to set up the optimization problem to get the best possible (or close to the best) bandwidth for that particular case.

– **Different estimators than KDE.** We can try to use other estimators for level sets, besides KDE. The nearest neighbor based estimator can give interesting results [14].

– **Integration with SVM.** Do we have to use densities to estimate the approximations defined in (1)? We showed that the estimation of the RV approximations (1) boils down to the estimation of level sets. We may explore the relation between SVM and level set estimation as has been done in [11,16,22]. On the other hand, there is a direct correspondence between principles of rough sets and SVM. The applications of rough sets in binary classification divide the domain into three sets, two certain regions for each class and one boundary region. SVM is doing something similar where it trains two margins which divide the space similarly as the rough sets: one boundary region and two regions for two classes. Thus, using the similarities between rough sets and SVM, we can try to integrate them in order to achieve better results.

6   Conclusion

We presented a new view on the definition of rough sets for the case when data are not necessarily categorical. From the statistical point of view, the calculation of rough set approximations is basically the estimation of the unknown RV (random value) approximations dependent on random variables that generate data. Such estimation under certain conditions (i.e., using radial kernels with bounded support) is equivalent to the definition of neighborhood based rough sets. We also showed a simple way how to calculate the neighborhood size using statistics. Moreover, we discussed several options for future research on the integration of rough sets and statistics. Of course, for each of the proposals it should be studied if it can be tailored to the main applications of rough sets: rule induction and attribute selection.

Acknowledgements. This work was supported by the Odysseus program of the Research Foundation-Flanders.

References

1. Asuncion, A., Newman, D.: UCI machine learning repository (2007)
2. Cadre, B.: Kernel estimation of density level sets. J. Multivar. Anal. **97**(4), 999–1023 (2006)
3. Chen, Y.C., Genovese, C.R., Wasserman, L.: Density level sets: asymptotics, inference, and visualization. J. Am. Stat. Assoc. **112**(520), 1684–1696 (2017)
4. Choubey, S.K., Deogun, J.S., Raghavan, V.V., Sever, H.: A comparison of feature selection algorithms in the context of rough classifiers. In: Proceedings of IEEE 5th International Fuzzy Systems, vol. 2, pp. 1122–1128. IEEE (1996)
5. Dubois, D., Prade, H.: Rough fuzzy sets and fuzzy rough sets. Int. J. Gen. Syst. **17**(2–3), 191–209 (1990)
6. Greco, S., Matarazzo, B., Słowiński, R.: Rough membership and bayesian confirmation measures for parameterized rough sets. In: Ślęzak, D., Wang, G., Szczuka, M., Düntsch, I., Yao, Y. (eds.) RSFDGrC 2005. LNCS (LNAI), vol. 3641, pp. 314–324. Springer, Heidelberg (2005). https://doi.org/10.1007/11548669_33
7. Grzymala-Busse, J.W., Stefanowski, J.: Three discretization methods for rule induction. Int. J. Intell. Syst. **16**(1), 29–38 (2001)
8. Grzymala-Busse, J.W., Werbrouck, P.: On the best search method in the LEM1 and LEM2 algorithms. In: Orłowska, E. (ed.) Incomplete Information: Rough Set Analysis, pp. 75–91. Springer, Heidelberg (1998). https://doi.org/10.1007/978-3-7908-1888-8_4
9. Hu, Q., Yu, D., Liu, J., Wu, C.: Neighborhood rough set based heterogeneous feature subset selection. Inf. Sci. **178**(18), 3577–3594 (2008)
10. Jensen, R.: Rough set-based feature selection: a review. In: Rough Computing: Theories, Technologies and Applications, pp. 70–107. IGI Global (2008)
11. Kloft, M., Nakajima, S., Brefeld, U.: Feature selection for density level-sets. In: Buntine, W., Grobelnik, M., Mladenić, D., Shawe-Taylor, J. (eds.) ECML PKDD 2009. LNCS (LNAI), vol. 5781, pp. 692–704. Springer, Heidelberg (2009). https://doi.org/10.1007/978-3-642-04180-8_62
12. Kulczycki, P.: Kernel estimators in industrial applications. In: Prasad, B. (ed.) Soft Computing Applications in Industry, pp. 69–91. Springer, Heidelberg (2008). https://doi.org/10.1007/978-3-540-77465-5_4
13. Muller, H.G., et al.: Smooth optimum kernel estimators of densities, regression curves and modes. Ann. Stat. **12**(2), 766–774 (1984)
14. Orava, J.: K-nearest neighbour kernel density estimation, the choice of optimal k. Tatra Mt. Math. Publ. **50**(1), 39–50 (2011)
15. Pawlak, Z.: Rough sets. Int. J. Comput. Inf. Sci. **11**(5), 341–356 (1982)
16. Rakotomamonjy, A., Davy, M.: One-class SVM regularization path and comparison with alpha seeding. In: ESANN, pp. 271–276. Citeseer (2007)
17. Rudin, W.: Real and Complex Analysis. Tata McGraw-Hill Education, New York (2006)
18. Scott, D.W.: Multivariate Density Estimation: Theory, Practice, and Visualization. Wiley, Hoboken (2015)
19. Silverman, B.W.: Density Estimation for Statistics and Data Analysis. Routledge, Abingdon (2018)
20. Singh, A., Scott, C., Nowak, R., et al.: Adaptive hausdorff estimation of density level sets. Ann. Stat. **37**(5B), 2760–2782 (2009)
21. Slowinski, R., Vanderpooten, D.: A generalized definition of rough approximations based on similarity. IEEE Trans. Knowl. Data Eng. **12**(2), 331–336 (2000)

22. Steinwart, I., Hush, D., Scovel, C.: Density level detection is classification. In: Advances in Neural Information Processing Systems, pp. 1337–1344 (2005)
23. Wand, M.P., Jones, M.C.: Kernel Smoothing. Chapman and Hall/CRC, Boca Raton (1994)
24. Węglarczyk, S.: Kernel density estimation and its application. In: ITM Web of Conferences, vol. 23, p. 00037. EDP Sciences (2018)
25. Yao, Y.: Decision-theoretic rough set models. In: Yao, J.T., Lingras, P., Wu, W.-Z., Szczuka, M., Cercone, N.J., Ślęzak, D. (eds.) RSKT 2007. LNCS (LNAI), vol. 4481, pp. 1–12. Springer, Heidelberg (2007). https://doi.org/10.1007/978-3-540-72458-2_1
26. Yao, Y., Yao, B.: Covering based rough set approximations. Inf. Sci. **200**, 91–107 (2012)
27. Ziarko, W.: Variable precision rough set model. J. Comput. Syst. Sci. **46**(1), 39–59 (1993)
28. Ziarko, W.: Probabilistic rough sets. In: Ślęzak, D., Wang, G., Szczuka, M., Düntsch, I., Yao, Y. (eds.) RSFDGrC 2005. LNCS (LNAI), vol. 3641, pp. 283–293. Springer, Heidelberg (2005). https://doi.org/10.1007/11548669_30

Three-Way Decision Theory

An Adjusted Apriori Algorithm to Itemsets Defined by Tables and an Improved Rule Generator with Three-Way Decisions

Zhiwen Jian[1], Hiroshi Sakai[1(✉)], Takuya Ohwa[1], Kao-Yi Shen[2],
and Michinori Nakata[3]

[1] Graduate School of Engineering, Kyushu Institute of Technology,
Tobata, Kitakyushu 804-8550, Japan
`sakai@mns.kyutech.ac.jp`
[2] Department of Banking and Finance, Chinese Culture University (SCE),
Taipei, Taiwan
`kyshen@sce.pccu.edu.tw`
[3] Faculty of Management and Information Science, Josai International University,
Gumyo, Togane, Chiba 283-0002, Japan
`nakatam@ieee.org`

Abstract. The NIS-Apriori algorithm, which is extended from the Apriori algorithm, was proposed for rule generation from non-deterministic information systems and implemented in SQL. The realized system handles the concept of certainty, possibility, and three-way decisions. This paper newly focuses on such a characteristic of table data sets that there is usually a fixed decision attribute. Therefore, it is enough for us to handle itemsets with one decision attribute, and we can see that one frequent itemset defines one implication. We make use of these characteristics and reduce the unnecessary itemsets for improving the performance of execution. Some experiments by the implemented software tool in Python clarify the improved performance.

Keywords: Rule generation · The Apriori algorithm · Frequent itemset · Incomplete information · Three-way decisions

1 Introduction

We are following rough set based rule generation from table data sets [10,14,22] and Apriori based rule generation from transaction data sets [1,2,9], and we are investigating a new framework of rule generation from table data sets with information incompleteness [17–21].

Table 1 is a standard table. We term such a table as a *Deterministic Information System* (DIS). In DISs, several rough set based rule generation methods are proposed [3,5,10,14,16,22,23]. Furthermore, *missing values '?'* [6,7,11] (Table 2) and a *Non-deterministic Information System* (NIS) [12,13,15] (Table 3) were also

© Springer Nature Switzerland AG 2020
R. Bello et al. (Eds.): IJCRS 2020, LNAI 12179, pp. 95–110, 2020.
https://doi.org/10.1007/978-3-030-52705-1_7

Table 1. An exemplary DIS ψ.

Object	P	Q	R	S	Dec
$x1$	3	1	2	2	a
$x2$	2	2	2	1	a
$x3$	1	2	2	1	b
$x4$	1	3	3	2	b
$x5$	3	2	3	1	c

Table 2. An exemplary NIS Φ with missing value '?', whose value is one of 1, 2, 3.

Object	P	Q	R	S	Dec
$x1$	3	?	2	2	a
$x2$	2	$\{2,3\}$	2	?	a
$x3$	?	2	2	$\{1,2\}$	b
$x4$	1	3	3	2	b
$x5$	3	2	3	?	c

Table 3. An exemplary NIS Φ. Each '?' is replaced with a set $\{1,2,3\}$ of possible attribute values.

Object	P	Q	R	S	Dec
$x1$	3	$\{1,2,3\}$	2	2	a
$x2$	2	$\{2,3\}$	2	$\{1,2,3\}$	a
$x3$	$\{1,2,3\}$	2	2	$\{1,2\}$	b
$x4$	1	3	3	2	b
$x5$	3	2	3	$\{1,2,3\}$	c

investigated to cope with information incompleteness. In [12], question-answering based on possible world semantics was investigated, and an axiom system was given for query translation to one equivalent normal form [12].

In NIS, some attribute values are given as a set of possible attribute values due to information incompleteness. In Tables 2, $\{2,3\}$ in $x2$ implies '*either 2 or 3 is the actual value, but there is no information to decide it*', and '?' does there is no information. We replace each '?' with all possible attribute values and have Table 3. Thus, we can handle '?' in NIS (some discretization may be necessary for continuous attribute values). Formerly in NISs, question-answering and information retrieval were investigated, and we are coping with rule generation from NISs.

The Apriori algorithm [1] was proposed by Agrawal for handling transaction data sets. We adjust this algorithm to DIS and NIS by using the characteristics of table data sets. The highlight of this paper is the following.

(1) A brief survey of Apriori based rule generation and a rule generator,
(2) Some improvements of the Apriori based algorithm and a rule generator,
(3) Experiment by the improved rule generator in Python.

This paper is organized as follows: Sect. 2 surveys our framework on NISs and the Apriori algorithm [1,2,9]. Section 3 connects table data sets to transaction data sets and copes with the manipulation of candidates of rules. Then, more effective manipulation is proposed in DISs and NISs. Section 4 describes a new NIS-Apriori based system in Python and presents the improved results. Section 5 concludes this paper.

2 Preliminary: An Overview of Rule Generation and Examples

This section briefly reviews rule generation from DISs and NISs.

2.1 Rules and Rule Generation from DISs

In Table 1, we consider implications like $[P, 3] \Rightarrow [Dec, a]$ from $x1$ and $[R, 2] \wedge [S, 1] \Rightarrow [Dec, b]$ from $x3$. Generally, a *rule* is defined as an implication satisfying some constraint. The following is one standard definition of rules [1,2,9,14,22]. We follow this definition and consider the following rule generation from DIS.

(A rule from DIS). A *rule* is an implication τ satisfying $support(\tau) \geq \alpha$ and $accuracy(\tau) \geq \beta$ ($0 < \alpha$, $\beta \leq 1.0$) for given threshold values α and β.
(Rule generation from DIS). If we fix α and β in DIS, the set of all rules is also fixed, but we generally do not know them. Rule generation is to generate all minimal rules (we term a rule with minimal condition part a *minimal rule*).

Here, $support(\tau)$ is an occurrence ratio of an implication τ for the total objects and $accuracy(\tau)$ is a consistency ratio of τ for the condition part of τ. For example, let us consider $\tau : [R, 2] \wedge [S, 1] \Rightarrow [Dec, b]$ from $x3$. Since τ occurs one time for five objects, we have $support(\tau) = 1/5$. Since $[R, 2] \wedge [S, 1]$ occurs two times, we have $accuracy(\tau) = 1/2$. Fig. 1 shows all minimal rules (redundant rules are not generated) from Table 1.

```
mysql> select * from rule1;
+-----------+------+------+-----------+---------+----------+
| att1      | val1 | deci | deci_value | support | accuracy |
+-----------+------+------+-----------+---------+----------+
| P         | 1    | Dec  | b          | 0.400   | 1.000    |
| P         | 2    | Dec  | a          | 0.200   | 1.000    |
| Q         | 1    | Dec  | a          | 0.200   | 1.000    |
| Q         | 3    | Dec  | b          | 0.200   | 1.000    |
| end_attrib | NULL | NULL | NULL      | NULL    | NULL     |
+-----------+------+------+-----------+---------+----------+
5 rows in set (0.00 sec)

mysql> select * from rule2;
+-----------+------+------+------+------+-----------+---------+----------+
| att1      | val1 | att2 | val2 | deci | deci_value | support | accuracy |
+-----------+------+------+------+------+-----------+---------+----------+
| P         | 3    | Q    | 2    | Dec  | c          | 0.200   | 1.000    |
| P         | 3    | R    | 2    | Dec  | a          | 0.200   | 1.000    |
| P         | 3    | R    | 3    | Dec  | c          | 0.200   | 1.000    |
| P         | 3    | S    | 1    | Dec  | c          | 0.200   | 1.000    |
| P         | 3    | S    | 2    | Dec  | a          | 0.200   | 1.000    |
| Q         | 2    | R    | 3    | Dec  | c          | 0.200   | 1.000    |
| R         | 2    | S    | 2    | Dec  | a          | 0.200   | 1.000    |
| R         | 3    | S    | 1    | Dec  | c          | 0.200   | 1.000    |
| R         | 3    | S    | 2    | Dec  | b          | 0.200   | 1.000    |
| end_attrib | NULL | NULL | NULL | NULL | NULL      | NULL    | NULL     |
+-----------+------+------+------+------+-----------+---------+----------+
10 rows in set (0.00 sec)
```

Fig. 1. The obtained all minimal rules ($support(\tau) \geq 0.2$, $accuracy(\tau) \geq 0.9$) from Table 1. Our system ensures that there is no other rule except them. In the table *rule1*, the first rule is $\tau : [P, 1] \Rightarrow [Dec, b]$. Even though $\tau' : [P, 1] \wedge [Q, 2] \Rightarrow [Dec, b]$ satisfies the constraint of rules, τ' is a redundant implication of τ and τ' is not minimal.

2.2 Rules and Rule Generation from NISs

From now, we employ the symbols Φ and ψ for expressing NIS and DIS, respectively. In NIS Φ, we replace a set of all possible values with an element of this set, and then we have one DIS. We term such a DIS a *derived DIS* from NIS, and let $DD(\Phi)$ denote a set of all derived DISs from NIS. Table 1 is a derived DIS from Table 3. In NISs like Table 3, we consider the following two types of rules,

(1) A rule which we certainly conclude from NIS (a certain rule),
(2) A rule which we may conclude from NIS (a possible rule).

These two types of rules seem to be natural for rule generation with information incompleteness. Yao recalls three-valued logic in rough sets and proposes *three-way decisions* [23,24]. These types of rules concerning missing values were also investigated in [6,11], and we coped with the following two types of rules based on possible world semantics [18,20]. The definition in [6,11] and the following definition are semantically different [18].

(A certain rule from NIS). An implication τ is a *certain rule*, if τ is a rule in each of derived DIS from NIS,
(A possible rule from NIS). An implication τ is a *possible rule*, if τ is a rule in at least one derived DIS from NIS.
(Rule generation from NIS). If we fix α and β in NIS, the set of all certain rules and the set of all possible rules are also fixed. Rule generation is to generate all minimal certain rules and all minimal possible rules.

Two types of rules depend on all derived DISs from NIS, and the number of them increases exponentially. For Table 3, the number is 324 ($=2^2 \times 3^4$), and the number is more than 10^{100} for the Mammographic data set [4]. Thus, the realization of a system to handle two types of rules was seemed to be hard, however, we gave one solution to this problem.

(Proved Property). For each implication τ, we developed some formulas to calculate the following,

(1) $minsupp(\tau) = \min_{\psi \in DD(\Phi)}\{support(\tau) \text{ in } \psi\}$,
(2) $minacc(\tau) = \min_{\psi \in DD(\Phi)}\{accuracy(\tau) \text{ in } \psi\}$,
(3) $maxsupp(\tau) = \max_{\psi \in DD(\Phi)}\{support(\tau) \text{ in } \psi\}$,
(4) $maxacc(\tau) = \max_{\psi \in DD(\Phi)}\{accuracy(\tau) \text{ in } \psi\}$.

This calculation employs the rough sets based concept and is independent of the number of derived DISs [18,20,21]. By using these formulas, we proved a method to examine 'τ is a certain rule or not' and 'τ is a possible rule or not'. This method is also independent of the number of all derived DISs [18,20,21].

We apply this property to the Apriori algorithm for realizing a rule generation system. The Apriori algorithm effectively enumerates itemsets (candidates of rules), and the *support* and *accuracy* values of every candidate are calculated by the Proved Property. Figures 2 and 3 show the obtained minimal certain rules and minimal possible rules from Table 3. As for the execution time, we discuss it in Sect. 4.

```
mysql> select * from c1rule;
```

att1	val1	deci	deci_value	minsupp	minacc
P	1	Dec	b	0.200	1.000
end_attrib	NULL	NULL	NULL	NULL	NULL

```
2 rows in set (0.00 sec)

mysql> select * from c2rule;
```

att1	val1	att2	val2	deci	deci_value	minsupp	minacc
P	3	R	3	Dec	c	0.200	1.000
Q	2	R	3	Dec	c	0.200	1.000
Q	3	R	3	Dec	b	0.200	1.000
end_attrib	NULL	NULL	NULL	NULL	NULL	NULL	NULL

```
4 rows in set (0.00 sec)
```

Fig. 2. The obtained all minimal certain rules $(support(\tau) \geq 0.2, accuracy(\tau) \geq 0.9)$ from Table 3. There is no rule except them.

```
mysql> select * from p1rule;
```

att1	val1	deci	deci_value	maxsupp	maxacc
P	1	Dec	b	0.400	1.000
P	2	Dec	a	0.200	1.000
Q	1	Dec	a	0.200	1.000
Q	3	Dec	b	0.200	1.000
S	1	Dec	a	0.200	1.000
S	1	Dec	b	0.200	1.000
S	1	Dec	c	0.200	1.000
S	3	Dec	a	0.200	1.000
S	3	Dec	c	0.200	1.000
end_attrib	NULL	NULL	NULL	NULL	NULL

```
10 rows in set (0.00 sec)

mysql> select * from p2rule;
```

att1	val1	att2	val2	deci	deci_value	maxsupp	maxacc
P	2	Q	2	Dec	b	0.200	1.000
P	2	S	2	Dec	b	0.200	1.000
P	3	Q	2	Dec	c	0.200	1.000
P	3	Q	3	Dec	a	0.200	1.000
P	3	R	2	Dec	a	0.200	1.000
P	3	R	3	Dec	c	0.200	1.000
P	3	S	2	Dec	a	0.200	1.000
Q	2	R	2	Dec	b	0.200	1.000
Q	2	R	3	Dec	c	0.200	1.000
Q	2	S	2	Dec	a	0.400	1.000
Q	2	S	2	Dec	b	0.200	1.000
Q	2	S	2	Dec	c	0.200	1.000
Q	3	R	2	Dec	a	0.400	1.000
R	2	S	2	Dec	a	0.400	1.000
R	3	S	2	Dec	b	0.200	1.000
end_attrib	NULL	NULL	NULL	NULL	NULL	NULL	NULL

```
16 rows in set (0.00 sec)
```

Fig. 3. The obtained all minimal possible rules $(support(\tau) \geq 0.2, accuracy(\tau) \geq 0.9)$ from Table 3. There is no rule except them.

2.3 A Relation Between Rules in DISs and Rules in NISs

Let ψ^{actual} be a derived DIS with actual information from NIS Φ (we cannot decide ψ^{actual} from Φ, but we suppose there is an actual ψ^{actual} for Φ), then we can easily have the next inclusion relation.

$$\{\tau \mid \tau \text{ is a certain rule in } \Phi\} \subseteq \{\tau \mid \tau \text{ is a rule in } \psi^{actual}\}$$
$$\subseteq \{\tau \mid \tau \text{ is a possible rule in } \Phi\}$$

Due to information incompleteness, we know lower and upper approximations of a set of rules in ψ^{actual}. This property follows the concept of rough sets based approximations.

2.4   The Apriori Algorithm for Transaction Data Sets

Let us consider Table 4, which shows four persons' purchase of items. Such structured data is termed a *transaction data set*. In this data set, let us focus on a set $\{ham, beer\}$. Such a set is generally termed an *itemset*. For this itemset, we consider two implications $\tau_1 : ham \Rightarrow beer$ and $\tau_2 : beer \Rightarrow ham$. In τ_1, $support(\tau_1) = 3/4$ and $accuracy(\tau_1) = 3/3$. In τ_2, $support(\tau_2) = 3/4$ and $accuracy(\tau_2) = 3/4$. For an itemset $\{ham, beer, corn\}$, we consider six implications, $ham \wedge beer \Rightarrow corn, \cdots, beer \Rightarrow corn \wedge ham$. Like this, Agrawal proposed a method to obtain rules from transaction data sets, which is known as the Apriori algorithm [1,2,9]. This algorithm makes use of the following.

Table 4. An exemplary transaction data set

Transaction	Items
1	bread, milk, ham, beer, corn
2	cheese, ham, beer
3	ham, beer, apple, potato, corn
4	cheese, cake, beer

(Monotonicity of *support*). For two itemsets P and Q, if $P \subseteq Q$, $support(Q) \leq support(P)$ holds.

By using this property, the Apriori algorithm enumerates all itemsets, which satisfy $support \geq \alpha$. Each of such itemsets is termed a *frequent itemset*. Let us consider the manipulation of itemsets in Table 4 under $support \geq 0.5$. Since there are four transactions, each itemset must occur more than two times. Let CAN_i and FI_i $(i \geq 0)$ denote a set of all candidates of itemsets and a set of all frequent itemsets consisting of $(i+1)$-items, respectively. We have the following.

$CAN_0 = \{\{bread\}(\text{Occurrence}=1), \{milk\}(1), \{ham\}(3), \{beer\}(4), \{corn\}(2),$
$\qquad \{cheese\}(2), \{apple\}(1), \{potato\}(1), \{cake\}(1)\},$
$FI_0 = \{\{ham\}(3), \{beer\}(4), \{corn\}(2), \{cheese\}(2)\},$
$CAN_1 = \{\{ham, beer\}, \{ham, corn\}, \{ham, cheese\}, \{beer, corn\},$
$\qquad \{beer, cheese\}, \{corn, cheese\}\},$
$FI_1 = \{\{ham, beer\}(3), \{ham, corn\}(2), \{beer, corn\}(2), \{beer, cheese\}(2)\},$
$CAN_2 = \{\{ham, beer, corn\}, \{ham, beer, cheese\}, \{ham, corn, cheese\},$
$\qquad \{beer, corn, cheese\}\},$
$FI_2 = \{\{ham, beer, corn\}(2)\}.$

Each element in CAN_i $(i \geq 1)$ is generated by the combination of two itemsets in FI_{i-1} [1,2]. Then, every itemset satisfying the *support* condition becomes the element of FI_i. For example, for A : $\{ham, corn\}$, B : $\{beer, cheese\} \in FI_1$, we add one element of B to A and have $\{ham, corn, beer\}$, $\{ham, corn, cheese\} \in CAN_2$. We also do the converse and have $\{beer, cheese, ham\}$, $\{beer, cheese, corn\} \in CAN_2$. Only one itemset $\{ham, corn, beer\}$ satisfies the *support* condition and becomes an element of FI_2. Like this, FI_1, FI_2, $\cdots$, FI_n are obtained at first, then the *accuracy* value of each implication defined by a frequent itemset is evaluated. In the subsequent sections, we change the above manipulation by using the characteristics of table data sets.

3 Some Improvements of the NIS-Apriori Based Rule Generator

We describe the improvements in our framework based on Sect. 2.

3.1 From Transaction Data Sets to Table Data Sets

We translate Table 1 to Table 5 and identify each descriptor with an item. Then, we can see that Table 5 is a transaction data set. Thus, we can apply the Apriori algorithm to rule generation.

Table 5. A transaction data set for DIS ψ in Table 1.

Object	Descriptors as items
$x1$	[P,3], [Q,1], [R,2], [S,2], [Dec,a]
$x2$	[P,2], [Q,2], [R,2], [S,1], [Dec,a]
$x3$	[P,1], [Q,2], [R,2], [S,1], [Dec,b]
$x4$	[P,1], [Q,3], [R,3], [S,2], [Dec,b]
$x5$	[P,3], [Q,2], [R,3], [S,1], [Dec,c]

We define the next sets IMP_1, IMP_2, $\cdots$, IMP_n.

$IMP_1 = \{[A, val_A] \Rightarrow [Dec, val]\}$,
$IMP_2 = \{[A, val_A] \wedge [B, val_B] \Rightarrow [Dec, val]\}$,
$IMP_3 = \{[A, val_A] \wedge [B, val_B] \wedge [C, val_C] \Rightarrow [Dec, val]\}$,

Here, IMP_i means a set of implications which consist of i-condition attributes. A minimal rule is an implication $\tau \in \cup_i IMP_i$, and we may examine each $\tau \in \cup_i IMP_i$. However, in the subsequent sections, we consider some effective manipulations to generate minimal rules in IMP_1, IMP_2, $\cdots$, sequentially.

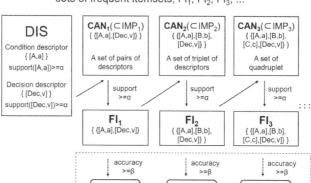

Fig. 4. The manipulation I for itemsets.

3.2   The Manipulation I for Frequent Itemsets by the Characteristics of Table Data Sets

Here, we make use of the characteristics of table data sets below.

(TA1). The decision attribute Dec is fixed. So, it is enough to consider each itemset including one descriptor whose attribute is Dec. For example, we do not handle any itemset like $\{[P,3],[Q,2]\}$ nor $\{[P,3],[Dec,a],[Dec,b]\}$ in Table 5.

(TA2). An attribute is related to each descriptor. So, we handle itemsets with different attributes. For example, we do not handle any itemset like $\{[P,3],[P,1],[Q,2],[Dec,b]\}$ in Table 5.

(TA3). To consider implications, we handle CAN_1, FI_1 ($\subseteq IMP_1$), CAN_2, FI_2 ($\subseteq IMP_2$), $\cdots$, which are defined in Sect. 2.4.

Based on the above characteristics, we can consider Fig. 4. In Fig. 4, itemsets satisfying (TA1) and (TA2) are enumerated. Generally, in the Apriori algorithm, the *accuracy* value is examined after obtaining all FI_i, because the decision attribute is not fixed. For each set in FI_i, there are plural implications. However, in a table data set, one implication corresponds to a frequent itemset. We employed this property and proposed the Apriori algorithm adjusted to table data sets [20,21] in Fig. 5. We term this algorithm the *DIS-Apriori algorithm*. Here, we calculate the *accuracy* value of every frequent itemset in each while loop (the rectangle area circled by the dotted line in Fig. 4 and lines 5-7 in Fig. 5). We can easily handle certain rules and possible rules in NISs by extending the DIS-Apriori algorithm.

Input: Table data set DIS ψ, decision attribute Dec, threshold values α, β.
Output: A set $Rule(\psi)$ of minimal rules.
1: $Rule(\psi) \leftarrow \{\}$; $i \leftarrow 1$;
2: create $FI_1 = \{\{[A, a], [Dec, v]\} | support([A, a] \Rightarrow [Dec, v]) \geq \alpha\}$ from CAN_1;
3: **while** $(|FI_i| \geq 1)$ **do**
4:      $Rest_i \leftarrow \{\}$; $Rule_i \leftarrow \{\}$;
5:      **for all** $\tau_{i,j} \in FI_i$ **do**
6:          **if** $accuracy(\tau_{i,j}) \geq \beta$ **then** <u>add $\tau_{i,j}$ to $Rule_i$</u>; **else** <u>add $\tau_{i,j}$ to $Rest_i$</u>;
7:          **end if**
8:      **end for**
9:      remove redundant implications from $Rule_i$;
10:     $i \leftarrow i + 1$; create FI_i;
11: **end while**
12: **return** $Rule(\psi) = \cup_{k < i} Rule_k$

Fig. 5. The Apriori algorithm adjusted to table data set DIS ψ. We can examine the *accuracy* value in each while loop (the rectangle area circled by the dotted line in Fig. 4). This examination is not done in the Apriori algorithm for transaction data sets.

Proposition 1. *[20, 21]*

(1) We replace DIS ψ with NIS Φ, support and accuracy with minsupp and minacc, respectively. Then, this algorithm generates all minimal certain rules.

(2) We replace DIS ψ with NIS Φ, support and accuracy with maxsupp and maxacc, respectively. Then, this algorithm generates all minimal possible rules.

(3) We term the algorithm consisting of (1) and (2) the NIS-Apriori algorithm.

Both DIS-Apriori and NIS-Apriori algorithms are logically sound and complete for rules. They generate rules without excess and deficiency.

Figures 1, 2 and 3 by the rule generator in SQL are based on the algorithm in Fig. 5 and Proposition 1.

3.3   The Manipulation II for Frequent Itemsets by the Characteristics of Table Data Sets

Now, we advance the manipulation I to the manipulation II. We focus on the statement 'create FI_i' in lines 2 and 10 in Fig. 5. In every while loop, we examine each $\tau \in FI_i \subseteq CAN_i \subseteq IMP_i$, so to reduce sets CAN_i and FI_i will influence the performance of execution. In Fig. 5, we at first need to remark the following.

(Rule generation). The purpose of rule generation is to generate each minimal implication $\tau \in \cup_i IMP_i$ satisfying $support(\tau) \geq \alpha$ and $accuracy(\tau) \geq \beta$. We obtain $Rule_1, Rest_1 \subseteq IMP_1$ in the 1st while loop, $Rule_2, Rest_2 \subseteq IMP_2$ in the 2nd while loop, and $Rule_3, Rest_3$ in the 3rd while loop, $\cdots$.

(Relation between sets in Fig. 5). We clarify the relation and the definition of $NOrule_i$ below.

(1) $Rule_i = \{\tau \in IMP_i \mid support(\tau) \geq \alpha,\ accuracy(\tau) \geq \beta\}$,
(2) $Rest_i = \{\tau \in IMP_i \mid support(\tau) \geq \alpha,\ accuracy(\tau) < \beta\}$,
(3) $FI_i = \{\tau \in IMP_i \mid support(\tau) \geq \alpha\}$,
(4) $NOrule_i = \{\tau \in IMP_i \mid support(\tau) < \alpha\}$,
(5) $IMP_i = FI_i \cup NOrule_i = (Rule_i \cup Rest_i) \cup NOrule_i$.

(A case of $\tau \in Rule_i$). If $\tau : \wedge_j[A_j, val_j] \Rightarrow [Dec, val] \in Rule_i$, we do not deal with any redundant implication $\tau' : (\wedge_j[A_j, val_j]) \wedge [B, b] \Rightarrow [Dec, val] \in IMP_{i+1}$, because τ' cannot be a minimal rule.
(A case of $\tau \in NOrule_i$). If $\tau : \wedge_j[A_j, val_j] \Rightarrow [Dec, val] \in NOrule_i$, any redundant implication $\tau' : (\wedge_j[A_j, val_j]) \wedge [B, b] \Rightarrow [Dec, val]$ satisfies $support(\tau') < \alpha$. So, $\tau' \in IMP_{i+1}$ cannot be a rule. Thus, we do not deal with any redundant implication τ'.
(A case of $\tau \in Rest_i$). In the $accuracy$ value, the monotonicity like $support$ does not hold (an example is in [20]). Thus, if $\tau : \wedge_j[A_j, val_j] \Rightarrow [Dec, val] \in Rest_i$, $accuracy(\tau') \geq \beta$ may hold for a redundant implication $\tau' : (\wedge_j[A_j, val_j]) \wedge [B, b] \Rightarrow [Dec, val] \in FI_{i+1}$.

Proposition 2. *Let us suppose that we had $Rule_i$ and $Rest_i$ ($IMP_i = Rule_i \cup Rest_i \cup NOrule_i$) in the i-th while loop in Fig. 5. Every candidate of a minimal rule in IMP_{i+1} is a redundant implication of $\tau \in Rest_i$.*

(Proof)
For every implication $\tau \notin FI_i \subseteq IMP_i$, its redundant implication τ' satisfies $support(\tau') \leq support(\tau) < \alpha$. Thus, τ' cannot be a minimal rule in IMP_{i+1}. Based on the Apriori algorithm, we need to combine two frequent itemsets in $FI_i = Rule_i \cup Rest_i$ (an example of this combination is described in Sect. 2.4). However, for the minimality condition of rules, we do not handle any redundant implication of $\tau \in Rule_i$. Thus, we conclude that every candidate of a minimal rule in IMP_{i+1} is a redundant implication of $\tau \in Rest_i$.

Definition 1. *We define a set $RCAN_i$ ($\subseteq CAN_i$), whose element is a candidate of a minimal rule in IMP_i w.r.t. rules $\cup_{j=1,\cdots,(i-1)} Rule_j$ and a set $RFI_i = \{\tau \in RCAN_i \mid support(\tau) \geq \alpha\}$ ($\subseteq FI_i \subseteq IMP_i$).*

In the Apriori algorithm, the concept of redundancy is not introduced, so that some redundant rules may be generated. The sets CAN_i and FI_i in Fig. 4 are generated from FI_{i-1} ($=Rule_{i-1} \cup Rest_{i-1}$). However, we can generate $RCAN_i (\subseteq CAN_i)$ and $RFI_i (\subseteq FI_i)$ from $Rest_{i-1}$. Furthermore, we previously generated itemsets $\{[A, a], [B, b], [Dec, v1]\}, \{[A, a], [B, b], [Dec, v2]\} \in RCAN_2$ from $\{[A, a], [Dec, v1]\}, \{[B, b], [Dec, v2]\} \in Rest_1$, and we removed this combination, because there is no object satisfying both $[Dec, v1]$ and $[Dec, v2]$. This combination formerly generated meaningless itemsets. This revision is another improvement in the manipulation of itemsets.

Proposition 3. *The set $RCAN_i$ and RFI_i are given as follows:*

$(i = 1) RCAN_1 = CAN_1$ *and* $RFI_1 = FI_1$,

$(i \geq 2) RCAN_i = \{\tau : (\wedge_j[A_j, val_j]) \wedge [B, b] \Rightarrow [Dec, val] \mid$

$\qquad \wedge_j[A_j, val_j] \Rightarrow [Dec, val] \in Rest_{i-1}, [B, b] \Rightarrow [Dec, val] \in Rest_1\}$,

$\qquad RFI_i = \{\tau \in RCAN_i \mid support(\tau) \geq \alpha\}.$

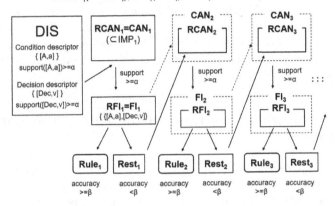

The manipulation II for generating
$RCAN_i(\subseteq CAN_i)$, $RFI_i(\subseteq FI_i)$, Rule$_i$, and Rest$_i$, :::

Fig. 6. New manipulation II of itemsets. We can handle $RCAN_i \subseteq CAN_i$ and $RFI_i \subseteq FI_i$ for generating minimal rules. In the Apriori algorithm, CAN_i and FI_i are employed, so redundant rules may be generated. By using $RCAN_i$ and RFI_i, the candidates of rules are reduced, and the performance of execution is improved.

(Proof)
(In case of $i = 1$) $RCAN_1 = CAN_1$ *and* $RFI_1 = FI_1$ *hold, because redundant rules occur after 2nd while loop.*
(In case of $i \geq 2$) We add one descriptor $[B, b]$ to $\wedge_j[A_j, val_j] \Rightarrow [Dec, val] \in Rest_{i-1}$ and have a redundant implication $\tau : (\wedge_j[A_j, val_j]) \wedge [B, b] \Rightarrow [Dec, val] \in IMP_i$ due to Proposition 2.

(1) In order to handle the same decision, $[B, b]$ must be the condition part of $\tau' : [B, b] \Rightarrow [Dec, val] \in RFI_1 = FI_1$. (If $\tau' \notin FI_1$, $support(\tau) < \alpha$ holds and τ cannot be a rule, because τ is a redundant implication of τ').
(2) $FI_1 = Rule_1 \cup Rest_1$ holds. If $\tau' \in Rule_1$, τ cannot be a minimal rule, because τ' is a minimal rule.

Based on the above discussion, we conclude $\tau' \in Rest_1$.

We propose the manipulation II in Fig. 6 due to the above propositions. In the Apriori algorithm, CAN_i is generated by FI_{i-1}, but we can remove redundant implications of $\tau \in Rule_{i-1}$. Thus, we can handle $RCAN_i$, which is a subset of CAN_i. If the number of elements in $Rule_{i-1}$ is large, the number of elements in $RCAN_i$ will be much smaller than that of CAN_i.

Proposition 4. *The DIS-Apriori algorithm with the manipulation II is sound and complete for minimal rules in DIS, and the NIS-Apriori algorithm with the manipulation II is also sound and complete for minimal certain rules and minimal possible rules in NIS. They do not miss any rule defined in DIS ψ or NIS Φ.*

(Sketch of Proof). We have proved that the DIS-Apriori and NIS-Apriori algorithms are sound and complete [20, 21]. We newly introduced sets $RCAN_i \subseteq CAN_i$ and $RFI_i \subseteq FI_i$ by using the redundancy of rules, and we extended the previous two algorithms to those with the manipulation II. The proposed algorithm does not examine each $\tau \in \cup_j IMP_j$, but examines each $\tau \in \cup_j RCAN_j$. As a result, this algorithm generates the same rules defined by the procedure 'to examine each $\tau \in \cup_j IMP_j$'.

4   An Improved Apriori Based Rule Generator and Some Experiments

This section compares the NIS-Apriori algorithm and the NIS-Apriori algorithm with the manipulation II. Of course, two algorithms generate the same rules due to Propositions 1 and 4, and the latter algorithm makes use of the redundancy concept. We newly implemented two systems in Python (Windows PC, CPU: Intel i7-4600U, 2.7 z). Table 6 shows the results on the Car Evaluation data set [4], and Table 7 does the results on the Phishing data set [4]. They are the cases of DISs, and the characteristic of $RCAN_i \subseteq CAN_i$ is effectively employed.

Now, we show two examples by the NIS-Apriori algorithm. The one is the Congressional Voting data set [4], and the other is the Lithology data set [8].

Table 6. The Car Evaluation data set (Objects: 1728, condition attributes: 6). A:$|Rule_1|$, B:$|CAN_2|$ or $|RCAN_2|$, C:$|Rule_2|$, D:$|CAN_3|$ or $|RCAN_3|$, E:$|Rule_3|$, F:$|CAN_4|$ or $|RCAN_4|$, G:$|Rule_4|$.

CASE	Manipulation	Time (sec)	A	B	C	D	E	F	G
support ≥ 0.2	I	0.037	5	24	0	0	0	0	0
accuracy ≥ 0.7	II	0.027	5	2	0	0	0	0	0
support ≥ 0.1	I	0.096	8	366	0	27	0	0	0
accuracy ≥ 0.7	II	0.059	8	74	0	0	0	0	0
support ≥ 0.05	I	0.189	8	366	0	1694	0	0	0
accuracy ≥ 0.7	II	0.123	8	176	0	572	0	0	0
support ≥ 0.01	I	0.621	8	732	0	3388	1	6588	0
accuracy ≥ 0.7	II	0.329	8	349	0	1172	1	1840	0

Table 7. The Phishing data set (Objects: 1353, condition attributes: 9). Here, A, B, $\cdots$, G are the same as Table 6.

CASE	Manipulation	Time (sec)	A	B	C	D	E	F	G
support ≥ 0.2	I	0.139	3	148	2	276	0	15	0
accuracy ≥ 0.7	II	0.083	3	25	2	30	0	0	0
support ≥ 0.1	I	0.847	6	426	13	2380	1	5774	0
accuracy ≥ 0.7	II	0.291	6	167	13	552	1	1101	0
support ≥ 0.05	I	1.409	7	831	23	5355	9	12438	2
accuracy ≥ 0.7	II	0.647	7	285	23	1259	9	3508	2
support ≥ 0.01	I	2.532	7	831	30	5355	25	22113	11
accuracy ≥ 0.7	II	1.522	7	583	30	3118	25	10611	11

Table 8. The Congressional Voting data set (Objects: 435, condition attributes: 16). There are 392 missing values, thus $|DD(\Phi)| = 2^{392} \geq 10^{100}$ (the number of derived DISs exceeds 10^{100}). A certain rule is a rule in each of more than 10^{100} derived DISs. A possible rule is a rule in at least one derived DISs. Here, A, B, $\cdots$, G are the same as Table 6.

CASE	Manipulation	Time (sec)	A	B	C	D	E	F	G
support ≥ 0.2 *accuracy* ≥ 0.6	I (certain rule)	23.73	23	900	6	8120	0	50960	0
	II (certain rule)	0.12	23	50	6	77	0	0	0
	I (possible rule)	23.56	28	960	3	8120	0	50960	0
	II (possible rule)	0.12	28	41	3	30	0	0	0
support ≥ 0.1 *accuracy* ≥ 0.6	I (certain rule)	26.35	23	960	6	8960	0	58240	0
	II (certain rule)	0.81	23	132	6	448	0	1064	0
	I (possible rule)	26.72	29	960	7	8960	2	58240	0
	II (possible rule)	0.52	29	100	7	290	2	580	0
support ≥ 0.05 *accuracy* ≥ 0.6	I (certain rule)	26.59	23	960	6	8960	0	58240	0
	II (certain rule)	1.79	23	220	6	949	0	2788	0
	I (possible rule)	27.29	29	960	7	8960	2	58240	0
	II (possible rule)	1.84	29	223	7	984	2	2967	0
support ≥ 0.01 *accuracy* ≥ 0.6	I (certain rule)	27.46	23	960	6	8960	0	58240	0
	II (certain rule)	4.28	23	354	6	1981	0	7630	0
	I (possible rule)	28.71	29	960	7	8960	2	58240	0
	II (possible rule)	3.59	29	296	7	1599	2	6141	0

As we described in Proposition 1, the NIS-Apriori algorithm (certain rule generation) is the DIS-Apriori algorithm with criterion values *minsupp* and *minacc*. Thus, the number of candidates of itemsets is also reduced by the manipulation II. The experiments easily examine the advancement of the manipulation II (Tables 8 and 9).

Table 9. The Lithology data set (Objects: 1923, condition attributes: 10). There are 519 missing values, therefore there are more than 10^{100} ($2^{519} \fallingdotseq (2^{10})^{50} > (10^3)^{50} > 10^{100}$) derived DISs. Here, A, B, $\cdots$, G are the same as Table 6.

CASE	Manipulation	Time (sec)	A	B	C	D	E	F	G
$support \geq 0.2\ accuracy \geq 0.5$	I (certain rule)	0.18	11	54	0	120	0	210	0
	II (certain rule)	0.06	11	0	0	0	0	0	0
	I (possible rule)	0.2	11	54	0	156	0	210	0
	II (possible rule)	0.07	11	0	0	0	0	0	0
$support \geq 0.1\ accuracy \geq 0.5$	I (certain rule)	0.43	17	127	0	464	0	985	0
	II (certain rule)	0.06	17	0	0	0	0	0	0
	I (possible rule)	0.51	17	127	0	549	0	1521	0
	II (possible rule)	0.06	17	0	0	0	0	0	0
$support \geq 0.05\ accuracy \geq 0.5$	I (certain rule)	0.84	18	900	0	1228	0	3657	0
	II (certain rule)	0.06	18	36	0	4	0	0	0
	I (possible rule)	1.26	19	1122	0	4128	0	4535	0
	II (possible rule)	0.08	19	76	0	97	0	0	0
$support \geq 0.01\ accuracy \geq 0.5$	I (certain rule)	17.05	23	6055	7	44940	21	222420	14
	II (certain rule)	4.18	23	1185	7	7772	21	36799	14
	I (possible rule)	48.87	39	8806	27	116466	37	755202	34
	II (possible rule)	6.45	39	1413	27	9804	37	48932	34

5 Concluding Remarks

We recently adjusted the Apriori algorithm to table data sets and proposed the DIS-Apriori and NIS-Apriori algorithms. This paper makes use of the characteristics of table data sets (one decision attribute Dec is fixed) and improved these algorithms. If we do not handle table data sets, there was no necessity for considering Fig. 6. The framework of the manipulation II (Fig. 6) is an improvement of Apriori based rule generation by using the characteristics of table data sets. We can generate minimal rules by using $RCAN_i \subseteq CAN_i$ and $RFI_i \subseteq FI_i$. This reduction causes to reduce the candidates of itemsets. We newly implemented the proposed algorithm in Python and examined the improvement of the performance of execution by experiments.

Acknowledgment. The authors would be grateful to the anonymous referees for their useful comments. This work is supported by JSPS (Japan Society for the Promotion of Science) KAKENHI Grant Number JP20K11954.

References

1. Agrawal, R., Srikant, R.: Fast algorithms for mining association rules in large databases. In: Proceedings of VLDB 1994, pp. 487–499. Morgan Kaufmann (1994)
2. Agrawal, R., Mannila, H., Srikant, R., Toivonen, H., Verkamo, A.I.: Fast discovery of association rules. In: Advances in Knowledge Discovery and Data Mining, pp. 307–328. AAAI/MIT Press (1996)

3. Ciucci, D., Flaminio, T.: Generalized rough approximations in PI 1/2. Int. J. Approx. Reason. **48**(2), 544–558 (2008)
4. Frank, A., Asuncion, A.: UCI machine learning repository. School of Information and Computer Science, University of California, Irvine (2010). http://mlearn.ics. uci.edu/MLRepository.html. Accessed 10 July 2019
5. Greco, S., Matarazzo, B., Słowiński, R.: Granular computing and data mining for ordered data: the dominance-based rough set approach. In: Meyers, R.A. (ed.) Encyclopedia of Complexity and Systems Science, pp. 4283–4305. Springer, New York (2009). https://doi.org/10.1007/978-0-387-30440-3
6. Grzymała-Busse, J.W., Werbrouck, P.: On the best search method in the LEM1 and LEM2 algorithms. In: Orłowska, E. (ed.) Incomplete Information: Rough Set Analysis. Studies in Fuzziness and Soft Computing, vol. 13, pp. 75–91. Springer, Heidelberg (1998). https://doi.org/10.1007/978-3-7908-1888-8_4
7. Grzymala-Busse, J.W.: Data with missing attribute values: generalization of indiscernibility relation and rule induction. In: Peters, J.F., Skowron, A., Grzymała-Busse, J.W., Kostek, B., Świniarski, R.W., Szczuka, M.S. (eds.) Transactions on Rough Sets I. LNCS, vol. 3100, pp. 78–95. Springer, Heidelberg (2004). https://doi.org/10.1007/978-3-540-27794-1_3
8. Hossain, T.M., Watada, J., Hermana, M., Shukri, S.R., Sakai, H.: A rough set based rule induction approach to geoscience data. In: Proceedings of UMSO 2018. IEEE (2018). https://doi.org/10.1109/UMSO.2018.8637237
9. Jovanoski, V., Lavrač, N.: Classification rule learning with APRIORI-C. In: Brazdil, P., Jorge, A. (eds.) EPIA 2001. LNCS (LNAI), vol. 2258, pp. 44–51. Springer, Heidelberg (2001). https://doi.org/10.1007/3-540-45329-6_8
10. Komorowski, J., Pawlak, Z., Polkowski, L., Skowron, A.: Rough sets: a tutorial. In: Pal, S.K., Skowron, A. (eds.) Rough Fuzzy Hybridization: A New Method for Decision Making, pp. 3–98. Springer, Heidelberg (1999)
11. Kryszkiewicz, M.: Rules in incomplete information systems. Inf. Sci. **113**(3–4), 271–292 (1999)
12. Lipski, W.: On databases with incomplete information. J. ACM **28**(1), 41–70 (1981)
13. Orłowska, E., Pawlak, Z.: Representation of nondeterministic information. Theoret. Comput. Sci. **29**(1–2), 27–39 (1984)
14. Pawlak, Z.: Rough sets. Int. J. Comput. Inf. Sci. **11**(5), 341–356 (1982)
15. Pawlak, Z.: Systemy Informacyjne: Podstawy Teoretyczne. WNT (1983). (in Polish)
16. Riza, L.S., et al.: Implementing algorithms of rough set theory and fuzzy rough set theory in the R package RoughSets. Inf. Sci. **287**(10), 68–89 (2014)
17. Sakai, H., Ishibashi, R., Koba, K., Nakata, M.: Rules and apriori algorithm in non-deterministic information systems. Trans. Rough Sets **9**, 328–350 (2008)
18. Sakai, H., Wu, M., Nakata, M.: Apriori-based rule generation in incomplete information databases and non-deterministic information systems. Fundam. Inf. **130**(3), 343–376 (2014)
19. Sakai, H.: Execution logs by RNIA software tools. http://www.mns.kyutech.ac.jp/ ~sakai/RNIA. Accessed 10 July 2019
20. Sakai, H., Nakata, M.: Rough set-based rule generation and Apriori-based rule generation from table data sets: a survey and a combination. CAAI Trans. Intell. Technol. **4**(4), 203–213 (2019)
21. Sakai, H., Nakata, M., Watada, J.: NIS-Apriori-based rule generation with three-way decisions and its application system in SQL. Inf. Sci. **507**, 755–771 (2020)

22. Skowron, A., Rauszer, C.: The discernibility matrices and functions in information systems. In: Słowiński, R. (ed.) Intelligent Decision Support - Handbook of Advances and Applications of the Rough Set Theory, pp. 331–362. Kluwer Academic Publishers, Berlin (1992)
23. Yao, Y.Y.: Three-way decisions with probabilistic rough sets. Inf. Sci. **180**, 314–353 (2010)
24. Hu, M., Yao, Y.: Structured approximations as a basis for three-way decisions in rough set theory. Knowl.-Based Syst. **165**, 92–109 (2019)

A Graph-Based Keyphrase Extraction Model with Three-Way Decision

Tianlei Chen[1,2], Duoqian Miao[1,2(✉)], and Yuebing Zhang[1,2]

[1] Department of Computer Science and Technology, Tongji University,
Shanghai 201804, China
ctlchentianlei@163.com, dqmiao@tongji.edu.cn
[2] Key Laboratory of Embedded System and Service Computing,
Ministry of Education, Tongji University, Shanghai 201804, China

Abstract. Keyphrase extraction has been a popular research topic in the field of natural language processing in recent years. But how to extract keyphrases precisely and effectively is still a challenge. The mainstream methods are supervised learning methods and graph-based methods. Generally, the effects of supervised methods are better than unsupervised methods. However, there are many problems in supervised methods such as the difficulty in obtaining training data, the cost of labeling and the limitation of the classification function trained by training data. In recent years, the development of the graph-based method has made great progress and its performance of extraction is getting closer and closer to the supervised method, so the graph-based method of keyphrase extraction has got a wide concern from researchers. In this paper, we propose a new model that applies the three-way decision theory to graph-based keyphrase extraction model. In our model, we propose algorithms dividing the set of candidate phrases into the positive domain, the boundary domain and the negative domain depending on graph-based attributes, and combining candidate phrases in the positive domain and the boundary domain qualified by graph-based attributes and non- graph-based attributes to get keyphrases. Experimental results show that our model can effectively improve the extraction precision compared with baseline methods.

Keywords: Keyphrase extraction · Three-way decision · Graph-based

1 Introduction

Keyphrase extraction has been a popular research topic in the natural language processing research field. Especially with the current increasing requirements for applications of texts, keyphrase extraction has attracted widespread attention from researchers. Although it has been greatly developed in recent years at home and abroad, the extracted results are far from the ideal.

With the rapid growth of text applications, the analysis of text data has become an important research area that has attracted much attention. Among them, how to extract keyphrases that reflect the subjects of texts has always been a research hotspot in the field of natural language processing, and its research results can be widely used in text retrieval, text summarization, text classification and question answering systems.

© Springer Nature Switzerland AG 2020
R. Bello et al. (Eds.): IJCRS 2020, LNAI 12179, pp. 111–121, 2020.
https://doi.org/10.1007/978-3-030-52705-1_8

Especially with the rise of research on unstructured big data of texts in recent years, the issue of keyphrase extraction has received in-depth research, and many researches have appeared in the international top conferences of artificial intelligence and natural language processing, such as the International Joint Conference on Artificial Intelligence (IJCAI) [1], The Annual Meeting of the Association for the Advance of Artificial Intelligence (AAAI) [2–4], International Computational Linguistics Association The Annual Meeting of the Association for Computational Linguistics (ACL) [5], The International Conference of World Wide Web (WWW) [6] and Conference on Empirical Methods in Natural Language Processing (EMNLP) [7], etc.

Researchers generally believe that the extracted keyphrases [8] should meet the following basic standards: (i) Keyphrases should be meaningful phrases. For example, "keyphrase extraction" is a meaningful phrase, but "and" does not meet the standard. (ii) Keyphrase extraction should meet the relevance standard that keyphrases must be closely related to the subjects of texts, which is the most essential requirement for keyphrase extraction. For example, the subtitle "Introduction" in this paper is not an appropriate keyphrase obviously. (iii) Keyphrase extraction should correspond to the coverage standard. Keyphrases should be able to cover various topics of the text and the main aspects of each topic, not just focus on only one topic and ignore others. (iv) Keyphrases extraction should meet the coherence standard. Several keyphrases of the text should be semantically and logically related. For an instance, a piece of academic paper that mainly introduces a graph-based keyphrase extraction model. The set of keyphrases is {"keyphrase extraction", "graph-based"}, which is more suitable than {"keyphrase extraction", "target detection"}. (v) Keyphrase extraction should correspond the conciseness standard. The number of keyphrases is limited, and the set of keyphrases should not contain any redundant phrase.

To meet any of the above standard, there is a huge challenge. Although there are many methods to solve this scientific problem such as statistical-based methods, supervised learning methods and graph-based methods, how to extract keyphrases precisely and efficiently is still a challenge.

In this paper, we propose a new model that applies the three-way decision theory to the graph-based keyphrase extraction model. In our model, we propose algorithms dividing the set of candidate phrases into the positive domain, the boundary domain and the negative domain depending on graph-based attributes, and combining candidate phrases in the positive domain and the boundary domain qualified by graph-based attributes and non-graph-based attributes to get keyphrases. Experimental results show that our model can effectively improve the extraction precision compared with baseline methods.

In Sect. 2, we briefly introduce the three-way decision theory and some related works in the field of keyphrase extraction. In Sect. 3, we describe the structure of our model and algorithms we proposed. In Sect. 4, we report the experimental results and analysis. Finally, we make a conclusion in Sect. 5.

2 Related Work

2.1 Statistical-Based Methods

Using statistical-based methods to extract keyphrases of texts is relatively simple, because it requires neither training data nor external knowledge. After the preprocessing of texts, simple statistical rules can be used to form a set of candidate phrases. The estimation of candidate phrases usually uses quantification of feature values. The main statistical-based keyphrase extraction method is TF-IDF (Term Frequency-Inverse Document Frequency) [9] and its improved methods. The advantage of the TF-IDF algorithm is that it is simple and fast. However, the traditional TF-IDF algorithm also has obvious shortcomings that it is not comprehensive enough to measure the importance of phrases based on the frequency. Sometimes important phrases may not appear frequently.

2.2 Graph-Based Methods

The graph-based keyphrase extraction method is the most effective and widely studied unsupervised keyphrase extraction method, because the method considers the co-occurrence relationship between phrases in the text. If there is a co-occurrence relationship between two phrases, it indicates that they are semantically related in the text. On the other hand, the graph-based method can incorporate more other features, so it has reached better effect of Extraction. The graph-based method has been widely concerned by researchers, from the TextRank method proposed by Mihalcea [10] to the PositionRank method proposed by Florescu [4]. In this paper, we propose a new model that applies the three-way decision theory to graph-based keyphrase extraction method.

2.3 Three-Way Decision

As generally considered, there are only acceptance and rejection in making a decision, which is a two-branch decision model, but it is often not the case in practical application. Based on the rough set theory proposed by Pawlak [11], Yao's three-way decision theory [12] provided a third alternative. The idea of three-way decision is based on three categories: acceptance, rejection and non-commitment. The goal is to divide a domain into three disjoint parts. Positive rules acquired from positive domain are used to accept something, negative rules acquired from negative domain are used to deny something, and rules that fall on boundary domain need further observation, which called delayed decision-making. Miao [13] has made some researches about three-way decision theory with multi-granularity, and Zhang [14] has applied it to the application of sentiment classification. The way of three-way decision describes the thinking mode of human beings in solving practical decision-making problems.

3 The Model with Three-Way Decision

3.1 Structure of the Model

We propose a graph-based keyphrase extraction model with three-way decision. As Fig. 1 illustrated, we could obtain candidate phrases through the preprocessing of texts from the raw, and then transform texts to text graphs with candidate phrases as nodes to get their graph-based attributes and non-graph-based attributes. With the support of the three-way decision theory, we divide the set of candidate phrases into the positive domain, the boundary domain and the negative domain depending on their graph-based attributes, and combine candidate phrases in the positive domain and the boundary domain qualified by their graph-based attributes and non-graph-based attributes to get keyphrases.

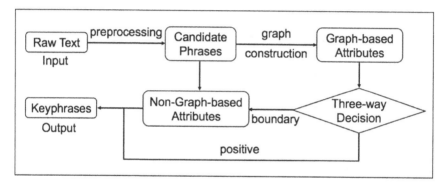

Fig. 1. The structure of the model with three-way decision

3.2 Preprocessing of Texts and Graph Construction

The step of preprocessing of texts from the raw plays an important role in the process of extracting keyphrases due to its output affecting the result deeply. The generic preprocessing way of graph-based keyphrases extraction: (i) Tokenizing: The process of tokenizing is to split strings into phrases. (ii) Tagging [15]: The task of tagging is to tag part-of-speech of phrases preparing for filtering. (iii) Filtering: Filter out phrases that do not meet the part-of-speech requirements according to the result of tagging. (iv) Stemming [16, 17]: Stemming phrases is in order to eliminate the effects of phrases forms that can get the main part of phrases. The differences between the phrases before stemming and after stemming are as follows (Table 1):

After preprocessing the raw, we construct the text graph to obtain graph-based attributes of candidate phrases. The text graph $G = (V, E, W)$, V is the set of nodes representing candidate phrases, E is the set of edges and W is the set of corresponding

Table 1. Examples for stemming results

Before stemming	After stemming
Harmonic	Harmon
Effective	Effect
Axiomatized	Axiom
Reality	Real
Validated	Valid

edge weights where weight w_{ij} for an edge e_{ij} indicates the frequency of two phrases v_i and v_j co-occurring in consecutive sentences, adopting the context-aware graph construction method from Duari [18] due to its simple construction method and well performance. The higher value of w_{ij} is, the stronger relationships between v_i and v_j are.

3.3 Keyphrase Extraction with Three-Way Decision

In our opinion, the three-way decision is making a delayed decision on uncertainty, and decides based on other information in the future. In this paper, we propose two Algorithms, which applies the three-way decision theory to the graph-based keyphrase extraction model (see Algorithm 1 and Algorithm 2). The main notations in this paper are listed in Table 2.

Table 2. The list of main notations

Variable	Explanation
c_i	Candidate phrases
ga_i	Graph-based attributes
nga_i	Non-graph-based attributes
r_i	Keyphrases extraction results
p	The positive domain
b	The boundary domain
n	The negative domain
C_i	The set of candidate phrases
G_i	The set of graph-based attributes
NG_i	The set of non-graph-based attributes
R	The set of keyphrases extraction results
Th	The threshold of the three-way decision

Algorithm 1 Classify the candidate phrases by graph-based attributes

---

Input: The set of candidate phrases $C = \{c_1, c_2, ..., c_n\}$ and its graph-based attributes
$G = \{ga_1, ga_2, ..., ga_n\}$
Output: $C_p = \{c_{(1)}, c_{(2)}, ..., c_{(Th*n)}\}$; $C_b = \{c_{(Th*n+1)}, c_{(Th*n+2)}, ..., c_{((1-Th)*n)}\}$
and $C_n = \{c_{((1-Th)*n+1)}, c_{((1-Th)*n+2)}, ..., c_{(n)}\}$
for i = 1 to n **do**
    if ga_i ranked in top Th * n **then**
        put c_i from C into C_p;
    else if ga_i ranked in bottom Th * n **then**
        put c_i from C into C_n;
    else
        put c_i from C into C_b;
    end if
end for

From Cohen's [19] Trusses theory, for a weighted, undirected and simple graph $G = (V, E, W)$, a k-truss subgraph of G is the maximal subgraph $G_k = (V_k, E_K, W_k)$, such that each edge $e_{ij} \in E_k$ belongs to at least $(k - 2)$ triangles. The truss level of an edge e_{ij} is k if it lies in k-truss but not in $(k + 1)$-truss. Kaur [20] expanded the concept of truss to nodes and defined truss level λ_i of node v_i as follows.

$$\lambda_i = max_{v_j \in N_i} \{l_{ij}\} \tag{1}$$

where N_i is the set of neighbours of node v_i and l_{ij} is the truss level of edge e_{ij}.

Based on the definition of the truss level of nodes, Duari [18] defined the semantic strength χ_i of node v_i and the semantic connectivity SC_i of node v_i as follows.

$$\chi_i = \sum_{v_j \in N_i} w_{ij} \times \lambda_j \tag{2}$$

$$SC_i = \frac{|\{\lambda_k : v_k \in N_i\}|}{maxtruss} \tag{3}$$

We take these attributes on the basis of the graph into account and define the graph-based attributes ga_i of node v_i as follows.

$$ga_i = \lambda_i \times \chi_i \times SC_i \tag{4}$$

In this paper, we propose Algorithm 1 to classify the candidate phrases by graph-based attributes and divide the set of candidate phrases C into the positive domain C_P, boundary domain C_b and negative domain C_n respectively.

Algorithm 2 Extract keyphrases with three-way decision

Input: Set of candidate phrases in the positive domain, boundary domain, negative domain

$C_p = \{c_{(1)}, c_{(2)}, ..., c_{(Th*n)}\}$, $C_b = \{c_{(Th*n+1)}, c_{(Th*n+2)}, ..., c_{((1-Th)*n)}\}$,

$C_n = \{c_{((1-Th)*n+1)}, c_{((1-Th)*n+2)}, ..., c_{(n)}\}$ and their corresponding attributes sets

$G_p = \{ga_{(1)}, ga_{(2)}, ..., ga_{(Th*n)}\}$, $G_b = \{ga_{(Th*n+1)}, ga_{(Th*n+2)}, ..., ga_{((1-Th)*n)}\}$,

$G_n = \{ga_{((1-Th)*n+1)}, ga_{((1-Th)*n+2)}, ..., ga_{(n)}\}$, $NG_p = \{nga_{(1)}, nga_{(2)}, ..., nga_{(Th*n)}\}$,

$NG_b = \{nga_{(Th*n+1)}, nga_{(Th*n+2)}, ..., nga_{((1-Th)*n)}\}$,

$NG_n = \{nga_{((1-Th)*n+1)}, nga_{((1-Th)*n+2)}, ..., nga_{(n)}\}$

Output: Set of keyphrases $R = \{r_1, r_2, ..., r_k\}$

if $Th * n \geq$ k **then**

    for i = 1 to $Th * n$ **do**

        put $ga_i * nga_i$ ranked top k from C_p into R;

    end for

else if $(1 - Th) * n \leq$ k **then**

    put C_p and C_b into R;

    for i = $(1 - Th) * n + 1$ to n **do**

        put $ga_i * nga_i$ ranked top $k - (1 - Th) * n$ from C_n into R;

    end for

else

    put C_p into R;

    for i = $Th * n + 1$ to $(1 - Th) * n$ **do**

        put $ga_i * nga_i$ ranked top $k - Th * n$ from C_b into R;

    end for

end if

Position information is an important factor in identifying keyphrases except for graph-based attributes. Florescu [4] proposed PositionRank and took the position of candidate phrases into account to identify keyphrases, we regard it as non-graph-based attributes nga_i of node v_i with the following definition.

$$nga_i = \sum_j^{n_i} \frac{1}{p_j} \tag{5}$$

In this paper, we propose Algorithm 2 taking graph-based and non-graph-based attributions of the candidate phrases into account in the boundary domain. Generally, both of the candidate phrases in the positive domain and the boundary domain are considered as the output of the Algorithm 2, where Th is the threshold of the three-way decision and the value of k represents the count of keyphrases to extract.

4   Experiments and Results

4.1   Benchmark Datasets and Baseline Methods

We evaluate the performance of the model with two widely used benchmark datasets, which are Hulth2003 and Krapivin2009. Hulth2003 is a dataset including about 2,000 abstracts of academic articles. Krapivin2009 consists of over 2,000 scientific papers from computer science domain published by ACM used for keyphrase extraction specially. We use the uncontrolled list of keyphrases of Hulth2003 and gold-standard keyphrases of Krapivin2009 for evaluation. We take Textrank [10], DegExt [21], k-core retention [22] and PositionRank [4] as baseline methods and evaluate our model against them.

4.2   Performance Results and Discussions

Duari [18] reported that values of k are 25 for Hulth2003 and 10 for Krapivin2009 that yield the highest F1-measure with all algorithms mentioned above, which correlate with the average number of labeled keyphrases in datasets, and we adopted the reported values of k and the results of baseline methods. In the experiment, we separate a part of data from data sets as validation sets to explore the most appropriate value of Th. The results show the value of Th is 0.1 for Hulth2003 and 0.4 for Krapivin2009 yields the best performance (see Table 3 and Table 4).

Table 3. The performance of Hulth2003 (k = 25)

Th	Precision	Recall	F1
0.1	**43.92**	**63.28**	**51.85**
0.2	43.20	62.25	51.01
0.3	42.62	61.40	50.31
0.4	42.90	61.81	50.65

Table 4. The performance of Krapivin2009 (k = 10)

Th	Precision	Recall	F1
0.1	27.57	29.69	28.60
0.2	39.07	42.07	40.52
0.3	41.78	44.99	43.32
0.4	**42.08**	**45.31**	**43.64**

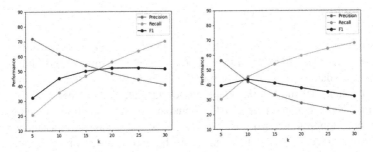

Fig. 2. The performance of Hulth2003 (threshold = 0.1) and Krapivin2009 (threshold = 0.4)

To verify the value of k yields the highest F1-measure mentioned above, we compared the F1-measure where the value of k was 5, 10, 15, 20, 25 and 30. The result shows that the F1-measure reaches the best when the value of k is 20 or 25 for Hulth2003 and 10 for Krapivin2009 (see Fig. 2). We find that the result of recall increases and the result of precision decreases when the value of k increases, which meets the fact.

The performance evaluation of keyphrase extraction can be divided into micro-statistical evaluation and macro-statistical evaluation. The micro one calculates the performance for each text first and then takes the average value. In comparison, the macro one statistics the result of extraction first and then calculates the performance at one time. We compared our model with Textrank [10], DegExt [21], k-core retention [22] and PositionRank [4] under the macro-statistical evaluation, where the value of k was 25 for Hulth2003 and 10 for Krapivin2009. The result shows that our model gets the best performance where the F1-measure reaches 51.85 for Hulth2003 and 43.64 for Krapivin2009 (see Table 5 and Fig. 3).

Table 5. The comparing performance with baseline methods

Dataset	DegExt	TextRank	K-core	PositionRank	Ours
Hulth2003	18.22	18.37	43.41	50.41	**51.85**
Krapivin2009	13.34	13.72	22.70	37.07	**43.64**

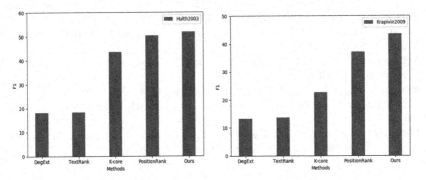

Fig. 3. The comparing performance with baseline methods

5 Conclusion

In this paper, we propose a new model that applies the three-way decision theory to graph-based keyphrase extraction model. In our model, we propose algorithms dividing the set of candidate phrases into the positive domain, the boundary domain and the negative domain depending on graph-based attributes, and combining candidate phrases in the positive domain and the boundary domain qualified by graph-based attributes and non-graph-based attributes to extract keyphrases. Experimental results show that our model can effectively improve the extraction accuracy compared with baseline methods. In future work, we will do more experiments to prove the performance of keyphrase extraction.

Acknowledgments. Authors would like to thank the anonymous reviewer for their critical and constructive comments and suggestions. This work was supported by National Natural Science Foundation of China (Grant No. 61976158). It was also supported by National Natural Science Foundation of China (Grant No. 61673301).

References

1. Zhang, W., Feng, W., Wang, J.Y.: Integrating semantic relatedness and words' intrinsic features for keyword extraction. In: Proceedings of the IJCAI, pp. 2225–2231. Morgan Kaufmann Publishers Inc., San Francisco (2013)
2. Gollapalli, S.D., Caragea, C.: Extracting keyphrases from research papers using citation networks. In: Proceedings of the AAAI, pp. 1629–1635. AAAI Press, Palo Alto (2014)
3. Gollapalli, S.D., Li, X.L., Yang, P.: Incorporating expert knowledge into keyphrase extraction. In: Proceedings of the AAAI, pp. 3180–3187. AAAI Press, Palo Alto (2017)
4. Florescu, C., Caragea, C.: A position-biased PageRank algorithm for keyphrase extraction, pp. 4923–4924. AAAI (2017)
5. Meng, R., Zhao, S.Q., Han, S.G., He, D.Q., Brusilovsky, P., Chi, Y.: Deep keyphrase generation. In: Proceedings of the ACL, pp. 582–592. ACL, Stroudsburg (2017)
6. Sterckx, L., Demeester, T., Deleu, J., Develder, C.: Topical word importance for fast keyphrase extraction. In: Proceedings of the WWW, pp. 121–122. ACM, New York (2015)
7. Sterckx, L., Caragea, C., Demeester, T., Develder, C.: Supervised keyphrase extraction as positive unlabeled learning. In: Proceedings of the EMNLP, pp. 1924–1929. ACL, Stroudsburg (2016)
8. Camacho, J.E.P., Ledeneva, Y., Hernández, R.A.G.: Comparison of automatic keyphrase extraction systems in scientific papers. Res. Comput. Sci. **115**, 181–191 (2016)
9. Salton, G., Buckley, C.: Term-Weighting approaches in automatic text retrieval. Inf. Process. Manag. **24**(5), 513–523 (1988)
10. Mihalcea, R., Tarau, P.: TextRank: bringing order into texts. In: Proceedings of the 2004 Conference on EMNLP, pp. 404–411. ACL (2004)
11. Pawlak, Z.: Rough sets. Int. J. Comput. Inform. Sci. **11**, 341–356 (1982)
12. Yao, Y.Y.: Three-way decisions with probabilistic rough sets. Inform. Sci. **180**, 341–353 (2010)
13. Miao, D.Q., Wei, Z.H., Wang, R.Z., Zhao, C.R., Chen, Y.M., Zhang, X.Y.: Uncertainty Analysis in Granular Computing. Science Press, Beijing (2019)

14. Zhang, Y.B., Miao, D.Q., Wang, J.Q., Zhang, Z.F.: A cost-sensitive three-way combination technique for ensemble learning in sentiment classification. Int. J. Approx. Reason. **105**, 85–97 (2019)
15. Toutanova, K., Klein, D., Manning, C.D., Singer, Y.: Feature-rich part-of-speech tagging with a cyclic dependency network. In: Proceedings of the ACL, pp. 173–180. ACL, Stroudsburg (2003)
16. Lovins, J.B.: Development of a stemming algorithm. Mech. Transl. Comput. Linguist. **11**, 22–31 (1968)
17. Bird, S.: NLTK: the natural language toolkit. In: Proceedings of the COLING/ACL on Interactive Presentation Sessions, pp. 69–72. ACL, Stroudsburg (2006)
18. Duari, S., Bhatnagar, V.: sCAKE: semantic connectivity aware keyword extraction. Inf. Sci. **477**, 100–117 (2019)
19. Cohen, J.: Trusses: cohesive subgraphs for social network analysis. National Security Agency Technical Report (2008)
20. Kaur, S., Saxena, R., Bhatnagar, V.: Leveraging hierarchy and community structure for determining influencers in networks. In: Bellatreche, L., Chakravarthy, S. (eds.) DaWaK 2017. LNCS, vol. 10440, pp. 383–390. Springer, Cham (2017). https://doi.org/10.1007/978-3-319-64283-3_28
21. Litvak, M., Last, M., Aizenman, H., Gobits, I., Kandel, A.: DegExt—a language-independent graph-based keyphrase extractor. In: Mugellini, E., Szczepaniak, P.S., Pettenati, M.C., Sokhn, M. (eds.) Advances in Intelligent Web Mastering – 3, pp. 121–130. Springer, Heidelberg (2011). https://doi.org/10.1007/978-3-642-18029-3_13
22. Rousseau, F., Vazirgiannis, M.: Main core retention on graph-of-words for single-document keyword extraction. In: Hanbury, A., Kazai, G., Rauber, A., Fuhr, N. (eds.) ECIR 2015. LNCS, vol. 9022, pp. 382–393. Springer, Cham (2015). https://doi.org/10.1007/978-3-319-16354-3_42

Modeling Use-Oriented Attribute Importance with the Three-Way Decision Theory

Xin Cui[✉], JingTao Yao, and Yiyu Yao

Department of Computer Science, University of Regina, Regina, SK S4S 0A2, Canada
{xcd325,jtyao,yyao}@cs.uregina.ca

Abstract. Ranking and measuring attribute importance is one of the key research topics in data mining and machine learning. Most of the existing attribute importance research relying on data-oriented approaches such as statistics and information theory perspectives. User preference, which involves a user specifying his or her preferential attitude towards a set of attributes, is another meaningful perspective. However, the research community has not paid much attention to this perspective. We adopt the three-way decision theory as a framework and concentrate on analyzing attribute importance based on user preference in this paper. In particular, we propose qualitative and quantitative analysis of attribute importance approaches that result a ranking list as well as a set of numerical weights of an attribute set. We then categorize attributes into different groups of importance using qualitative and quantitative analysis results. Finally, a unified model to analyze user-oriented attribute importance is constructed.

Keywords: Three-way decision · Attribute importance · User preference

1 Introduction

The main task of data mining is to derive valuable and representative patterns or knowledge from a dataset. Usually, a dataset is represented as a set of objects described by a set of attributes. In some clustering and classification problems, we treat attributes with equal importance. However, in real-world situations, different importance of attributes should be considered. There are various methods to analyze attribute importance from different perspectives, such as entropy based methods [2,11], maximizing deviation methods [17], and rough set based methods [18].

Generally speaking, analysis of attribute importance can be categorized into two classes, data-oriented and user-oriented. For data-oriented methods, we care more about the inner data structure by different attributes, which is also called

This work is partially supported by NSERC Discovery Grants, Canada.

internal information based analysis. In contrast, user-oriented methods emphasize the preferential attitude of a user towards an attribute set, which is called external information based analysis.

Attribute importance analysis based on data-oriented research is an objective approach. Using statistics or information theory, data-oriented analysis focuses on predictive ability or objects distribution of different attributes. One good example is entropy based methods, whose basic idea is that attributes leading to more entropy reduction would have a higher predictive ability, therefore, they are considered to be relatively more important. Apart from the data-oriented approach, user-oriented attribute importance analysis is subjective. It underlines the user's preferential judgment towards a set of attributes. Most current studies only focus on the former one, and user-oriented analysis has not received its due attention. In fact, researches from both perspectives are meaningful. This paper concentrates on the user-oriented approach, as a part of integrated attribute importance analysis, this is shown in Fig. 1.

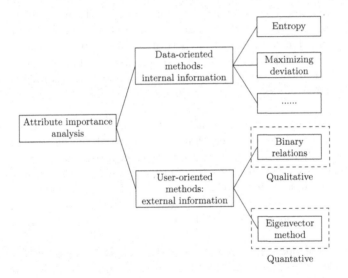

Fig. 1. A framework of attribute importance analysis

The primary purpose of this paper is to provide a general framework and adopt concrete methods to analyze user-oriented attribute importance. The content of this paper is generally arranged into three parts, qualitative attribute importance analysis using binary relations, quantitative attribute importance analysis using the eigenvector method, and evaluation based analysis.

2 Three-Way Decision in Processing User-Oriented Attribute Importance

Three-way decision theory, which was proposed by Yao [20], aims at providing a unified framework for thinking, problem solving, and information processing in three. It provides us with a practical framework for modeling real-world problems. Three-way decision has been expanded in various fields and acquired fruitful results, for example, three-way conflict analysis [9,22], three-way clustering [1,16,25,28], three-way recommender systems [5,26,27], three-way concept analysis [14,21], three-way granular computing [19,20], three-way face recognition [10] etc.

To analyze user-oriented attribute importance, we utilize three-way decision theory as our basic framework. Three models, namely, the Trisecting-Acting-Outcome (TAO) model, the three-level computing model, and the evaluation based model, are used to analyze attribute importance. We conduct qualitative and quantitative analysis of the attribute importance as shown in Fig. 1.

In qualitative analysis, we aim at ranking a set of attributes. The ranking is induced by considering the order relations of attribute pairs. We use TAO mode in three-way decision to model the structure and analyze attribute importance. We first compare and determine the relative importance of all attributes in pairs. Then, we trisect all these pairs into three classes, preferred, indifferent, and less preferred. Finally, by adopting a certain binary relation, we rank attributes in order.

In quantitative analysis, we utilize the eigenvector method to derive attributes' weights. It is reported that a drawback of the eigenvector method is when the number of objects is over 9, significant error could be introduced in the calculation [15]. The three-level computing model is used to overcome this problem by building a three-level structure. And then, weights calculation is applied from top to bottom using the eigenvector method several times, so that we can derive a large number of attributes' weights without losing too much accuracy.

The results of qualitative and quantitative analysis are a ranking list and a set of numerical weights considering attributes' importance. For the purpose of understanding and representing these results in a more clear way, we further categorize attributes into three groups with different importance levels, three-way evaluation based model is used in this analysis.

3 Three-Way Qualitative User-Oriented Attribute Importance Analysis

An important implication of binary relations is order relations, which is an intuitive notion ranking element against one another. For example, (x, y) is an ordered pair of two elements, we can determine order relations between x and y, which could be x is larger than y, x is worse than y, or x is a part of y considering different situations. In decision theory, order relations are commonly used in representing user preference, we write an order relation as $\succeq$ or $\succ$. If $x \succeq y$, we say x is at least as good as y, if $x \succ y$, we say x is strictly better than y. In this paper, to define user preference later on in a more straightforward way based on the property of trichotomy, we only concentrate on the strict order relation $\succ$.

3.1 User Preference and the Property of Trichotomy

The theory of user preference, also named individual choice behavior, is widely studied in different user-oriented researches, such as information retrieval [8, 29], economics [3], and social sciences [6]. The idea of user preference theory can also be applied in qualitative attribute importance analysis and the property of trichotomy plays an essential role. Order relations having this property are suitable to model a user's preferential attitude towards a set of attributes.

In our daily lives, human beings are good at making a relative comparison between numbers, products, strategies, etc. In number theory, given two arbitrary real numbers n and m, it is easy for us to conclude that exactly one of $n < m$, $n = m$, or $n > m$ must hold, this is called the trichotomy property of real numbers. Similarly, by comparing a pair of objects x and y under a specific criterion, an individual can determine the ordering relation between x and y as one of the followings, x is preferred than y, x is indifferent with y, or x is less preferred than y. Obviously, an individual's preferential attitude towards a pair of objects is three. This idea can also be generalized into order relations.

If we use an order relation $\succ$ to represent the meaning "preferred", the indifference relation $\sim$ is defined as an absence of $\succ$, which is defined as:

$$x \sim y \iff \neg(x \succ y) \wedge \neg(y \succ x). \tag{1}$$

Give an ordered pair (x, y), if an order relation $\succ$ expresses the first element is preferred than the second element. Its converse relation, written as $\overleftarrow{\succ}$, is called a less preferred relation, which is defined as:

$$x \overleftarrow{\succ} y \iff (y \succ x). \tag{2}$$

We usually write $\overleftarrow{\succ}$ as $\prec$ if it does not cause any ambiguity.

Definition 1. *An order relation $\succ$ on a set A is called trichotomous if $\forall(x, y)$, $x, y \in A$, exactly one of $x \succ y, x \sim y$ or $x \prec y$ holds.*

From the perspective of a decision maker, the goal of user preference related studies is to find optimal choice by analyzing the order relations among elements of a nonempty set, this primitive characteristic of a user is summarized as preference relation. The process of user preference theory can be viewed as to bring up rational axioms based on the decision maker's preference first, then analyzing a user's choice behavior based on preference [3]. From the perspective of mathematics, we model a preference relation using the property of trichotomy and transitivity.

Definition 2. *A preference relation, denoted as $\succ$, is a special type of binary relations on the set of elements A, that satisfies the following two rationality properties.* $\forall x, y, z \in A$,

$$\text{Trichotomous}: (x \succ y) \vee (x \sim y) \vee (x \prec y),$$
$$\text{Transitive}: x \succ y \wedge y \succ z \implies x \succ z. \tag{3}$$

If we use an order relation $\succ$ as a preference relation, user preference is represented as:

$$x \succ y \iff x \ is \ preferred \ than \ y$$
$$x \sim y \iff x \ is \ indifferent \ with \ y$$
$$x \prec y \iff x \ is \ less \ preferred \ than \ y \tag{4}$$

For an attribute set At, we divide all attribute pairs into three classes. Based on this trisection, attribute ranking can be induced. This process is shown in Fig. 2:

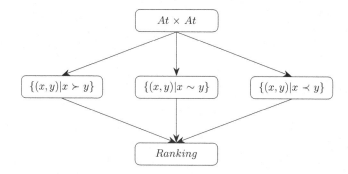

Fig. 2. The property of trichotomy

There are three kinds of order relations, namely linear orders, weak orders, and semiorders, which all equip the property of trichotomy and transitivity. In this paper, these three order relations are used in representing user preference for attribute importance analysis.

3.2 Modeling User Preference as a Linear Order

Given an attribute set At, a linear order $\succ$ enables us to arrange attributes in the form $At = \{a_1, a_2, ..., a_n\}$, such that $a_i \succ a_j$ if and only if $i < j$, for this reason, a linear order is also called a chain.

Definition 3. *Given a set At, a binary relation $\succ$ is a linear order on At, if it satisfies for any $x, y, z \in At$:*

$$\text{Asymmetric}: x \succ y \implies \neg(y \succ x),$$
$$\text{Transitive}: x \succ y \wedge y \succ z \implies x \succ z,$$
$$\text{Weakly Complete}: x \neq y \implies (x \succ y) \vee (y \succ x). \quad (5)$$

The property of asymmetric excludes the situation of a_i is better than a_j, as well as a_j better than a_i happen at the same time. The transitive property ensures that reasonable inference can be applied. The property of weakly complete ensures that all attributes are comparable with others.

Example 1. *Given a set of attributes $At = \{a_1, a_2, a_3, a_4, a_5\}$, a user's preference on At is defined by a linear order $\succ$. Suppose the ordering between attributes is specified by a user as:*

$$a_1 \succ a_5, a_1 \succ a_4, a_1 \succ a_2, a_3 \succ a_1, a_3 \succ a_2,$$
$$a_3 \succ a_4, a_3 \succ a_5, a_5 \succ a_4, a_5 \succ a_2, a_4 \succ a_2.$$

Then, attributes are ranked as:

$$a_3 \succ a_1 \succ a_5 \succ a_4 \succ a_2.$$

3.3 Modeling User Preference as a Weak Order

Weak orders are widely used in representing user preference relations in different fields [3,6,8,29]. Different from a linear order arranges elements in a chain, which is pretty strong in modeling real-world problems, a weak order allows ties in the ranking results. In other wolds, some attributes in a set could be considered as indifferent.

Definition 4. *A weak order $\succ$ is a binary relation on set At, if it satisfies for any $x, y \in At$:*

$$\text{Asymmetric}: x \succ y \implies \neg(y \succ x),$$
$$\text{Negative transitive}: \neg(x \succ y) \wedge \neg(y \succ z) \implies \neg(x \succ z). \quad (6)$$

Example 2. *Given a set of attributes $At = \{a_1, a_2, a_3, a_4, a_5\}$, a user's preference on At is defined by a weak order $\succ$. Suppose the ordering between attributes is specified as:*

$$a_1 \succ a_3, a_1 \succ a_4, a_1 \succ a_5, a_2 \succ a_3, a_2 \succ a_4, a_2 \succ a_5, a_3 \succ a_4, a_3 \succ a_5.$$

Because the user neither preferences a_1 to a_2, nor prefer a_2 to a_1, so a_1 is indifferent with a_2, written $a_1 \sim a_2$. Similarly, $a_4 \sim a_5$. By considering the above ordering, we can rank attributes like:

$$a_1 \sim a_2 \succ a_3 \succ a_4 \sim a_5.$$

3.4 Modeling User Preference as a Semiorder

Actually, an indifference relation does not necessarily be transitive. One good example is after reading three books, a reader might believe book A and B are equally good, so does book B and C, while he can tell that he prefers A to C based on his intuition. In other words, from an individual's preferential attitude, he can not distinguish the preference between A and B, and he also can not distinguish the preference between B and C, however, he can tell apart his preference between A and C. Luce [12] introduced semiorders to model this kind of problems.

Definition 5. *A semiorder $\succ$ on a set At is a binary relation which satisfies for any $x, x', x'', y, y' \in At$:*

$$\text{Asymmetric}: \ x \succ y \implies \neg(y \succ x),$$
$$\text{Ferrers}: \ (x \succ x') \land (y \succ y') \implies (x \succ y') \lor (y \succ x'),$$
$$\text{Semitransitive}: \ (x \succ x') \land (x' \succ x'') \implies (x \succ y) \lor (y \succ x''). \quad (7)$$

Example 3. *Given a set of attributes $At = \{a_1, a_2, a_3, a_4, a_5\}$, a user's preference on At is defined by a semiorder $\succ$. Suppose the ordering between attributes is specified as:*

$$a_1 \succ a_2, a_1 \succ a_3, a_1 \succ a_4, a_1 \succ a_5, a_2 \succ a_4, a_2 \succ a_5, a_3 \succ a_5, a_4 \succ a_5.$$

The user neither prefers a_2 to a_3, nor prefer a_3 to a_2, so $a_2 \sim a_3$, similarly we can get $a_3 \sim a_4$, however, the indifference is intransitive, because $a_2 \succ a_4$. So, we can not rank all attributes in one order but several, like below:

$$a_1 \succ a_2 \succ a_4 \succ a_5,$$

$$a_1 \succ a_2 \sim a_3 \succ a_5,$$

$$a_1 \succ a_3 \sim a_4 \succ a_5.$$

4 Three-Way Quantitative User-Oriented Attribute Importance Analysis

Mathematically, quantitative attribute importance analysis can be considered as a process of mapping each attribute to a numerical value,

$$w : At \longrightarrow \mathcal{R}, \quad (8)$$

where At is a set of attributes, $\mathcal{R}$ is a real number set, and w is a mapping function that calculates or assigns a numerical value to each attribute. For an attribute $a \in At, w(a)$ represents its weight from the perspective of a user.

4.1   A Three-Level Structure

To calculate or assign numerical weights for each attribute, this paper proposed two approaches. The first one is weights calculation using the eigenvector method, this is described in Sect. 4.2. The second approach is weights assignment. To be more specific, we derive an importance scale with numerical weights using the eigenvector method first, and then each attribute will be compared with this scale to get its weight, this approach is described in Sect. 4.3. Obviously, the eigenvector method plays an important role in both approaches, while, it has a drawback that not applicable in the situation when the number of objects is more than 9, because significant errors would be introduced in the calculation [15]. To overcome this problem, we introduce the three-way decision theory. More explicitly, the problem is arranged into a three-level structure, then, we apply the eigenvector method for weights calculation from top to bottom. The three-level structure enables us to control the number of objects in weights calculation is no more than 9, so that we can use the eigenvector method to calculate weights without losing too much accuracy.

4.2   Three-Way Quantitative Attribute Weighting Method Based on Eigenvector Method

The framework of the quantitative attribute weighting model is shown in Fig. 3. Suppose we have an attribute set At, a_{ij} at the bottom level represents an attribute. By grouping attributes into different clusters concerning semantic meaning, we build a three-level structure.

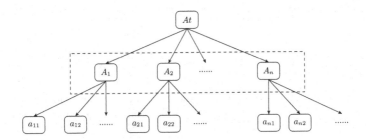

Fig. 3. The structure of the three-level attribute weighting method

Once we have this three-level structure, the calculation using the eigenvector method is applied from top to bottom as the second step. We calculate clusters' weights based on user preference, then, for each cluster, the weights of attributes within this cluster are calculated based on this cluster' weight.

Weights calculation using the eigenvector method is described as follows. Suppose an attribute set At has been grouped it into n clusters, $n \leq 9$ and the number of attributes in each cluster is also no more than 9. To derive a weight vector $w = (w_1, w_2, \cdots, w_n)$ for clusters, we build a comparison matrix M as defined in Definition 6, where element m_{ij} represents the relative importance of a cluster A_i compared with a cluster A_j.

Definition 6. *A comparison matrix M is a square matrix of order n, whose elements are m_{ij}, M is a positive reciprocal matrix if M is:*

$$\text{Positive}: \forall i, j < n, m_{ij} > 0,$$
$$\text{Reciprocal}: \forall i, j < n, m_{ij} = 1/m_{ji}. \tag{9}$$

A comparison matrix M is in the form like below, and in a perfect situation, m_{ij} should exactly be the weights ratio of a cluster A_i compared with A_j.

$$M = \begin{pmatrix} m_{11} & m_{12} & \cdots & m_{1n} \\ m_{21} & m_{22} & \cdots & m_{2n} \\ \vdots & \vdots & \ddots & \vdots \\ m_{n1} & m_{n2} & \cdots & m_{nn} \end{pmatrix} = \begin{pmatrix} w_1/w_1 & w_1/w_2 & \cdots & w_1/w_n \\ w_2/w_1 & w_2/w_2 & \cdots & w_2/w_n \\ \vdots & \vdots & \ddots & \vdots \\ w_n/w_1 & w_n/w_2 & \cdots & w_n/w_n \end{pmatrix} \tag{10}$$

In a real situation, the values of elements in a comparison matrix is given by a user based on his or her preference. For the purpose of determining two clusters' weight ratio w_1/w_2 precisely, we utilize the 9-points rating scale proposed by Saaty [15], see Table 1.

Table 1. The Saaty's 9-points rating scale [15]

Intensity of importance	Definition	Explanation
1	Equal importance	Two activities contribute equally to the objective
3	Weak importance of one over another	Experience and judgment slightly favor one activity over another
5	Essential or strong importance	Experience and judgment strongly favor one activity over another
7	Demonstrated importance	An activity is strongly favored and its dominance demonstrated in practice
9	Absolute importance	The evidence favoring one activity over another is of the highest possible order of affirmation
2, 4, 6, 8	Intermediate values between the two adjacent judgments	When compromise is needed

In Table 1, we can learn that the value 1 means two clusters are equally important. For an arbitrary cluster, it should be equally important with itself, thus in a comparison matrix, the value m_{ii} of the main diagonal must be 1. Besides, for two clusters a and b, if a is preferred than b, the weight ratio w_a/w_b should be greater than 1, otherwise, it should be equal or less than 1.

Under ideal conditions, we can get the matrix equation as follows:

$$Mw = \begin{pmatrix} w_1/w_1 & w_1/w_2 & \cdots & w_1/w_n \\ w_2/w_1 & w_2/w_2 & \cdots & w_2/w_n \\ \vdots & \vdots & \ddots & \vdots \\ w_n/w_1 & w_n/w_2 & \cdots & w_n/w_n \end{pmatrix} \times \begin{pmatrix} w_1 \\ w_2 \\ \vdots \\ w_n \end{pmatrix} = nw, \tag{11}$$

where M is multiplied on the left by the vector of weights w, and the result of this multiplication is nw. The problem we are dealing with has been transformed into solving $Mw = nw$, or $(M - nI)w = 0$. Ideally, M is consistent if and only if its principle eigenvalue $\lambda_{max} = n$ [15]. However, elements in a comparison matrix are personal judgments which are estimated by a user and inconsistency is inevitable. Under this circumstance, perturbations in the matrix imply perturbations in the eigenvalues. Now, we need to solve a new problem:

$$M'w' = \lambda_{max}w', \tag{12}$$

where $M' = (m'_{ij})$ is the perturbed matrix of $M = (m_{ij})$, w' is the principal eigenvector and λ_{max} is the principal eigenvalue of M'. What we want to learn is how good the principal eigenvector w' represents w. Consistency ratio C.R. is used to determine whether an inconsistency is acceptable:

$$\text{C.R.} = \frac{\lambda_{max} - n}{(n-1) \times \text{R.I.}}, \tag{13}$$

where R.I. is an average random consistency index, see Table 2. These indexes are derived from a sample of randomly generated reciprocal matrices using 9-point rating scale [15].

Table 2. Average random consistency index (R.I.)

n	1	2	3	4	5	6	7	8	9	10
R.I	0	0	0.52	0.89	1.11	1.25	1.35	1.40	1.45	1.49

If C.R. $< 10\%$, eigenvector w can be used as weights of clusters, else, the comparison matrix is need to be revised until C.R. $< 10\%$.

Example 4. *Suppose we have an attribute set At which has been categorized into six clusters, that is $At = \{A_1, A_2, A_3, A_4, A_5, A_6\}$. A comparison matrix has been built based on a user's judgment, the process of weights calculation is shown as follows (Table 3):*

Table 3. Weights calculation of six clusters

	A_1	A_2	A_3	A_4	A_5	A_6	Weight
A_1	1	3	1/2	4	2	1/3	0.140
A_2	1/3	1	1/7	1	1/2	1/9	0.041
A_3	2	7	1	9	5	1/2	0.290
A_4	1/4	1	1/9	1	1/2	1/9	0.038
A_5	1/2	2	1/5	2	1	1/6	0.071
A_6	3	9	2	9	6	1	0.420
$\lambda_{max} = 6.048$
C.I. $= 0.010$
C.R. $= 0.762\% < 10\%$

Since C.R. $\leq 10\%$, *which satisfies consistency checking, the eigenvector of comparison matrix can be used as the weights of* $\{A_1, A_2, A_3, A_4, A_5, A_6\}$, *that is:*

$$w = (0.140, 0.041, 0.290, 0.038, 0.071, 0.420).$$

Once we have clusters' weights, we use them the as a basic and calculate the weights of attributes in each cluster using the same idea. Then, we normalize the weights of attributes by dividing by their cluster' weights. Finally, we can derive weights of all attributes.

4.3 A Quantitative Attribute Weighting Method Using an Importance Scale

It is a straightforward way for a user to directly assign numerical values for attributes as their weights based on his or her preferential attitude. However, when the number of attribute is large, a fluctuated performance in judgment is inevitable, and this causes low accuracy in the result. Considering this situation, an importance scale is used to overcome this problem [15].

The process of attribute weighting method using an importance scale can be described as the following three steps. First, from a user's perspective, intensities of preferential degree of attributes are categorized into different levels, like: significantly important, very important, moderately important, weakly important, not important. Next, using the eigenvector method described in Sect. 4.2, we can derive weights for each intensity degree. When the number of intensity degrees is over 9, a three-level structure is a necessity. By doing this, we build an importance scale to assist our judgment. Finally, attributes are compared with this scale to determine their weights.

Example 5. *Suppose we set five intensities of preferential degree, which are A : significantly important, B : very important, C : moderately important, D : weakly important, E : not important. A user builds a comparison matrix of these intensities, and the weights calculation of intensities is described as (Table 4):*

Table 4. A pairwise comparison matrix of intensity levels

	A	B	C	D	E	Weight
A	1	2	3	5	9	0.450
B	1/2	1	2	4	6	0.277
C	1/3	1/2	1	2	3	0.147
D	1/5	1/4	1/2	1	2	0.081
E	1/9	1/6	1/3	1/2	1	0.046

$\lambda_{max} = 5.024$

C.I. = 0.006

C.R. = 0.533%

Since consistency checking is satisfied, an importance scale is built using these intensities' weights. Then, we compare attributes with this scale one by one, different weights will be assigned to different attributes from the perspective of a user.

5   Three-Way Evaluation Based Attribute Importance Analysis

Dividing the universe into three regions and applying different strategies separately are the main idea of the three-way decision [23]. In qualitative or quantitative attribute importance analysis, the result we get is a ranking list or a set of numerical weights, they are meaningful but relatively impractical for a user to make a decision. In this section, these results will be processed and categorized into three pair-wise disjoint classes with different importance levels, which are high importance, median importance, and low importance. In the rest of this paper, We use H, M and L to represent these three classes. The reason we select three as the number of classes is that human cognition and problem solving rely on such a three-way division, which enables us turns complexity into simplicity in many situations [24].

5.1   Trisecting an Attribute Set Based on Percentiles

One approach to trisect an attribute set is by using two percentiles. The first step is to transform the result of qualitative or quantitative analysis into a linear order $\succ$. For the result of the qualitative approach based on a linear order, this procedure can be skipped. The second step is using a pair of percentiles to determine the three regions.

There are several ways to transform the qualitative and quantitative results in a linear order. The first one is topological sorting, whose basic idea is an element will not appear in a ranking list until all elements preferred than this element have been listed [7]. By using a topological sorting, we can get a ranking list in descending order. Another way is using an evaluation function to transform the results of qualitative and quantitative analysis into a set of attributes' evaluation status values (ESVs). The ESV of attribute a can be defined as:

$$v(a) = \frac{|\{x \in At | a \succ x\}|}{|At|}. \tag{14}$$

Attributes will be ranked based on their ESVs in descending order, attributes having a same ESV will be ranked in any order.

Now, we have a list of ESVs, which is in the form of $v_1, v_2, ..., v_n$, where v_1 is the largest value and v_n is the smallest value. Using the ranking lists of the above two methods, we then adopt two ESVs at α^{th} and β^{th} percentiles with $0 < \beta < \alpha < 100$ to calculate a pair of thresholds l and h as:

$$h = v_{\lceil \alpha n/100 \rceil},$$
$$l = v_{\lfloor \beta n/100 \rfloor}, \tag{15}$$

where the ceiling function $\lceil x \rceil$ gives the smallest integer that is not less than x, and the floor function $\lfloor x \rfloor$ gives the largest integer that not greater than x. The floor and ceiling functions are necessary for the reason that $\alpha n/100$ and $\beta n/100$ may not be integers [23].

Based on the descending ranking list and a pair of thresholds, three regions H, M and L can be constructed. Attributes in H region have high importance, attributes in M region are with median importance and attributes in L region have low importance.

5.2   Trisecting an Attribute Set Based on a Statistical Method

Yao and Gao [23] discussed the process of constructing and interpreting three regions from a statistical view. Mean and standard deviation are useful statistical tools for analyzing numerical values, which are applicable to the results of quantitative user-oriented attribute importance analysis. Suppose $w(a_1), w(a_2), ..., w(a_n)$ are the weights of attributes in At, n is the cardinality of At, the mean and standard deviation are calculated by:

$$\mu = \frac{1}{n} \sum_{i=1}^{n} w(a_i), \tag{16}$$

$$\sigma = \left(\frac{1}{n} \sum_{i=1}^{n} (w(a_i) - \mu)^2 \right)^{\frac{1}{2}}. \tag{17}$$

We use two non-negative number k_1 and k_2 to represent the position of thresholds away from the mean, then a pair of thresholds is determined as [23]:

$$h = \mu + k_1\sigma, k_1 \geq 0,$$
$$l = \mu - k_2\sigma, k_2 \geq 0. \tag{18}$$

Based on thresholds h and l, three regions of an attribute set can be constructed as:

$$
\begin{aligned}
H_{(k_1,k_2)}(w) &= \{x \in At | w(x) \geq h\} \\
&= \{x \in At | w(x) \geq \mu + k_1\sigma\}, \\
M_{(k_1,k_2)}(w) &= \{x \in At | l < w(x) < h\} \\
&= \{x \in At | \mu - k_2\sigma < w(x) < \mu + k_1\sigma\}, \\
L_{(k_1,k_2)}(w) &= \{x \in At | w(x) \leq l\} \\
&= \{x \in At | w(x) \leq \mu - k_2\sigma\}.
\end{aligned}
\tag{19}
$$

Attributes can be categorized into three regions H, M and L considering their weights.

6 Conclusions

Attribute importance analysis includes two perspectives: data-oriented and user-oriented. Using the three-way decision as a framework, we propose a unified model for user-oriented attribute importance analysis which consists of three parts, quantitative analysis, quantitative analysis, and evaluation based analysis. In the qualitative analysis, by using binary relations and the TAO model, we rank attributes considering their importance. In quantitative analysis, the three-level computing model is adopted, numerical weights are assigned to attributes using the eigenvector method. Finally, we trisect the results of qualitative and quantitative analysis into three classes of different importance in advance.

References

1. Afridi, M.K., Azam, N., Yao, J.T., Alanazi, E.: A three-way clustering approach for handling missing data using GTRS. Int. J. Approx. Reason. **98**, 11–24 (2018)
2. Chen, Y., Li, B.: Dynamic multi-attribute decision making model based on triangular intuitionistic fuzzy numbers. Sci. Iranica **18**, 268–274 (2011)
3. Evren, Ö., Ok, E.A.: On the multi-utility representation of preference relations. J. Math. Econ. **47**, 554–563 (2011)
4. Figueira, J., Greco, S., Ehrgott, M.: Multiple Criteria Decision Analysis: State of the Art Surveys. Springer, New York (2005). https://doi.org/10.1007/b100605
5. Huang, J.J., Wang, J., Yao, Y.Y., Zhong, N.: Cost-sensitive three-way recommendations by learning pair-wise preferences. Int. J. Approx. Reason. **86**, 28–40 (2017)
6. Hogg, T.: Inferring preference correlations from social networks. Electron. Commer. Res. Appl. **9**, 29–37 (2010)
7. Kahn, A.B.: Topological sorting of large networks. Commun. ACM **5**, 558–562 (1962)
8. Kelly, D., Teevan, J.: Implicit feedback for inferring user preference: a bibliography. In: Proceedings of the ACM SIGIR Forum, pp. 18–28 (2003)
9. Lang, G.M., Miao, D.Q., Cai, M.J.: Three-way decision approaches to conflict analysis using decision-theoretic rough set theory. Inf. Sci. **406**, 185–207 (2017)

10. Li, H.X., Zhang, L.B., Huang, B., Zhou, X.Z.: Sequential three-way decision and granulation for cost-sensitive face recognition. Knowl.-Based Syst. **91**, 241–251 (2016)

11. Liu, P.D., You, X.L.: Some linguistic neutrosophic Hamy mean operators and their application to multi-attribute group decision making. PLoS ONE **13**, e0193027 (2018)

12. Luce, R.D.: Semiorders and a theory of utility discrimination. Econometrica **24**, 178–191 (1956)

13. Pomerol, J.C., Barba-Romero, S.: Multicriterion Decision in Management: Principles and Practice. Springer, New York (2012). https://doi.org/10.1007/978-1-4615-4459-3

14. Ren, R., Wei, L.: The attribute reductions of three-way concept lattices. Knowl.-Based Syst. **99**, 92–102 (2016)

15. Saaty, T.L., Vargas, L.G.: Models, Methods, Concepts & Applications of the Analytic Hierarchy Process. Springer, New York (2012). https://doi.org/10.1007/978-1-4615-1665-1

16. Wang, P.X., Yao, Y.Y.: CE3: a three-way clustering method based on mathematical morphology. Knowl.-Based Syst. **155**, 54–65 (2018)

17. Xu, Z.S., Zhang, X.L.: Hesitant fuzzy multi-attribute decision making based on TOPSIS with incomplete weight information. Knowl.-Based Syst. **52**, 53–64 (2013)

18. Xie, G., Zhang, J.L., Lai, K.K., Yu, L.: Variable precision rough set for group decision-making: an application. Int. J. Approx. Reason. **49**, 331–343 (2008)

19. Yang, X., Li, T.R., Liu, D., Fujita, H.: A temporal-spatial composite sequential approach of three-way granular computing. Inf. Sci. **486**, 171–189 (2017)

20. Yao, Y.Y.: Three-way decision and granular computing. Int. J. Approx. Reason. **103**, 107–123 (2018)

21. Yao, Y.: Interval sets and three-way concept analysis in incomplete contexts. Int. J. Mach. Learn. Cybernet. **8**(1), 3–20 (2016). https://doi.org/10.1007/s13042-016-0568-1

22. Yao, Y.Y.: Three-way conflict analysis: Reformulations and extensions of the Pawlak model. Knowl.-Based Syst. **180**, 26–37 (2019)

23. Yao, Y., Gao, C.: Statistical interpretations of three-way decisions. In: Ciucci, D., Wang, G., Mitra, S., Wu, W.-Z. (eds.) RSKT 2015. LNCS (LNAI), vol. 9436, pp. 309–320. Springer, Cham (2015). https://doi.org/10.1007/978-3-319-25754-9_28

24. Yao, Y.Y.: Three-way decisions and cognitive computing. Cogn. Comput. **8**, 543–554 (2016)

25. Yu, H., Zhang, C., Wang, G.: A tree-based incremental overlapping clustering method using the three-way decision theory. Knowl.-Based Syst. **91**, 189–203 (2016)

26. Zhang, H.R., Min, F.: Three-way recommender systems based on random forests. Knowl.-Based Syst. **91**, 275–286 (2016)

27. Zhang, H.R., Min, F., Shi, B.: Regression-based three-way recommendation. Inf. Sci. **378**, 444–461 (2017)

28. Zhang, Y., Yao, J.T.: Gini objective functions for three-way classifications. Int. J. Approx. Reason. **81**, 103–114 (2017)

29. Zhou, B., Yao, Y.Y.: Evaluating information retrieval system performance based on user preference. J. Intell. Inf. Syst. **34**, 227–248 (2010)

Three-Way Decision for Handling Uncertainty in Machine Learning: A Narrative Review

Andrea Campagner, Federico Cabitza, and Davide Ciucci$^{(\boxtimes)}$ (iD)

Dipartimento di Informatica, Sistemistica e Comunicazione,
University of Milano-Bicocca, viale Sarca 336, 20126 Milan, Italy
davide.ciucci@unimib.it

Abstract. In this work we introduce a framework, based on three-way decision (TWD) and the trisecting-acting-outcome model, to handle uncertainty in Machine Learning (ML). We distinguish between handling uncertainty affecting the input of ML models, when TWD is used to identify and properly take into account the uncertain instances; and handling the uncertainty lying in the output, where TWD is used to allow the ML model to abstain. We then present a narrative review of the state of the art of applications of TWD in regard to the different areas of concern identified by the framework, and in so doing, we will highlight both the points of strength of the three-way methodology, and the opportunities for further research.

1 Introduction

Three-way decision (TWD) is a recent paradigm emerged from rough set theory (RST) that is acquiring its own status and visibility [46]. This paradigm is based on the simple idea of *thinking in three "dimensions"* (rather then in binary terms) when considering how to represent computational objects. This idea leads to the so-called *trisecting-acting-outcome* (TAO) model [82]: *Trisecting* addresses the question of how to divide the universe under investigation in three partitions; *Acting* explains how to deal with the three parts identified; and *Outcome* gives methodological indications on how to evaluate the adopted strategy.

Based on the TAO model, we propose a framework to handle uncertainty in Machine Learning: this model can be applied both to the input and the output of the Learning algorithm. Obviously, these two latter aspects are strictly related and they mutually affect each other in real applications. Schematically, the framework looks as illustrated in Table 1.

With reference to the table, we distinguish between applications that handle uncertainty in the input and those that handle uncertainty with respect to the output. By *uncertainty in the input* we mean different forms of uncertainty that are already explicitly present in the training datasets used by ML algorithms. By *uncertainty in the output* we mean mechanisms adopted by the ML algorithm in order to create more robust models or making the (inherent and partly insuppressible) predictive uncertainty more explicit.

© Springer Nature Switzerland AG 2020
R. Bello et al. (Eds.): IJCRS 2020, LNAI 12179, pp. 137–152, 2020.
https://doi.org/10.1007/978-3-030-52705-1_10

Table 1. TAO model applied to Machine Learning

	Trisecting	Acting	Outcome
Input	The dataset contains different forms of uncertainty and it can be split in certain/uncertain instances	The ML-algorithm should take into account the dataset uncertainty and handle it	Ad-hoc measures should be introduced to quantify the dataset uncertainty, which should also be considered in the algorithm evaluation
Output	The output can contain instances with no decision (classification, clustering, etc.)	The ML algorithm abstains from giving the result on uncertain instances	New measures to evaluate ML algorithms with abstention should be introduced

In the following Sections, we will explain in more detail the different parts of the framework outlined in Table 1, and discuss the recent advances and current research in the framework areas by means of a narrative review of the literature indexed by the Google Scholar database. In particular, in Sect. 2, we describe the different steps of the proposed model with respect to the handling of uncertainty in the input, while in Sect. 3 we do the same for the handling of the uncertainty in the output. In Sect. 4, we will then discuss the advantages of incorporating TWD and the TAO model for uncertainty handling into Machine Learning, and some relevant future directions.

2   Handling Uncertainty in the Input

Real-world datasets are far from being perfect: typically they are affected by different forms of uncertainty (often missingness) that can be mainly related to either the data acquisition process or the complexity (e.g, in terms of volatility) of the phenomena under consideration or for both these factors.

These forms of uncertainty are usually distinguished in three common variants:

1. *Missing data*: this is usually the most common type of uncertainty in the input [6]. The dataset could contain missing values in its predictive features either because the original value was not recorded (e.g. the data was collected in two separate times, and the instrumentation to measure the feature was available only at one time), was subsequently lost or considered irrelevant (e.g. a doctor decided not to measure the BMI of a seemingly healthy person). This type of uncertainty has been the most studied, typically under the *data imputation* perspective, that is the task in which missing values are *filled in* before any subsequent ML process. This can be done in various ways, with techniques based on clustering [34,65], statistical or regression approaches [7], rough set or fuzzy rough set methods [4,51,67];

2. *Weak supervision*: in the case of supervised problems, the supervision (i.e. the target or decision variable) is only given in an imprecise form or only partially specified. This type of uncertainty has seen some increase in interest in the recent years [105], with a growing literature focusing specifically on *superset learning* [17,29]; this is a specific type of weak supervision in which instances are associated with sets of *possible but mutually exclusive labels* that are guaranteed to contain the true value of the decision label;

3. *Multi-rater annotation*: this form of uncertainty is getting more and more impact due to the increasing use of *crowdsourcing* [5,23,69] for data annotation purposes, but it is also inherent in many domains where it is common (and in fact recommended) practice to involve multiple experts to increase the reliability of the Ground Truth, which is a crucial requirement in many situations where ML models are applied for sensitive or critical tasks (like in medicine for diagnostic tasks). Involving multiple raters who annotate the dataset independently of each others often results in multiple and conflicting decision labels for a given instance [9], for a common phenomenon that has been denoted with many expressions, like *observer variability* or *inter-rater reliability*.

While superficially similar (e.g. weak supervision could be seen as a form of missing data), the problems inherent to and the methods to handle these types of uncertainty are such that they should be distinguished. In the case of missing data, the main problem is to build reliable models of knowledge despite the incomplete information, and the *completion* of the dataset is but a means to an end, often under assumptions that are difficult to attain (or verify). In the case of weak supervision, on the other hand, the task of completion (which is usually called *disambiguation*) is of fundamental importance and the goal is, usually, to simultaneously build ML models and disambiguate the uncertain instances. Finally, in the case of multi-rater annotations, while the task of disambiguation is obviously present, there is also the problem of inferring the extent each single rater can be trusted (i.e., how accurate they are) and how to meaningfully aggregate the information they provide in order to build a consensus which is to be used to build the ground truth by which to train the ML model.

2.1 Trisecting and Acting Steps

In all three uncertainty forms, the *trisecting* act is at the basis of the process of uncertainty handling, as the uncertain instances (e.g., the instances missing some feature values, or those for which the provided annotations are only weak) must be necessarily recognised for any action to be considered: this also means that the trisecting act usually amounts to simply dividing the certain instances from the uncertain ones, and the bulk of the work is usually performed in the *acting* step in order to decide how differently handle the two kinds of instances. According to the three kinds of problems described at the beginning of the section, we present the following solutions.

Missing Data. Missing data is the type of uncertainty for which a TWD methodology to handle this kind of uncertainty is more mature, possibly because the problem has been well studied in RST and other theories for the management of uncertainty that are associated with TWD [21,22]. Most approaches in this direction have been based on the notion of *incomplete information table*, which is typically found in RST: Liu et al. [42] introduced a TWD model based on an incomplete information table augmented with interval-valued loss functions; Luo et al. [45] proposed a multi-step approach by which to distinguish different types of missing data (e.g. "don't know", "don't care") and similarity relations; Luo et al. [44] focused on how to update TWD in incomplete and multi-scale information systems using decision-theoretic rough sets; Sakai et al. [57–59] described an approach based on TWD to construct certain and possible rules using an algorithm which combines the classical A-priori algorithm [3] and possible world semantics [30]. Other approaches (not directly based on the incomplete information table notion) have also been considered: Nowicki et al. [52] proposed a TWD algorithm for classification with missing or interval-valued data based on rough sets and SVM; Yang et al. [75] proposed a method for TWD based on intuitionistic fuzzy sets that are construed based on a similarity relation of instances with missing values.

While all the above approaches propose techniques based on TWD with missing data for *classification* problems, there have also been proposals to deal with this type of uncertainty in *clustering*, starting from the original approach proposed by Yu [85,87], to deal with missing data in clustering using TWD: Afridi et al. [2] described an approach which is based, as for the classification case, on a simple trisecting step in which complete instances are used to produce an initial clustering and then use an approach based on game-theoretic rough sets to cluster the instances with missing values; Yang et al. [74] proposed a method for three-way clustering with missing data based on clustering density.

Weak Supervision. With respect to the case of weak supervision, the application of three-way based strategies is more recent and different techniques have been proposed in the recent years. Most of the work in this sense has focused on the specific case of *semi-supervised learning*, in which the uncertain instances have no supervision, and *active learning*, in which the missing labels can be requested to an external oracle (usually a human user) at some cost: Miao et al. [48] proposed a method for semi-supervised learning based on TWD; Yu et al. [88] proposed a three-way clustering approach for semi-supervised learning that uses an active learning approach to obtain labels for instances that are considered as uncertain after the initial clustering; Triff et al. [66] proposed an evolutionary semi-supervised algorithm based on rough sets and TWD and compare it with other algorithms obtaining interesting results when only the certainly classified objects are considered; Dai et al. [18] introduced a co-training technique for cost-sensitive semi-supervised learning based on sequential TWD and apply it to different standard ML algorithms (k-NN, PCA, LDA) in order to obtain a multi-view dataset; Campagner et al. [10,13] introduced a three-way Decision Tree model for semi-supervised learning and show that this model achieves good performance with respect to standard ML algorithms for semi-supervised learning; Wang et al. [70,71] proposed a cost-sensitive three-way active learning algorithm based on the computation of label error statistics; Min et al. [49] proposed a cost-sensitive active learning strategy based on k-nearest neighbours and a tripartition of the instances in certain and uncertain ones.

In the case of more general weakly supervised learning, Campagner et al. [12] proposed a collection of approaches based on TWD and standard ML algorithms in order to take into account this type of uncertainty in the setting of *classification*. In particular, the authors considered an algorithm for Decision Tree (and ensemble-based extensions, such as Random Forest) learning, in which the trisecting and acting steps are dynamically and iteratively performed during the Decision Tree induction process on the basis of TWD and generalized information theory [33], and a generalized stochastic gradient descent algorithm based on interval analysis and TWD, in order to take into account the fact that the uncertain instances naturally determine interval-valued information with respect to the loss function to be optimized. In both cases, promising results were reported, showing that they outperform standard superset learning and semi-supervised techniques. A different approach, which is based on treating weakly supervision as a type of missing data, proposed by Sakai et al. [58], employs a three-way rule extraction algorithm that could also be applied in the case of weakly supervised data: this approach is of particular interest in that it suggests an integrated end-to-end approach to simultaneously handle missing data and weakly supervised data.

Multi-rater Annotation. With respect to the third type of uncertainty, that is multi-rater annotation, in [12] we noted that the issue has largely been ignored in the ML community. With respect to the application of TWD methodologies to handle this type of uncertainty, there has been some recent works with respect to aggregation methods and information fusion using TWD, mainly under the perspective of group decision making [25,39,53,96] and the modelling of multi-agent systems [76]. However, there has been so far a lack of studies concerning the application of these TWD based techniques to ML problems. Some related approaches have been explored under the perspective of *multi-source information tables* in RST, in which the multi-rater, and possibly conflicting, information is available not only for the decision variable but also for the predictor ones: Huang et al. [28] proposed a three-way concept learning method for multi-source data; Sang et al. [60] studied the application of decision-theoretic rough sets for TWD in multi-source information systems; Sang et al. [61] proposed an alternative approach which is not directly based on merging different information systems but instead it employs multi-granulation double-quantitative decision-theoretic rough set, which the authors show to be more fault tolerant with respect to traditional approaches. Campagner et al. [8,15] proposed a novel aggregation strategy, based on TWD, which can be applied to implement the trisecting step to handle the multi-rater annotation uncertainty type. In this case, the instances are categorized as certain or uncertain depending on the distribution of labels given by the raters and a set of parameters that have a cost-theoretic interpretation. After the aggregation step, the problem is converted into a weakly supervised one and a learning algorithm is proposed that is shown to be significantly more effective than the traditional approach of simply assigning the most frequent labels (among the multi-rater annotations) to the uncertain instances.

2.2 Outcome Step: Evaluating the Results

All of the articles considered for this review mainly deal with the trisecting and acting step in the TAO model that we propose. The outcome step has rarely been considered and is usually addressed as it would be for traditional ML models: that is by simply considering the accuracy of the trained models, sometimes even in naive ways [14]. According to the framework that we propose, the main goal of employing TWD for ML is the handling of uncertainty. In this light, attention should also be placed on how much the TWD approach allows to reduce the initial uncertainty in the input data or at least to which degree the TWD-based algorithm is able to obtain good performances despite of the uncertainty. For example, with respect to the missing data problem, the outcome step should also consider the amount of missing values that have been correctly imputed (for imputation-based approaches), or the *robustness* of the induced ML algorithm with respect to different values that could be present in the missing features, for instance using interval-valued accuracy or information-theoretic metrics [11,14], or by distinguishing which predictions made by the algorithm are certain (i.e., robust with respect to the missing values or the weakly supervised instance) or only possible. Similarly, with respect to the multi-rater annotation uncertainty type, besides the accuracy of the proposed approaches with respect to a known ground truth (when available), the outcome step should also consider the robustness of the proposed approach when varying the degree of conflicting information, and the level of noise of the raters who annotate the datasets, as we considered in [15]. In this sense, we believe that more attention should be put on the outcome step of the proposed framework, and further research in this sense should be performed.

3 Handling Uncertainty in the Output

The application of TWD to handle uncertainty in the output of the ML is a mature research area, and has possibly been considered since the original proposal of TWD, both for classification [80,103] and for clustering [40]. In both cases, the uncertainty in the output of the ML model refers to the inability of the ML model to properly discriminate the instances and assign them a certain, precisely known, label. This could be due to a variety of issues: the chosen data representation (i.e., the selected features and/or their level of granularity) is not informative enough; the inability to distinguish different instances that are either identical or "too near" in the sample space, but are associated with different decision labels; the selected model class is not powerful enough to properly represent the concept to be learned. All these issues have been widely studied, both under the perspective of RST with the notion of *indiscernibility* [54,55], and of more traditional ML approaches, with the notion of *decision boundary*. The approach suggested by TWD in this setting consists in allowing the classifier to abstain [81], even partially, that is excluding some of the possible alternative classifications. In so doing, the focus is on the trisecting step, which involves deciding on which instances the ML model (both for classification or clustering) should be considered uncertain, and hence the model should abstain on.

3.1 Trisecting and Acting Steps for Classification

With respect to classification, the traditional model of TWD applies only to binary classification cases, for which a third "uncertain" category is added, for which extensions of the most traditional ML methods are available. In all of the cases, the trisecting step is performed in a similar manner, on the basis of the original decision-theoretic rules proposed by Yao [81]; these rules are often embedded in different models, and the main variation relates to how the *acting step* is implemented. This step has usually been based on Bayesian decision analysis under the decision-theoretic rough set paradigm [31,36,79,103,104]. However, also other approaches to implement the acting step have been proposed, such as *structured approximations* in RST [27], or the combination of TWD with more traditional ML techniques, for instance, Deep Learning [37,100,101], optimization-based learning [41,43,95] or frequent pattern mining [38,50]: all of these implementations of the TWD model for the handling of uncertainty have been successfully applied to different fields, such as face recognition, spam filtering or recommender systems.

A particularly interesting use, with respect to the acting outcome, consists of integrating TWD in active learning methodologies: Chen et al. [16] proposed a three-way rule-based decision algorithm that employs active learning to re-classify the uncertain instances; Zhang et al. [94] proposed a random forest-based recommender systems with the capability to ask for user supervision on uncertain objects; Yao et al. [78] proposed a TWD model based on a game-theoretic rough set for medical decision systems that distinguish *certain rules* (for acceptance and rejection) from *deferment rules* which require intervention from the user.

In recent years, different proposals have also been considered for the extension to the multi-class case, mainly under two major approaches. The first one is based on sequential TWD [83], which essentially implements a hierarchical one-vs-all learning scheme; Yang et al. [77] considered a Bayesian extension of multi-class decision theoretic rough sets [102]; Savchenko [62,63] proposed sequential TWD and granular computing for speed-up of image classification when the number of classes is large; Zhang et al. [98] proposed a sequential TWD model based on the use of autoencoders for granular feature extraction. The second approach, which can be defined as natively multi-class, has been proposed by some authors (e.g., in [11,12]): it employs a decision-theoretic procedure to convert every standard probabilistic classifier into a multi-class TWD classifier. A similar approach, but based on decision-theoretic rough sets, have also been developed by Jia et al. [32].

While all the approaches mentioned above consider the combination of TWD and ML models in a *a posteriori* strategy in which the trisecting step is performed after, or as a consequence of, the standard ML training procedure, in [11,12] we also considered how to directly embed TWD in the training algorithm of a wide class of standard ML models, either by a direct modification of the learning algorithm (for decision trees and related methods), or by adopting ad-hoc regularized loss functions (for optimization-based procedures such as SVM or logistic regression).

3.2 Trisecting and Acting Steps for Clustering

In regards clustering, various approaches have been proposed to implement the TWD-based handling of the uncertainty in the output, hence to construct clusterings in which the assignment of some instances to clusters is uncertain, mainly under the frameworks of *rough clustering* [40], *interval-set clustering* [84] and *three-way clustering* [90]. In all of the above approaches, the trisecting step is implemented as a modification of standard clustering assignment criteria, and it allows instances to be considered as *uncertain* with respect to their assignment to one or more clusters: Yu [90] proposed a three-way clustering algorithm that also works with incomplete data; Wang et al. [73] proposed a three-way clustering method based on mathematical morphology; Yu et al. [91] considered a flexible tree-based incremental three-way clustering algorithm; Yu et al. [86] proposed an optimized ensemble-based three-way clustering algorithm for large-scale datasets; Afridi et al. [1] proposed a variance-based three-way clustering algorithm; Zhang et al. [99] proposed a novel improvement on the original rough k-means based on a weighted Guassian distance function; Li et al. [35] extended standard rough k-means with an approach based on decision-theoretic rough sets, Yu et al. [89] proposed an hybrid clustering/active learning based on TWD for multi-view data; Zhang [97] proposed a three-way c-means algorithm; Wang et al. [72] proposed a refinement three-way clustering algorithm based on the re-clustering of ensemble of traditional hard clustering algorithms; Yu et al. [93] proposed a density three-way clustering algorithm based on DBscan; Yu et al. [92] proposed a three-way clustering algorithm optimized for high-dimensionality datasets based on a modification of the k-medoids algorithm and the random projection method; Hu et al. [26] proposed a sequential TWD model for consensus clustering based on the notion of co-association matrix.

3.3 Outcome Step: Evaluating the Results

With respect to the outcome step, both clustering and classification techniques based on TWD have been shown to significantly improve the performance in comparison to traditional ML algorithms (see the referenced literature). Despite this promising assessment, one should also consider that the evaluation of ML algorithms using TWD to handle the uncertainty in output, at least in principle, cannot be made on the same grounds of traditional ML models (i.e., only on the basis of accuracy metrics). Indeed, since these models are allowed to *abstain* on uncertain instances, metrics for their evaluation should take into account the trade-off between the accuracy on the classified/clustered instances but also the *coverage* of the algorithm, that is on how many instances the model defers its decision. As an example of this issue, suffice it to consider that a three-way classifier that abstains on all the instances but one, which is correctly classified/clustered, has perfect accuracy but it is hardly a useful predictive model. However, attention towards this trade-off has emerged only recently, where the majority of the surveyed papers only focus on the accuracy of the models on the classified/clustered instances: Peters [56] proposed a modified Davis-Bouldin index for evaluation of three-way clustering; Depaolini et al. [19] proposed generalizations of Rand, Jaccard and Fowlkes-Mallows indices; similarly, we proposed a generalization of information-theoretic measures of clustering quality [14] and generalization of accuracy metrics for classification [11]. Promisingly, the superior performance of TWD techniques for the handling of output uncertainty can be observed also under these more robust, and conservative, metrics.

4   Discussion

In this article, we proposed a TAO model for the management of uncertainty in Machine Learning that is based on TWD. After describing the proposed framework, we have reviewed the current state of the art for the different areas of concern identified by our framework, and discussed about the strengths, limitations and areas requiring further investigation of the main works considered.

In what follows, we emphasise both what we believe are the main advantages of adopting this methodology in ML and also delineate some topics that in our opinion are particularly in need of further study.

4.1   Advantages of Three-Way ML

It is undeniable that in the recent years, the application of TWD and the TAO model to ML applications has been growing and showing promising results. In this Section, we will emphasise the advantages of TWD under the perspective of uncertainty handling for ML. In this perspective, TWD and the TAO model look promising as a means to provide a principled way to handle uncertainty in the ML process in an end-to-end fashion, by directly using the information obtained in the *trisecting* act (i.e., the splitting of instances into certain/uncertain ones), in the subsequent *acting* and *outcome* steps, without the need to address and "correct" the uncertainty in a separate pre-processing step. This is particularly clear in our discussion about the *handling of the uncertainty in the input*: in this case, the TAO model enables one to directly deal with different forms of data uncertainty in a theoretically-sound, robust and *non-invasive* manner [20], while also obtaining higher predictive accuracy than with traditional ML methodologies. The same holds true also with respect to the handling of uncertainty in the output. In this case, the TAO model allows to obtain classifiers that are both more accurate and robust, thanks to the possibility of abstention that allows more conservative decision boundaries. Abstention is a more *informative* strategy also from a decision-support perspective, in that the model that is enhanced with TWD can expose its predictive uncertainty by abstaining as a sign that the situation needs more information, or the careful consideration of the human decision maker.

4.2   Future Directions

Despite the increasing popularity of TWD to handle the uncertainty in ML pipelines, and the relative maturity of the application of this methodology with respect to the trisecting and acting steps of our framework (see Table 1), we believe that some specific aspects merit further investigations. Then, as already discussed in Sects. 2 and 3, the outcome step has not been sufficiently explored, especially with respect to the handling of uncertainty in the input. As discussed in Sect. 2.2, we believe that conceiving appropriate metrics to assess the robustness of TWD methods represents a particularly promising strand of research, which would also enable *counterfactual*-like reasoning [47] for ML models, a topic that has recently been considered important in the light of *eXplainable AI* [68]. For instance, this can be done by analyzing the robustness and performance of the ML models with respect to specific counterfactual instantiations of the instances affected by uncertainty that would most likely alter the learnt decision boundary. Similarly, while there have been more proposals for the outcome step for the output part of our framework, we believe that further work should be done towards the general adoption of these measures in the application of TWD-based ML. Similarly, a second promising direction of research regards the acting step for the management of the uncertainty in the output: as we previously discussed, active learning and *human-in-the-loop* [24] techniques to handle the instances recognized as uncertain by the ML algorithms are of particular interest. Similarly, it would be interesting to study the connection between the TWD model to handle uncertainty in the output and the *conformal prediction* paradigm [64], as both are based on the idea of providing set-valued predictions on uncertain instances. A third research direction regards the fact that the different steps have currently been studied mostly in isolation: so far, most studies applying TWD in ML focused either on the input *or* the output part of our framework. While some initial works with respect to a unified treatment of both types of uncertainty have recently been considered [12], we believe that further work toward such a uniform methodology would be particularly promising. Finally, missing data is usually understood as a problem of completeness: this is missing data at feature level, for instances at least partly observed. But there is also a "missingness" at row level, that is a source of uncertainty (which makes the data we have uncertain and less reliable) that regards instances that we have not observed or whose characteristics are not well represented in the data collected: more research is due to how TWD can tackle this important source of bias, which is usually called sampling bias.

References

1. Afridi, M.K., Azam, N., Yao, J.: Variance based three-way clustering approaches for handling overlapping clustering. IJAR **118**, 47–63 (2020)
2. Afridi, M.K., Azam, N., Yao, J., et al.: A three-way clustering approach for handling missing data using GTRS. IJAR **98**, 11–24 (2018)
3. Agrawal, R., Srikant, R., et al.: Fast algorithms for mining association rules. In: Proceedings of the 20th International Conference on Very Large Data Bases, VLDB, vol. 1215, pp. 487–499 (1994)
4. Amiri, M., Jensen, R.: Missing data imputation using fuzzy-rough methods. Neurocomputing **205**, 152–164 (2016)
5. Awasthi, P., Blum, A., Haghtalab, N., et al.: Efficient PAC learning from the crowd. arXiv preprint arXiv:1703.07432 (2017)
6. Brown, M.L., Kros, J.F.: Data mining and the impact of missing data. Ind. Manag. Data Syst. **103**(8), 611–621 (2003)
7. Buuren, S.V., Groothuis-Oudshoorn, K.: Mice: multivariate imputation by chained equations in R. J. Stat. Softw. **45**(3), 1–67 (2010)
8. Cabitza, F., Campagner, A., Ciucci, D.: New frontiers in explainable AI: understanding the GI to interpret the GO. In: Holzinger, A., Kieseberg, P., Tjoa, A.M., Weippl, E. (eds.) CD-MAKE 2019. LNCS, vol. 11713, pp. 27–47. Springer, Cham (2019). https://doi.org/10.1007/978-3-030-29726-8_3
9. Cabitza, F., Locoro, A., Alderighi, C., et al.: The elephant in the record: on the multiplicity of data recording work. Health Inform. J. **25**(3), 475–490 (2019)
10. Campagner, A., Cabitza, F., Ciucci, D.: Exploring medical data classification with three-way decision tree. In: Proceedings of BIOSTEC 2019 - Volume 5: HEALTHINF, pp. 147–158. SCITEPRESS (2019)
11. Campagner, A., Cabitza, F., Ciucci, D.: Three-way classification: ambiguity and abstention in machine learning. In: Mihálydeák, T., et al. (eds.) IJCRS 2019. LNCS (LNAI), vol. 11499, pp. 280–294. Springer, Cham (2019). https://doi.org/10.1007/978-3-030-22815-6_22
12. Campagner, A., Cabitza, F., Ciucci, D.: The three-way-in and three-way-out framework to treat and exploit ambiguity in data. IJAR **119**, 292–312 (2020)
13. Campagner, A., Ciucci, D.: Three-way and semi-supervised decision tree learning based on orthopartitions. In: Medina, J., et al. (eds.) IPMU 2018. CCIS, vol. 854, pp. 748–759. Springer, Cham (2018). https://doi.org/10.1007/978-3-319-91476-3_61
14. Campagner, A., Ciucci, D.: Orthopartitions and soft clustering: soft mutual information measures for clustering validation. Knowl.-Based Syst. **180**, 51–61 (2019)
15. Campagner, A., Ciucci, D., Svensson, C.M., et al.: Ground truthing from multi-rater labelling with three-way decisions and possibility theory. IEEE Trans. Fuzzy Syst. (2020, submitted)
16. Chen, Y., Yue, X., Fujita, H., et al.: Three-way decision support for diagnosis on focal liver lesions. Knowl.-Based Syst. **127**, 85–99 (2017)
17. Cour, T., Sapp, B., Taskar, B.: Learning from partial labels. J. Mach. Learn. Res. **12**, 1501–1536 (2011)
18. Dai, D., Zhou, X., Li, H., et al.: Co-training based sequential three-way decisions for cost-sensitive classification. In: 2019 IEEE 16th ICNSC, pp. 157–162 (2019)
19. Depaolini, M.R., Ciucci, D., Calegari, S., Dominoni, M.: External indices for rough clustering. In: Nguyen, H.S., Ha, Q.-T., Li, T., Przybyła-Kasperek, M. (eds.) IJCRS 2018. LNCS (LNAI), vol. 11103, pp. 378–391. Springer, Cham (2018). https://doi.org/10.1007/978-3-319-99368-3_29

20. Düntsch, I., Gediga, G.: Rough set data analysis–a road to non-invasiveknowledge discovery. Methodos (2000)
21. Greco, S., Matarazzo, B., Slowinski, R.: Dealing with missing data in rough set analysis of multi-attribute and multi-criteria decision problems. In: Zanakis, S.H., Doukidis, G., Zopounidis, C. (eds.) Decision Making: Recent Developments and Worldwide Applications, pp. 295–316. Springer, Boston (2000). https://doi.org/10.1007/978-1-4757-4919-9_20
22. Grzymala-Busse, J.W., Hu, M.: A comparison of several approaches to missing attribute values in data mining. In: Ziarko, W., Yao, Y. (eds.) RSCTC 2000. LNCS (LNAI), vol. 2005, pp. 378–385. Springer, Heidelberg (2001). https://doi.org/10.1007/3-540-45554-X_46
23. Heinecke, S., Reyzin, L.: Crowdsourced PAC learning under classification noise. In: Proceedings of AAAI HCOMP 2019, vol. 7, pp. 41–49 (2019)
24. Holzinger, A.: Interactive machine learning for health informatics: when do we need the human-in-the-loop? Brain Inf. 3(2), 119–131 (2016)
25. Hu, B.Q., Wong, H., Yiu, K.F.C.: The aggregation of multiple three-way decision spaces. Knowl.-Based Syst. 98, 241–249 (2016)
26. Hu, M., Deng, X., Yao, Y.: A sequential three-way approach to constructing a co-association matrix in consensus clustering. In: Nguyen, H.S., Ha, Q.-T., Li, T., Przybyła-Kasperek, M. (eds.) IJCRS 2018. LNCS (LNAI), vol. 11103, pp. 599–613. Springer, Cham (2018). https://doi.org/10.1007/978-3-319-99368-3_47
27. Hu, M., Yao, Y.: Structured approximations as a basis for three-way decisions in rough set theory. Knowl.-Based Syst. 165, 92–109 (2019)
28. Huang, C., Li, J., Mei, C., et al.: Three-way concept learning based on cognitive operators: an information fusion viewpoint. IJAR 83, 218–242 (2017)
29. Hüllermeier, E., Cheng, W.: Superset learning based on generalized loss minimization. In: Appice, A., Rodrigues, P.P., Santos Costa, V., Gama, J., Jorge, A., Soares, C. (eds.) ECML PKDD 2015. LNCS (LNAI), vol. 9285, pp. 260–275. Springer, Cham (2015). https://doi.org/10.1007/978-3-319-23525-7_16
30. Imieliński, T., Lipski Jr., W.: Incomplete information in relational databases. J. ACM 31(4), 761–791 (1984)
31. Jia, X., Deng, Z., Min, F., Liu, D.: Three-way decisions based feature fusion for chinese irony detection. IJAR 113, 324–335 (2019)
32. Jia, X., Li, W., Shang, L.: A multiphase cost-sensitive learning method based on the multiclass three-way decision-theoretic rough set model. Inf. Sci. 485, 248–262 (2019)
33. Klir, G.J., Wierman, M.J.: Uncertainty-based information: elements of generalized information theory, vol. 15. Physica (2013)
34. Li, D., Deogun, J., Spaulding, W., Shuart, B.: Towards missing data imputation: a study of fuzzy k-means clustering method. In: Tsumoto, S., Słowiński, R., Komorowski, J., Grzymała-Busse, J.W. (eds.) RSCTC 2004. LNCS (LNAI), vol. 3066, pp. 573–579. Springer, Heidelberg (2004). https://doi.org/10.1007/978-3-540-25929-9_70
35. Li, F., Ye, M., Chen, X.: An extension to rough c-means clustering based on decision-theoretic rough sets model. IJAR 55(1), 116–129 (2014)
36. Li, H., Zhang, L., Huang, B., et al.: Sequential three-way decision and granulation for cost-sensitive face recognition. Knowl.-Based Syst. 91, 241–251 (2016)
37. Li, H., Zhang, L., Zhou, X., et al.: Cost-sensitive sequential three-way decision modeling using a deep neural network. IJAR 85, 68–78 (2017)

38. Li, Y., Zhang, Z.H., Chen, W.B., et al.: TDUP: an approach to incremental mining of frequent itemsets with three-way-decision pattern updating. IJMLC **8**(2), 441–453 (2017). https://doi.org/10.1007/s13042-015-0337-6

39. Liang, D., Pedrycz, W., Liu, D., Hu, P.: Three-way decisions based on decision-theoretic rough sets under linguistic assessment with the aid of group decision making. Appl. Soft Comput. **29**, 256–269 (2015)

40. Lingras, P., West, C.: Interval set clustering of web users with rough k-means. Technical report 2002-002, Department of Mathematics and Computing Science, St. Mary's University, Halifax, NS, Canada (2002)

41. Liu, D., Li, T., Liang, D.: Incorporating logistic regression to decision-theoretic rough sets for classifications. IJAR **55**(1), 197–210 (2014)

42. Liu, D., Liang, D., Wang, C.: A novel three-way decision model based on incomplete information system. Knowl.-Based Syst. **91**, 32–45 (2016). Three-way Decisions and Granular Computing

43. Liu, J., Li, H., Zhou, X., et al.: An optimization-based formulation for three-way decisions. Inf. Sci. **495**, 185–214 (2019)

44. Luo, C., Li, T., Huang, Y., et al.: Updating three-way decisions in incomplete multi-scale information systems. Inf. Sci. **476**, 274–289 (2019)

45. Luo, J., Fujita, H., Yao, Y., Qin, K.: On modeling similarity and three-way decision under incomplete information in rough set theory. Knowl.-Based Syst. **191**, 105251 (2020)

46. Ma, M.: Advances in three-way decisions and granular computing. Knowl.-Based Syst. **91**, 1–3 (2016)

47. Mandel, D.R.: Counterfactual and causal explanation: from early theoretical views to new frontiers. In: The Psychology of Counterfactual Thinking, pp. 23–39. Routledge (2007)

48. Miao, D., Gao, C., Zhang, N.: Three-way decisions-based semi-supervised learning. In: Theory and Applications of Three-Way Decisions, pp. 17–33 (2012)

49. Min, F., Liu, F.L., Wen, L.Y., et al.: Tri-partition cost-sensitive active learning through kNN. Soft. Comput. **23**(5), 1557–1572 (2019)

50. Min, F., Zhang, Z.H., Zhai, W.J., et al.: Frequent pattern discovery with tri-partition alphabets. Inf. Sci. **507**, 715–732 (2020)

51. Nelwamondo, F.V., Marwala, T.: Rough set theory for the treatment of incomplete data. In: 2007 IEEE International Fuzzy Systems Conference, pp. 1–6. IEEE (2007)

52. Nowicki, R.K., Grzanek, K., Hayashi, Y.: Rough support vector machine for classification with interval and incomplete data. J. Artif. Intell. Soft Comput. Res. **10**(1), 47–56 (2020)

53. Pang, J., Guan, X., Liang, J., Wang, B., Song, P.: Multi-attribute group decision-making method based on multi-granulation weights and three-way decisions. IJAR **117**, 122–147 (2020)

54. Pawlak, Z.: Rough Sets: Theoretical Aspects of Reasoning About Data. Kluwer (1991)

55. Pawlak, Z., Skowron, A.: Rough sets: some extensions. Inf. Sci. **177**(1), 28–40 (2007)

56. Peters, G.: Rough clustering utilizing the principle of indifference. Inf. Sci. **277**, 358–374 (2014)

57. Sakai, H., Nakata, M.: Rough set-based rule generation and apriori-based rule generation from table data sets: a survey and a combination. CAAI Trans. Intell. Technol. **4**(4), 203–213 (2019)

58. Sakai, H., Nakata, M., Watada, J.: NIS-apriori-based rule generation with three-way decisions and its application system in SQL. Inf. Sci. **507**, 755–771 (2020)
59. Sakai, H., Nakata, M., Yao, Y.: Pawlak's many valued information system, non-deterministic information system, and a proposal of new topics on information incompleteness toward the actual application. In: Wang, G., Skowron, A., Yao, Y., Ślęzak, D., Polkowski, L. (eds.) Thriving Rough Sets. SCI, vol. 708, pp. 187–204. Springer, Cham (2017). https://doi.org/10.1007/978-3-319-54966-8_9
60. Sang, B., Guo, Y., Shi, D., et al.: Decision-theoretic rough set model of multi-source decision systems. IJMLC **9**(11), 1941–1954 (2018)
61. Sang, B., Yang, L., Chen, H., et al.: Generalized multi-granulation double-quantitative decision-theoretic rough set of multi-source information system. IJAR **115**, 157–179 (2019)
62. Savchenko, A.V.: Fast multi-class recognition of piecewise regular objects based on sequential three-way decisions and granular computing. Knowl.-Based Syst. **91**, 252–262 (2016)
63. Savchenko, A.V.: Sequential three-way decisions in multi-category image recognition with deep features based on distance factor. Inf. Sci. **489**, 18–36 (2019)
64. Shafer, G., Vovk, V.: A tutorial on conformal prediction. J. Mach. Learn. Res. **9**(Mar), 371–421 (2008)
65. Tian, J., Yu, B., Yu, D., Ma, S.: Missing data analyses: a hybrid multiple imputation algorithm using gray system theory and entropy based on clustering. Appl. Intell. **40**(2), 376–388 (2014)
66. Triff, M., Wiechert, G., Lingras, P.: Nonlinear classification, linear clustering, evolutionary semi-supervised three-way decisions: a comparison. FUZZ-IEEE **2017**, 1–6 (2017)
67. W. Grzymala-Busse, J.: Rough set strategies to data with missing attribute values. In: Proceedings of ISMIS 2005, vol. 542, pp. 197–212 (2005)
68. Wachter, S., Mittelstadt, B., Russell, C.: Counterfactual explanations without opening the black box: automated decisions and the GDPR. Harv. JL Tech. **31**, 841 (2017)
69. Wang, L., Zhou, Z.H.: Cost-saving effect of crowdsourcing learning. In: IJCAI, pp. 2111–2117 (2016)
70. Wang, M., Fu, K., Min, F., Jia, X.: Active learning through label error statistical methods. Knowl.-Based Syst. **189**, 105140 (2020)
71. Wang, M., Lin, Y., Min, F., Liu, D.: Cost-sensitive active learning through statistical methods. Inf. Sci. **501**, 460–482 (2019)
72. Wang, P., Liu, Q., Yang, X., Xu, F.: Ensemble re-clustering: refinement of hard clustering by three-way strategy. In: Sun, Y., Lu, H., Zhang, L., Yang, J., Huang, H. (eds.) IScIDE 2017. LNCS, vol. 10559, pp. 423–430. Springer, Cham (2017). https://doi.org/10.1007/978-3-319-67777-4_37
73. Wang, P., Yao, Y.: Ce3: a three-way clustering method based on mathematical morphology. Knowl.-Based Syst. **155**, 54–65 (2018)
74. Yang, L., Hou, K.: A method of incomplete data three-way clustering based on density peaks. In: AIP Conference Proceedings, vol. 1967, p. 020008. AIP Publishing LLC (2018)
75. Yang, X., Tan, A.: Three-way decisions based on intuitionistic fuzzy sets. In: Polkowski, L., Yao, Y., Artiemjew, P., Ciucci, D., Liu, D., Ślęzak, D., Zielosko, B. (eds.) IJCRS 2017. LNCS (LNAI), vol. 10314, pp. 290–299. Springer, Cham (2017). https://doi.org/10.1007/978-3-319-60840-2_21
76. Yang, X., Yao, J.: Modelling multi-agent three-way decisions with decision-theoretic rough sets. Fundam. Inform. **115**(2–3), 157–171 (2012)

77. Yang, X., Li, T., Fujita, H., Liu, D.: A sequential three-way approach to multi-class decision. IJAR **104**, 108–125 (2019)

78. Yao, J., Azam, N.: Web-based medical decision support systems for three-way medical decision making with game-theoretic rough sets. IEEE Trans. Fuzzy Syst. **23**(1), 3–15 (2014)

79. Yao, Y.: Three-way decision: an interpretation of rules in rough set theory. In: Wen, P., Li, Y., Polkowski, L., Yao, Y., Tsumoto, S., Wang, G. (eds.) RSKT 2009. LNCS (LNAI), vol. 5589, pp. 642–649. Springer, Heidelberg (2009). https://doi.org/10.1007/978-3-642-02962-2_81

80. Yao, Y.: Three-way decisions with probabilistic rough sets. Inf. Sci. **180**(3), 341–353 (2010)

81. Yao, Y.: An outline of a theory of three-way decisions. In: Yao, J.T., et al. (eds.) RSCTC 2012. LNCS (LNAI), vol. 7413, pp. 1–17. Springer, Heidelberg (2012). https://doi.org/10.1007/978-3-642-32115-3_1

82. Yao, Y.: Three-way decision and granular computing. Int. J. Approx. Reason. **103**, 107–123 (2018)

83. Yao, Y., Deng, X.: Sequential three-way decisions with probabilistic rough sets. In: Proceedings of IEEE ICCI-CC 2011, pp. 120–125. IEEE (2011)

84. Yao, Y., Lingras, P., Wang, R., Miao, D.: Interval set cluster analysis: a re-formulation. In: Sakai, H., Chakraborty, M.K., Hassanien, A.E., Ślęzak, D., Zhu, W. (eds.) RSFDGrC 2009. LNCS (LNAI), vol. 5908, pp. 398–405. Springer, Heidelberg (2009). https://doi.org/10.1007/978-3-642-10646-0_48

85. Yu, H.: A framework of three-way cluster analysis. In: Polkowski, L., et al. (eds.) IJCRS 2017. LNCS (LNAI), vol. 10314, pp. 300–312. Springer, Cham (2017). https://doi.org/10.1007/978-3-319-60840-2_22

86. Yu, H., Chen, Y., Lingras, P., et al.: A three-way cluster ensemble approach for large-scale data. IJAR **115**, 32–49 (2019)

87. Yu, H., Su, T., Zeng, X.: A three-way decisions clustering algorithm for incomplete data. In: Miao, D., Pedrycz, W., Ślęzak, D., Peters, G., Hu, Q., Wang, R. (eds.) RSKT 2014. LNCS (LNAI), vol. 8818, pp. 765–776. Springer, Cham (2014). https://doi.org/10.1007/978-3-319-11740-9_70

88. Yu, H., Wang, X., Wang, G.: A semi-supervised three-way clustering framework for multi-view data. In: Polkowski, L., et al. (eds.) IJCRS 2017. LNCS (LNAI), vol. 10314, pp. 313–325. Springer, Cham (2017). https://doi.org/10.1007/978-3-319-60840-2_23

89. Yu, H., Wang, X., Wang, G., et al.: An active three-way clustering method via low-rank matrices for multi-view data. Inf. Sci. **507**, 823–839 (2020)

90. Yu, H., Wang, Y.: Three-way decisions method for overlapping clustering. In: Yao, J.T., et al. (eds.) RSCTC 2012. LNCS (LNAI), vol. 7413, pp. 277–286. Springer, Heidelberg (2012). https://doi.org/10.1007/978-3-642-32115-3_33

91. Yu, H., Zhang, C., Wang, G.: A tree-based incremental overlapping clustering method using the three-way decision theory. Knowl.-Based Syst. **91**, 189–203 (2016)

92. Yu, H., Zhang, H.: A three-way decision clustering approach for high dimensional data. In: Flores, V., et al. (eds.) IJCRS 2016. LNCS (LNAI), vol. 9920, pp. 229–239. Springer, Cham (2016). https://doi.org/10.1007/978-3-319-47160-0_21

93. Yu, H., Chen, L., Yao, J., et al.: A three-way clustering method based on an improved dbscan algorithm. Phys. A **535**, 122289 (2019)

94. Zhang, H.R., Min, F.: Three-way recommender systems based on random forests. Knowl.-Based Syst. **91**, 275–286 (2016)

95. Zhang, H.R., Min, F., Shi, B.: Regression-based three-way recommendation. Inf. Sci. **378**, 444–461 (2017)

96. Zhang, H.Y., Yang, S.Y.: Three-way group decisions with interval-valued decision-theoretic rough sets based on aggregating inclusion measures. IJAR **110**, 31–45 (2019)

97. Zhang, K.: A three-way c-means algorithm. Appl. Soft Comput. **82**, 105536 (2019)

98. Zhang, L., Li, H., Zhou, X., et al.: Sequential three-way decision based on multi-granular autoencoder features. Inf. Sci. **507**, 630–643 (2020)

99. Zhang, T., Ma, F.: Improved rough k-means clustering algorithm based on weighted distance measure with gaussian function. Int. J. Comput. Math. **94**(4), 663–675 (2017)

100. Zhang, Y., Miao, D., Wang, J., et al.: A cost-sensitive three-way combination technique for ensemble learning in sentiment classification. IJAR **105**, 85–97 (2019)

101. Zhang, Y., Zhang, Z., Miao, D., et al.: Three-way enhanced convolutional neural networks for sentence-level sentiment classification. Inf. Sci. **477**, 55–64 (2019)

102. Zhou, B.: Multi-class decision-theoretic rough sets. IJAR **55**(1), 211–224 (2014)

103. Zhou, B., Yao, Y., Luo, J.: A three-way decision approach to email spam filtering. In: Farzindar, A., Kešelj, V. (eds.) AI 2010. LNCS (LNAI), vol. 6085, pp. 28–39. Springer, Heidelberg (2010). https://doi.org/10.1007/978-3-642-13059-5_6

104. Zhou, B., Yao, Y., Luo, J.: Cost-sensitive three-way email spam filtering. JIIS **42**(1), 19–45 (2014)

105. Zhou, Z.H.: A brief introduction to weakly supervised learning. Natl. Sci. Rev. **5**(1), 44–53 (2018)

Three-Way Decisions Community Detection Model Based on Weighted Graph Representation

Jie Chen[1], Yang Li[1], Shu Zhao[1(✉)], Xiangyang Wang[2], and Yanping Zhang[1]

[1] School of Computer Science and Technology, Anhui University,
Hefei 230601, Anhui, People's Republic of China
`zhaoshuzs2002@hotmail.com`
[2] Anhui Electrical Engineering Professional Technique College,
Hefei 230051, Anhui, People's Republic of China

Abstract. Community detection is of great significance to the study of complex networks. Community detection algorithm based on three-way decisions (TWD) forms a multi-layered community structure by hierarchical clustering and then selects a suitable layer as the community detection result. However, this layer usually contains overlapping communities. Based on the idea of TWD, we define the overlapping part in the communities as boundary region (BND), and the non-overlapping part as positive region (POS) or negative region (NEG). How to correctly divide the nodes in the BND into the POS or NEG is a challenge for three-way decisions community detection. The general methods to deal with boundary region are modularity increment and similarity calculation. But these methods only take advantage of the local features of the network, without considering the information of the divided communities and the similarity of the global structure. Therefore, in this paper, we propose a method for three-way decisions community detection based on weighted graph representation (WGR-TWD). The weighted graph representation (WGR) can well transform the global structure into vector representation and make the two nodes in the boundary region more similar by using frequency of appearing in the same community as the weight. Firstly, the multi-layered community structure is constructed by hierarchical clustering. The target layer is selected according to the extended modularity value of each layer. Secondly, all nodes are converted into vectors by WGR. Finally, the nodes in the BND are divided into the POS or NEG based on cosine similarity. Experiments on real-world networks demonstrate that WGR-TWD is effective for community detection in networks compared with the state-of-the-art algorithms.

Keywords: Community detection · Hierarchical clustering · Three-way decisions · Weighted graph representation

© Springer Nature Switzerland AG 2020
R. Bello et al. (Eds.): IJCRS 2020, LNAI 12179, pp. 153–165, 2020.
https://doi.org/10.1007/978-3-030-52705-1_11

1   Introduction

Nowadays, there are all kinds of complex systems with specific functions in the real world such as online social systems, medical systems and computer systems. These systems can be abstracted into networks with complex internal structures, called complex networks. The research of complex networks has received more and more attention due to the development of the Internet. Community structure [6,23] is a common feature of complex networks, which means that a network consists of several communities, the connections between communities are sparse and the connections within a community are dense [10]. Mining the community structure in the network is of great significance to understand the network structure, analyze the network characteristics and predict the network behavior. Thus, community detection has become one of the most important issues in the study of complex networks.

In recent years, a great deal of research is devoted to community detection in networks. Most community detection methods are used to identify non-overlapping communities (i.e., a node belongs to only one community). The main approaches include graph partitioning and clustering [9,10,13], modularity maximization [1,20], information theory [12,25] and non-negative matrix factorization [16,27]. The Kernighan-Lin algorithm [13] is a heuristic graph partitioning method that detects communities by optimizing the edges within and between communities. GN algorithm [10] is a representative hierarchical clustering method, which can find communities by removing the links between communities. Blondel et al. proposed the Louvain algorithm [1], which is a well-known optimization method based on modularity. It is used to handle large-scale networks due to low time complexity. Liu et al. [16] put forward a community detection method by using non-negative matrix factorization. Zhao et al. [33] introduced the idea of granular computing into the community detection of network and proposed a community detection method based on clustering granulation.

The existing non-overlapping community detection algorithms have made great achievements, but these algorithms only use the traditional two-way decisions [29,30] method (the acceptance or rejection decision) to deal with the overlapping nodes between communities. Compared with the two-way decisions method, the three-way decisions theory (TWD) [28] adds a non-commitment decision. The main idea of TWD is to divide an entity set into three disjoint regions, which are denoted as positive region (POS), negative region (NEG) and boundary region (BND) respectively. The POS adopts the acceptance decision, the NEG adopts the rejection decision, and the BND adopts the non-commitment decision (i.e., entities that cannot make a decision based on the current information are placed in the BND). For entities in the BND, we can further mine more information to realize their final partition. The introduction of non-commitment decision can effectively solve the decision-making errors caused by insufficient information, which is more flexible and closer to the actual situation.

How to deal with the boundary region has become a key issue for three-way decisions community detection. At present, the commonly used methods to process the boundary region include modularity increment [20] and similarity calculation [2,8]. But these methods only take advantage of the local features of the network, without considering the information of the divided communities and the similarity of the global structure. Therefore, how to tackle the boundary region effectively is a challenge.

In this paper, we propose a three-way decisions community detection model based on weighted graph representation (WGR-TWD). The graph representation can well transform the global structure of the network into vector representation and make the two nodes in the boundary region that appear in the same community more similar by using the weight. Firstly, the multi-layered community structure is constructed by hierarchical clustering. The target layer is selected according to the extended modularity value of each layer. Secondly, all nodes are converted into vectors by weighted graph representation. Finally, nodes in the boundary region are divided into positive or negative region based on cosine similarity. Thus, non-overlapping community detection is realized.

The key contributions of this paper can be summarized as follows:

(1) We use weighted graph representation to obtain the global structure information of the network to guide the processing of the boundary region, which gets a better three-way decisions community detection method.
(2) Based on the knowledge of the communities in the target layer, we make the two nodes connected by a direct edge in the boundary region more similar by using frequency of appearing in the same community as the weight. Then the walk sequences are constructed according to the weight of the edge. Finally, the Skip-Gram model is used to obtain the vector representation of nodes. Therefore, the weighted graph representation method is realized.
(3) We evaluate the effectiveness the proposed model WGR-TWD on real-world networks compared with the baseline methods. The experimental results show the superior performance of our model.

The rest of this paper is organized as follows. We introduce related work in Sect. 2. We give the detailed description of our algorithm in Sect. 3. Experiments on real-world networks are reported in Sect. 4. Finally, we conclude the paper in Sect. 5.

2   Related Work

2.1   Community Detection of Hierarchical Clustering

Hierarchical clustering method has been widely used in community detection due to the hierarchical nature of the network structure. This approach can be divided into two forms: divisive method and agglomerative method. The divisive method removes the link with the lowest similarity index repeatedly, while the agglomerative method merges the pair of clusters with the highest similarity index repeatedly. These two methods eventually form a dendrogram, and communities are detected by cutting the tree.

The research of community detection based on hierarchical clustering has received widespread attention from scholars. Girvan and Newman proposed the GN algorithm [10], which is a typical divisive method. Clauset et al. [5] proposed a community detection algorithm based on data analysis, which is a representative agglomerative method. Fortunato et al. [9] presented an algorithm to find community structures based on node information centrality. Chen et al. proposed the LCV algorithm [4] which detects communities by finding local central nodes. Zhang et al. [32] introduced a hierarchical community detection algorithm based on partial matrix convergence using random walks.

Combining hierarchical clustering with granular computing, we introduce an agglomerative method based on variable granularity to build a dendrogram. Given an undirected and unweighted graph $G = (V, E)$, where V is the set of nodes, E denotes the set of edges. The set of neighbor nodes to a node v_i is denoted as $N(v_i) = \{v_j \in V \mid (v_i, v_j) \in E\}$, the set of initial granules is defined as $H^1 = \{C_1^1, C_2^1, ..., C_p^1\}$. The formation process of the initial granules is as follows. First, we calculate the local importance of each node in the network. The local importance of a node v_i is defined as follows:

$$I(v_i) = \frac{|Z|}{|N(v_i)|}, \tag{1}$$

where $Z = \{v_j \in N(v_i) \mid d(v_j) \le d(v_i)\}$, $d(v_i)$ is the degree of node v_i, and $|\cdot|$ denotes the number of elements in a set. Second, all important nodes are found according to the local importance of nodes. The node v_i is an important node if $I(v_i) > 0$. Finally, for any important node, an initial granule is composed of all neighbor nodes of the important node and the important node itself. After all the initial granules are obtained, the hierarchical clustering method based on variable granularity is described. The clustering coefficient between the two granules is defined as

$$f\left(C_i^m, C_j^m\right) = \frac{\left|C_i^m \cap C_j^m\right|}{min\left\{|C_i^m|, |C_j^m|\right\}}, C_i^m, C_j^m \in H^m \tag{2}$$

where $\left|C_i^m \cap C_j^m\right|$ denotes the number of common nodes in granules C_i^m and C_j^m, $min\left\{|C_i^m|, |C_j^m|\right\}$ is the smaller number of nodes in granules C_i^m and C_j^m, H^m is the granules set of the mth ($m = 1, 2, ...$) layer. The collection of clustering thresholds is denoted as $\lambda = \{\lambda_m, m = 1, 2, ...\}$, where λ_m is the clustering threshold of the mth layer. In order to automatically obtain the clustering threshold of each layer, λ_m is defined as

$$\lambda_m = med\left\{f\left(C_i^m, C_j^m\right) |\forall C_i^m, C_j^m \in H^m, C_i^m \cap C_j^m \neq \varnothing \wedge i \neq j\right\} \tag{3}$$

where $med\{\}$ is a median function. The clustering process is as follows. Firstly, for $\forall C_i^m, C_j^m \in H^m$, the clustering coefficient between them is calculated. Then the clustering threshold λ_m of the current layer is calculated. And the maximum clustering coefficient is found, which is denoted as $f\left(C_\alpha^m, C_\beta^m\right)$.

If $f\left(C_\alpha^m, C_\beta^m\right) \geqslant \lambda_m$, the two granules C_α^m and C_β^m are merged to form a new granule and the new granule is added to H^{m+1}. Otherwise, all the granules in H^m are added to H^{m+1} and H^m is set to empty. For each layer, repeat above clustering process until all nodes in the network are in a granule. Therefore, a dendrogram is built.

2.2 DeepWalk

Traditional network representation usually uses high-dimensional sparse vectors, which takes more running time and computational space in statistical learning. Network representation learning (NRL) is proposed to address the problem. NRL aims to learn the low-dimensional potential representations of nodes in networks. The learned representations can be used as features of the graph for various graph-based tasks, such as classification, clustering, link prediction, community detection, and visualization.

DeepWalk [24] is the first influential NRL model in recent years, which adopts the approach of natural language processing by using the Skip-Gram model [18,19] to learn the representation of nodes in the network. The goal of Skip-Gram is to maximize the probability of co-occurrence among the words that appear within a window. DeepWalk first generates a large number of random walk sequences by sampling from the network. These walk sequences can be analogized to the sentences of the article, and the nodes are analogized to the words in the sentence. Then Skip-Gram can be applied to these walk sequences to acquire network embedding. DeepWalk can express the connection of the network well, and has high efficiency when the network is large.

3 The Proposed Algorithm

3.1 Weighted Graph Representation

To effectively deal with overlapping communities in the target layer, a weighted graph representation approach is proposed. At first, a weighted graph is constructed according to the community structure of the target layer. The weights of edges in an unweighted graph are defined as follows

$$W_{ij} = 1.0 + \sigma_{ij}/N_c \tag{4}$$

where σ_{ij} is the number of communities in which nodes v_i and v_j appear in a community at the same time, N_c is the total number of communities in the target layer. After that, an improved DeepWalk (IDW) model is used to acquire the vector representation of all nodes in the graph. Unlike DeepWalk, the IDW model constructs the walk sequences according to the weight of the edge. The greater the weight, the higher the walk probability. Assume that the current walk node is v_i, if $v_j \in N(v_i)$, then the walk probability from node v_i to node v_j is

$$P(v_i \to v_j) = \frac{W_{ij}}{\sum\limits_{v_k \in N(v_i)} W_{ik}}. \tag{5}$$

After obtaining all the walk sequences, the Skip-Gram model is used to learn the vector representation of nodes from the walk sequences. The objective function of IDM is as follows

$$\min_{R} \sum_{-\omega \leqslant j \leqslant \omega, j \neq 0} -logP\left(v_{i+j}|R\left(v_i\right)\right) \tag{6}$$

where $R\left(v_i\right)$ is the vector representation of node v_i, ω is the window size which is maximum distance between the current and predicted node within a walk sequence. Thus, the vector representation of all nodes in the network is obtained.

3.2 The WGR-TWD Algorithm

We will present the proposed WGR-TWD algorithm in this section. Figure 1 shows the overall framework of the proposed algorithm. Our algorithm consists of two parts: the construction of multi-layered community structure and boundary region processing.

The first part, we employ the hierarchical clustering method based on variable granularity to construct a multi-layered community structure according to Sect. 2.1. Some overlapping communities exist in the multi-layered community structure because of clustering mechanism, so we use the extended modularity (EQ) [26] to measure the partition quality of each layer. It is defined as follows

$$EQ = \frac{1}{2m} \sum_{i} \sum_{u \in C_i, v \in C_i} \frac{1}{O_u O_v} \left(A_{uv} - \frac{d_u d_v}{2m}\right) \tag{7}$$

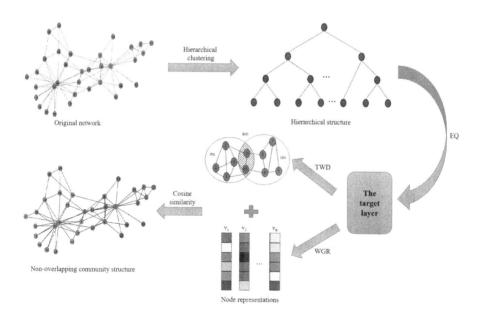

Fig. 1. The framework of the proposed method.

where m is the number of edges in the network, C_i represents a community, O_u is the number of communities that node u belongs to, A_{uv} is the element of adjacent matrix, and d_u is the degree of node u. A larger EQ value means better performance for overlapping community division. Thus, we select the layer corresponding to the largest EQ value as the target layer.

Algorithm 1. WGR-TWD

Input: An undirected and unweighted graph $G = (V, E)$.
Output: Non-overlapping community structure $POS(G)$, $NEG(G)$.
1: construct the muti-layered community structure: $H = \{H^1, H^2, ..., H^m\}$
2: calculate EQ for each layer in H
3: $H^t \leftarrow$ find the layer corresponding to the maximum EQ
4: divide three regions according to H^t: $POS(G)$, $NEG(G)$, $BND(G)$
5: get a weighted graph $G' = (V, E, W)$ according to Equation (4)
6: $walks \leftarrow$ generate the walk sequences by W
7: learn node representation from $walks$ by using Skip-Gram model
8: **for** $\forall v \in BND(G)$ **do**
9:     $s(v, POS) \leftarrow$ calculate the similarity between node v and $POS(G)$
10:    $s(v, NEG) \leftarrow$ calculate the similarity between node v and $NEG(G)$
11:    **if** $s(v, POS) > s(v, NEG)$ **then**
12:        $POS(G) \leftarrow POS(G) + \{v\}$
13:    **else if** $s(v, POS) < s(v, NEG)$ **then**
14:        $NEG(G) \leftarrow NEG(G) + \{v\}$
15:    **end if**
16: **end for**
17: **return** $POS(G)$, $NEG(G)$

The second part introduces the method of dealing with overlapping communities in the target layer. Since there are overlapping communities in the target layer, we need to further divide the target layer to achieve non-overlapping community detection. Therefore, the three-way decisions theory (TWD) is introduced to handle overlapping communities. Based on the idea of TWD, we define the overlapping part in the communities as boundary region (BND), and the non-overlapping part as positive region (POS) or negative region (NEG). And our goal is to process nodes in the BND. First of all, we adopt the weighted graph representation method to learn the vector representation of all nodes in the network. After that, the nodes in the BND are divided into the POS or NEG by using cosine similarity. Suppose the vector of node u is $u = (x_1, x_2, ..., x_n)$, node v is $v = (y_1, y_2, ..., y_n)$, then the cosine similarity is defined as

$$S(u, v) = \frac{\sum\limits_{i=1}^{n} x_i y_i}{\sqrt{\sum\limits_{i=1}^{n} (x_i)^2} \cdot \sqrt{\sum\limits_{i=1}^{n} (y_i)^2}}. \tag{8}$$

For arbitrary node v_i in the BND, find out all communities containing node v_i in the target layer, calculate the average value of cosine similarity between node v_i and non-overlapping nodes in each community as the similarity between node v_i and this community, then join node v_i into the community corresponding to the maximum similarity and update the community structure of the target layer. Repeat the above operation until all nodes in the BND are processed. The WGR-TWD algorithm is described in Algorithm 1.

4   Experiments

4.1   Datasets

We test the performance of our method on eight real-world datasets in which each dataset is described as follows, and the main information of those datasets are shown in Table 1.

Zachary's karate club [31]. This is a social network of friendships between 34 members of a karate club at a US university in the 1970s.

Dolphin social network [17]. It is an undirected social network of frequent associations between 62 dolphins in a community living off Doubtful Sound, New Zealand.

Books about US politics [22]. A network of books about US politics published around the time of the 2004 presidential election and sold by the online bookseller Amazon.com. Edges between books represent frequent co-purchasing of books by the same buyers.

American college football [10]. A network of American football games between Division IA colleges in 2000.

Email communication network [21]. It is a complex network which indicates the email communications of a university. The network was composed by Alexandre Arenas.

Facebook [15]. The network was collected from survey participants using Facebook app.

Geom [11]. The authors collaboration network in computational geometry.

Collaboration [14]. The network is from the e-print arXiv and covers scientific collaborations between authors papers submitted to High Energy Physics Theory category.

4.2   Baseline Methods

In this paper, two representative algorithms are chosen to compare with the proposed WGR-TWD, as shown below:

- Modularity increment (MI) [3]. A hierarchical clustering method based on variable granularity, and the overlapping nodes between communities are divided according to modularity optimization.
- DeepWalk [24]. It is a network representation learning method. This approach is used to handle the overlapping communities in the target layer.

Table 1. Information of datasets.

Network	Nodes	Edges	Real clusters
Karate	34	78	2
Dolphin	62	159	2
Polbooks	105	441	3
Football	115	613	12
Email	1133	5451	Unknown
Facebook	4039	88234	Unknown
Geom	7343	11898	Unknown
Collaboration	9877	25998	Unknown

4.3 Evaluation Metrics

We employ two widely used criteria to evaluate the performance of community detection algorithms.

The first index is modularity (Q) [5], which is often used when the real community structure is not known. Q is defined as follows

$$Q = \frac{1}{2m} \sum_{i,j} \left[A_{ij} - \frac{d_i d_j}{2m} \right] \delta(c_i, c_j) \tag{9}$$

where m is the number of edges in the network, A is the adjacent matrix, d_i is the degree of node i, c_i represents the community to which node i belongs, and $\delta(c_i, c_j) = 1$ when $c_i = c_j$, else $\delta(c_i, c_j) = 0$. The higher the modularity value, the better the result of community detection.

Another index is normalized mutual information (NMI) [7], which is defined as follows

$$NMI = \frac{-2 \sum_{i=1}^{C_A} \sum_{j=1}^{C_B} C_{ij} \log \frac{C_{ij} n}{C_{i.} C_{.j}}}{\sum_{i=1}^{C_A} C_{i.} \log \frac{C_{i.}}{n} + \sum_{j=1}^{C_B} C_{.j} \log \frac{C_{.j}}{n}} \tag{10}$$

where C_A (C_B) denotes the number of communities in partition A (B), C_{ij} is the number of nodes shared by community i in partition A and by community j in partition B, $C_{i.}$ $(C_{.j})$ represents the sum of elements of matrix C in row i (column j), and n is the number of nodes in the network. A higher value of NMI indicates the detected community structure is closer to the real community structure.

4.4 Experimental Results

In the networks with known real partition (the first four small networks), we use two indicators (Q and NMI) to evaluate our algorithm. Table 2 presents the community detection results of the proposed algorithm and the baseline algorithms on networks with known real partition. We can see that our method

Table 2. Experimental results on networks with known real partition (under EQ criterion).

Network	Index	MI	DeepWalk	WGR-TWD
Karate	Q	0.360	0.360	**0.371**
	NMI	0.837	0.837	**1.000**
Dolphin	Q	**0.385**	0.379	**0.385**
	NMI	0.814	**0.889**	0.830
Polbooks	Q	0.439	**0.442**	0.441
	NMI	**0.469**	0.448	0.453
Football	Q	0.480	0.545	**0.558**
	NMI	0.626	0.726	**0.732**

Table 3. Experimental results on networks with known real partition (under Q criterion).

Network	Index	MI	DeepWalk	WGR-TWD
Karate	Q	0.391	0.391	**0.401**
	NMI	0.606	0.620	**0.700**
Dolphin	Q	0.510	0.520	**0.523**
	NMI	**0.619**	0.613	0.618
Polbooks	Q	**0.514**	0.492	0.509
	NMI	**0.501**	0.463	0.489
Football	Q	0.582	**0.596**	**0.596**
	NMI	0.904	0.899	**0.911**

obtains best results on the Karate and Football datasets. On the Dolphin dataset, the Q value of our method is better and the NMI value is second to the DeepWalk method. On the Polbooks dataset, our method performs not well because the connections between nodes are sparse which is difficult to mine the structural information of the network.

To further verify the effectiveness of the proposed algorithm, the MI method is used to deal with each layer in the multi-layered community structure. And we select the layer corresponding to the maximum Q value as the target layer. The experimental results are shown in Table 3. Compared with Table 2, Table 3 can obtain higher Q value. Combined with Tables 2 and 3, our method can get better community detection results compared with the baseline methods.

We also conducted experiments on four large networks. On these networks, the real partition is unknown. Therefore, we only use modularity to evaluate the performance of different methods. Table 4 shows the community detection results of the proposed method and baseline methods. On the first three networks, we can see that our method obtains better results compared with the

Table 4. Modularity values on networks with unknown real partition.

Network	MI	DeepWalk	WGR-TWD
Email	0.537	0.538	**0.542**
Facebook	0.771	0.770	**0.772**
Geom	0.702	0.710	**0.711**
Collaboration	**0.723**	0.718	0.720

two baseline methods. On the Collaboration dataset, the MI method achieves the best performance which is a little bit higher than our method. The main reason is that the Collaboration network is very sparse which leads to poor vector representation of the nodes. In conclusion, the proposed method effectively addresses the problem of non-overlapping community detection in networks.

5 Conclusion

In this paper, we propose a method for three-way decisions community detection based on weighted graph representation. The target layer in multi-layered community structure is selected according to the extended modularity value of each layer. For the overlapping communities in the target layer, the weighted graph representation can well transform the global structure into vector representation and make the two nodes in the boundary region more similar by using frequency of appearing in the same community as the weight. Finally, the nodes in the boundary region are divided according to cosine similarity. Experiments on real-world networks demonstrate that the proposed method is effective for community detection in networks.

Acknowledgments. This work is supported by the National Natural Science Foundation of China (Grants Numbers 61876001) and the Major Program of the National Social Science Foundation of China (Grant No. 18ZDA032).

References

1. Blondel, V.D., Guillaume, J.L., Lambiotte, R., Lefebvre, E.: Fast unfolding of communities in large networks. J. Stat. Mech: Theory Exp. **2008**(10), P10008 (2008)
2. Chen, J., et al.: Three-way dicision community detection algorithm based on local group information. In: Polkowski, L., et al. (eds.) IJCRS 2017. LNCS (LNAI), vol. 10314, pp. 171–182. Springer, Cham (2017). https://doi.org/10.1007/978-3-319-60840-2_12
3. Chen, J., Li, Y., Yang, X., Zhao, S., Zhang, Y.: VGHC: a variable granularity hierarchical clustering for community detection. Granular Comput. **4**, 1–10 (2019)
4. Chen, Q., Wu, T.T.: A method for local community detection by finding maximal-degree nodes. In: 2010 International Conference on Machine Learning and Cybernetics, vol. 1, pp. 8–13. IEEE (2010)

5. Clauset, A., Newman, M.E., Moore, C.: Finding community structure in very large networks. Phys. Rev. E **70**(6), 066111 (2004)
6. Cui, Y., Wang, X., Eustace, J.: Detecting community structure via the maximal sub-graphs and belonging degrees in complex networks. Phys. A **416**, 198–207 (2014)
7. Danon, L., Diaz-Guilera, A., Duch, J., Arenas, A.: Comparing community structure identification. J. Stat. Mech: Theory Exp. **2005**(09), P09008 (2005)
8. Fang, L., Zhang, Y., Chen, J., Wang, Q., Liu, F., Wang, G.: Three-way decision based on non-overlapping community division. CAAI Trans. Intell. Syst. **12**(3), 293–300 (2017)
9. Fortunato, S., Latora, V., Marchiori, M.: Method to find community structures based on information centrality. Phys. Rev. E **70**(5), 056104 (2004)
10. Girvan, M., Newman, M.E.: Community structure in social and biological networks. Proc. Nat. Acad. Sci. **99**(12), 7821–7826 (2002)
11. Gong, M., Cai, Q., Chen, X., Ma, L.: Complex network clustering by multiobjective discrete particle swarm optimization based on decomposition. IEEE Trans. Evol. Comput. **18**(1), 82–97 (2013)
12. Hajek, B., Wu, Y., Xu, J.: Information limits for recovering a hidden community. In: 2016 IEEE International Symposium on Information Theory (ISIT), pp. 1894–1898. IEEE (2016)
13. Kernighan, B.W., Lin, S.: An efficient heuristic procedure for partitioning graphs. Bell Syst. Tech. J. **49**(2), 291–307 (1970)
14. Leskovec, J., Kleinberg, J., Faloutsos, C.: Graph evolution: densification and shrinking diameters. ACM Trans. Knowl. Discov. Data (TKDD) **1**(1), 2 (2007)
15. Leskovec, J., Mcauley, J.J.: Learning to discover social circles in ego networks. In: Advances in Neural Information Processing Systems, pp. 539–547 (2012)
16. Liu, X., Wang, W., He, D., Jiao, P., Jin, D., Cannistraci, C.V.: Semi-supervised community detection based on non-negative matrix factorization with node popularity. Inf. Sci. **381**, 304–321 (2017)
17. Lusseau, D.: The emergent properties of a dolphin social network. Proc. R. Soc. Lond. Ser. B: Biol. Sci. **270**(Suppl. 2), S186–S188 (2003)
18. Mikolov, T., Chen, K., Corrado, G., Dean, J.: Efficient estimation of word representations in vector space. arXiv preprint arXiv:1301.3781 (2013)
19. Mikolov, T., Sutskever, I., Chen, K., Corrado, G.S., Dean, J.: Distributed representations of words and phrases and their compositionality. In: Advances in Neural Information Processing Systems, pp. 3111–3119 (2013)
20. Newman, M.E.: Fast algorithm for detecting community structure in networks. Phys. Rev. E **69**(6), 066133 (2004)
21. Newman, M.E.: Finding community structure in networks using the eigenvectors of matrices. Phys. Rev. E **74**(3), 036104 (2006)
22. Newman, M.E.: Modularity and community structure in networks. Proc. Nat. Acad. Sci. **103**(23), 8577–8582 (2006)
23. Newman, M.E., Girvan, M.: Finding and evaluating community structure in networks. Phys. Rev. E **69**(2), 026113 (2004)
24. Perozzi, B., Al-Rfou, R., Skiena, S.: Deepwalk: online learning of social representations. In: Proceedings of the 20th ACM SIGKDD International Conference on Knowledge Discovery and Data Mining, pp. 701–710 (2014)
25. Rosvall, M., Bergstrom, C.T.: An information-theoretic framework for resolving community structure in complex networks. Proc. Natl. Acad. Sci. **104**(18), 7327–7331 (2007)

26. Shen, H., Cheng, X., Cai, K., Hu, M.B.: Detect overlapping and hierarchical community structure in networks. Phys. A **388**(8), 1706–1712 (2009)
27. Wu, W., Kwong, S., Zhou, Y., Jia, Y., Gao, W.: Nonnegative matrix factorization with mixed hypergraph regularization for community detection. Inf. Sci. **435**, 263–281 (2018)
28. Yao, Y.: Three-way decision: an interpretation of rules in rough set theory. In: Wen, P., Li, Y., Polkowski, L., Yao, Y., Tsumoto, S., Wang, G. (eds.) RSKT 2009. LNCS (LNAI), vol. 5589, pp. 642–649. Springer, Heidelberg (2009). https://doi.org/10.1007/978-3-642-02962-2_81
29. Yao, Y.: Three-way decisions with probabilistic rough sets. Inf. Sci. **180**(3), 341–353 (2010)
30. Yao, Y.: Two semantic issues in a probabilistic rough set model. Fundam. Inform. **108**(3–4), 249–265 (2011)
31. Zachary, W.W.: An information flow model for conflict and fission in small groups. J. Anthropol. Res. **33**(4), 452–473 (1977)
32. Zhang, W., Kong, F., Yang, L., Chen, Y., Zhang, M.: Hierarchical community detection based on partial matrix convergence using random walks. Tsinghua Sci. Technol. **23**(1), 35–46 (2018)
33. Zhao, S., Wang, K., Chen, J., Zhang, Y.: Community detection algorithm based on clustering granulation. J. Comput. Appl. **34**(10), 2812–2815 (2014)

Attribute Reduction

Quick Maximum Distribution Reduction in Inconsistent Decision Tables

Baizhen Li[1], Wei Chen[2(✉)], Zhihua Wei[1], Hongyun Zhang[1], Nan Zhang[3], and Lijun Sun[1]

[1] Tongji University, Shanghai 201804, China
[2] Shanghai Institute of Criminal Science and Technology, Shanghai 200003, China
weichen_82@163.com
[3] Yantai University, Yantai 264005, Shandong, China

Abstract. Attribute reduction is a key issue in rough set theory, and this paper focuses on the maximum distribution reduction for complete inconsistent decision tables. It is quite inconvenient to judge the maximum distribution reduct directly according to its definition and the existing heuristic based judgment methods are inefficient due to the lack of acceleration mechanisms that mainstream heuristic judgment methods have. In this paper, we firstly point out the defect of judgment method proposed by Li et al. [15]. After analyzing the root cause of the defect, we proposed two novel heuristic attribute reduction algorithms for maximum distribution reduction. The experiments show that proposed algorithms are more efficient.

Keywords: Rough sets · Attribute reduction · Maximum distribution reduction · Heuristic algorithm

1 Introduction

Rough set theory, introduced by Z. Pawlak [1] in 1982, is an efficient tool to imprecise, incomplete and uncertain information processing [2–5]. Currently, rough set theory has been successfully applied to many practical problems, including machine learning [6,7], pattern recognition [8,9], data mining [10], decision support systems [11], etc.

Attribute reduction, the process of obtaining a minimal set of attributes that can preserve the same ability of classification as the entire attribute set, is one of the core concepts in rough set theory [12]. Maximum distribution reduction, proposed as a compromise between the capability of generalized decision preservation reduction and the complexity of distribution preservation reduction [13] by Zhang et al. [14] in 2003, guarantees the decision value with maximum probability of object in inconsistent decision tables unchanged. Subsequently, Pei et al. proposed a theorem for maximum distribution reduct judgment in 2005. Next, Li et al. [15] paid attention to the computational efficiency of reduction definition and designed a new definition of maximum distribution reduction to

© Springer Nature Switzerland AG 2020
R. Bello et al. (Eds.): IJCRS 2020, LNAI 12179, pp. 169–182, 2020.
https://doi.org/10.1007/978-3-030-52705-1_12

speed up attribute reduction. Taking into consideration the general reduction on inconsistent decision tables, Ge et al. [16] proposed new definition of maximum distribution reduction.

Heuristic approaches is one of import method in attribute reduction. The heuristic approach is composed of two parts: the attribute reduction heuristic and the search strategy [17]. The attribute reduction heuristic is the fitness function of a heuristic approach. Existing definitions of heuristics are mainly based on three aspects: dependency degree [18], entropy [19–21], and consistency [22,23]. The search strategy is the control structure of the heuristic approach. Speaking loosely, the search strategy mainly includes three kinds of methods: the deletion method, the addition method, and the addition-deletion method [24].

Existing methods for the judgment of maximum distribution were weak association with mainstream heuristics. As a result, the efficiency of heuristic maximum distribution reduction algorithm was limited due to lack of the support of acceleration policies that mainstream heuristics have. This paper focuses on the quick reduction algorithms for maximum distribution reduction. At first, we analyze the defect of the quick maximum distribution reduction algorithm (Q-MDRA) proposed in [15] and explore the root cause of its defect. Next, based on the existing mainstream heuristic function, we develop three heuristic maximum distribution reduction algorithms. Finally, we conduct some experiments to evaluate the effectiveness and efficiency of proposed algorithms.

The rest of this paper is organized as follows. In Sect. 2, we review some basic notions related to maximum distribution reduction and three classic heuristic functions. In Sect. 3, we show the defect of Q-MDRA with a calculation example of maximum distribution reduction. After exploring the root cause of its defect, we present three novel algorithms for maximum distribution reduction. In Sect. 4, we evaluate the efficiency of proposed algorithms through algorithm complexity analysis and comparison experiments.

2   Preliminary

In this section, we review some basic notions related to maximum distribution reduction and three classic heuristic functions.

The research object of the rough set theory is called the information system. The information system IS can be expressed as four tuple, $i.e.$ $< U, A, V, f >$, where U stands for the universe of discourse, a non-empty finite set of objects. A is the set of attributes, $V = \bigcup_{a \in A} V_a$ is the set of all attribute values, and $f : U \times A \to V$ is an information function that maps an object in U to exactly one value in V_a. For $\forall x \in U, \forall a \in A$, we have $f(x, a) \in V_a$. Specifically in the classification problem, the information table contains two kinds of attributes, which can be characterized by a decision table $DT = (U, C \cup D, V, f)$ with $C \cap D = \emptyset$, where an element of C is called a condition attribute, C is called a condition attribute set, an element D is called a decision attribute, and D is called a decision attribute set.

For the condition attribute set $B \subseteq C$, the indiscernibility relation and discernibility relation of B is respectively defined by $IND(B) = \{< x,y > \in U \times U | \forall a \in B, f(x,a) = f(y,a)\}$ and $DIS(B) = \{< x,y > \in U \times U | \exists a \in B, f(x,a) \neq f(y,a)\}$. For an object $x \in U$, the equivalence class of x, denoted by $[x]_B$, is defined by $[x]_B = \{y \in U| < x,y > \in IND(B)\}$. The family of all equivalence classes of $IND(B)$, $i.e.$, the partition determined by B, is denoted by $U/IND(B)$ or simply U/B. Obviously, $IND(B)$ is reflexive, symmetric and transitive. Meanwhile, $DIS(C)$ is irreflexive, symmetric, but not transitive. Something else needed to be reminded of is that $DIS(C) \cup IND(C) = U \times U, DIS(C) \cap IND(C) = \emptyset$.

One the basis of above notions, the concept of maximum distribution reduction was proposed by Zhang et al. [14] in 2003.

Definition 1. *Let $DT = (U, C \cup D, V, f)$ be a decision table, $B \subseteq C$ is a maximum distribution reduct of C if and only if B satisfies*

$$\forall x \in U, \gamma_B(x) = \gamma_C(x);$$
$$\forall B' \subset B, \exists x \in U, \gamma_{B'}(x) \neq \gamma_C(x),$$

where $\gamma_C(x) = \{P_i : P_i \in U/D \wedge |P_i \cap [x]_C| = max_{P_j \in U/D}(|P_j \cap [x]_C|)\}$.

It is said that B is a maximum distribution consistent attribute set if B satisfies condition (1) mentioned above only. There are two methods of maximum distribution reduction: the discernibility matrix based methods and the heuristic methods. For that the discernibility matrix based methods are low-efficiency, heuristic methods are the more reasonable choice for processing the larger scale data. The heuristic attribute reduction algorithms comprises two parts: the heuristic function and the control strategy. We take the addition strategy based heuristic algorithms as the research object of paper. For the heuristic functions, we take three classic heuristic functions, $i.e.$, the dependency degree, the condition entropy, and the consistency as the alternatives for the construction of improved algorithms.

Definition 2. *Given a decision table $DT = (U, C \cup D, V, f)$ and $B \subseteq C$, $U/B = \{X_1, X_2, \cdots, X_m\}$, $U/D = \{Y_1, Y_2, \cdots, Y_n\}$, three classic heuristic functions (dependency degree, the consistency and conditional entropy) are defined by:*

(1) $\Gamma_B(D) = \frac{|POS_B(D)|}{|U|}$;

(2) $\delta_B(D) = |\{D_j|\frac{|[x]_B \cap D_j|}{|[x]_B|} = \max\limits_{k=1}^{|U/D|}\{\frac{|[x]_B \cap D_k|}{|[x]_B|}\}\}|/|U|$;

(3) $H(D|B) = -\sum_{i=1}^{m} P(X_i) \sum_{j=1}^{n} P(Y_j|X_i) \log P(Y_j|X_i)$, where $P(Y_j|X_i) = |X_i \cap Y_j|/|X_i|$, where $H(B) = -\sum_{i=1}^{m} P(X_i) \log P(X_i)$, $P(X_i) = |X_i|/|U|$.

3 Novel Heuristic Maximum Distribution Reduction Algorithms

In this section, we present two defects in Q-MDRA firstly. After analyzing its cause, we construct two quick heuristic maximum distribution reduction algorithms based on classic heuristic functions.

At first, we want to review the quick maximum distribution reduction algorithm (Q-MDRA) proposed by Li et al. Here. Based upon Definition 1, Li et al. [15] proposed following theorem for the judgment of the maximum distribution reduct.

Theorem 1. *Let $DT = (U, C \cup D, V, f)$ be a decision table and $B \subseteq C$, B is a maximum distribution reduct of C if and only if B satisfies*

$$\forall x \in U, \gamma_B^{Md}(D) = \gamma_C^{Md}(D);$$
$$\forall B' \subset B, \gamma_{B'}^{Md}(D) \neq \gamma_C^{Md}(D),$$

where $\gamma_{B'}^{Md}(D) = \sum_{X \in U/B} \frac{|X \cap P_i : argmax_{P_i \in U/D}|X \cap P_i||}{|U|}$.

This theorem is expressed by the Theorem 6.11 of Ref. [15]. $\gamma_B^{MD}(D) = \gamma_C^{MD}(D)$ maintains unchanged the scale of the maximum decision classes instead of the maximum decision classes for all of the objects in decision tables. That is to say, B may be not a maximum distribution reduct of C in some special conditions. We present the detail information in Sect. 3.1. Based on the variant of dependency degree heuristic function in Theorem 1, Algorithm 1 was constructed by the way of the addition strategy. Something needed to be reminded of in Algorithm 1 is that we denote the assignment operation as ":=" and use the "=" to represent that two items are on equal term.

Algorithm 1. Quick Maximum Distribution Reduction Algorithm (Q-MDRA)

Require: Decision table $DT = (U, C \cup D, V, f)$
Ensure: A maximum distribution reduct of DT
1: $red := \phi$
2: **while** $True$ **do**
3:    $T := red$
4:    **for** $a \in C - red$ **do**
5:        **if** $\gamma_{red \cup \{a\}}^{Md}(D) > \gamma_T^{Md}(D)$ **then**
6:            $T := red \cup \{a\}$
7:        **end if**
8:    **end for**
9:    **if** $red = T$ **then**
10:        break
11:    **else**
12:        $red := T$
13:    **end if**
14: **end while**
15: **return** red

3.1 The Defects of Q-MDRA

Here, in the way of calculation example, we show the detail information about that Q-MDRA may not perform well as our expectation. Assume that there is a decision table given as Table 1, we are assigned to get the maximum distribution reduct of Table 1.

Table 1. A decision table

U	a_1	a_2	a_3	d
x_1	0	0	0	0
x_2	1	0	0	1
x_3	1	1	0	0
x_4	1	1	0	1
x_5	1	1	1	0
x_6	1	1	1	1
x_7	1	1	1	1

For Table 1, we know that $U = \{x_1, x_2, \cdots, x_7\}$, $C = \{a_1, a_2, a_3\}$, $D = \{d\}$, and obviously we have $U/C = \{X_1, X_2, X_3, X_4\} = \{\{x_1\}, \{x_2\}, \{x_3, x_4\}, \{x_5, x_6, x_7\}\}$ and $U/D = \{P_1, P_2\} = \{\{x_1, x_3, x_5\}, \{x_2, x_4, x_6, x_7\}\}$. According to Definition 1, we know that $\gamma_C(x_1) = \{P_1\}$, $\gamma_C(x_2) = \{P_2\}$; for $x \in X_3$, $\gamma_C(x) = \{P_1, P_2\}$; for $x \in X_4$, we have $\gamma_C(x) = \{P_2\}$.

The process of Q-MDRA for obtaining maximum distribution reduct of Table 1 is shown as follows.

Step 1. $red := \emptyset$.

Step 2. $T := red$, $\gamma_T^{Md}(D) = |P_2|/|U| = 4/7$; $\gamma_{T\cup\{a_1\}}^{Md}(D) = (|P_1\cap X_1| + |\{x_2, x_3,$ $\cdots, x_7\} \cap P_2|)/|U| = 5/7$; $T := T \cup \{a_1\}$; $\gamma_{T\cup\{a_2\}}^{Md}(D) = 4/7$; $\gamma_{T\cup\{a_3\}}^{Md}(D) = 4/7$. Because of $T \neq red$, we operate the assignment of $red := T = \{a_1\}$.

Step 3. $T := red$, $\gamma_T^{Md}(D) = |P_2|/|U| = 5/7$; $\gamma_{T\cup\{a_2\}}^{Md}(D) = 5/7$; $\gamma_{T\cup\{a_3\}}^{Md}(D) = 5/7$. Because T is equal to red, program is over.

Using Q-MDRA we get a collection of attributes $\{a_1\}$. According to Theorem 1, $\{a_1\}$ is a maximum distribution reduct of Table 1 for that $\{a_1\}$ satisfies $\gamma_{\{a_1\}}^{Md}(D) = \gamma_{\{C\}}^{Md}(D) = 5/7$ and $\gamma_\phi^{Md}(D) \neq 5/7$. But checking it with original Definition 1, we know that $\{a_1\}$ is not a maximum distribution reduct for Table 1 because $\gamma_{\{a_1\}}(x_3) = \{P_2\} \neq \gamma_C(x_3) = \{P_1, P_2\}$. Consequently, Theorem 1 is incorrect.

Here we analyze the root of the defect of Theorem 1. Given a decision table $DT = (U, C \cup D, V, f)$, $U/C = \{X_1, X_2, \cdots, X_n\}$, $U/D = \{P_1, P_2, \cdots, P_m\}$. Let $mxcf(X_i) = max_{P_j \in U/D}(|P_j \cap X_i|)$, we have $\gamma_C^{Md}(D) = \sum_{X_i \in U/C} \frac{mxcf(X_i)}{|U|}$.

Assume that $x_1 \in X_1$, $x_2 \in X_2$, $\gamma_C(x_1) \neq \gamma_C(x_2)$, $|\gamma_C(x_1)| > 1$, $|\gamma_C(x_2)| > 1$, $|\gamma_C(x_1) \cap \gamma_C(x_1)| \geq 1$ and $B \subseteq C$, $U/B = \{X_1 \cup X_2, X_3, \cdots, X_n\}$, it is obvious that $mxcf(X_1) + mxcf(X_2) = mxcf(X_1 \cup X_2)$ and $\gamma_C^{Md}D = \gamma_B^{Md}(D)$. But for $x \in X_1 \cup X_2$, $\gamma_B(x) = \gamma_C(x_1) \cap \gamma_C(x_2)$, it is not equal to $\gamma_C(x_1)$ or $\gamma_C(x_2)$. The measure $\gamma_C^{Md}(D)$, used in Theorem 1, is not sensitive to the change of the maximum decision classes of objects that have two or more than two maximum decision classes.

On the other side, an attribute set red outputted by Q-MDRA does not always satisfy $\gamma_{red}^{Md}(D) = \gamma_C^{Md}(D)$. The reason is that $\forall a \in C - red$, $\gamma_{red\cup\{a\}}^{Md}(D) = \gamma_{red}^{Md}(D)$ does not guarantee $\gamma_{red}^{Md}(D) = \gamma_C^{Md}(D)$. That is to say, $\forall a \in C - red$, $\gamma_{red\cup\{a\}}^{Md}(D) = \gamma_{red}^{Md}(D)$ is not conflicted with $\exists B \subseteq C - red$, $\gamma_{red\cup B}^{Md}(D) > \gamma_{red}^{Md}(D)$.

3.2   Novel Maximum Distribution Reduction Algorithms

To solve the problems identified in Q-MDRA, the concept of indiscernibility relation and discernibility relation of maximum distribution with respect to the specific attribute set are defined. Firstly. Next, the maximum distribution reduct is defined using the indiscernibility relation of maximum distribution. Finally, we construct heuristic maximum distribution reduction algorithms with classic heuristic functions.

Definition 3. *Given a decision table $DT = (U, C \cup D, V, f)$, the indiscernibility relation of maximum distribution of U with respect to $B \subseteq C$ is defined as $IND_{md}(B) = \{< x, y > | x, y \in U, \gamma_B(x) = \gamma_B(y)\}$, and the discernibility relation of maximum distribution of U with respect to B stands for $DIS_{md}(B) = \{< x, y > | x, y \in U, \gamma_B(x) \neq \gamma_B(y)\}$.*

Obviously, $IND_{md}(C)$ is reflexive, symmetric and transitive; $DIS_{md}(C)$ is irreflexive, symmetric, but not transitive. It is worth noting that $IND_{md}(C) \cup DIS_{md}(C) = U \times U$, $IND_{md}(C) \cap DIS_{md}(C) = \emptyset$.

Theorem 2. *Given $DT = (U, C, D, V, f)$, B is a maximum distribution consistent attribute set of C if and only if B satisfies $IND(C) \subseteq IND(B) \subseteq IND_{md}(C)$, $DIS_{md}(C) \subseteq DIS(B) \subseteq DIS(C)$.*

Proof. It is apparent that $DIS(B) \subseteq DIS(C)$, $IND(C) \subseteq IND(B)$, and based on $IND(B) \cap DIS(B) = \phi$, $IND(B) \cup DIS(B) = U \times U$, $IND_{md}(C) \cap DIS_{md}(C) = \phi$, $IND_{md}(C) \cup DIS_{md}(C) = U \times U$, we know that $DIS(C) \subseteq DIS(B) \subseteq DIS_{md}(C)$ is equal to $IND(C) \subseteq IND(B) \subseteq IND_{md}(C)$. Thus what all we need is to prove that $DIS_{md}(C) \subseteq DIS(B)$ is true.

- Sufficiency($\Rightarrow$): Assume that if B is a maximum distribution consistent attribute set then $DIS_{md}(C) \nsubseteq DIS(B)$. $DIS_{md}(C) \nsubseteq DIS(B)$ means $\exists < x, y > \in DIS_{md}(C)$, $< x, y > \notin DIS(B)$. Then we know $\gamma_C(x) \neq \gamma_C(y)$ and $\gamma_B(x) = \gamma_B(y)$. It is conflicted with our assumption. So if B is a maximum distribution consistent attribute set, then $DIS_{md}(C) \subseteq DIS(B)$.

- Neccessity($\Leftarrow$): Assume that if B satisfies $DIS_{md}(C) \subseteq DIS(B)$ then $\exists x \in U, \gamma_B(x) \neq \gamma_C(x)$. According to the assumption, we know $\exists y \in [x]_B - [x]_C, \gamma_C(y) \neq \gamma_C(x)$. That is to say, $< x, y > \in DIS_{md}(C)$, $< x, y > \notin DIS(B)$. It is conflicted with $DIS_{md}(C) \subseteq DIS(B)$. Consequently we know if B satisfies $DIS_{md}(C) \subseteq DIS(B)$ then $\forall x \in U, \gamma_B(x) = \gamma_C(x)$.

As mentioned above, Theorem 2 is true.   □

Above theorem is good for understanding but it is not friendly in computing. So we represent maximum distribution reduction in the way of classic heuristic functions. According to Definition 2, we can present the definition of the maximum distribution reduct by conditional entropy.

Theorem 3. *Given a decision table* $DT = (U, C, D, V, f)$, *Let TGran stands for* $U/IND_{md}(C)$, $B \subseteq C$ *is a maximum distribution reduct if and only if B satisfies*

(1) $H(TGran|B) = 0$;
(2) $\forall B' \subset B$, B' *doesn't satisfy condition (1).*

Proof. On the basis of Theorem 2, we can prove this theorem by explaining the equivalence relation between $H(TGran|B) = 0$ and $DIS_{md}(C) \subseteq DIS(B)$.

- Sufficiency($\Rightarrow$): According to the definition of $H(Q|P)$, it is easy to know that $H(TGran|B) = 0 \Leftrightarrow \forall Y \in TGran, \exists \{X : X \in U/B \wedge X \cap Y \neq \phi\}, \bigcup_{X_i \in X} X_i = Y$. Therefore, we conclude that $DIS_{md}(C) \subseteq DIS(B)$, $IND_{md}(C) \supseteq IND(B)$. As a result, $H(TGran|B) = 0 \Rightarrow B$ is a maximum distribution consistent attribute set.
- Neccessity($\Leftarrow$): Assume that B is a maximum distribution consistent attribute set, and B satisfies $H(TGran|B) \neq 0$. According to the definition of conditional etropy, we know $H(TGran|B) \neq 0$ means $\exists Y \in TGran, X \in U/B$ satisfies $X \cap Y \neq \phi \wedge X \not\subseteq Y$. That is to say, $\exists p \in X - X \cap Y, q \in X \cap Y, \gamma_C(p) \neq \gamma_C(q), \gamma_B(p) = \gamma_B(q)$. This concludes a conflict with B is a maximum distribution reduct. That is to say, if B is a maximum distribution consistent attribute set, then $H(TGran|B) = 0$.

As a result, Theorem 3 is true.   □

According to Definition 2, we can use dependency degree for the presentation of the maximum distribution reduct.

Theorem 4. *Given a decision table* $DT = (U, C, D, V, f)$, *Let TGran stands for* $U/IND_{md}(C)$, $B \subseteq C$ *is a maximum distribution reduct if and only if B satisfies (1)* $\Gamma_B(TGran) = 1$; *(2)* $\forall B' \subset B$, B' *doesn't satisfy condition (1).*

Proof. According to Theorem 2 and Theorem 3, the conclusion is clearly established.

For that $\Gamma_C(D) = 1 \Leftrightarrow \delta_C(D) = 1$, we have $\Gamma_B(TGran) = 1 \Leftrightarrow \delta_B(TGran) = 1$. As a result, there is no need to construct a theorem for maximum distribution reduction with $\delta_B(TGran)$. Based on upon theorems, the significance functions for maximum distribution reduction can be defined as follows.

(1) $Sig_1^{outer}(a, B, TGran) = H(TGran|B) - H(TGran|B \cup \{a\})$, $a \notin B$;
    $Sig_2^{outer}(a, B, TGran) = \Gamma_B(TGran) - \Gamma_{B \cup \{a\}}(TGran)$, $a \notin B$.
(2) $Sig_1^{inner}(a, B, TGran) = H(TGran|B - \{a\}) - H(TGran|B)$, $a \in B$;
    $Sig_2^{inner}(a, B, TGran) = \Gamma_B(TGran) - \Gamma_{B-\{a\}}(TGran)$, $a \in B$.

For convenience of algorithm description, we denote $Sig_i^j(a, B, TGran, U')$, $i \in \{1, 2\}$, $j \in \{inner, outer\}$ as the significance value computed in U'. Using Theorem 3 and Theorem 4, we can construct Algorithms 2 and 3 for maximum distribution reduction. Algorithms 2 and 3, indeed, are the variant of the discernibility matrix based reduction algorithms. The difference of two algorithms to the discernibility matrix based algorithms is the focus paid toward the indiscernibility relation instead of the discernibility relation. It can be proved by extending the relation of $IND(B) \cup DIS(B) = U \times U$ to the reduction algorithms. As a result, in intuition, the correctness of two algorithms can be transmitted from the discernibility matrix based algorithm for obtaining maximum distribution reducts.

Algorithm 2. Maximum Distribution Reduction Algorithm Using Condition Entropy (MDRAUCE)

Require: Decision table $DT = (U, C \cup D, V, f)$
Ensure: A maximum distribution reduct of DT
 1: $red := \emptyset$;
 2: $TGran := U/IND_{md}(C)$;
 3: $U' := U$;
 4: **while** $U_r \neq \emptyset$ **do**
 5:      Calculate $a_{max} : a_{max} = argmax_{a \in C - red} Sig_1^{outer}(a, red, TGran, U_r)$;
 6:      $red := red \cup \{a_{max}\}$;
 7:      $U' := U' - POS_{red}(TGran)$;
 8:      $TGran := TGran - POS_{red}(TGran)$;
 9: **end while**
10: **return** red

---

4   Correctness Analysis and Experiments Results

The objective of this section is to present the correctnes and the efficiency of the attribute reduction algorithms proposed in this paper, *i.e.* MDRAUCE and MDRAUDD. To show the correctness of two algorithms, we calculate the maximum distribution reduct of Table 1 using MDRAUCE and MDRAUDD, and check outputs of two algorithms with the definition of maximum distribution

Algorithm 3. Maximum Distribution Reduction Algorithm Using Dependency Degree (MDRAUDD)

Require: Decision table $DT = (U, C \cup D, V, f)$
Ensure: A maximum distribution reduct of DT
1: $red := \emptyset$;
2: $TGran := U/IND_{md}(C)$;
3: $U' := U$;
4: **while** $U' \neq \emptyset$ **do**
5:     Calculate $a_{max} : a_{max} = argmax_{a \in C - red} Sig_2^{outer}(a, red, TGran, U')$;
6:     $red := red \cup \{a_{max}\}$;
7:     $U' := U' - POS_{red}(TGran)$;
8:     $TGran := TGran - POS_{red}(TGran)$;
9: **end while**
10: **return** red

---

reduction for validation. On the other side, we employed 12 UCI data sets to verify the performance of time consumption of MDRAUCE, MDRAUDD, and existing maximum distribution reduction algorithms.

4.1 The Validation of Correctness

In this part, we show the correctness of two algorithms proposed in Sect. 3 through presenting the process of calculating the maximum distribution reduct for Table 1 using Algorithm 2 and Algorithm 3. After that, we check the outputs of two algorithms according to the maximum distribution definition.

The process of MDRAUCE for finding the maximum distribution reduct of Table 1 is presented here. In the following description of calculation process, "item1=item2" denotes that the relationship of two are on equal item, and ":=" stands for the assignment operation.

Step 1. $red := \emptyset$, $TGran = U/IND_{md}(C) = \{\{x_1\}, \{x_2, x_5, x_6, x_7\}, \{x_3, x_4\}\}$, $U' = \{x_1, x_2, \cdots, x_7\}$.

Step 2. $Sig_1^{outer}(a_1, red, TGran, U') = H^{U'}(TGran|red) - H^{U'}(TGran|red \cup \{a_1\}) = 1.38 - 0.79 = 0.59$, $Sig_{md}^{outer}(a_2, red, TGran, U') = H^{U'}(TGran|red) - H^{U'}(TGran|red \cup \{a_2\}) = 1.38 - 0.98 = 0.40$, $Sig_{md}^{outer}(a_3, red, TGran, U') = H^{U'}(TGran|red) - H^{U'}(TGran|red \cup \{a_3\}) = 1.38 - 0.86 = 0.52$. So $a_{max} = a_1$, $red := red \cup \{a_1\} = \{a_1\}$. We have $POS_{red}(TGran) = \{x_1\}$. U' and $TGran$ are updated as follows, $U' := U' - POS_{red}(TGran) = \{x_2, x_3, \cdots, x_7\}$, $TGran = TGran - POS_{red}(TGran) = \{\{x_2, x_5, x_6, x_7\}, \{x_3, x_4\}\}$.

Step 3. $Sig_1^{red}(a_2, red, TGran, U') = H^{U'}(TGran|red) - H^{U'}(TGran|red \cup \{a_2\}) = 0.92 - 0.81 = 0.11$, $Sig_{md}^{outer}(a_3, red, TGran, U') = H^{U'}(TGran|red) - H^{U'}(TGran|red \cup \{a_3\}) = 0.92 - 0.46 = 0.46$. So $a_{max} = a_3$, $red := red \cup \{a_3\} = \{a_1, a_3\}$. We have $POS_{red}(TGran) = \{x_5, x_6, x_7\}$. U' and $TGran$ are updated as follows, $U' := U' - POS_{red}(TGran) = \{x_2, x_3, x_4\}$, $TGran = TGran - POS_{red}(TGran) = \{\{x_2\}, \{x_3, x_4\}\}$.

Step 4. $Sig_{md}^{outer}(a_2, red, TGran, U') = H^{U'}(TGran|red) - H^{U'}(TGran|red \cup \{a_2\}) = 0.92 - 0 = 0.92$. So $a_{max} = a_2$, $red := red \cup \{a_2\} = \{a_1, a_3, a_2\}$. We have $POS_{red}(TGran) = \{x_2, x_3, x_4\}$. U' and $TGran$ are updated as follows, $U' := U' - POS_{red}(TGran) = \emptyset$, $TGran = TGran - POS_{red}(TGran) = \emptyset$.

Step 5. Because $U' = \emptyset$, program is over. Algorithm outputs $red = \{a_1, a_3, a_2\}$ as the result.

The process of MDRAUDD for obtaining the maximum distribution reduct of Table 1 is presented as follows.

Step 1. $red := \emptyset$, $TGran = U/IND_{md}(C) = \{\{x_1\}, \{x_2, x_5, x_6, x_7\}, \{x_3, x_4\}\}$, $U' = \{x_1, x_2, \cdots, x_7\}$.

Step 2. $Sig_1^{outer}(a_1, red, TGran, U') = \Gamma_{red \cup \{a_1\}}^{U'}(TGran) - \Gamma_{red}^{U'}(TGran) = \frac{1}{7} - 0 = \frac{1}{7}$, $Sig_1^{outer}(a_2, red, TGran, U') = \Gamma_{red \cup \{a_2\}}^{U'}(TGran) - \Gamma_{red}^{U'}(TGran) = 0 - 0 = 0$, $Sig_1^{outer}(a_3, red, TGran, U') = \Gamma_{red \cup \{a_3\}}^{U'}(TGran) - \Gamma_{red}^{U'}(TGran) = \frac{3}{7} - 0 = \frac{3}{7}$. So $a_{max} = a_3$, $red := red \cup \{a_3\} = \{a_3\}$. We have $POS_{red}(TGran) = \{x_5, x_6, x_7\}$. U' and $TGran$ are updated as follows, $U' := U' - POS_{red}(TGran) = \{x_1, x_2, x_3, x_4\}$, $TGran = TGran - POS_{red}(TGran) = \{\{x_1\}, \{x_2\}, \{x_3, x_4\}\}$.

Step 3. $Sig_1^{outer}(a_1, red, TGran, U') = \Gamma_{red \cup \{a_1\}}^{U'}(TGran) - \Gamma_{red}^{U'}(TGran) = \frac{1}{4} - 0 = \frac{1}{4}$, $Sig_1^{outer}(a_2, red, TGran, U') = \Gamma_{red \cup \{a_2\}}^{U'}(TGran) - \Gamma_{red}^{U'}(TGran) = \frac{2}{4} - 0 = 0.5$. So $a_{max} = a_2$, $red := red \cup \{a_2\} = \{a_3, a_2\}$. We have $POS_{red}(TGran) = \{x_3, x_4\}$. U' and $TGran$ are updated as follows, $U' := U' - POS_{red}(TGran) = \{x_1, x_2\}$, $TGran = TGran - POS_{red}(TGran) = \{\{x_1\}, \{x_2\}\}$.

Step 4. $Sig_1^{outer}(a_1, red, TGran, U') = \Gamma_{red \cup \{a_1\}}^{U'}(TGran) - \Gamma_{red}^{U'}(TGran) = 1 - 0 = 1$. So $a_{max} = a_1$, $red := red \cup \{a_12\} = \{a_3, a_2, a_1\}$. We have $POS_{red}(TGran) = \{x_1, x_2\}$. U' and $TGran$ are updated as follows, $U' := U' - POS_{red}(TGran) = \emptyset$, $TGran = TGran - POS_{red}(TGran) = \emptyset$.

Step 5. Because $U' = \emptyset$, program is over. Algorithm outputs $red = \{a_3, a_2, a_1\}$ as the result.

According to Definition 1, we know $\gamma_{red}(x_1) = \{P_1\}$ and $\gamma_{red}(x_2) = \{P_2\}$; for $x \in \{x_3, x_4\}$, we have $\gamma_{red}(x) = \{P_1, P_2\}$; for $x \in \{x_5, x_6, x_7\}$, we know $\gamma_{red}(x) = \{P_2\}$. Meanwhile, we know $\gamma_C(x_1) = \{P_1\}$ and $\gamma_C(x_2) = \{P_2\}$; for $x \in \{x_3, x_4\} \gamma_C(x) = \{P_1, P_2\}$; for $x \in \{x_5, x_6, x_7\}$, we have $\gamma_C(x) = \{P_2\}$. It is obvious that for $\forall x \in U$, $\gamma_{red}(x) = \gamma_C(x)$. Finally we know that MDRAUCE and MDRAUDD are correct.

4.2   The Efficiency of Proposed Algorithms

In this part, we employed 12 data sets to verify the performance of time consumption of MDRAUDD, MDRAUDD, Q-MDRA [15] and QGARA-FS [16]. We carried out all the attribute reduction algorithms in experiments on a personal computer with Windows 10, Intel(R) Core(TM) CPU i5-8265U 1.60GHZ and 8GB RAM memory. The software used was Visual Studio Code 1.3.8, and the programming language was python 3.7.

The data sets used in experiments are all downloaded from UCI repository of machine learning data sets [25] whose basic information is outlined in Table 2. For the sake that reduction algorithms can address only symbolic data, data sets containing continuous attributes were preprocessed by CAIM [26] discretization algorithm. For each data sets, the positive region dependency degree, $i.e.$ $\gamma_C(D)$, is listed in the last column of Table 2. As we know, the data set is consistent if $\gamma_C(D) = 1$; otherwise, it is inconsistent. As shown in Table 2, Wpbc, Wine, and Sonar are consistent. Taking into consideration the value of $\gamma_C(D)$, we take Sat, Segment, Wdbc, and Wave as consistent data sets whose value of $\gamma_C(D)$ satisfies $0.981 \leq \gamma_C(D) \leq 1$. The other 5 data sets (Vehicle, Ion, Glass, Heart, and Pid) are inconsistent.

Table 2. Description of data sets

ID	Data sets	Cases	Attributes	Classes	$\gamma_C(D)$
1	Wpdc	198	34	2	1
2	Wine	178	13	3	1
3	Sat	6435	86	6	0.993
4	Segment	2310	19	7	0.991
5	Wdbc	569	30	2	0.989
6	Waveform	5000	21	3	0.981
7	Vehicle	846	18	4	0.946
8	Ions	351	34	2	0.940
9	Glass	214	9	7	0.937
10	Heart	270	6	2	0.935
11	Sonar	208	60	2	1
12	Pid	768	8	2	0.519

Table 3 indicate the computational time of MDRAUCE, MDRADD, Q-MDRA, and QGARA-FS for obtaining maximum distribution reduct on 12 data sets. We can see that MDRADD was the fastest of four attribute reduction algorithms for that it was the best on 11 data sets, and MDRAUCE was faster than QGARA-FS. MDRAUCE performed better than Q-MDRA in obtaining the

Table 3. Time consumption of maximum distribution reduction algorithms

Data sets	Time consumption(s)			
	MDRAUCE	MDRADD	Q-MDRA	QGARA-FS
Wpdc	0.253	**0.208**	0.447	1.110
Wine	0.051	0.051	**0.045**	0.166
Sat	5.395	**3.220**	59.683	16.604
Segment	1.027	**0.546**	1.743	1.213
Wdbc	0.591	**0.452**	0.915	2.411
Waveform	3.607	**1.634**	11.820	2.551
Vehicle	0.463	**0.304**	1.587	0.633
Ions	0.612	**0.300**	0.572	1.582
Glass	**0.034**	**0.034**	0.064	0.092
Heart	0.087	**0.040**	0.121	0.105
Sonar	0.649	**0.411**	0.621	1.832
Pid	0.101	**0.081**	0.100	0.137
Average	1.073	**0.607**	6.477	2.370

reduct of 9 data sets. Q-MDRA performed better than MDRAUCE, MDRAUCE on small data sets,*i.e.* Wine data set. However, in processing the large scale data, Q-MDRA consumed more time than MDRAUCE, MDRADD. From results of experiments on both consistent and inconsistent decision tables, the computational times of four algorithms in obtaining the maximum distribution reduct followed this order: MDRADD ≥ MDRAUCE, Q-MDRA > QGARA-FS. For most of the cases in experiments, the computational time of MDRAUDD can reduce half of the computation time of QGARA-FS and Q-MDRA, such as data sets Wpdc, Glass, Heart, etc. In the same condition. from the row of average time consumption in obtaining reduct of 12 data sets, we know that MDRAUCE and MDRADD are more efficient and steady in time consumption of maximum distribution reduction than existing maximum distribution reduction algorithms.

5   Conclusion

In this paper, we focus on the maximum distribution reduction for complete inconsistent decision tables. The problems in Li's algorithm for obtaining the maximum distribution reduct were pointed out, and based on classic heuristic functions, we designed two novel heuristic algorithms, *i.e.* MDRAUCE and MDRADD, to efficiently finding a maximum distribution reduct. Because the scale of data processed becomes larger and larger, the efficiency of attribute reduction algorithms is still our focus of future researches.

Acknowledgement. A special thank is owed to professor Li Min, the discussing with him on Q-MDRA contributes a lot to this paper. The work is partially supported by the National Key Research and Development Project (No. 213), the National Nature Science Foundation of China (No. 61976160, No. 61673301) and Key Lab of Information Network Security, Ministry of Public Security (No. C18608).

References

1. Pawlak, Z.: Rough sets. Int. J. Comput. Inf. Sci. **11**(5), 341–356 (1982)
2. Huang, K.Y., Li, I.-H.: A multi-attribute decision-making model for the robust classification of multiple inputs and outputs datasets with uncertainty. Appl. Soft Comput. **38**, 176–189 (2016)
3. Dai, J., Xu, Q.: Approximations and uncertainty measures in incomplete information systems. Inf. Sci. **198**, 62–80 (2012)
4. Shi, J., Lei, Y., Zhou, Y., Gong, M.: Enhanced rough-fuzzy c-means algorithm with strict rough sets properties. Appl. Soft Comput. **46**, 827–850 (2016)
5. Zhan, J., Ali, M.I., Mehmood, N.: On a novel uncertain soft set model: Z-soft fuzzy rough set model and corresponding decision making methods. Appl. Soft Comput. **56**, 446–457 (2017)
6. Das, R.T., Ang, K.K., Quek, C.: ieRSPOP: a novel incremental rough set-based pseudo outer-product with ensemble learning. Appl. Soft Comput. **46**, 170–186 (2016)
7. Xie, X., Qin, X., Yu, C., Xu, X.: Test-cost-sensitive rough set based approach for minimum weight vertex cover problem. Appl. Soft Comput. **64**, 423–435 (2018)
8. Hu, Y.C.: Flow-based tolerance rough sets for pattern classification. Appl. Soft Comput. **27**, 322–331 (2015)
9. Huang, K.Y.: An enhanced classification method comprising a genetic algorithm, rough set theory and a modified PBMF-index function. Appl. Soft Comput. **12**(1), 46–63 (2012)
10. Wang, F., Liang, J., Dang, C.: Attribute reduction for dynamic data sets. Appl. Soft Comput. **13**(1), 676–689 (2013)
11. Kaya, Y., Uyar, M.: A hybrid decision support system based on rough set and extreme learning machine for diagnosis of hepatitis disease. Appl. Soft Comput. **13**(8), 3429–3438 (2013)
12. Thangavel, K., Pethalakshmi, A.: Dimensionality reduction based on rough set theory: a review. Appl. Soft Comput. **9**(1), 1–12 (2009)
13. Kryszkiewicz, M.: Comparative study of alternative types of knowledge reduction in inconsistent systems. Int. J. Intell. Syst. **16**(1), 105–120 (2001)
14. Zhang, W.X., Mi, J.S., Wu, W.Z.: Knowledge reductions in inconsistent information systems. Chin. J. Comput.-Chin. Ed. **26**(1), 12–18 (2003)
15. Li, M., Shang, C.X., Feng, S.Z., Fan, J.P.: Quick attribute reduction in inconsistent decision tables. Inf. Sci. **254**, 155–180 (2014)
16. Ge, H., Li, L.S., Xu, Y., Yang, C.J.: Quick general reduction algorithms for inconsistent decision tables. Int. J. Approx. Reason. **82**, 56–80 (2017)
17. Yao, Y., Zhao, Y., Wang, J.: On reduct construction algorithms. In: Gavrilova, M.L., Tan, C.J.K., Wang, Y., Yao, Y., Wang, G. (eds.) Transactions on Computational Science II. LNCS, vol. 5150, pp. 100–117. Springer, Heidelberg (2008). https://doi.org/10.1007/978-3-540-87563-5_6
18. Hu, X.H., Cercone, N.: Learning in relational databases: a rough set approach. Comput. Intell. **11**(2), 323–338 (1995)

19. Tian, J., Wang, Q., Yu, B., Yu, D.: A rough set algorithm for attribute reduction via mutual information and conditional entropy. In: 2013 10th International Conference on Fuzzy Systems and Knowledge Discovery (FSKD), pp. 567–571. IEEE (2013)
20. Sun, H.R., Wang, R., Xie, B.X., Tian, Y.: Continuous attribute reduction method based on an automatic clustering algorithm and decision entropy. In: Control Conference (2017)
21. Yan, T., Han, C.Z.: Entropy based attribute reduction approach for incomplete decision table. In: International Conference on Information Fusion (2017)
22. Dash, M., Liu, H.: Consistency-based search in feature selection. Artif. Intell. **151**(1), 155–176 (2003)
23. Hu, Q., Zhao, H., Xie, Z., Yu, D.: Consistency based attribute reduction. In: Zhou, Z.-H., Li, H., Yang, Q. (eds.) PAKDD 2007. LNCS (LNAI), vol. 4426, pp. 96–107. Springer, Heidelberg (2007). https://doi.org/10.1007/978-3-540-71701-0_12
24. Wang, J., Miao, D.Q.: Analysis on attribute reduction strategies of rough set. J. Comput. Sci. Technol. **13**(2), 189–192 (1998)
25. Asuncion, A., Newman, D.J.: UCI machine learning repository (2007)
26. Feng, S.Z., Li, M., Deng, S.B., Fan, J.P.: An effective discretization based on class-attribute coherence maximization. Pattern Recogn. Lett. **32**, 1962–1973 (2011)

Attribute Reduction from Closure Operators and Matroids in Rough Set Theory

Mauricio Restrepo[1]([⊠])[iD] and Chris Cornelis[2][iD]

[1] Universidad Militar Nueva Granada, Bogotá, Colombia
mauricio.restrepo@unimilitar.edu.co
[2] Ghent University, Ghent, Belgium
chris.cornelis@ugent.be
https://www.umng.edu.co, https://www.ugent.be

Abstract. In this paper, we present a new closure operator defined on the set of attributes of an information system that satisfies the conditions for defining a matroid. We establish some basic relationships between equivalence classes and approximation operators where different sets of attributes are used. It is shown that the reducts of an information system can be obtained from dependent sets of a matroid. Finally, we show that the closure operator can be defined at least in three different ways.

Keywords: Attribute reduction · Closure operators · Matroids · Rough sets · Approximation operators

1 Introduction

The main concept of rough set theory, proposed by Z. Pawlak, is the indiscernibility between objects given by an equivalence relation in a non-empty set U, called *Universe*. This theory has been used for the study of information systems and it has been successfully applied in artificial intelligence fields such as machine learning, pattern recognition, decision analysis, process control, knowledge discovery in databases, and expert systems.

Matroids were introduced in 1935 by Whitney and have been used as a generalization of the concept of independence in different mathematical theories like linear spaces, graph theory, field theory, and fuzzy sets [5,11]. In relation to rough sets, many papers have shown interesting connections with matroids [6–10,14–17]. Zhu et al., presented the concept of rough matroids based on a relation [21,22] and rough matroids based on coverings [19]. An interesting generalization of rough matroids for a general approximation operator was proposed in [12].

The problem of attribute reduction is a problem of high computational complexity and turns out to be fundamental in the construction of machine learning models.

This work was supported by Universidad Militar Nueva Granada's VICEIN Special Research Fund, under project CIAS 2548-2018, and by the Odysseus program of the Research Foundation-Flanders.

R. Bello et al. (Eds.): IJCRS 2020, LNAI 12179, pp. 183–192, 2020.
https://doi.org/10.1007/978-3-030-52705-1_13

This problem has attracted the attention of many researchers and has been addressed from the perspective of different theories. Some relationships of attribute reduction with matroids and rough sets, can be found in [4, 17, 18, 20]. However, in most publications related to matroids and the attribute reduction problem, matroidal structures have been defined in terms of the set of objects and not the set of attributes. This paper proposes a matroidal structure defined in terms of an attribute set whose maximal independent elements, matching reducts. Therefore the idea of independence in a set given by a matroid can also be applied to the set of attributes A, in order to address the problem of attribute reduction in a better way.

The paper is organized as follows: Sect. 2 presents preliminary concepts regarding rough set theory, lower and upper approximations, closure operators and matroids. Section 3 establishes some basic relationships between equivalence classes and approximation operators where different sets of attributes are used. Section 4 presents three definitions of closure operators and the necessary conditions for them to define a matroidal structure. Finally, Sect. 5 presents the main conclusions of the paper and describes future work.

2   Preliminaries

2.1   Rough Sets

Rough set theory involves a data table composed of a finite set U of objects described by a finite set A of attributes. The basic notions of rough set theory are: indiscernibility relation on U, lower and upper approximation operators, dependence among attributes, and decision rules derived from lower approximations [3].

The pair (U, A) is called an information system. A simple example can be seen in Table 1. In this case $U = \{1, 2, 3, 4, 5, 6\}$ and $A = \{a_1, a_2, a_3, a_4\}$. From each subset of attributes $P \subseteq A$, it is possible to define a binary relation E_P: $x E_P y$ if and only if $f_a(x) = f_a(y)$ for all $a \in P$. In this case, $f_a(x)$ is the value of an object x for the attribute $a \in A$. According to the same table, we have that $f_{a_1}(2) = B$, while $f_{a_2}(1) = G$. Clearly this is an equivalence relation, i.e., a reflexive, symmetric and transitive relation.

If $X \subseteq U$, the operators:

$$\underline{apr}(X) = \{x \in U : [x]_P \subseteq X\} \tag{1}$$

$$\overline{apr}(X) = \{x \in U : [x]_P \cap X \neq \emptyset\} \tag{2}$$

are called the lower and upper approximations of X. $[x]_P$ represents the equivalence class of x for the set of attributes P, and $I_P = U/E_P$ represents the equivalence classes, defined from the equivalence relation E_P. Each subset of attributes $P \subseteq A$ defines a partition of U. In particular, for $P = \{a\}$, the partition is represented as $\mathbb{P}_a$.

The pair $(\underline{apr}, \overline{apr})$ is a dual pair of approximation operators, that is, for $X \subseteq U$, $\underline{apr}(\sim X) = \sim \overline{apr}(X)$, where $\sim X$ represents the complement of X, i.e., $\sim X = U \setminus X$.

Example 1. In Table 1 we have an information system with six objects and four condition attributes $\{a_1, a_2, a_3, a_4\}$.

Table 1. An information system.

Object	Attributes			
	a_1	a_2	a_3	a_4
1	G	G	B	M
2	B	B	B	G
3	M	G	A	M
4	M	B	B	G
5	M	B	B	M
6	G	B	A	B

For $P = \{a_1, a_2\}$ and $Q = \{a_1, a_2, a_3\}$, we have that $[x]_P = [x]_Q$ for all $x \in U$, therefore $\underline{apr}_P(X) = \underline{apr}_Q(X)$ for all $X \subseteq U$.

Also, it is easy to see that:

- $\mathbb{P}_{a_1} = \{\{1,6\}, \{2\}, \{3,4,5\}\}$
- $\mathbb{P}_{a_2} = \{\{1,3\}, \{2,4,5,6\}\}$
- $\mathbb{P}_{a_3} = \{\{3,6\}, \{1,2,4,5\}\}$
- $\mathbb{P}_{a_4} = \{\{1,3,5\}, \{2,4\}, \{6\}\}$

2.2 Reducts in Rough Set Theory

Attribute reduction in rough set theory involves the removal of attributes that have no significance to the classification problem. An attribute reduct set (or simply reduct) is a subset P of the set of attributes A such that the quality of classification is the same [13].

According to the definition of superfluous attribute given in [3] we can say that if $[x]_P = [x]_{P \cup \{a\}}$ for all $x \in U$, then a is called a superfluous attribute of P; otherwise, a is called indispensable in P. The set P is independent if all of its attributes are indispensable. The subset Q of P is a reduct of P (denoted as $Red(P)$) if Q is independent and $[x]_Q = [x]_P$ for all $x \in U$.

2.3 Matroids

Matroids are related to the notion of linear independence. They can be introduced from an elementary point of view as a collection of sets of linearly independent vectors. Let us suppose that $v_1 = \begin{bmatrix} 1 \\ 1 \\ 2 \end{bmatrix}$, $v_2 = \begin{bmatrix} 0 \\ 0 \\ -1 \end{bmatrix}$, $v_3 = \begin{bmatrix} 2 \\ 2 \\ 0 \end{bmatrix}$ and $v_4 = \begin{bmatrix} 1 \\ 2 \\ 0 \end{bmatrix}$ are vectors in $\mathbb{R}^3$. It is easy to see that $S = \{v_1, v_2, v_3, v_4\}$ is a set of linearly dependent vectors, because the matrix A with columns v_i has free variables.

$$A = \begin{bmatrix} 1 & 0 & 2 & 1 \\ 1 & 0 & 2 & 2 \\ 2 & -1 & 0 & 0 \end{bmatrix} \tag{3}$$

However, there are several subsets of S whose vectors are linearly independent. For example, $S_1 = \{v_1, v_2\}$ and $S_2 = \{v_1, v_2, v_4\}$. A collection of sets with linearly independent vectors determines a structure that we will call a matroid.

Additionally, we know that any subset of linearly independent vectors is also linearly independent. In this case, a collection of independent sets is:

$$\mathbb{I} = \{\emptyset, \{v_1\}, \{v_2\}, \{v_4\}, \{v_1, v_2\}, \{v_1, v_4\}, \{v_2, v_4\}, \{v_1, v_2, v_4\}\}.$$

There are different definitions of a matroid. In this case, we consider the following definition in terms of independence.

Definition 1. *Let U be a finite set. A matroid on U is an ordered pair $M = (U, \mathbb{I})$, where $\mathbb{I}$ is a collection of subsets of U with the following properties:*

1. $\emptyset \in \mathbb{I}$.
2. *If $I \in \mathbb{I}$ and $I' \subseteq I$ then $I' \in \mathbb{I}$.*
3. *If $I_1, I_2 \in \mathbb{I}$ and $|I_1| < |I_2|$, then there exists $x \in I_2 - I_1$ such that $I_1 \cup \{x\} \in \mathbb{I}$. Here $|I|$ denotes the cardinality of the set I.*

If $I \in \mathbb{I}$, then I is called a independent set. If a subset of U is not an independent set, then it is called a dependent set. By property 3, every two maximal independent sets in a matroid have the same cardinal number.

Rank Function. The rank function of a matroid $M = (U, \mathbb{I})$ is a function $r : \mathscr{P}(U) \to \mathbb{N}$ defined as:

$$r(X) = \max\{|Y| : Y \subseteq X, Y \in \mathbb{I}\} \tag{4}$$

This function satisfies:

1. $r(\emptyset) = 0$
2. $0 \leq r(X) \leq |X|$.
3. If $Y \subseteq X$, then $r(Y) \leq r(X)$
4. $r(X \cup Y) + r(X \cap Y) \leq r(X) + r(Y)$

2.4 Closure Operators

The notion of closure operator usually is applied to ordered sets and topological spaces. Some relations between closure operators with upper approximation and matroids are presented in [1,6].

We present some concepts about ordered structures, according to Blyth [2].

Definition 2. *A map $c : \mathscr{P}(U) \to \mathscr{P}(U)$ is a **closure operator** on U if it is such that, for all $A, B \subseteq U$:*

P_1. $A \subseteq c(A)$, *(extensive)*.
P_2. $A \subseteq B$ implies $c(A) \subseteq c(B)$, *(order preserving)*.
P_3. $c(A) = c[c(A)]$, *(idempotent)*.

If $M = (U, \mathbb{I})$ is a matroid and r its rank function, then

$$c_M(A) = \{x \in E : r(A) = r(A \cup \{x\})\} \tag{5}$$

is a closure operator. Also, the operator c_M satisfies the following property:

P_4. For any $x, y \in U$ and any $A \subseteq U$, it follows from $y \in c(A \cup \{x\})$ and $y \notin c(A)$ that $x \in c(A \cup \{y\})$.

Any closure operator which satisfies property P_4 defines a matroidal structure according to the following proposition.

Proposition 1. *[11] If c is a closure operator on a finite set U that satisfied property P_4, and*

$$\mathbb{I} = \{I \subseteq U : x \notin c(I - x) \text{ for all } x \in I\}$$

then $(U, \mathbb{I})$ is a matroid.

3 Relationships on Attribute Sets

The following propositions establish useful relationships between equivalence classes and approximation operators for different sets of attributes.

Proposition 2. *If $P, Q \subseteq A$ and $P \subseteq Q$, then $[x]_P \supseteq [x]_Q$, for all $x \in U$.*

Proof. If $w \in [x]_Q$ then $f_a(x) = f_a(w)$ for all $a \in Q$, in particular for all $a \in P$, since $P \subseteq Q$. Therefore $f_a(x) = f_a(w)$ for all $a \in P$, and so $w \in [x]_P$.

Proposition 3. *If $P, Q \subseteq A$ and $P \subseteq Q$, then $\underline{apr}_P(X) \subseteq \underline{apr}_Q(X)$, for all $X \subseteq U$.*

Proof. If $w \in \underline{apr}_P(X)$, then $[w]_P \subseteq X$. From Proposition 2, we have that $[w]_Q \subseteq [w]_P$. Therefore $[w]_Q \subseteq X$ and so, $w \in \underline{apr}_Q(X)$.

Proposition 4. *If $P \subseteq A$ is a set of attributes, $[x]_{P \cup \{a\}} = [x]_P \cap [x]_a$, for all $x \in U$.*

Proof. We have that $w \in [x]_{P \cup \{a\}}$ if and only if $f_b(x) = f_b(w)$ for all $b \in P \cup \{a\}$, i.e., if and only if for all $a \in P$ and for $b = a$, if and only if $w \in [x]_P$ and $w \in [x]_a$, if and only if $w \in [x]_P \cap [x]_a$.

Using the order relation above it is possible to establish the following:
For all $P, Q \subseteq A$ and $X \subseteq U$ we have:

$$\underline{apr}_{P \cap Q}(X) \leq \underline{apr}_P(X) \cap \underline{apr}_Q(X) \leq \underline{apr}_P(X) \cup \underline{apr}_Q(X) \leq \underline{apr}_{P \cup Q}(X) \tag{6}$$

Example 2. In Table 2 we have the equivalence classes of each element $x \in U$, using some sets of attributes. For the particular cases of $P = \{a_i\}$, we have the partitions given in Example 1.

The sets of attributes $Q_1 = \{a_1, a_2, a_4\}$, $Q_2 = \{a_1, a_3, a_4\}$, and $A = \{a_1, a_2, a_3, a_4\}$ have the same equivalence class for each $x \in U$. It is easy to see that lower and upper approximations of each $X \subseteq U$ are also the same. In this case, Q_1 and Q_2 are reducts of A.

Table 2. Equivalence classes $[x]_P$ for different sets of attributes P.

Object	$\{a_1\}$	$\{a_1,a_2\}$	$\{a_1,a_3\}$	$\{a_1,a_4\}$	$\{a_1,a_2,a_3\}$	$\{a_1,a_2,a_4\}$	$\{a_1,a_3,a_4\}$	A
1	$\{1,6\}$	$\{1\}$	$\{1\}$	$\{1\}$	$\{1\}$	$\{1\}$	$\{1\}$	$\{1\}$
2	$\{2\}$	$\{2\}$	$\{2\}$	$\{2\}$	$\{2\}$	$\{2\}$	$\{2\}$	$\{2\}$
3	$\{3,4,5\}$	$\{3\}$	$\{3\}$	$\{3,5\}$	$\{3\}$	$\{3\}$	$\{3\}$	$\{3\}$
4	$\{3,4,5\}$	$\{4,5\}$	$\{4,5\}$	$\{4\}$	$\{4,5\}$	$\{4\}$	$\{4\}$	$\{4\}$
5	$\{3,4,5\}$	$\{4,5\}$	$\{4,5\}$	$\{3,5\}$	$\{4,5\}$	$\{5\}$	$\{5\}$	$\{5\}$
6	$\{1,6\}$	$\{6\}$	$\{6\}$	$\{6\}$	$\{6\}$	$\{6\}$	$\{6\}$	$\{6\}$

4   Closure Operators on Attribute Sets

Let (U, A) be a decision system, where U is a finite set and A is a finite set of attributes. For each $P \subseteq A$, we can define at least three closure operators, as follows.

Definition 3. *For each $P \subseteq A$ subset of attributes, we define:*

$$c_1(P) = \{a \in A : \underline{apr}_{P\cup\{a\}}(X) = \underline{apr}_P(X) \text{ for all } X \subseteq U\} \tag{7}$$

Proposition 5. *The operator c_1 is a closure on A.*

Proof. We will show the three properties:

1. $P \subseteq c_1(P)$. *If $a \in P$, then $P \cup \{a\} = P$ and $\underline{apr}_{P\cup a}(X) = \underline{apr}_P(X)$.*
2. *If $P \subseteq Q$, then $c_1(P) \subseteq c_1(Q)$. If $a \in c_1(P)$, we have that $\underline{apr}_{P\cup\{a\}}(X) = \underline{apr}_P(X)$ for all $X \subseteq U$. We will show that $a \in c(Q)$, i.e., $\underline{apr}_{Q\cup\{a\}}(X) = \underline{apr}_Q(X)$. If $w \in \underline{apr}_{Q\cup\{a\}}(X)$, then $[w]_{Q\cup\{a\}} \subseteq [w]_{P\cup\{a\}} = [w]_P$. Therefore, $\underline{apr}_{Q\cup\{a\}}(X) \subseteq \underline{apr}_Q(X)$. According to Proposition 2, $\underline{apr}_Q(X) \subseteq \underline{apr}_{Q\cup\{a\}}(X)$. So, $\underline{apr}_{Q\cup\{a\}}(X) = \underline{apr}_Q(X)$ and $a \in c_2(P)$.*
3. $c_1(c_1(P)) = c_1(P)$. *From properties 1 and 2 we have that $c_1(P) \subseteq c_1(c_1(P))$. For the other relation let a be such that $a \in c_1(c_1(P))$. We have that $\underline{apr}_{c_1(P)\cup\{a\}}(X) = \underline{apr}_{c_1(P)}(X)$ for all $X \subseteq U$. We need to show that $\underline{apr}_{P\cup\{a\}}(X) = \underline{apr}_P(X)$. The relation $\underline{apr}_{c_1(P)}(X) \subseteq \underline{apr}_{c_1(c_1(P))}(X)$ always holds. For $X \in \mathbb{P}_a$, we have that $\underline{apr}_{c_1(P)}(X) = X$. Therefore $\underline{apr}_{c_1(c_1(P))}(X) = X$, so $\in c_1(P)$, (see Definition 5).*

There are at least other two definitions for the closure operator.

Definition 4. *For each $P \subseteq A$ subset of attributes, we define:*

$$c_2(P) = \{a \in A : [x]_P = [x]_{P\cup\{a\}} \text{ for all } x \in U\} \tag{8}$$

The following proposition establishes the equivalence between c_1 and c_2 operators.

Proposition 6. $c_2 = c_1$.

Proof. We will see that $c_1(P) \subseteq c_2(P)$ and $c_2(P) \subseteq c_1(P)$, for all $P \subseteq A$.

1. $c_2 \leq c_1$. If $a \in c_2(P)$, then $[x]_P = [x]_{P \cup \{a\}}$ for all $x \in U$. $w \in \underline{apr}_P(X)$ iff $[w]_P \subseteq X$ iff $[w]_{P \cup \{a\}} \subseteq X$ iff $w \in \underline{apr}_{P \cup \{a\}}(X)$. So $\underline{apr}_P(X) = \underline{apr}_{P \cup \{a\}}(X)$ for all $X \subseteq U$, and $a \in c_1(P)$.

2. $c_1 \leq c_2$. Suppose that $a \in c_1(P)$, and $a \notin c_2(P)$. For some $a \in A$, $P \subseteq A$. Since $a \notin c_2(P)$, there exists $x \in U$ such that $[x]_P \neq [x]_{P \cup \{a\}}$. Hence, there exists $y \in U$ for which $y \in [x]_P$ but $y \notin [x]_{P \cup \{a\}}$. Suppose $X = [x]_{P \cup \{a\}}$, then $\underline{apr}_{P \cup \{a\}}(X) = X$ and since $a \in c_1(P)$, $\underline{apr}_P(X) = X$. Since $x \in X$, therefore $[x]_P \subseteq X$. But this means that $y \in X = [x]_{P \cup \{a\}}(X) = X$, a contradiction. So $c_1 \leq c_2$.

Proposition 7. *Operator c_2 satisfies property P_4. This is: for any $a, b \in A$ and any $P \subseteq A$, if $b \in c_2(P \cup \{a\})$ and $b \notin c_2(P)$, then $a \in c_2(P \cup \{b\})$.*

Proof. The expression $b \in c_2(P \cup \{a\})$ means that $[x]_{P \cup \{a\}} = [x]_{P \cup \{a,b\}}$ for all $x \in U$, and we have to show that $[x]_{P \cup \{b\}} = [x]_{P \cup \{a,b\}}$ for all $x \in U$. From a preliminary proposition, we have that $[x]_{P \cup \{b\}} \supseteq [x]_{P \cup \{a,b\}}$. Suppose that $y \in [x]_{P \cup \{b\}} = [x]_P \cap [x]_b$ and $y \notin [x]_{P \cup \{a,b\}}$. Then $y \notin [x]_{P \cup \{a\}} = [x]_P \cap [x]_a$. Thus, $y \notin [x]_a$ nor $y \notin [x]_{P \cup \{a,b\}}$. Therefore, $y \notin [x]_{P \cup \{b\}}$. A contradiction. □

From Propositions 6 and 7, follows this corollary.

Corollary 1. *Operator c_1 satisfies property P_4.*

Definition 5. *For each $P \subseteq A$ subset of attributes, we define:*

$$c_3(P) = \{a \in A : \underline{apr}_P(X) = X \text{ for all } X \in \mathbb{P}_a\} \tag{9}$$

Proposition 8. $c_3 = c_2$.

Proof. We will see that $c_2(P) \subseteq c_3(P)$, and $c_3(P) \subseteq c_2(P)$ for all $P \subseteq A$.

1. $c_2 \leq c_3$. Let X be such that $X \in \mathbb{P}_a$. $X = [w]_a$ for some $w \in U$. Since $[x]_P = [x]_{P \cup \{a\}}$, we have that $[x]_P \subseteq [x]_a = X$, so $w \in \underline{apr}_P(X)$. Therefore, $a \in c_3(P)$.

2. $c_3 \leq c_2$. Suppose $a \in c_3(P)$ for some $a \in A$, $P \subseteq A$. Take $x \in U$, then $[x]_{P \cup \{a\}} \subseteq [x]_P$. Since $[x]_a \in \mathbb{P}_a$, $\underline{apr}([x]_a) = [x]_a$, so $[x]_P \subseteq [x]_a$. To see that $[x]_P \subseteq [x]_{P \cup \{a\}}$, let $y \in [x]_P$. Then $(\forall b \in P)(b(x) = b(y))$, and we also have that $a(x) = a(y)$. Therefore, $y \in [x]_{P \cup \{a\}}$. So $[x]_P \subseteq [x]_{P \cup \{a\}}$ and $a \in c_2(P)$.

From Propositions 1 and 6, follows this corollary.

Corollary 2. *The operator c_3 is a closure on A.*

Example 3. The values of the closure operator c_1 for some sets of attributes in Example 1, are the following:

- $c_1(\{a_1\}) = \{a_1\}$
- $c_1(\{a_2\}) = \{a_2\}$
- $c_1(\{a_3\}) = \{a_3\}$
- $c_1(\{a_4\}) = \{a_4\}$
- $c_1(\{a_1, a_2\}) = \{a_1, a_2, a_3\}$

- $c_1(\{a_2,a_4\}) = \{a_2,a_4\}$
- $c_1(\{a_3,a_4\}) = \{a_1,a_2,a_3,a_4\}$
- $c_1(\{a_1,a_2,a_4\}) = \{a_1,a_2,a_3,a_4\}$

According to Propositions 1 and 7, operator c_1 defines a matroidal structure, given by:

$$\mathbb{I} = \{P \subseteq A : a \notin c_1(P-a), \text{ for all } a \in A\} \tag{10}$$

Figure 1 shows the matroidal structure obtained from the closure operator on the dataset in Example 1. The connecting lines represent the inclusion relation.

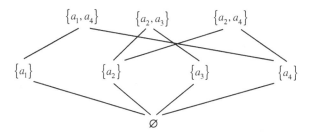

Fig. 1. Matroidal structure of the closure operator.

According to Eq. 10, the set $P = \{a_2,a_4\}$ belongs to $\mathbb{I}$, because for all $a_i \in A$, we have that $a_i \notin c_1(P-a_i)$. On the other hand, the set $P = \{a_3,a_4\}$ does not belong to $\mathbb{I}$, because for $a_1 \in A$, we have that $a_1 \in c_1(P-a_1) = c_1(P) = \{a_1,a_2,a_3,a_4\}$.

As we can see, the maximal independent sets $\{a_1,a_4\}$, $\{a_2,a_3\}$ and $\{a_2,a_4\}$ have the same number of elements.

If P is different to $c_1(P)$, i.e. if $P \subsetneq c_1(P)$, then P is a dependent set, as is shown in the following proposition.

Proposition 9. *If $P \subsetneq c_1(P)$, then P is a dependent set*

Proof. *If $P \subsetneq c_1(P)$, then there exists $a \in c_1(P)$ such that $a \notin P$. Since $a \notin P$, we have $P - a = P$, therefore $a \in c_1(P) = c_1(P-a)$, so $P \notin \mathbb{I}$.*

Figure 2 shows the structure obtained from the dependent set related with the matroidal structure.

In this collection of dependent sets we can find all the possible reducts. For example, $P_1 = \{a_1,a_2\}$ and $P_2 = \{a_1,a_3\}$ produce the same partition as $Q = \{a_1,a_2,a_3\}$, i.e., $I_{P_1} = I_Q$ and $I_{P_2} = I_Q$. Also, we have that $P_1 = \{a_1,a_3,a_4\}$ and $P_4 = \{a_2,a_3,a_4\}$ produce the same partition as A. In this case, $P_1 = \{a_1,a_3,a_4\}$ and $P_2 = \{a_1,a_2,a_4\}$ are reducts of A.

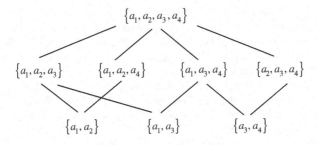

Fig. 2. Dependent sets of the matroidal structure.

5 Conclusions

This paper introduces a new methodology to address the problem of attribute reduction of an information system in Rough Set Theory. Specifically, a closure operator is introduced into the set of attributes that satisfies the conditions for defining a matroid. It is shown that the dependent sets of the matroid contain the reducts. Two definitions of the closure operator are additionally presented and it is proved that they coincide with the proposed operator.

The novelty of the proposal to define new structures on the set of attributes undoubtedly opens the possibility of analyzing the problem of attribute reduction from a new perspective. As future work, it is planned to extend the definition to the different generalizations of Rough Set Theory, in particular relation-based rough sets and covering-based rough sets.

References

1. Bian, X., Wang, P., Yu, Z., Bai, X., Chen, B.: Characterization of coverings for upper approximation operators being closure operators. Inf. Sci. **314**, 41–54 (2015)
2. Blyth, T.S.: Lattices and Ordered Algebraic Structures. Springer, London (2005). https://doi.org/10.1007/b139095
3. Greco, S., Matarazzo, B., Slowinski, R.: Rough sets theory for multicriteria decision analysis. Eur. J. Oper. Res. **129**, 1–47 (2001)
4. Huang, A., Zhao, H., Zhu, W.: Nullity-based matroid of rough sets and its application to attribute reduction. Inf. Sci. **263**, 153–165 (2014)
5. Lai, W.: Matroid Theory. Higher Education Press, Beijing (2001)
6. Li, X., Liu, S.: Matroidal approaches to rough sets via closure operators. Int. J. Approx. Reason. **53**, 513–527 (2012)
7. Li, X., Huangjian, Y., Sanyang, L.: Rough sets and matroids from a lattice-theoretic viewpoint. Inf. Sci. **342**, 37–52 (2016)
8. Li, Y., Wang, Z.: The relationships between degree rough sets and matroids. Ann. Fuzzy Math. Inf. **12**(1), 139–153 (2012)
9. Liu, Y., Zhu, W.: Relation matroid and its relationship with generalized rough set based on relations. CoRR., abs 1209.5456 (2012)
10. Liu, Y., Zhu, W., Zhang, Y.: Relationship between partition matroids and rough sets through k-rank matroids. J. Inf. Comput. Sci. **8**, 2151–2163 (2012)

11. Oxley, J.G.: Matroid Theory. Oxford University Press, Oxford (1992)
12. Restrepo, M., Cornelis, C.: Rough matroids based on dual approximation operators. In: Mihálydeák, T., Min, F., Wang, G., Banerjee, M., Düntsch, I., Suraj, Z., Ciucci, D. (eds.) IJCRS 2019. LNCS (LNAI), vol. 11499, pp. 118–129. Springer, Cham (2019). https://doi.org/10.1007/978-3-030-22815-6_10
13. Shen, Q., Alexios, C.A.: Rough set-based dimensionality reduction for supervised and unsupervised learning. Int. J. Appl. Math. Comput. Sci. **11**(3), 583–601 (2001)
14. Tang, J., She, K., Min, F., Zhu, W.: A matroidal approach to rough set theory. Theoret. Comput. Sci. **47**(1), 1–11 (2013)
15. Tsang, E., Chen, D., Lee, J., Yeung, D.S.: On the upper approximations of covering generalized rough sets. In: Proceedings of the 3rd International Conference on Machine Learning and Cybernetics, pp. 4200–4203 (2004)
16. Wang, S., Zhu, W., Min, F.: Transversal and function matroidal structures of covering-based rough sets. In: Yao, J.T., Ramanna, S., Wang, G., Suraj, Z. (eds.) RSKT 2011. LNCS (LNAI), vol. 6954, pp. 146–155. Springer, Heidelberg (2011). https://doi.org/10.1007/978-3-642-24425-4_21
17. Wang, S., Zhu, Q., Zhu, W., Min, F.: Matroidal structure of rough sets and its characterization to attribute reduction. Knowl.-Based Syst. **54**, 155–161 (2012)
18. Yang, T., Li, Q.: Reduction about approximation spaces of covering generalized rough sets. Int. J. Approx. Reason. **51**, 335–345 (2010)
19. Yang, B., Zhao, H., Zhu, W.: Rough matroids based on covering, vol. 407–411 (2013)
20. Wang Z, Feng Q, Wang H: The lattice and matroid representations of definable sets in generalized rough sets based on relations. Inf. Sci. **155 & 485**, 505–520 (2019)
21. Zhu, W., Wang, S.: Rough matroids. In: IEEE International Conference on Granular Computing, pp. 817–8221 (2011)
22. Zhu, W., Wang, S.: Rough matroids based on relation. Inf. Sci. **232**, 241–252 (2013)

The Reduct of a Fuzzy β-Covering

Lingling Mao[1], Jingqian Wang[2($\boxtimes$)], and Peiqiu Yu[3]

[1] Department of Basic Education, Xi'an Traffic Engineering University, Xi'an 710300, China
[2] College of Electrical and Control Engineerings,
Shaanxi University of Science and Technology, Xi'an 710021, China
wangjingqianw@163.com, 81157@sust.edu.cn
[3] School of Mathematics and Statistics, Minnan Normal University,
Zhangzhou 363000, Fujian, China

Abstract. This paper points some mistakes of three algorithms of updating the reduct in fuzzy β-covering via matrix approaches while adding and deleting some objects of the universe, and gives corrections of these mistakes. Moreover, we study the reduct of a fuzzy β-covering while adding and deleting objects further.

Keywords: Covering-based rough sets · Fuzzy sets · Matrix · Reduct

1 Introduction

Recently, fuzzy covering approximation spaces [1–3] were generalized to fuzzy β-covering approximation spaces by Ma [4] by replacing 1 with a parameter β, where 1 is a condition in fuzzy covering approximation spaces. Inspired by Ma's work, many researches were done. For example, some fuzzy covering-based rough set models were constructed by Yang and Hu [5–7], D'eer et al. [8,9] studied fuzzy neighborhood operators, and Huang et al. [10] presented a matrix approach for computing the reduct of a fuzzy β-covering.

The research idea of Ref. [10] is very good, but we find that Algorithms 1, 2 and 3 are incorrect after checking the paper carefully. Moreover, the result of a fuzzy β-covering can be studied further while adding and deleting objects. Hence, a further study about Ref. [10] can be done in this paper. Firstly, we explain the mistakes about Algorithms 1, 2 and 3 in Huang et al. (2020) [10]. Then, we give corresponding corrections of them. Finally, we present some new definitions and properties for updating the reduct while adding and deleting objects of a universe. The concepts about a fuzzy β-covering approximation space after adding and deleting objects are presented, respectively. Some new properties about the fuzzy β-covering approximation space and the new fuzzy β-covering approximation space after adding and deleting objects are given.

The rest of this paper is organized as follows. Section 2 reviews some fundamental definitions about fuzzy covering-based rough sets. In Sect. 3, we show some mistakes in [7]. Moreover, we give corresponding corrections of them. In Sect. 4, we present some new definitions and properties for updating the reduct while adding and deleting objects. This paper is concluded and further work is indicated in Sect. 5.

© Springer Nature Switzerland AG 2020
R. Bello et al. (Eds.): IJCRS 2020, LNAI 12179, pp. 193–203, 2020.
https://doi.org/10.1007/978-3-030-52705-1_14

2 Basic Definitions

This section recalls some fundamental definitions related to fuzzy covering-based rough sets. Supposing U is a nonempty and finite set called universe.

For any family $\gamma_i \in [0,1], i \in I, I \subseteq \mathbb{N}^+$ ($\mathbb{N}^+$ is the set of all positive integers), we write $\vee_{i \in I} \gamma_i$ for the supremum of $\{\gamma_i : i \in I\}$, and $\wedge_{i \in I} \gamma_i$ for the infimum of $\{\gamma_i : i \in I\}$. Some basic operations on $F(U)$ are shown as follows [11]: $A, B \in F(U)$,

(1) $A \subseteq B$ iff $A(x) \leq B(x)$ for all $x \in U$;

(2) $A = B$ iff $A \subseteq B$ and $B \subseteq A$;

(3) $A \cup B = \{\langle x, A(x) \vee B(x) \rangle : x \in U\}$;

(4) $A \cap B = \{\langle x, A(x) \wedge B(x) \rangle : x \in U\}$;

(5) $A' = \{\langle x, 1 - A(x) : x \in U\}$.

Ma [4] presented the notion of fuzzy β-covering approximation space.

Definition 1. *([4]) Let U be an arbitrary universal set and $F(U)$ be the fuzzy power set of U. For each $\beta \in (0,1]$, if $(\bigcup_{i=1}^{m} C_i)(x) \geq \beta$ for each $x \in U$, then we call $\widehat{C} = \{C_1, C_2, ..., C_m\}$ a fuzzy β-covering of U with $C_i \in F(U)$ $(i = 1, 2, ..., m)$. We also call $(U, \widehat{C})$ a fuzzy β-covering approximation space.*

The concept of reducible elements is important for us to deal with some problems in fuzzy covering-based rough sets [5]. Let $\widehat{C}$ be a fuzzy β-covering of U and $C \in \widehat{C}$. If C can be expressed as a union of some elements in $\widehat{C} - \{C\}$, then C is called a reducible element in $\widehat{C}$; otherwise C is called an irreducible element in $\widehat{C}$.

As shown in [5], if all reducible elements are deleted from a fuzzy β-covering $\widehat{C}$, then the remainder is still a fuzzy β-covering and this new fuzzy β-covering does not have any reducible element. We call this new fuzzy β-covering the reduct of the original fuzzy β-covering $\widehat{C}$. The following definition presents its concept.

Definition 2. *([5]) Let $(U, \widehat{C})$ be a fuzzy β-covering approximation space. Then the family of all irreducible elements of $\widehat{C}$ is called the reduct of $\widehat{C}$, denoted as $\Gamma(\widehat{C})$.*

To calculate the result of a fuzzy β-covering by matrix, Huang et al. gave the following definition.

Definition 3. *([10]) Let $(U, \widehat{C})$ be a fuzzy β-covering approximation space. The containing relation character matrix on U is denoted by $Q^U = (q_{ij}^U)_{m \times m}$, where*

$$q_{ij}^U = \begin{cases} 1, & C_i \subseteq C_j \wedge i \neq j; \\ 0, & otherwise; \end{cases} \quad i, j \in \{1, 2, \cdots, m\}$$

3   Some Corrections on the Reduct of a Fuzzy β-covering

In [10], we find that Algorithms 1, 2 and 3 have mistakes after checking the paper carefully. Then we give corresponding corrections of the paper in this section.

Algorithm 1: Algorithm to compute the reduct of a fuzzy β-covering.

Input: $(1)\widehat{C} = \{C_1, C_2, \cdots, C_m,\}$ $(2)U = \{x_1, x_2, \cdots, x_n\}$. $(3)\beta \in (0, 1]$.
Output: $\Gamma(\widehat{C}), Q^U$.

```
1  for i = 1, 2, · · · , m do
2  │   C(x_i) ← 0;
3  │   for j = 1, 2, · · · , m do
4  │   │   C(x_i) ← C(x_i) ∨ C_j(x_i)
5  │   if C(x_i) < β then
6  │   │   return Ĉ is not a fuzzy β-covering
7  Γ(Ĉ) ← Ĉ;
8  for k = 1, 2, · · · , m do
9  │   T ← ∅;
10 │   for l = 1, 2, · · · , m do
11 │   │   if C_k ⊆ C_l then
12 │   │   │   q^U_{kl} ← 1;
13 │   │   if q^U_{kl} = 1 then
14 │   │   │   T ← T ∪ C_k;
15 │   if T = C_l then
16 │   │   Γ(Ĉ) ← Γ(Ĉ) − {C_l};
17 return Γ(Ĉ)
```

Fig. 1. Algorithm 1 (In [10])

By Algorithm 1 (In [10]), we know that $\Gamma(\widehat{C}) = \emptyset$ for any fuzzy β-covering, which is incorrect. To explain the incorrect results in Algorithm 1, we show the Algorithm 1 (In [10]) in Fig. 1:

In Algorithm 1 (In [10]), $U = \{x_1, x_2, \cdots, x_n\}$. By Step 2, $C(x_i) \leftarrow 0$. Hence,

- Step 1: "$i = 1, 2, \cdots, m$" should be changed as "$i = 1, 2, \cdots, n$".

According to Definition 3 (Definition 5 in [10]) and Step 12, we find Step 11 of Algorithm 1 (In [10]) is incorrect. By Steps 11 and 12, if $C_k \subseteq C_l$ then $q^U_{kl} \leftarrow 1$. But according to Definition 3 (Definition 5 in [10]), if $C_k \subseteq C_l$ and $k \neq l$ then $q^U_{kl} \leftarrow 1$. Hence,

- Step 11: "if $C_k \subseteq C_l$ then" should be changed as "if $C_k \subseteq C_l$ and $k \neq l$ then".

From Steps 11 to 12 of Algorithm 1 (In [10]), it is to find all $C_k \in \widehat{C} - \{C_l\}$ which satisfy $C_k \subseteq C_l$ for any $C_l \in \widehat{C}$. From Steps 13 to 16 of Algorithm 1 (In [10]), if $\bigcup_{C_k \in \widehat{C} - \{C_l\}} = C_l$ then C_l is a reducible element in $\widehat{C}$. Hence, Steps 8 and 10 should be swapped places. That is to say,

- Step 8: "for $k = 1, 2, \cdots, m$ do" should be changed as "for $l = 1, 2, \cdots, m$ do".
- Step 10: "for $l = 1, 2, \cdots, m$ do" should be changed as "for $k = 1, 2, \cdots, m$ do".

The result of Algorithm 2 (In [10]) will be $\widehat{\mathcal{G}}$ all the time, which is incorrect. To explain the incorrect results in Algorithm 2, we show the Algorithm 2 (In [10]):

By Algorithm 2, we find:

- Step 15: "$T \leftarrow T \cup C_k$;" should be changed as "$T \leftarrow T \cup \mathcal{G}_k$".

From Steps 14 to 17 of Algorithm 2 (In [10]), if $\bigcup_{\mathcal{G}_k \in \widehat{\mathcal{G}} - \{\mathcal{G}_l\}} = \mathcal{G}_l$ then $\mathcal{G}_l$ is a reducible element in $\widehat{\mathcal{G}}$. Hence, Steps 11 and 13 should be swaped places. That is to say,

- Step 11: "**for** $k = 1, 2, \cdots, m$ **do**" should be changed as "**for** $l = 1, 2, \cdots, m$ **do**".
- Step 13: "**for** $l = 1, 2, \cdots, m$ **do**" should be changed as "**for** $k = 1, 2, \cdots, m$ **do**".

The result of Algorithm 3 (In [10]) will be $\widehat{\mathcal{G}}$ all the time, which is incorrect. To explain the incorrect results in Algorithm 3, we show the Algorithm 3 (In [10]):

By Algorithm 3, we find:

- Step 15: "$T \leftarrow T \cup C_k$;" should be changed as "$T \leftarrow T \cup \mathcal{G}_k$".

From Steps 14 to 17 of Algorithm 3 (In [10]), if $\bigcup_{\mathcal{G}_k \in \widehat{\mathcal{G}} - \{\mathcal{G}_l\}} = \mathcal{G}_l$ then $\mathcal{G}_l$ is a reducible element in $\widehat{\mathcal{G}}$. Hence, Steps 11 and 13 should be swaped places. That is to say,

- Step 11: "**for** $k = 1, 2, \cdots, m$ **do**" should be changed as "**for** $l = 1, 2, \cdots, m$ **do**".
- Step 13: "**for** $l = 1, 2, \cdots, m$ **do**" should be changed as "**for** $k = 1, 2, \cdots, m$ **do**".

4   New Properties of Reducts of Fuzzy β-Coverings While Adding and Deleting Some Objects

This section presents some new properties of reducts in fuzzy β-coverings while adding and deleting some objects, respectively. In this section, t denotes an integer which is more than 1.

Firstly, we give some new properties on reducts of fuzzy β-coverings while adding some objects of a universe. The concept of increasing fuzzy β-covering approximation space is presented in the following definition.

Definition 4. *Let* $(U, \widehat{C})$ *be a fuzzy* β-*covering approximation space of* U, *where* $U = \{x_1, x_2, \cdots, x_n\}$ *and* $\widehat{C} = \{C_1, C_2, \cdots, C_m\}$. *We call* $(U^+, \widehat{C}^+)$ *an increasing fuzzy* β-*covering approximation space from* $(U, \widehat{C})$, *where* $U^+ = \{x_1, x_2, \cdots, x_n, x_{n+1}, \cdots, x_{n+t}\}$, $\widehat{C}^+ = \{C_1^+, C_2^+, \cdots, C_m^+\}$, *and for any* $1 \leq j \leq m$,

$$\begin{cases} C_j^+(x_i) = C_j(x_i), & 1 \leq i \leq n; \\ (\bigcup_{j=1}^m C_j^+)(x_i) \geq \beta, & n+1 \leq i \leq n+t. \end{cases}$$

The following proposition shows that an increasing fuzzy β-covering approximation space from a fuzzy β-covering approximation space is also a fuzzy β-covering approximation space.

Algorithm 2: Algorithm to update the reduct of a fuzzy β-covering while adding some objects into the universe.

Input: $(1)\widehat{\mathcal{G}} = \{\mathcal{G}_1, \mathcal{G}_2, \cdots, \mathcal{G}_m\}$ $(2)\mathcal{U} = \{x_1, x_2, \cdots, x_n, x_{n+1}, \cdots, x_{n+t}\}$. $(3)\beta \in (0,1]$,$(4)Q^{\mathcal{U}}$.

Output: $\Gamma(\widehat{\mathcal{G}})$.

1   $Q^{\mathcal{U}} \leftarrow Q^U$;
2   **for** $i = 1, 2, \cdots, m$ **do**
3   **for** $j = 1, 2, \cdots, m$ **do**
4   **if** $q_{ij}^{\mathcal{U}} = 1$ **then**
5         $s \leftarrow 1$;
6   **for** $k = n+1, n+2, \cdots, n+t$ **do**
7   **if** $\mathcal{G}_i(x_k) > \mathcal{G}_j(x_k)$ **then**
8             $s \leftarrow 0$;
9         $q_{ij}^{\mathcal{U}} \leftarrow s$;
10   $\Gamma(\widehat{\mathcal{G}}) \leftarrow \widehat{\mathcal{G}}$;
11   **for** $k = 1, 2, \cdots, m$ **do**
12     $T \leftarrow \varnothing$;
13   **for** $l = 1, 2, \cdots, m$ **do**
14   **if** $q_{kl}^{\mathcal{U}} = 1$ **then**
15         $T \leftarrow T \cup C_k$;
16   **if** $T = \mathcal{G}_l$ **then**
17       $\Gamma(\widehat{\mathcal{G}}) \leftarrow \Gamma(\widehat{\mathcal{G}}) - \{\mathcal{G}_l\}$;
18   **return** $\Gamma(\widehat{\mathcal{G}})$

Fig. 2. Algorithm 2 (In [10])

Proposition 1. *Let* $(U, \widehat{C})$ *be a fuzzy* β-*covering approximation space of* U, *where* $U = \{x_1, x_2, \cdots, x_n\}$ *and* $\widehat{C} = \{C_1, C_2, \cdots, C_m\}$. *Then* $(U^+, \widehat{C}^+)$ *is also a fuzzy* β-*covering approximation space of* U^+.

Proof. By Definition 4, $(\bigcup_{j=1}^{m} C_j^+)(x_i) = (\bigcup_{j=1}^{m} C_j)(x_i) \geq \beta$ for any $i \in \{1, 2, \cdots, n\}$, and $(\bigcup_{j=1}^{m} C_j^+)(x_i) \geq \beta$ for each $i \in \{n+1, \cdots, n+t\}$. Hence, $(U^+, \widehat{C}^+)$ is also a fuzzy β-covering approximation space of U^+ by Definition 1.

Example 1. Let $U = \{x_1, x_2, x_3, x_4, x_5\}$ and $\widehat{C} = \{C_1, C_2, C_3, C_4\}$, where

$$C_1 = \frac{0.7}{x_1} + \frac{0.8}{x_2} + \frac{0.6}{x_3} + \frac{0.6}{x_4} + \frac{0.7}{x_5},$$

$$C_2 = \frac{0.3}{x_1} + \frac{0.8}{x_2} + \frac{0.3}{x_3} + \frac{0.5}{x_4} + \frac{0.6}{x_5},$$

$$C_3 = \frac{0.7}{x_1} + \frac{0.6}{x_2} + \frac{0.6}{x_3} + \frac{0.6}{x_4} + \frac{0.7}{x_5},$$

$$C_4 = \frac{0.4}{x_1} + \frac{0.6}{x_2} + \frac{0.3}{x_3} + \frac{0.2}{x_4} + \frac{0.5}{x_5}.$$

According to Definition 1, we know $\widehat{C}$ is a fuzzy β-covering of U $(0 < \beta \leq 0.6)$. Suppose $\beta = 0.5$. Let $U^+ = \{x_1, x_2, x_3, x_4, x_5, x_6\}$ and $\widehat{\mathbf{C}}^+ = \{C_1^+, C_2^+, C_3^+, C_4^+\}$, where

$$C_1^+ = \frac{0.7}{x_1} + \frac{0.8}{x_2} + \frac{0.6}{x_3} + \frac{0.6}{x_4} + \frac{0.7}{x_5} + \frac{0.6}{x_6},$$

$$C_2^+ = \frac{0.3}{x_1} + \frac{0.8}{x_2} + \frac{0.3}{x_3} + \frac{0.5}{x_4} + \frac{0.6}{x_5} + \frac{0.5}{x_6},$$

$$C_3^+ = \frac{0.7}{x_1} + \frac{0.6}{x_2} + \frac{0.6}{x_3} + \frac{0.6}{x_4} + \frac{0.7}{x_5} + \frac{0.5}{x_6},$$

$$C_4^+ = \frac{0.4}{x_1} + \frac{0.6}{x_2} + \frac{0.3}{x_3} + \frac{0.2}{x_4} + \frac{0.5}{x_5} + \frac{0.7}{x_6}.$$

Algorithm 3: Algorithm to update the reduct of a fuzzy β-covering while deleting objects from the universe

Input: (1)$\widehat{\mathcal{G}} = \{\mathcal{G}_1, \mathcal{G}_2, \cdots, \mathcal{G}_m\}$ (2)$\mathcal{U} = \{x_1, x_2, \cdots, \cdots, x_{n-t}\}$. (3)$\beta \in (0,1]$,(4)$Q^{\mathcal{U}}$.
Output: $\Gamma(\widehat{\mathcal{G}})$.

1   $Q^{\mathcal{U}} \leftarrow Q^{\mathcal{U}}$;
2   **for** $i = 1, 2, \cdots, m$ **do**
3   **for** $j = 1, 2, \cdots, m$ **do**
4   **if** $q_{ij}^{\mathcal{U}} = 0$ **then**
5         $s \leftarrow 1$;
6   **for** $k = 1, 2, \cdots, n - t$ **do**
7   **if** $\mathcal{G}_i(x_k) > \mathcal{G}_j(x_k)$ **then**
8             $s \leftarrow 0$;
9         $q_{ij}^{\mathcal{U}} \leftarrow s$;
10   $\Gamma(\widehat{\mathcal{G}}) \leftarrow \widehat{\mathcal{G}}$;
11   **for** $k = 1, 2, \cdots, m$ **do**
12     $T \leftarrow \varnothing$;
13   **for** $l = 1, 2, \cdots, m$ **do**
14   **if** $q_{kl}^{\mathcal{U}} = 1$ **then**
15         $T \leftarrow T \cup C_k$;
16   **if** $T = \mathcal{G}_l$ **then**
17       $\Gamma(\widehat{\mathcal{G}}) \leftarrow \Gamma(\widehat{\mathcal{G}}) - \{\mathcal{G}_l\}$;
18   **return** $\Gamma(\widehat{\mathcal{G}})$

Fig. 3. Algorithm 3 (In [10])

According to Definitions 1 and 4, we know $\widehat{C}^+$ is a fuzzy 0.5-covering of U.

We give a relationship about the relation character matrices between a fuzzy β-covering approximation space and it's increasing fuzzy β-covering approximation space in the following proposition.

Proposition 2. *Let* $(U, \widehat{C})$ *and* $(U^+, \widehat{C}^+)$ *be two fuzzy β-covering approximation spaces, where* $U = \{x_1, x_2, \cdots, x_n\}$ *and* $\widehat{C} = \{C_1, C_2, \cdots, C_m\}$. *If* $q_{ij}^U = 0$, *then* $q_{ij}^{U^+} = 0$ *for any* $i, j \in \{1, 2, \cdots, m\}$.

Proof. For any $i, j \in \{1, 2, \cdots, m\}$, we have the following two conditions:

For $i = j$: if $i = j$, then $q_{ij}^U = 0$ and $q_{ij}^{U^+} = 0$;

For $i \neq j$: by Definition 3, if $q_{ij}^U = 0$, then there exists $k \in \{1, 2, \cdots, n\}$ such that $C_i(x_k) > C_j(x_k)$. Hence, there exists $k \in \{1, 2, \cdots, n\}$ such that $C_i^+(x_k) > C_j^+(x_k)$ according to Definition 4. Therefore, C_i^+ is not contained in C_j^+. That is to say, $q_{ij}^{U^+} = 0$.

Example 2. (Continued from Example 1)

$$Q^U = (q_{ij}^U)_{4\times 4} = \begin{array}{c} \\ C_1 \\ C_2 \\ C_3 \\ C_4 \end{array} \begin{pmatrix} \begin{array}{cccc} C_1 & C_2 & C_3 & C_4 \\ 0 & 0 & 0 & 0 \\ 1 & 0 & 0 & 0 \\ 1 & 0 & 0 & 0 \\ 1 & 0 & 1 & 0 \end{array} \end{pmatrix},$$

$$Q^{U^+} = (q_{ij}^{U^+})_{4\times 4} = \begin{array}{c} \\ C_1^+ \\ C_2^+ \\ C_3^+ \\ C_4^+ \end{array} \begin{pmatrix} \begin{array}{cccc} C_1^+ & C_2^+ & C_3^+ & C_4^+ \\ 0 & 0 & 0 & 0 \\ 1 & 0 & 0 & 0 \\ 1 & 0 & 0 & 0 \\ 0 & 0 & 0 & 0 \end{array} \end{pmatrix}.$$

Hence, if $q_{ij}^U = 0$, then $q_{ij}^{U^+} = 0$ for any $i, j \in \{1, 2, \cdots, 4\}$.

We give a relationship about reducible elements between a fuzzy β-covering approximation space and it's increasing fuzzy β-covering approximation space in the following proposition.

Proposition 3. *Let $(U, \widehat{C})$ and $(U^+, \widehat{C}^+)$ be two fuzzy β-covering approximation spaces, where $U = \{x_1, x_2, \cdots, x_n\}$ and $\widehat{C} = \{C_1, C_2, \cdots, C_m\}$. If C_i^+ is a reducible element in $\widehat{C}^+$, then C_i is a reducible element in $\widehat{C}$ for any $i \in \{1, 2, \cdots, m\}$.*

Proof. It is immediate by Definition 4 and the concept of reducible element.

The converse of Proposition 3 is not true, i.e., "If C_i is a reducible element in $\widehat{C}$, then C_i^+ is a reducible element in $\widehat{C}^+$ for any $i \in \{1, 2, \cdots, m\}$." is not true. Example 1 can explain this. In Example 1, since $C_1 = C_2 \bigcup C_3$, C_1 is a reducible element in $\widehat{C}$. However, C_1^+ is not a reducible element in $\widehat{C}^+$. Based on Proposition 3, we give the following corollary.

Corollary 1. *Let $(U, \widehat{C})$ and $(U^+, \widehat{C}^+)$ be two fuzzy β-covering approximation spaces, where $U = \{x_1, x_2, \cdots, x_n\}$ and $\widehat{C} = \{C_1, C_2, \cdots, C_m\}$. If C_i is a irreducible element in $\widehat{C}$, then C_i^+ is a irreducible element in $\widehat{C}^+$ for any $i \in \{1, 2, \cdots, m\}$.*

Proof. By Proposition 3, it is immediate.

Example 3. (Continued from Example 1) C_2, C_3 and C_4 are irreducible elements in $\widehat{C}$. C_2^+, C_3^+ and C_4^+ are irreducible elements in $\widehat{C}^+$.

The converse of Corollary 1 is not true, i.e., "If C_i^+ is a irreducible element in $\widehat{C}^+$, then C_i is a irreducible element in $\widehat{C}$ for any $i \in \{1, 2, \cdots, m\}$." is not true. Example 1 can explain this. In Example 1, C_1^+ is a irreducible element in $\widehat{C}^+$. But C_1 is not a irreducible element in $\widehat{C}$. Inspired by Corollary 1, we give the following theorem.

Theorem 1. *Let $(U, \widehat{C})$ and $(U^+, \widehat{C}^+)$ be two fuzzy β-covering approximation spaces. Then $|\Gamma(\widehat{C})| \le |\Gamma(\widehat{C}^+)|$.*

Proof. By Definition 2, $\Gamma(\widehat{C})$ and $\Gamma(\widehat{C}^+)$ are families of all irreducible elements of $\widehat{C}$ and $\widehat{C}^+$, respectively. Hence, it is immediate by Corollary 1.

Note that $|\Gamma(\widehat{C})|$ and $|\Gamma(\widehat{C}^+)|$ denote the cardinality of $\Gamma(\widehat{C})$ and $\Gamma(\widehat{C}^+)$, respectively.

Example 4. (Continued from Example 1) $\Gamma(\widehat{C}) = \{C_2, C_3, C_4\}$, $\Gamma(\widehat{C}^+) = \{C_1^+, C_2^+, C_3^+, C_4^+\}$. Hence, $|\Gamma(\widehat{C})| = 3$ and $|\Gamma(\widehat{C}^+)| = 4$. That is to say, $|\Gamma(\widehat{C})| \leq |\Gamma(\widehat{C}^+)|$.

Then, we give some new properties on reducts of fuzzy β-coverings while deleting some objects of a universe. The concept of declining fuzzy β-covering approximation space is presented in the following definition.

Definition 5. *Let* $(U, \widehat{C})$ *be a fuzzy β-covering approximation space of U, where $U = \{x_1, x_2, \cdots, x_n\}$ and $\widehat{C} = \{C_1, C_2, \cdots, C_m\}$. We call $(U^-, \widehat{C}^-)$ a declining fuzzy β-covering approximation space from $(U, \widehat{C})$, where $U^+ = \{x_1, x_2, \cdots, x_{n-t}\}$, $\widehat{C}^- = \{C_1^-, C_2^-, \cdots, C_m^-\}$ and $C_j^-(x_i) = C_j(x_i)$ for any $1 \leq i \leq n - t$, $1 \leq j \leq m$.*

The following proposition shows that a declining fuzzy β-covering approximation space from a fuzzy β-covering approximation space is also a fuzzy β-covering approximation space.

Proposition 4. *Let* $(U, \widehat{C})$ *be a fuzzy β-covering approximation space of U, where $U = \{x_1, x_2, \cdots, x_n\}$ and $\widehat{C} = \{C_1, C_2, \cdots, C_m\}$. Then $(U^-, \widehat{C}^-)$ is also a fuzzy β-covering approximation space of U^-.*

Proof. By Definition 5, $(\bigcup_{j=1}^{m} C_j^-)(x_i) = (\bigcup_{j=1}^{m} C_j)(x_i) \geq \beta$ for any $i \in \{1, 2, \cdots, n - t\}$. Hence, $(U^-, \widehat{C}^-)$ is also a fuzzy β-covering approximation space of U^- by Definition 1.

Example 5. Let $U = \{x_1, x_2, x_3, x_4, x_5\}$ and $\widehat{C} = \{C_1, C_2, C_3, C_4\}$, where

$$C_1 = \frac{0.7}{x_1} + \frac{0.8}{x_2} + \frac{0.6}{x_3} + \frac{0.6}{x_4} + \frac{0.7}{x_5},$$
$$C_2 = \frac{0.3}{x_1} + \frac{0.8}{x_2} + \frac{0.3}{x_3} + \frac{0.8}{x_4} + \frac{0.6}{x_5},$$
$$C_3 = \frac{0.7}{x_1} + \frac{0.6}{x_2} + \frac{0.6}{x_3} + \frac{0.6}{x_4} + \frac{0.7}{x_5},$$
$$C_4 = \frac{0.4}{x_1} + \frac{0.6}{x_2} + \frac{0.3}{x_3} + \frac{0.2}{x_4} + \frac{0.5}{x_5}.$$

According to Definition 1, we know $\widehat{C}$ is a fuzzy β-covering of U ($0 < \beta \leq 0.6$). Suppose $\beta = 0.5$. Let $U^- = \{x_1, x_2, x_3\}$ and $\widehat{\mathbf{C}}^- = \{C_1^-, C_2^-, C_3^-, C_4^-\}$, where

$$C_1^- = \frac{0.7}{x_1} + \frac{0.8}{x_2} + \frac{0.6}{x_3},$$
$$C_2^- = \frac{0.3}{x_1} + \frac{0.8}{x_2} + \frac{0.3}{x_3},$$
$$C_3^- = \frac{0.7}{x_1} + \frac{0.6}{x_2} + \frac{0.6}{x_3},$$
$$C_4^- = \frac{0.4}{x_1} + \frac{0.6}{x_2} + \frac{0.3}{x_3}.$$

According to Definitions 1 and 5, we know $\widehat{C}^-$ is a fuzzy 0.5-covering of U.

We give a relationship about the relation character matrices between a fuzzy β-covering approximation space and it's declining fuzzy β-covering approximation space in the following proposition.

Proposition 5. *Let $(U, \widehat{C})$ and $(U^-, \widehat{C}^-)$ be two fuzzy β-covering approximation spaces, where $U = \{x_1, x_2, \cdots, x_n\}$ and $\widehat{C} = \{C_1, C_2, \cdots, C_m\}$. If $q_{ij}^{U^-} = 0$, then $q_{ij}^U = 0$ for any $i, j \in \{1, 2, \cdots, m\}$.*

Proof. For any $i, j \in \{1, 2, \cdots, m\}$, we have the following two conditions:

For $i = j$: if $i = j$, then $q_{ij}^U = 0$ and $q_{ij}^{U^-} = 0$;

For $i \neq j$: by Definition 3, if $q_{ij}^{U^-} = 0$, then there exists $k \in \{1, 2, \cdots, n - t\}$ such that $C_i(x_k) > C_j(x_k)$. Hence, there exists $k \in \{1, 2, \cdots, n - t\}$ such that $C_i(x_k) > C_j(x_k)$ according to Definition 5, i.e., there exists $k \in \{1, 2, \cdots, n\}$ such that $C_i(x_k) > C_j(x_k)$. Therefore, C_i is not contained in C_j. That is to say, $q_{ij}^U = 0$.

Example 6. (Continued from Example 5)

$$Q^U = (q_{ij}^U)_{4 \times 4} = \begin{array}{c} \\ C_1 \\ C_2 \\ C_3 \\ C_4 \end{array} \begin{pmatrix} \begin{array}{cccc} C_1 & C_2 & C_3 & C_4 \end{array} \\ \begin{array}{cccc} 0 & 0 & 0 & 0 \\ 0 & 0 & 0 & 0 \\ 1 & 0 & 0 & 0 \\ 1 & 0 & 1 & 0 \end{array} \end{pmatrix},$$

$$Q^{U^-} = (q_{ij}^{U^-})_{4 \times 4} = \begin{array}{c} \\ C_1^- \\ C_2^- \\ C_3^- \\ C_4^- \end{array} \begin{pmatrix} \begin{array}{cccc} C_1^- & C_2^- & C_3^- & C_4^- \end{array} \\ \begin{array}{cccc} 0 & 0 & 0 & 0 \\ 1 & 0 & 0 & 0 \\ 1 & 0 & 0 & 0 \\ 1 & 0 & 1 & 0 \end{array} \end{pmatrix}.$$

Hence, if $q_{ij}^{U^-} = 0$, then $q_{ij}^U = 0$ for any $i, j \in \{1, 2, \cdots, 4\}$.

Huang et al. [10] gave a relationship about reducible elements between a fuzzy β-covering approximation space and it's declining fuzzy β-covering approximation space in the following proposition.

Lemma 1. *([10]) Let $(U, \widehat{C})$ and $(U^-, \widehat{C}^-)$ be two fuzzy β-covering approximation spaces, where $U = \{x_1, x_2, \cdots, x_n\}$ and $\widehat{C} = \{C_1, C_2, \cdots, C_m\}$. If C_i is a reducible element in $\widehat{C}$, then C_i^- is a reducible element in $\widehat{C}^-$ for any $i \in \{1, 2, \cdots, m\}$.*

The converse of Lemma 1 is not true, i.e., "If C_i^- is a reducible element in $\widehat{C}$, then C_i is a reducible element in $\widehat{C}$ for any $i \in \{1, 2, \cdots, m\}$." is not true. Example 5 can explain this. In Example 5, since $C_1^- = C_2^- \bigcup C_3^-$, C_1^- is a reducible element in $\widehat{C}^-$. But C_1 is not a reducible element in $\widehat{C}$. Based on Lemma 1, we give the following corollary.

Corollary 2. *Let $(U, \widehat{C})$ and $(U^-, \widehat{C}^-)$ be two fuzzy β-covering approximation spaces, where $U = \{x_1, x_2, \cdots, x_n\}$ and $\widehat{C} = \{C_1, C_2, \cdots, C_m\}$. If C_i^- is a irreducible element in $\widehat{C}^-$, then C_i is a irreducible element in $\widehat{C}$ for any $i \in \{1, 2, \cdots, m\}$.*

Proof. By Lemma 1, it is immediate.

Example 7. (Continued from Example 5) C_2^-, C_3^- and C_4^- are irreducible elements in $\widehat{C}^-$. C_2, C_3 and C_4 are irreducible elements in $\widehat{C}$.

Based on Corollary 2, we give the following theorem.

Theorem 2. *Let* $(U, \widehat{C})$ *and* $(U^-, \widehat{C}^-)$ *be two fuzzy β-covering approximation spaces. Then* $|\Gamma(\widehat{C})| \geq |\Gamma(\widehat{C}^-)|$.

Proof. By Definition 2, $\Gamma(\widehat{C})$ and $\Gamma(\widehat{C}^-)$ are families of all irreducible elements of $\widehat{C}$ and $\widehat{C}^-$, respectively. Hence, it is immediate by Corollary 2.

Example 8. (Continued from Example 5) $\Gamma(\widehat{C}) = \{C_1, C_2, C_3, C_4\}$, $\Gamma(\widehat{C}^-) = \{C_2^-, C_3^-, C_4^-\}$. Hence, $|\Gamma(\widehat{C})| = 4$ and $|\Gamma(\widehat{C}^-)| = 3$. That is to say, $|\Gamma(\widehat{C})| \geq |\Gamma(\widehat{C}^-)|$.

5   Conclusions

In this paper, we explain the mistakes about Algorithms 1, 2 and 3 in Huang et al. (2020) [10]. Moreover, we present some new definitions and properties for updating the reduct while adding and deleting objects of a universe. It is helpful for others to investigate the work further. In future, updating the reduct while adding and deleting objects at the same time will be done. Neutrosophic sets and related algebraic structures [12–15] will be connected with the research content of this paper in further research.

Acknowledgments. This work is supported by the Natural Science Foundation of Education Department of Shaanxi Province, China, under Grant No. 19JK0506, the National Natural Science Foundation of China under Grant No. 61976130.

References

1. Feng, T., Zhang, S., Mi, J.: The reduction and fusion of fuzzy covering systems based on the evidence theory. Int. J. Approximate Reasoning **53**, 87–103 (2012)
2. Li, T., Leung, Y., Zhang, W.: Generalized fuzzy rough approximation operators based on fuzzy coverings. Int. J. Approximate Reasoning **48**, 836–856 (2008)
3. Šešelja, B.: L-fuzzy covering relation. Fuzzy Sets Syst. **158**, 2456–2465 (2007)
4. Ma, L.: Two fuzzy covering rough set models and their generalizations over fuzzy lattices. Fuzzy Sets Syst. **294**, 1–17 (2016)
5. Yang, B., Hu, B.: A fuzzy covering-based rough set model and its generalization over fuzzy lattice. Inf. Sci. **367–368**, 463–486 (2016)
6. Yang, B., Hu, B.: On some types of fuzzy covering-based rough sets. Fuzzy Sets Syst. **312**, 36–65 (2017)
7. Yang, B., Hu, B.: Fuzzy neighborhood operators and derived fuzzy coverings. Fuzzy Sets Syst. **370**, 1–33 (2019)
8. D'eer, L., Cornelis, C., Godo, L.: Fuzzy neighborhood operators based on fuzzy coverings. Fuzzy Sets Syst. **312**, 17–35 (2017)

9. D'eer, L., Cornelis, C.: A comprehensive study of fuzzy covering-based rough set models: definitions, properties and interrelationships. Fuzzy Sets Syst. **336**, 1–26 (2018)

10. Huang, J., Yu, P., Li, W.: Updating the reduct in fuzzy β-covering via matrix approaches while adding and deleting some objects of the universe. Information **11**, 3 (2020)

11. Klir, G., Yuan, B.: Fuzzy Sets and Fuzzy Logic: Theory and Applications. Prentice Hall, New Jersey (1995)

12. Zhang, X., Wu, X., Mao, X.: On neutrosophic extended triplet groups (loops) and Abel-Grassmann's groupoids (AG-groupoids). J. Intell. Fuzzy Syst. **37**, 5743–5753 (2019)

13. Zhang, X., Wang, X., Smarandache, F., Jaiyeola, T.G., Lian, T.: Singular neutrosophic extended triplet groups and generalized groups. Cogn. Syst. Res. **57**, 32–40 (2019)

14. Konecny, J., Krajča, P.: On attribute reduction in concept lattices: experimental evaluation shows discernibility matrix based methods inefficient. Inf. Sci. **467**, 431–445 (2018)

15. Konecny, J.: On attribute reduction in concept lattices: methods based on discernibility matrix are outperformed by basic clarification and reduction. Inf. Sci. **415**, 199–212 (2017)

The Problem of Finding the Simplest Classifier Ensemble is NP-Hard – A Rough-Set-Inspired Formulation Based on Decision Bireducts

Dominik Ślęzak[1($\boxtimes$)] and Sebastian Stawicki[2]

[1] Institute of Informatics, University of Warsaw, Warsaw, Poland
slezak@mimuw.edu.pl
[2] QED Software, Warsaw, Poland

Abstract. We investigate decision bireducts which extend the notion of a decision reduct developed in the theory of rough sets. For a decision table $\mathbb{A} = (U, A \cup \{d\})$, a decision bireduct is a pair (X, B), where $B \subseteq A$ is a subset of attributes which allows to distinguish between all pairs of objects in $X \subseteq U$ labeled with different values of decision attribute d, and where B and X cannot be made, respectively, smaller and bigger without losing this property. We refer to our earlier studies on deriving bireducts (X, B) from decision tables and utilizing them to construct families of rule-based classifiers, where $X \subseteq U$ is equal to total support of decision rules built using attributes in $B \subseteq A$. We introduce the notion of a correct ensemble of decision bireducts $(X_1, B_1), ..., (X_m, B_m)$, where each $u \in U$ must be validly classified by more than 50% of the corresponding models. We show that the problem of finding a correct ensemble of bireducts with the lowest cardinalities of subsets $B_i \subseteq A$ is NP-hard.

Keywords: Decision bireducts · Classifier ensembles · Rule-based classifiers · Decision model simplification · NP-hardness

1 Introduction

There are a number of approaches based on ensembles of classifiers – or decision models, more generally – in the areas of knowledge discovery and data classification [1, 4]. One can name several reasons to use ensembles. First, we can count on stability and robustness of the collective. Moreover, it is expected that each of classifiers – a part of a bigger ensemble – can be simpler than a single, not ensemble-based decision model that would yield a similar level of accuracy. On the other hand, if we look at all ensemble parts as a whole, they often lose something out of their interpretability. It may be hard to set up their cooperation. Finally, a lot of computing power is needed to derive them from the data.

Several optimization tendencies repeat in case of numerous methods for learning classifier ensembles. For instance, it is widely assumed that components of

© Springer Nature Switzerland AG 2020
R. Bello et al. (Eds.): IJCRS 2020, LNAI 12179, pp. 204–212, 2020.
https://doi.org/10.1007/978-3-030-52705-1_15

the ensemble should not misclassify too often the same training cases. Each single classifier is expected to make mistakes (in other words, each single part can be relatively weak with respect to its classification power) but for each training case, a (possibly weighted) majority of models in the ensemble should be correct. To address this aspect while building classifier ensembles, popular machine learning meta-algorithms such as boosting or bagging can be used.

One more aspect – besides tending to simplicity (and to some extent weakness) and complementarity (avoiding repeatable mistakes) in the ensemble – corresponds to diversification of attributes (features) that are used as inputs to learn particular models. For example, in case of rough-set-inspired approaches to knowledge discovery, it refers to computation of diverse decision reducts, i.e., irreducible (the smaller – the better) subsets of attributes that are sufficient to determine decision labels. If we want to put this idea together with diversification of cases which are classified correctly/wrongly by particular models, we can rely on ensembles of so-called decision bireducts [7,8], whereby irreducibility of a subset of attributes is combined with non-extendability of a subset of objects (cases), for which those attributes let us form valid classification rules.

Rough-set-based approaches are also a good reference while considering formal optimization problems behind construction of decision models. Indeed, starting from fundamental works on NP-hardness of the problem of finding minimal (in terms of the number of attributes) decision reduct, a lot of attention is paid in the rough set literature to develop mathematical and algorithmic methods for operating with the simplest yet sufficiently accurate (and thus the most powerful) classifiers [2,3]. However, even in the realm of rough sets, there are no studies on formulation of optimization problems related to classifier ensembles. In other words, optimization goals and their complexity characteristics are investigated only at the level of single models, rather than the whole ensembles.

Accordingly, in this paper we propose how to define the optimization problem related to searching for the simplest possible ensembles of decision models that meet specific accuracy constraints. We realize that there are plenty of ways of stating such constraints, with respect to various aspects of an ensemble as a whole or its single components. Similarly, there may be many ways of understanding the simplicity of an ensemble. Nevertheless, we believe that the introduced formulation – based on collections of decision bireducts that include the minimal amounts of attributes (the optimization goal) and in the same time, sufficiently cover all considered objects using the corresponding decision rules (the accuracy constraint) – can be a good starting point for further investigations.

The paper is organized as follows: Sect. 2 recalls decision bireducts (X, B), whereby – for the training data set (referred as so-called decision table) $\mathbb{A} = (U, A \cup \{d\})$ – $B \subseteq A$ is a subset of attributes which allows to distinguish between all pairs of objects in $X \subseteq U$ labeled with different values of decision attribute d, and where B and X cannot be reduced and extended, respectively. Section 3 introduces *correct* ensembles of bireducts $(X_1, B_1), ..., (X_m, B_m)$, $m > 0$, whereby there is inequality $|\{i = 1, ..., m : u \in X_i\}| > m/2$ for each $u \in U$. Section 4 presents our idea of expressing simplicity (which is the optimization goal) of correct ensembles in terms of cardinalities of subsets $B_1, ..., B_m$. Section 5 shows our main theoretical result, i.e., NP-hardness of finding the simplest correct ensembles of decision bireducts. Section 6 concludes the paper.

2   Decision Bireducts

We use tabular data representation by means of decision tables [3] – pairs $\mathbb{A} = (U, A \cup \{d\})$ of finite sets U and $A \cup \{d\}$, where U is the universe of objects and $A \cup \{d\}$ is the set consisting of attributes such that every $a \in A \cup \{d\}$ is associated with function $a : U \to V_a$, where V_a is called the value set of a. The distinguished attribute $d \notin A$ is called the decision. Elements of A are called conditional attributes. Values $v_d \in V_d$ correspond to decision classes.

Definition 1. *Let $\mathbb{A} = (U, A \cup \{d\})$ and $B \subseteq A$, $X \subseteq U$ be given. We say that B determines d within X, further denoted as $B \Rightarrow_X d$, if and only if B discerns all pairs $u_i, u_j \in X$ such that $d(u_i) \neq d(u_j)$. Further, we say that the pair (X, B) is a decision bireduct, if and only if the following holds:*

1. *There is $B \Rightarrow_X d$,*
2. *There is no proper subset $B' \subsetneq B$ such that $B' \Rightarrow_X d$,*
3. *There is no proper superset $X' \supsetneq X$ such that $B \Rightarrow_{X'} d$.*

We will say that objects in X are covered by $B \Rightarrow_X d$.

Every decision bireduct (X, B) may be understood as a pair consisting of an irreducible subset of attributes that can be evaluated by means of an non-extendable subset of objects for which it provides good classification. It was shown in [8] that X is actually the set-theoretic sum of objects supporting deterministic rules using the values of attributes in B to describe the values of d.

Figure 1 displays some bireducts derived for decision table $\mathbb{A}$ with objects $U = \{u_1, ..., u_6\}$ and conditional attributes $A = \{a_1, a_2, a_3\}$. For instance, consider the pair $B = \{a_1, a_2\}$ and $X = \{u_2, ..., u_6\}$. It corresponds to rules "if $a_1 = 0$ and $a_2 = 0$ then $d = 0$" (supported by u_2), "if $a_1 = 0$ and $a_2 = 1$ then $d = 0$" (supported by u_3), "if $a_1 = 1$ and $a_2 = 0$ then $d = 1$" (supported by u_4 and u_5) and "if $a_1 = 1$ and $a_2 = 1$ then $d = 0$" (supported by u_6). Neither a_1 nor a_2 would be sufficient by itself to cover X with shorter rules. Moreover, u_1 cannot be added to X because it is inconsistent with the first rule.

The first algorithms aimed at deriving decision bireducts from the data were proposed in [7]. They were based on random generation of mixed orderings of attributes and objects. Such orderings were utilized to encode sequences of attempts to remove attributes from B (starting with $B = A$) and add objects to X (starting with $X = \emptyset$) in order to obtain pairs (X, B) such that $B \Rightarrow_X d$, with possibly minimal B and maximal X. By using an appropriate process of generation of families of diverse orderings, one could derive collections of bireducts with quite different subsets of attributes and objects involved.

Probably the most efficient known technique for deriving decision bireducts consists of four phases: 1) Choose randomly a subset $C \subseteq A$; 2) For each block of objects (so-called indiscernibility class) with the same values on C, add randomly one of its elements to $Y \subseteq U$; 3) For table $\mathbb{A} = (Y, C \cup \{d\})$, run one of classical methods for finding standard decision reducts [2]; 4) For a decision reduct $B \subseteq C$, construct superset $X \supseteq Y$ by adding to Y all objects that are consistent with decision rules generated from table $\mathbb{A} = (Y, B \cup \{d\})$.

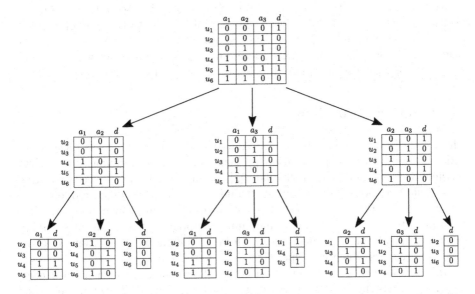

Fig. 1. Examples of decision bireducts for table $\mathbb{A} = (\{u_1, ..., u_6\}, \{a_1, a_2, a_3\} \cup \{d\})$. Bireducts in the middle layer form a correct ensemble (each object is validly classified by at least two bireducts out of three). For each "middle" bireduct treated as a new decision table, its corresponding correct ensemble is provided in the lowest layer.

3   Bireduct Ensembles

The notion of a decision bireduct allows us to operate with subsets of conditional attributes treated as classification descriptions, and with the associated subsets of objects for which those descriptions are valid. This gives us an elegant way to investigate complementarity of bireducts interpreted as classifiers in the ensemble. For instance, the following formulation expresses the idea of majority voting between ensemble components which – if properly tuned on the training data – gives us a chance of efficient performance over new cases too.

Definition 2. Let $\mathbb{A} = (U, A \cup \{d\})$ and the ensemble of decision bireducts $\mathcal{B} = (X_1, B_1), ..., (X_m, B_m)$ be given. We say that $\mathcal{B}$ is correct, if and only if there is inequality $|\{i = 1, ..., m : u \in X_i\}| > m/2$ for each $u \in U$.

The above inequality means that more than 50% of decision rules triggered for $u \in U$ point at the valid decision $d(u)$. Figure 1 illustrates a kind of hierarchy of correct bireduct ensembles for $m = 3$. Alternatively, one can work with a "flat" collection of decision bireducts that are supposed to vote correctly on each of objects, even if some single bireducts are wrong for some single cases.

Figure 1 implicitly suggests a top-down way of constructing correct ensembles, whereby each of m bireducts is derived in the same time, with an option of further decompositions on even smaller pieces. Such algorithms have been already considered in [5] for another type of (bi)reducts, i.e., so-called generalized decision reducts. On the other hand, one can proceed with the aforementioned ordering-based methods [7], whereby – somewhat reflecting the mechanisms of bagging and boosting – each next ordering may take into account which objects were covered least frequently by decision bireducts derived up to now.

4 Ensemble Simplicity

The rough literature provides a great number of theoretical works on computational complexity of optimization problems focused on deriving the simplest possible decision models from the data [2]. Let us refer to a recent comparative study reflecting both decision bireducts and so-called approximate reducts with this respect [8]. By "the simplest" one can mean (bi)reducts involving the minimal amounts of attributes, generating minimal amounts of decision rules, having the minimal information entropy, etc. However, all those formulations refer to single (bi)reducts which correspond to single classifiers.

In other words, as it was emphasized in [3], simplicity is a crucial aspect of decision models, in relation to paradigms such as Occam's Razor or the Minimum Description Length Principle. However, there is no clear guidance how to understand simplicity of ensembles. Thus, if we want to define optimization problems for ensembles, we need to know how to aggregate "complexities" of particular ensemble components (e.g.: the number of attributes in a single decision bireduct, the number of leaves in a single decision tree, etc.).

Intuitively, in case of ensembles of decision bireducts, the corresponding optimization problem should be stated by means of finding the smallest subsets $B_1, ..., B_m$ that satisfy – together with their counterparts $X_1, ..., X_m$ – the constraints of Definition 2. The question remains what we should mean by "the smallest" in case of a collection of subsets. In [6], for the analogous task related to the already-mentioned generalized decision reducts, it was proposed to look at it from the perspective of the maximum cardinality out of all involved subsets. For the purpose of bireducts it can be phrased as follows:

Definition 3. *Let decision table* $\mathbb{A} = (U, A \cup \{d\})$ *and two correct ensembles of decision bireducts* $\mathcal{B} = \{(X_1, B_1), ..., (X_m, B_m)\}$ *and* $\mathcal{C} = \{(Y_1, C_1), ..., (Y_n, C_n)\}$, $m, n \geq 0$, *be given. We say that* $\mathcal{B}$ *is simpler than* $\mathcal{C}$, *denoted as* $\mathcal{B} \prec \mathcal{C}$, *if and only if the following procedure yields it:*

1. *Sort sequences of cardinalities of attribute subsets in a descending order.*
2. *Add a sentinel item with the value -1 at the end of each of sequences.*
3. *Find the first position for which the sorted sequences differ from each other.*
4. *If the value in the above-found position is lower for B than for C, then $B \prec C$.*

The above procedure – illustrated additionally by Fig. 2 – induces a linear order over ensembles of bireducts for a given $\mathbb{A}$. We therefore propose to search through a space of all correct ensembles $B = \{(X_1, B_1), ..., (X_m, B_m)\}$, paying special attention to cardinalities of their largest components along a kind of cardinality-based lexicographic order. This is because the largest subsets of attributes correspond to the largest collections of the longest rules, i.e., they affect complexity of the model more significantly than other subsets.

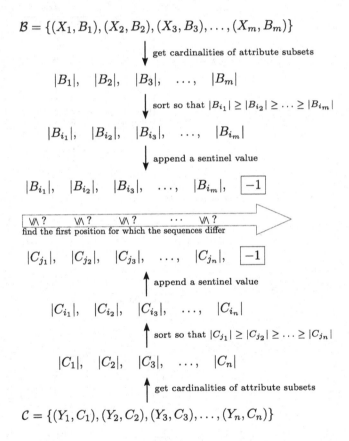

Fig. 2. Illustration of the procedure in Definition 3.

5   Main Result

Let us formalize the optimization goal that we drafted in the previous section:

Definition 4. *By the Simplest Correct Decision Bireduct Ensemble Problem (SCDBEP) we mean the task of finding – for each input decision table $\mathbb{A}$ – the correct ensemble of decision bireducts $\mathcal{B}$ such that there is no other correct ensemble for $\mathbb{A}$ that would be simpler than $\mathcal{B}$ according to Definition 3.*

Theorem 1. *SCDBEP is NP-hard.*

Before we present the proof, let us refer to Fig. 3. The proof is based on polynomial reduction of the problem of finding the smallest dominating sets in undirected graphs to SCDBEP. It requires encoding of each input graph $\mathbb{G}$ to its corresponding decision table $\mathbb{A}_{\mathbb{G}}$. This encoding is analogous to those that were utilized for other (bi)redect-related optimization problems [2,8].

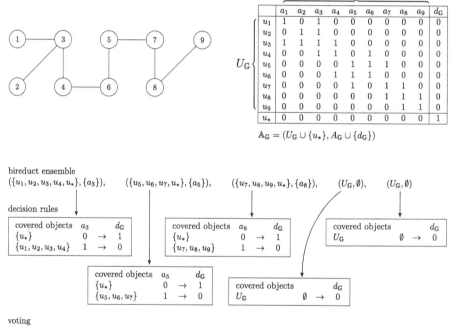

$A_{\mathbb{G}}$

	a_1	a_2	a_3	a_4	a_5	a_6	a_7	a_8	a_9	$d_{\mathbb{G}}$
u_1	1	0	1	0	0	0	0	0	0	0
u_2	0	1	1	0	0	0	0	0	0	0
u_3	1	1	1	1	0	0	0	0	0	0
u_4	0	0	1	1	0	1	0	0	0	0
u_5	0	0	0	0	1	1	1	0	0	0
u_6	0	0	0	1	1	1	0	0	0	0
u_7	0	0	0	0	1	0	1	1	0	0
u_8	0	0	0	0	0	0	1	1	1	0
u_9	0	0	0	0	0	0	0	1	1	0
u_*	0	0	0	0	0	0	0	0	0	1

$U_{\mathbb{G}}$ brackets rows u_1–u_9.

$$\mathbb{A}_{\mathbb{G}} = (U_{\mathbb{G}} \cup \{u_*\}, A_{\mathbb{G}} \cup \{d_{\mathbb{G}}\})$$

bireduct ensemble

$(\{u_1,u_2,u_3,u_4,u_*\},\{a_3\})$, $(\{u_5,u_6,u_7,u_*\},\{a_5\})$, $(\{u_7,u_8,u_9,u_*\},\{a_8\})$, $(U_{\mathbb{G}},\emptyset)$, $(U_{\mathbb{G}},\emptyset)$

decision rules

covered objects	a_3		$d_{\mathbb{G}}$
$\{u_*\}$	0	→	1
$\{u_1,u_2,u_3,u_4\}$	1	→	0

covered objects	a_8		$d_{\mathbb{G}}$
$\{u_*\}$	0	→	1
$\{u_7,u_8,u_9\}$	1	→	0

covered objects		$d_{\mathbb{G}}$
$U_{\mathbb{G}}$	$\emptyset$ →	0

covered objects	a_5		$d_{\mathbb{G}}$
$\{u_*\}$	0	→	1
$\{u_5,u_6,u_7\}$	1	→	0

covered objects		$d_{\mathbb{G}}$
$U_{\mathbb{G}}$	$\emptyset$ →	0

voting

	$(\{u_1,u_2,u_3,u_4,u_*\},\{a_3\})$	$(\{u_5,u_6,u_7,u_*\},\{a_5\})$	$(\{u_7,u_8,u_9,u_*\},\{a_8\})$	$(U_{\mathbb{G}},\emptyset)$	$(U_{\mathbb{G}},\emptyset)$	voting	
u_1	$1 \to 0$	$0 \to 1$	$0 \to 1$	$\emptyset \to 0$	$\emptyset \to 0$	$d_{\mathbb{G}}=0$	$(3 \times 0, 2 \times 1)$
u_2	$1 \to 0$	$0 \to 1$	$0 \to 1$	$\emptyset \to 0$	$\emptyset \to 0$	$d_{\mathbb{G}}=0$	$(3 \times 0, 2 \times 1)$
u_3	$1 \to 0$	$0 \to 1$	$0 \to 1$	$\emptyset \to 0$	$\emptyset \to 0$	$d_{\mathbb{G}}=0$	$(3 \times 0, 2 \times 1)$
u_4	$1 \to 0$	$0 \to 1$	$0 \to 1$	$\emptyset \to 0$	$\emptyset \to 0$	$d_{\mathbb{G}}=0$	$(3 \times 0, 2 \times 1)$
u_5	$0 \to 1$	$1 \to 0$	$0 \to 1$	$\emptyset \to 0$	$\emptyset \to 0$	$d_{\mathbb{G}}=0$	$(3 \times 0, 2 \times 1)$
u_6	$0 \to 1$	$1 \to 0$	$0 \to 1$	$\emptyset \to 0$	$\emptyset \to 0$	$d_{\mathbb{G}}=0$	$(3 \times 0, 2 \times 1)$
u_7	$0 \to 1$	$1 \to 0$	$1 \to 0$	$\emptyset \to 0$	$\emptyset \to 0$	$d_{\mathbb{G}}=0$	$(4 \times 0, 1 \times 1)$
u_8	$0 \to 1$	$0 \to 1$	$1 \to 0$	$\emptyset \to 0$	$\emptyset \to 0$	$d_{\mathbb{G}}=0$	$(3 \times 0, 2 \times 1)$
u_9	$0 \to 1$	$0 \to 1$	$1 \to 0$	$\emptyset \to 0$	$\emptyset \to 0$	$d_{\mathbb{G}}=0$	$(3 \times 0, 2 \times 1)$
u_*	$0 \to 1$	$0 \to 1$	$0 \to 1$	$\emptyset \to 0$	$\emptyset \to 0$	$d_{\mathbb{G}}=1$	$(2 \times 0, 3 \times 1)$

Fig. 3. Illustration for the proof of Theorem 1.

Figure 3 can also serve as one more illustration of creation of correct ensembles of decision bireducts. It displays how to interpret those bireducts rule-based classifiers. In particular, as it could be already noticed earlier in Fig. 1, some bireducts can correspond to empty sets of attributes. We can interpret them as "dummy" classifiers which point always at the same decision class. They may help to tune the majority voting mechanism in the ensemble.

Proof. As already stated, we intend to show NP-hardness of SCDBEP by polynomial reduction of the minimum dominating set problem. Let us consider an undirected graph $\mathbb{G} = (V, E)$ and create decision table $\mathbb{A}_{\mathbb{G}} = (U_{\mathbb{G}} \cup \{u_*\}, A_{\mathbb{G}} \cup \{d_{\mathbb{G}}\})$, where $a_v \in A_{\mathbb{G}}$ corresponding to $v \in V$ takes 1 on $u_{v'} \in U_{\mathbb{G}}$ corresponding to $v' \in V$, i.e., $a_v(u_{v'}) = 1$, if and only if $v = v'$ or $(v, v') \in E$, and where $a_v(u_*) = 0$. Let us also put $d_{\mathbb{G}}(u_{v'}) = 0$ and $d_{\mathbb{G}}(u_*) = 1$ (see Fig. 3).

Clearly, any $B \subseteq V$ is a dominating set in $\mathbb{G}$, if and only if it corresponds to a decision bireduct $(U_{\mathbb{G}}, B_{\mathbb{G}})$. It is obvious that a single-element bireduct ensemble $\{(U_{\mathbb{G}}, B_{\mathbb{G}})\}$ is correct according to Definition 2. However, we can always construct a simpler (or equally simple if $B_{\mathbb{G}}$ is a singleton) correct ensemble.

Assuming that $B_{\mathbb{G}} = \{a_{v_1}, a_{v_2}, ..., a_{v_n}\}$, let us define new subsets of attributes as $B_{\mathbb{G},1} = \{a_{v_1}\}$, ..., $B_{\mathbb{G},n} = \{a_{v_n}\}$, $B_{\mathbb{G},n+1} = \emptyset$, ..., $B_{\mathbb{G},2n-1} = \emptyset$ and new subsets of objects as $X_{\mathbb{G},1} = \{u_*\} \cup \{u \in U_{\mathbb{G}} | a_{v_1}(u) = 1\}$, ..., $X_{\mathbb{G},n} = \{u_*\} \cup \{u \in U_{\mathbb{G}} | a_{v_n}(u) = 1\}$, $X_{\mathbb{G},n+1} = U_{\mathbb{G}}$, ..., $X_{\mathbb{G},2n-1} = U_{\mathbb{G}}$. Then, the proposed simpler ensemble would be equal to $\{(X_{\mathbb{G},1}, B_{\mathbb{G},1}), ..., (X_{\mathbb{G},2n-1}, B_{\mathbb{G},2n-1})\}$ and it would be still correct according to Definition 2 (see Fig. 3 again).

The consequence of the above is that the simplest correct ensemble of decision bireducts corresponds to the smallest dominating set in the graph $\mathbb{G}$.   □

6   Conclusions

We investigated ensembles of so-called decision bireducts, which can be interpreted as rule-based classifiers. We introduced the notion of a correct ensemble, which means that every object (training case) must be validly recognized using the corresponding rules by more than majority of classifiers. We discussed how to specify a kind of simplicity criterion for such ensembles and we formulated an example of optimization problem related to extracting possibly simplest correct ensembles of decision bireducts from the input data. The main mathematical result of our paper is the NP-hardness of the considered problem.

In future, given such a sound theoretical framework, more attention should be paid to further extensions of our previous algorithmic approaches [7,8] to deriving and applying decision bireducts for the real-life data. Moreover, some alternative formulations of optimization problems should be discussed as well, possibly referring to ensembles of other types of classifiers.

References

1. Dietterich, T.G.: An experimental comparison of three methods for constructing ensembles of decision trees: bagging, boosting, and randomization. Mach. Learn. **40**(2), 139–157 (2000). https://doi.org/10.1023/A:1007607513941

2. Pawlak, Z., Skowron, A.: Rough sets and Boolean reasoning. Inf. Sci. **177**(1), 41–73 (2007)
3. Pawlak, Z., Skowron, A.: Rudiments of rough sets. Inf. Sci. **177**(1), 3–27 (2007)
4. Polikar, R., DePasquale, J., Mohammed, H.S., Brown, G., Kuncheva, L.I.: Learn++.MF: a random subspace approach for the missing feature problem. Pattern Recogn. **43**(11), 3817–3832 (2010)
5. Ślęzak, D.: Decomposition and synthesis of decision tables with respect to generalized decision functions. In: Pal, S.K., Skowron, A. (eds.) Rough Fuzzy Hybridization - A New Trend in Decision Making, pp. 110–135. Springer, Singapore (1999)
6. Ślęzak, D.: On generalized decision functions: reducts, networks and ensembles. In: Yao, Y., Hu, Q., Yu, H., Grzymala-Busse, J.W. (eds.) RSFDGrC 2015. LNCS (LNAI), vol. 9437, pp. 13–23. Springer, Cham (2015). https://doi.org/10.1007/978-3-319-25783-9_2
7. Ślęzak, D., Janusz, A.: Ensembles of bireducts: towards robust classification and simple representation. In: Kim, T., Adeli, H., Slezak, D., Sandnes, F.E., Song, X., Chung, K., Arnett, K.P. (eds.) FGIT 2011. LNCS, vol. 7105, pp. 64–77. Springer, Heidelberg (2011). https://doi.org/10.1007/978-3-642-27142-7_9
8. Stawicki, S., Ślęzak, D., Janusz, A., Widz, S.: Decision bireducts and decision reducts - a comparison. Int. J. Approximate Reasoning **84**, 75–109 (2017)

On Positive-Correlation-Promoting Reducts

Joanna Henzel[1], Andrzej Janusz[2], Marek Sikora[1], and Dominik Ślęzak[2(✉)]

[1] Department of Computer Networks and Systems, Faculty of Automatic Control, Electronics and Computer Science, Silesian University of Technology, Gliwice, Poland
[2] Institute of Informatics, Faculty of Mathematics, Informatics and Mechanics, University of Warsaw, Warsaw, Poland
slezak@mimuw.edu.pl

Abstract. We introduce a new rough-set-inspired binary feature selection framework, whereby it is preferred to choose attributes which let us distinguish between objects (cases, rows, examples) having different decision values according to the following mechanism: for objects $u1$ and $u2$ with decision values $dec(u1) = 0$ and $dec(u2) = 1$, it is preferred to select attributes a such that $a(u1) = 0$ and $a(u2) = 1$, with the secondary option – if the first one is impossible – to select a such that $a(u1) = 1$ and $a(u2) = 0$. We discuss the background for this approach, originally inspired by the needs of the genetic data analysis. We show how to derive the sets of such attributes – called positive-correlation-promoting reducts (PCP reducts in short) – using standard calculations over appropriately modified rough-set-based discernibility matrices. The proposed framework is implemented within the RoughSets R package which is widely used for the data exploration and knowledge discovery purposes.

Keywords: Rough sets · Feature selection · Discernibility · Rule induction · Positive-correlation-promoting reducts · RoughSets R package

1 Introduction

Rough set approaches are successfully utilized in the areas of machine learning and knowledge discovery, particularly for feature selection and classifiers simplification, as well as for deriving easily interpretable decision models from the data [1,9]. There are a number of generalizations and hybridizations of rough set methods available in the form of software toolkits, including dominance-based rough set algorithms [3], fuzzy-rough set algorithms [10], and others. There is plenty of research connecting rough sets with other knowledge representation methodologies such as e.g. formal concept analysis [4], as well as application-oriented studies such as e.g. extensions of standard rough set techniques aimed at handling high-dimensional data sets [7]. Finally, it is worth noting that rough set approaches can be combined in a natural way with various symbolic machine learning methods, in particular those designed for rule induction [5,12].

© Springer Nature Switzerland AG 2020
R. Bello et al. (Eds.): IJCRS 2020, LNAI 12179, pp. 213–221, 2020.
https://doi.org/10.1007/978-3-030-52705-1_16

```
library(RoughSets)

## Loading required package: Rcpp

data(RoughSetData)
binary.dt <- RoughSetData$binary.dt

# our exemplary decision table
binary.dt
```

V1 <fctr>	V2 <fctr>	V3 <fctr>	V4 <fctr>	V5 <fctr>	V6 <fctr>	V7 <fctr>	V8 <fctr>	V9 <fctr>	V10 <fctr>	dec <fctr>
1	0	1	1	1	1	0	0	1	1	0
1	0	0	1	0	0	0	0	0	0	0
1	0	1	1	0	1	1	1	0	1	0
1	0	1	0	0	1	0	1	1	0	0
0	1	1	1	1	0	1	1	0	1	0
1	0	1	0	0	0	1	1	0	1	1
0	1	0	1	0	0	1	1	0	0	1
0	1	0	1	0	1	0	0	1	1	1
0	1	1	0	1	0	1	0	1	0	1
0	1	1	1	1	1	0	0	1	0	1

1-10 of 10 rows

Fig. 1. Example of a binary decision table with 10 objects, 10 attributes $V1, ..., V10$, as well as decision dec, displayed using the RoughSets R package [10].

In this study, we are interested in inducing rules which follow a specific pattern of selecting conditions pointing at particular decisions. Using an example of decision table in Fig. 1, we seek for rules describing the case $dec = 1$ with conditions $Vi = 1$ (e.g. $V2 = 1 \wedge V6 = 1 \Rightarrow dec = 1$ supported by rows 8 and 10) and $dec = 0$ with conditions $Vi = 0$ (e.g. $V8 = 0 \wedge V9 = 0 \Rightarrow dec = 0$ supported by row 2). Only if there is no other choice, we would allow additional descriptors of the form $Vi = 0$ for $dec = 1$ and $Vi = 1$ for $dec = 0$ (e.g. it is impossible to construct a rule covering row 7 without using conditions $Vi = 0$).

We propose a new approach to feature selection, aimed at finding attributes which are suitable for constructing such rules. In order to do this, we modify the rough-set-based notion of a reduct [9]. For the binary data, our *positive-correlation-promoting (PCP) reducts* will prefer to contain attributes Vi such that – for objects $u1$ and $u2$ with decisions $dec(u1) = 0$ and $dec(u2) = 1$ – there is $Vi(u1) = 0$ and $Vi(u2) = 1$, or else – but only if the former option does not hold for any attribute – there is $Vi(u1) = 1$ and $Vi(u2) = 0$. (On the contrary, both those options of discernibility – i.e. $Vi(u1) = 0$, $Vi(u2) = 1$ versus $Vi(u1) = 1$, $Vi(u2) = 0$ – have the same importance for standard reducts.)

Going further, in Sect. 2 we recall the RoughSets R package [10], whereby we implement our new approach. In Sect. 3 we present the background for PCP reducts. In Sect. 4 we show that they are derivable using a modification of

```
## building a classical decision-relative discernibility matrix
disc.matrix <- BC.discernibility.mat.RST(binary.dt, return.matrix = TRUE)
head(disc.matrix$disc.list)
```

```
## [[1]]
## [1] "V4" "V5" "V6" "V7" "V8" "V9"
##
## [[2]]
## [1] "V1"  "V2"  "V3"  "V5"  "V6"  "V7"  "V8"  "V9"  "V10"
##
## [[3]]
## [1] "V1"  "V2"  "V3"  "V5"
##
## [[4]]
## [1] "V1"  "V2"  "V4"  "V6"  "V7"  "V10"
##
## [[5]]
## [1] "V1"  "V2"  "V10"
##
## [[6]]
## [1] "V3"  "V4"  "V7"  "V8"  "V10"
```

	1	2	3	4	5
6	V4, V5, V6, V7, V8, V9	V3, V4, V7, V8, V10	V4, V6	V6, V7, V9, V10	V1, V2, V4, V5
7	V1, V2, V3, V5, V6, V7, V8, V9, V10	V1, V2, V7, V8	V1, V2, V3, V6, V10	V1, V2, V3, V4, V6, V7, V9	V3, V5, V10
8	V1, V2, V3, V5	V1, V2, V6, V9, V10	V1, V2, V3, V7, V8, V9	V1, V2, V3, V4, V8, V10	V3, V5, V6, V7, V8, V9
9	V1, V2, V4, V6, V7, V10	V1, V2, V3, V4, V5, V7, V9	V1, V2, V4, V5, V6, V8, V9, V10	V1, V2, V5, V6, V7, V8	V4, V8, V9, V10
10	V1, V2, V10	V1, V2, V3, V5, V6, V9	V1, V2, V5, V7, V8, V9, V10	V1, V2, V4, V5, V8	V6, V7, V8, V9, V10

Fig. 2. Example continued: [Top] An excerpt from the standard discernibility matrix computed for decision table in Fig. 1 using the RoughSets R package; [Bottom] Full standard discernibility matrix for the considered decision table.

rough-set-based discernibility matrices. In Sect. 5 we discuss future relevant extensions of the package. In Sect. 6 we conclude the paper.

2   About the RoughSets R Package

The RoughSets package is available in CRAN (http://cran.r-project.org/web/packages/RoughSets/index.html). Its newest version can be found also in GitHub (https://github.com/janusza/RoughSets). It provides implementations of classical rough-set-based methods and their fuzzy-related extensions for data modeling and analysis. In particular, it includes tools for feature selection and attribute reduction, as well as rule induction and rule-based classification.

Figures 2 and 3 illustrate two out of the most fundamental functionalities of the package – calculation of a discernibility matrix from the input decision table, and calculation of all decision reducts from the input discernibility matrix. Let us recall that decision tables stand for standard representation of the labeled tabular data in the rough set framework. Discernibility matrices assign the pairs of objects (rows) having different decision values with attributes which are able to distinguish between them. Decision reducts are irreducible attribute subsets which distinguish between all such pairs (for decision table in Fig. 1, one needs to distinguish rows 1, 2, 3, 4, 5 from 6, 7, 8, 9, 10), i.e., those which have non-empty intersection with every cell of the corresponding matrix.

```
# computation of all classical reducts
classic.reducts <- FS.all.reducts.computation(disc.matrix)

# top 3 classical reducts
head(classic.reducts$decision.reduct, 3)
```

```
## $reduct1
## A feature subset consisting of 3  attributes:
## V1, V6, V10
##
## $reduct2
## A feature subset consisting of 3  attributes:
## V2, V6, V10
##
## $reduct3
## A feature subset consisting of 4  attributes:
## V1, V3, V4, V10
```

```
# a total number of found reducts
cat("A total number of reducts found: ",
    length(classic.reducts$decision.reduct), "\n", sep = "")
```

```
## A total number of reducts found: 38
```

```
# a decision core of the data table is empty...
classic.reducts$core
```

```
## character(0)
```

Fig. 3. Example continued: Standard reducts for decision table displayed in Fig. 1, calculated using the all-reducts function in the RoughSets R package.

The considered package contains also other, more modern methods of decision reduct calculation. Some of them work on far more efficient data structures than discernibility matrices. Some of them search heuristically for single reducts or small groups of reducts instead of all of them. Nevertheless, referring to functions in Figs. 2 and 3 is a good starting point for further investigations.

3   Inspiration for PCP Reducts

The idea of operating with rules exemplified in Sect. 1 comes from our earlier studies on the data produced in the cancer genome atlas project (https://en. wiki\discretionary-pedia.org/wiki/The_Cancer_Genome_Atlas) [11] and other gene-related data sets [7]. Let us consider the copy number variation pipeline (https://en.wikipedia.org/wiki/Copy-number_variation) [6] which uses the Affymetrix SNP 6.0 array data [2] to identify the repeating genomic regions and to infer the copy number of those repeats. Imagine that attributes in Fig. 1 represent some of the protein coding genes and rows represent patient samples. For each patient, a gene can be characterized by 0 (no change) or 1 (change in the copy number for that gene). Assume that dec takes value 1 for patients with short survival time. Then, we would like to describe decision class $dec = 1$ by genes for which a change in the copy number was registered i.e. using conditions $Vi = 1$.

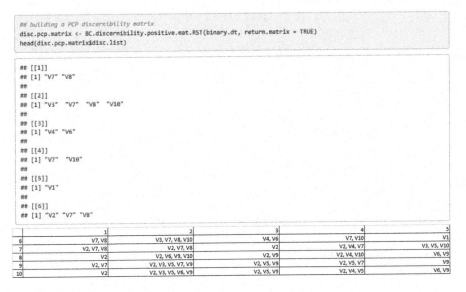

Fig. 4. Example continued: [Top] An excerpt from the PCP discernibility matrix computed for decision table in Fig. 1 using the new functionality of the RoughSets R package; [Bottom] Full PCP discernibility matrix for the considered table.

If we were interested only in such rules, then they could be modeled using formal concept analysis [4]. However, we also need rules describing $dec = 0$ by $Vi = 0$. In such a case, one might suggest that it is worth using the dominance-based rough set framework [3]. However, it is not so strict that we should use only conditions $Vi = 1$ for $dec = 1$ and $Vi = 0$ for $dec = 0$. Such conditions are preferred and should be *promoted* by the rule generation process. However, if it is impossible to form rules using only such conditions, then the other ones ($Vi = 0$ for $dec = 1$ and $Vi = 1$ for $dec = 0$) are allowed too.

4 Discernibility Characteristics of PCP Reducts

Discussion in the previous section leads us toward the following:

Definition 1. *Let a binary decision table* $\mathbb{A} = (U, A \cup \{d\})$ *be given. (In Fig. 1:* $U = \{u1, ..., u10\}$, $A = \{V1, ..., V10\}$, $d = dec$.) *Consider object* $u \in U$ *and subset* $B \subseteq A$. *We say that rule* $\bigwedge_{a \in B} a = a(u) \Rightarrow d = d(u)$ *is a positive-correlation-promoting (PCP) rule, if and only if it holds irreducibly in* $\mathbb{A}$ *and there is* $\forall_{a \in B} a(u) = d(u)$ *or else, one cannot replace conditions* $a = a(u)$ *such that* $a(u) \neq d(u)$ *with any conditions* $b = b(u)$, $b(u) = d(u)$, $b \notin B$.

Definition 2. *Subset* $B \subseteq A$ *is a positive-correlation-promoting (PCP) reduct, if and only if each* $u \in U$ *can be covered by a PCP rule* $\bigwedge_{a \in B^u} a = a(u) \Rightarrow d = d(u)$, $B^u \subseteq B$, *and there is no proper* $B' \subsetneq B$ *with this property.*

```
# analogically, computation of all PCP reducts
pcp.reducts <- FS.all.reducts.computation(disc.pcp.matrix)

# top 3 PCP reducts - notice how large they are relative to the classical ones
head(pcp.reducts$decision.reduct, 3)
```

```
## $reduct1
## A feature subset consisting of 6  attributes:
## V1, V2, V3, V4, V7, V9
##
## $reduct2
## A feature subset consisting of 6  attributes:
## V1, V2, V3, V6, V7, V9
##
## $reduct3
## A feature subset consisting of 6  attributes:
## V1, V2, V4, V5, V7, V9
```

```
# a total number of reducts found
cat("A total number of PCP reducts found: ",
    length(pcp.reducts$decision.reduct), "\n", sep = "")
```

```
## A total number of PCP reducts found: 8
```

```
# a decision core for PCP reducts contains attributes V1 and V2,
# even though they provide the same information in a classical sense:
pcp.reducts$core
```

```
## [1] "V1" "V2" "V9"
```

Fig. 5. Example continued: The all-reducts function in the RoughSets R package, now executed on the PCP discernibility matrix for decision table in Fig. 1.

The following characteristics can be shown in straightforward way:

Proposition 1. *For binary* $\mathbb{A} = (U, A \cup \{d\})$, *for every* $u1, u2 \in U$ *such that* $d(u1) \neq d(u2)$, *define* $M^+(u1, u2) = \{a \in A : a(u1) = d(u1) \land a(u2) = d(u2)\}$ *and* $M^-(u1, u2) = \{a \in A : a(u1) \neq d(u1) \land a(u2) \neq d(u2)\}$. *Consider the PCP discernibility matrix which labels the pairs of objects as follows:*

$$M(u1, u2) = \begin{cases} M^+(u1, u2) \ if \ M^+(u1, u2) \neq \emptyset \\ M^-(u1, u2) \ otherwise \end{cases} \tag{1}$$

Then, a given $B \subseteq A$ *is a PCP reduct, if and only if it is an irreducible subset such that* $B \cap M(u1, u2) \neq \emptyset$ *for every* $u1, u2 \in U$, $d(u1) \neq d(u2)$.

5   Heuristic Search of PCP Reducts and Rules

Definition 2 reflects the requirements of the feature selection process if the ultimate goal is to induce the rules of the form discussed in previous sections and considered earlier in [11]. Moreover, Proposition 1 provides us with an easy way to derive PCP reducts. Namely, it is enough to modify classical discernibility matrices and then apply the same techniques as those outlined in [8].

```
# decision tables reduced based on the top 2 classical reducts and one PCP reduct
dt.classic1 = SF.applyDecTable(binary.dt, classic.reducts$decision.reduct[[1]])
dt.classic2 = SF.applyDecTable(binary.dt, classic.reducts$decision.reduct[[2]])
dt.pcp = SF.applyDecTable(binary.dt, pcp.reducts$decision.reduct[[1]])

# computing decision rules using the LEM2 algorithm
rules.classic1 <- RI.LEM2Rules.RST(dt.classic1)
rules.classic1
```

```
## A set consisting of  6  rules:
## 1. IF V1 is 1 and V6 is 1 THEN dec is 0;
##      (supportSize=3; laplace=0.8)
## 2. IF V6 is 0 and V1 is 0 and V10 is 1 THEN dec is 0;
##      (supportSize=1; laplace=0.6667)
## 3. IF V1 is 1 and V10 is 0 THEN dec is 0;
##      (supportSize=2; laplace=0.75)
## 4. IF V1 is 0 and V10 is 0 THEN dec is 1;
##      (supportSize=3; laplace=0.8)
## 5. IF V1 is 0 and V6 is 1 THEN dec is 1;
##      (supportSize=2; laplace=0.75)
## 6. IF V1 is 1 and V6 is 0 and V10 is 1 THEN dec is 1;
##      (supportSize=1; laplace=0.6667)
```

```
rules.classic2 <- RI.LEM2Rules.RST(dt.classic2)
rules.classic2
```

```
## A set consisting of  6  rules:
## 1. IF V2 is 0 and V6 is 1 THEN dec is 0;
##      (supportSize=3; laplace=0.8)
## 2. IF V2 is 0 and V10 is 0 THEN dec is 0;
##      (supportSize=2; laplace=0.75)
## 3. IF V2 is 1 and V6 is 0 and V10 is 1 THEN dec is 0;
##      (supportSize=1; laplace=0.6667)
## 4. IF V2 is 1 and V10 is 0 THEN dec is 1;
##      (supportSize=3; laplace=0.8)
## 5. IF V10 is 1 and V2 is 0 and V6 is 0 THEN dec is 1;
##      (supportSize=1; laplace=0.6667)
## 6. IF V2 is 1 and V6 is 1 THEN dec is 1;
##      (supportSize=2; laplace=0.75)
```

```
rules.pcp <- RI.LEM2Rules.RST(dt.pcp)
rules.pcp
```

```
## A set consisting of  5  rules:
## 1. IF V2 is 0 and V7 is 0 THEN dec is 0;
##      (supportSize=3; laplace=0.8)
## 2. IF V9 is 0 and V3 is 1 and V4 is 1 THEN dec is 0;
##      (supportSize=2; laplace=0.75)
## 3. IF V2 is 1 and V9 is 1 THEN dec is 1;
##      (supportSize=3; laplace=0.8)
## 4. IF V2 is 1 and V3 is 0 THEN dec is 1;
##      (supportSize=2; laplace=0.75)
## 5. IF V4 is 0 and V7 is 1 THEN dec is 1;
##      (supportSize=2; laplace=0.75)
```

Fig. 6. Decision rules derived using the LEM2 [5] algorithm's version available in the RoughSets R package [10]. The rules are derived for two examples of standard reducts and one example of a PCP reduct (the last one). This means that only attributes contained in the given reduct are considered as input to LEM2. Although PCP reducts are designed to promote attributes which let us construct rules including more descriptors of the form $Vi = 1$ pointing at decision $dec = 1$, as well as more descriptors of the form $Vi = 0$ pointing at $dec = 0$, this information is lost during the phase of rule shortening. This is because – in its current implementation – this phase does not distinguish between positively (M^+) and negatively (M^-) correlated discernibility cases.

This fact allowed us to extend the RoughSets package [10], as visible in Figs. 4 and 5. When comparing the PCP matrix (Fig. 4) with its classical counterpart (Fig. 2), one can see that the attribute sets are now smaller. (The only unchanged cells are $M(u3, u6)$ and $M(u5, u7)$ – this is because $M^+ = \emptyset$ in both cases.) Consequently, PCP reducts are bigger than standard ones. In particular, PCP reducts can include both attributes $V1$ and $V2$ which are mutually interchangeable [7], so they would never co-occur in a standard reduct.

Still, there is a lot left to be done in the area of heuristic extraction of PCP reducts. One might expect that the corresponding algorithms should seek for PCP reducts which yield rules with maximum number of descriptors $Vi = 1$ for $dec = 1$ (and $Vi = 0$ for $dec = 0$). Unfortunately, classical methods cannot distinguish between the cases $M^+ \neq \emptyset$ and $M^+ = \emptyset$ in equation (1), so their heuristic optimization functions do not work properly. The same happens with standard rule induction methods [5,10] as further outlined in Fig. 6.

6   Further Research Directions

The newly introduced PCP reducts require further study in many aspects. Besides the aforementioned need of better heuristic search methods, we shall design algorithms working on more efficient data structures than PCP matrices. Herein, we will attempt to adapt some of modern data structures which are used to derive classical reducts and rules in rough-set-based toolkits [3,5].

Another future direction may refer to PCP reducts for non-binary data sets. In this paper, the nature of *promoting positive correlations* was expressed in terms of selecting these attributes which share – if possible – the same value differences as observed for the decision column. An analogous idea could be considered e.g. for numerical data sets, whereby one may think about appropriate modifications of fuzzy-rough discernibility characteristics [4,10].

References

1. Bello, R., Falcon, R.: Rough sets in machine learning: a review. In: Wang, G., Skowron, A., Yao, Y., Ślęzak, D., Polkowski, L. (eds.) Thriving Rough Sets. SCI, vol. 708, pp. 87–118. Springer, Cham (2017). https://doi.org/10.1007/978-3-319-54966-8_5
2. Bhaskar, H., Hoyle, D.C., Singh, S.: Machine learning in bioinformatics: a brief survey and recommendations for practitioners. Comp. Bio. Med. **36**(10), 1104–1125 (2006)
3. Błaszczyński, J., Greco, S., Matarazzo, B., Słowiński, R., Szeląg, M.: jMAF - dominance-based rough set data analysis framework. In: Skowron, A., Suraj, Z. (eds.) Rough Sets and Intelligent Systems, vol. 1, pp. 185–209. Springer, Heidelberg (2013). https://doi.org/10.1007/978-3-642-30344-9_5
4. Cornejo Piñero, M.E., Medina-Moreno, J., Ramírez-Poussa, E.: Fuzzy-attributes and a method to reduce concept lattices. In: Cornelis, C., et al. (eds.) RSCTC 2014. LNCS (LNAI), vol. 8536, pp. 189–200. Springer, Cham (2014). https://doi.org/10.1007/978-3-319-08644-6_20

5. Grzymała-Busse, J.W.: A comparison of rule induction using feature selection and the LEM2 algorithm. In: Stańczyk, U., Jain, L.C. (eds.) Feature Selection for Data and Pattern Recognition. SCI, vol. 584, pp. 163–176. Springer, Heidelberg (2015). https://doi.org/10.1007/978-3-662-45620-0_8
6. Jakobsson, M., et al.: Genotype, haplotype and copy-number variation in worldwide human populations. Nature **451**(7181), 998–1003 (2008)
7. Janusz, A., Ślęzak, D.: Rough set methods for attribute clustering and selection. Appl. Artif. Intell. **28**(3), 220–242 (2014)
8. Pawlak, Z., Skowron, A.: Rough sets and Boolean reasoning. Inf. Sci. **177**(1), 41–73 (2007)
9. Pawlak, Z., Skowron, A.: Rudiments of rough sets. Inf. Sci. **177**(1), 3–27 (2007)
10. Riza, L.S., Janusz, A., Bergmeir, C., Cornelis, C., Herrera, F., Ślęzak, D., Benítez, J.M.: Implementing algorithms of rough set theory and fuzzy rough set theory in the R package, "RoughSets". Inf. Sci. **287**, 68–89 (2014)
11. Sikora, M., Gruca, A.: Induction and selection of the most interesting gene ontology based multiattribute rules for descriptions of gene groups. Pattern Recognit. Lett. **32**(2), 258–269 (2011)
12. Sikora, M., Wróbel, L., Gudyś, A.: GuideR: a guided separate-and-conquer rule learning in classification, regression, and survival settings. Knowl. Based Syst. **173**, 1–14 (2019)

Granular Computing

Feature and Label Association Based on Granulation Entropy for Deep Neural Networks

Marilyn Bello[1,2(✉)], Gonzalo Nápoles[2,3], Ricardo Sánchez[1], Koen Vanhoof[2], and Rafael Bello[1]

[1] Computer Science Department, Central University of Las Villas, Santa Clara, Cuba
mbgarcia@uclv.cu
[2] Faculty of Business Economics, Hasselt University, Hasselt, Belgium
[3] Department of Cognitive Science and Artificial Intelligence, Tilburg University, Tilburg, The Netherlands

Abstract. Pooling layers help reduce redundancy and the number of parameters before building a multilayered neural network that performs the remaining processing operations. Usually, pooling operators in deep learning models use an explicit topological organization, which is not always possible to obtain on multi-label data. In a previous paper, we proposed a pooling architecture based on association to deal with this issue. The association was defined by means of Pearson's correlation. However, features must exhibit a certain degree of correlation with each other, which might not hold in all situations. In this paper, we propose a new method that replaces the correlation measure with another one that computes the entropy in the information granules that are generated from two features or labels. Numerical simulations have shown that our proposal is superior in those datasets with low correlation. This means that it induces a significant reduction in the number of parameters of neural networks, without affecting their accuracy.

Keywords: Granular computing · Rough sets · Association-based pooling · Deep learning · Multi-label classification

1 Introduction

Multi-Label Classification (MLC) is a type of classification where each of the objects in the data has associated a vector of outputs, instead of being associated with a single value [8, 20]. Formally speaking, suppose $X = R^d$ denotes the d-dimensional instance space, and $L = \{l_1, l_2, \ldots, l_k\}$ denotes the label space with k being the possible class labels. The task of multi-label learning is to estimate a function $h : X \longrightarrow 2^L$ from the multi-label training set $\{(x_i, L_i) \mid 1 \leq i \leq n\}$. For each multi-label example (x_i, L_i), $x_i \in X$ is a d-dimensional feature vector $(x_{i1}, x_{i2}, \ldots, x_{id})$ and $L_i \subseteq L$ is the set of labels associated with x_i. For any unseen instance $x \in X$, the multi-label classifier $h(\cdot)$ predicts $h(x) \subseteq L$ as the

R. Bello et al. (Eds.): IJCRS 2020, LNAI 12179, pp. 225–235, 2020.
https://doi.org/10.1007/978-3-030-52705-1_17

set of proper labels for x. This particular case of classification requires additional efforts in extracting relevant features describing both input and decision domains, since the boundaries regions of decisions usually overlap with each other. This often causes the decision space to be quite complex.

Deep learning [6,10] is a promising avenue of research into the automated extraction of complex data representations at high levels of abstraction. Such algorithms develop a layered, hierarchical architecture of learning and representing data, where higher-level (more abstract) features are defined in terms of lower-level (less abstract) features. For example, pooling layers [6,11,12] provide an approach to down sampling feature maps by summarizing the presence of features in patches of the feature map. Two common pooling methods are average pooling and max pooling, which compute the average presence of a feature and the most activated presence of a feature, respectively.

In the case of MLC, this must be done for both features and labels. Several authors [5,15,17,21] have proposed MLC solutions inspired on deep learning techniques. All these solutions are associated with application domains in which the data have a topological organization (i.e. recognizing faces, coloring black and white images or classifying objects in photographs). In [1] the authors introduced the *association-based pooling* that exploits the correlation among neurons instead of exploiting the topological information as typically occurs when using standard pooling operators. Despite of the relatively good results reported by this model, the function used to quantify the association between problem variables does not seem to be suitable for datasets having poor correlation among their features or labels. An alternative to deal with this issue consists in replacing the correlation measure with a more flexible association estimator.

In this paper, we compute the entropy of the granules that are generated from two problem features or labels. Several methods based on Granular Computing use granules as basic elements of analysis [7,18], so that from two similar granulations of the universe of discourse, similar results must be achieved. One way to measure this similarity between the granulations is to measure the entropy in the data that they generate [19]. The rationale of our proposal suggests that two features (or labels) can be associated if the generated granulations from them have equal entropy. Therefore, the proposal consists in obtaining a universe granulation, where each feature (or label) defines an indiscernibility relation. In this method, the information granules are the set of indiscernible objects with respect to the feature (or label) under consideration.

The rest of the paper is organized as follows. Section 2 presents the theoretical background related to our proposal. Section 3 introduces the new measure to quantify the association between features and labels, and Sect. 4 is dedicated to evaluating its performance in the model on synthetic datasets. Finally, in Sect. 5 we provide relevant concluding remarks.

2   Theoretical Background

In this section, we briefly describe the bidirectional neural network to be modified, and the granulation approach used in our proposal.

2.1 Bidirectional Deep Neural Network

Recently, in [1] the authors introduced a new bidirectional network architecture that is composed of stacked association-based pooling layers to extract high-level features and labels in MLC problems. This approach, unlike the classic use of pooling, does not pool pixels but problem features or labels.

The first pooling layer is composed of neurons denoting the problem features and labels (i.e. low-level features and labels), whereas in deeper pooling layers the neurons denote high-level features and labels extracting during the construction process. Each pooling layer uses a function that detects pairs of highly associated neurons, while performing an aggregation operation to derive the pooled neurons. Such neurons are obtained from neurons belonging to the previous layer such that they fulfil a certain association threshold. Figure 1 shows an example where two pooling layers are running for both features (left figure) and labels (right figure). In this example, five high-level neurons were formed from the association of the feature pairs (f_1, f_2) and (f_3, f_4), and the label pairs (l_1, l_2) and (l_3, l_4). The f_5 feature is not associated with any other feature, so it is transferred directly to the $t+1$ pooling layer. In this pooling architecture, $\oplus$ and $\odot$ are the aggregation operators used to conform the pooled neurons.

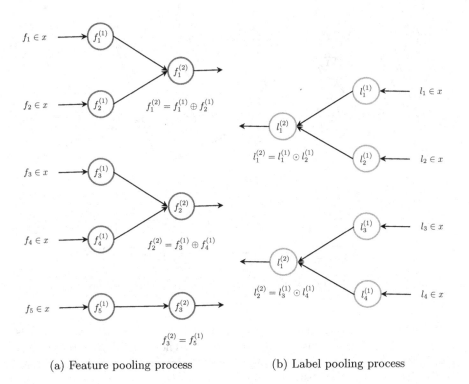

(a) Feature pooling process   (b) Label pooling process

Fig. 1. Bidirectional association-based pooling.

This model uses Pearson's correlation to estimate the association degree between two neurons. Overall, the authors computed the correlation matrix among features and labels, and derive the degree of association of the pooled neurons from the degree of association between each pair of neurons in the previous layer. The pooling process is repeated over aggregated features and labels until a maximum number of pooling layers is reached.

Once the high-level features and labels have been extracted from the dataset, they are connected together with one or several hidden processing layers. Finally, a decoding process [9] is performed to connect the high-level labels to the original ones by means of one or more hidden processing layers. Figure 2 depicts the network architecture resulting five high-level neurons that emerge from the association-based pooling layers. These hidden layers are equipped with either ReLU, sigmoid or hyperbolic tangent transfer functions, therefore conferring the neural system with prediction capabilities.

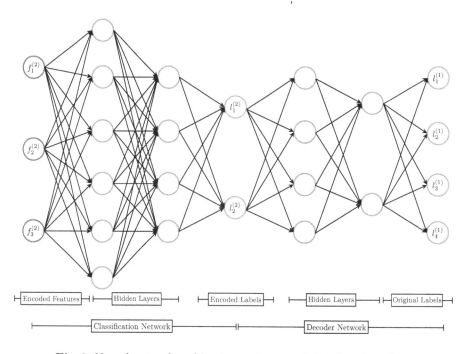

Fig. 2. Neural network architecture using association-based pooling.

It is worth reiterating that this model is aimed at pooling features and labels in traditional MLC problems where neither features nor labels have a topological organization. For example, when using numerical descriptors to encode a protein, it might happen that two distant positions in the sequence are actually close two each other in the tri-dimensional space.

2.2   Universe Granulation

The underlying notion for granulation in classic rough sets [2,13,14] relies on equivalence relations or partitions. Let U be a finite and non-empty universe, and A is a finite non-empty set of features that describe each object. Given a subset of attributes $B \subseteq A$, an indiscernibility relation is defined as $IND = \{(x,y) \in U \times U | \forall b \in B, x(b) = y(b)\}$. This relation is reflexive, symmetric and transitive. The equivalence class $[x]_{IND}$ consists of all elements equivalent to x according to relation IND. The family of equivalence classes $U/IND = \{[x]_{IND} | x \in U\}$ is a partition of the universe.

The indiscernibility relation seems to be excessively restrictive. In presence of numerical attributes, two inseparable objects (according to some similarity relation R [16]) will be gathered together in the same set of non-identical (but reasonably similar) objects. The definition of R may admit that a small difference between features values is considered as unsignificant. This relation delimits whether two objects x and y are inseparable or not, and defines a similarity class where $\bar{R}(x) = \{y \in U | yRx\}$. Equation (1) shows the similarity relation, assuming that $0 \leq \varphi(x,y) \leq 1$ is a similarity function,

$$R : yRx \Leftrightarrow \varphi(x,y) \geq \xi. \tag{1}$$

This weaker binary relation states that objects x and y are deemed inseparable as long as their similarity degree $\varphi(x,y)$ exceeds a similarity threshold $0 \leq \xi \leq 1$. It is worth mentioning that the similarity relation R does not induce a partition of U into a set of equivalence classes but rather a covering [3] of U into multiple similarity classes $\bar{R}(x)$.

3   Feature Association Using the Granulation Entropy

The granular approach in [19] uses the Shannon entropy to characterize partitions of a universe. Two granulations with the same (or similar) entropy value could be considered equivalent. Similarly, the degree of association between two features (or labels) could be determined using the entropy of the granulations they generate. Our method verifies if the coverings (or partitions) generated by two features (or labels) induce similar entropy values.

Let us assume that the problem feature f_1 generates the covering $Cf_1 = \{GF_1, GF_2, \ldots, GF_s\}$ that contains s granules, i.e. the family of similarity classes when only the f_1 feature is considered. Thus, we define $\varphi(x,y) = 1 - |x(f_1) - y(f_1)|$ as a similarity function used in Eq. (1), where $x(f_1)$ and $y(f_1)$ are the values of the feature f_1 in objects x and y. In addition, the l_1 label generates the partition $Pl_1 = \{GL_1, GL_2, \ldots, GL_t\}$ with t granules, i.e. the family of equivalence classes where all objects have exactly the same value on the l_1 label. Since the domain of the label is $\{0,1\}$, this partition will only contain two equivalence classes. Equations (2) and (3) define the probability distributions for the partitions Cf_1 and Pl_1, respectively,

$$D_{Cf_1} = \left\{ \frac{|GF_1|}{|U|}, \frac{|GF_2|}{|U|}, \ldots, \frac{|GF_s|}{|U|} \right\} \tag{2}$$

$$D_{Pl_1} = \left\{ \frac{|GL_1|}{|U|}, \frac{|GL_2|}{|U|}, \dots, \frac{|GL_t|}{|U|} \right\} \tag{3}$$

where $|\cdot|$ denotes the cardinality of a set. The Shannon entropy function of the probability distributions is defined by Eqs. (4) and (5),

$$H(Cf_1) = -\sum_{i=1}^{s} \left(\frac{|GF_i|}{|U|} \right) log \left(\frac{|GF_i|}{|U|} \right) \tag{4}$$

$$H(Pl_1) = -\sum_{j=1}^{t} \left(\frac{|GL_j|}{|U|} \right) log \left(\frac{|GL_j|}{|U|} \right). \tag{5}$$

The similarities of the granulations generated by two features f_1 and f_2, or two labels l_1 and l_2 can be defined according to the measures $GSE_1(f_1, f_2)$ and $GSE_2(l_1, l_2)$ defined in the Eqs. (6) and (7) respectively,

$$GSE_1(f_1, f_2) = \frac{(1 + E1)}{(1 + E2)} \tag{6}$$

$$GSE_2(l_1, l_2) = \frac{(1 + N1)}{(1 + N2)} \tag{7}$$

where $E1 = min\{H(Cf_1), H(Cf_2)\}$, $E2 = max\{H(Cf_1), H(Cf_2)\}$, $N1 = min\{H(Pl_1), H(Pl_2)\}$, and $N2 = max\{H(Pl_1), H(Pl_2)\}$.

In this way, two features can be associated if $GSE_1(f_1, f_2) \geq \alpha_1$, where α_1 is the association threshold regulating the aggregation of features. In the same way, two labels will be associated if $GSE_2(l_1, l_2) \geq \alpha_2$, where α_2 is the association threshold regulating the aggregation of labels.

In our approach, we estimate the association degree between pairs of pooled neurons from the values of the association matrix calculated for the original features and labels of the problem. Then, the association between two pooled neurons would be performed as the average of the values determined by GSE_1 and GSE_2 for each pair of features (or labels) in these neurons.

Equations (8) and (9) define the association between $f_1^{(v)}$ and $f_2^{(v)}$ (i.e. neurons in the v-th pool of features), and between $l_1^{(w)}$ and $l_2^{(w)}$ (i.e. neurons in the w-th pool of labels), respectively,

$$SP_1(f_1^{(v)}, f_2^{(v)}) = \frac{1}{k_1} \sum_{i=1}^{k_1} GSE_1(p_i^f) \tag{8}$$

$$SP_2(l_1^{(w)}, l_2^{(w)}) = \frac{1}{k_2} \sum_{j=1}^{k_2} GSE_2(p_j^l) \tag{9}$$

where k_1, k_2 are the number of pairs of features and labels that can be formed from the aggregation of $f_1^{(v)}$ and $f_2^{(v)}$, and $l_1^{(w)}$ and $l_2^{(w)}$, respectively. Similarly, p_i^f and p_j^l denote the ith and jth pairs of features and labels. In this way, we say that two pooled neurons $f_1^{(v)}$ and $f_2^{(v)}$, or $l_1^{(w)}$ and $l_2^{(w)}$ can be associated in the current layer if $SP_1 \geq \alpha_1$, or $SP_2 \geq \alpha_2$, respectively.

4   Simulations

In this section, we evaluate the ability of our proposal to estimate the association between problem variables (low-level features and labels), and between pooled neurons (high-level features and labels).

To perform the simulations, we use 10 multi-label datasets taken from the RUMDR repository [4]. In these problems (see Table 1), the number of instances ranges from 207 to 10,491, the number of features goes from 72 to 635, and the number of labels from 6 to 400. Also, the average maximal correlation of Pearson according to both features and labels is reported.

Table 1. Characterization of datasets used for simulations.

Dataset	Name	Instances	Features	Labels	Correlation-F	Correlation-L
D1	Emotions	593	72	6	0.62	0.39
D2	Scene	2,407	294	6	0.74	0.22
D3	Yeast	2,417	103	14	0.49	0.57
D4	Stackex-chemistry	6,961	540	175	0.18	0.13
D5	Stackex-chess	1,675	585	227	0.27	0.24
D6	Stackex-cooking	10,491	577	400	0.14	0.14
D7	Stackex-cs	9,270	635	274	0.18	0.18
D8	GnegativePseAAC	1,392	440	8	0.29	0.22
D9	GpositivePseAAC	519	440	4	0.33	0.34
D10	VirusPseAAC	207	440	6	0.40	0.22

The simulations aim at comparing our approach with the correlation-based method proposed in [1]. In order to make fair comparisons, we will use the same network architecture proposed by the authors. Similarly, as far as the pooling process is concerned, we set the maximum number of pooling layers to 5 for the features and 3 for the labels. The association thresholds α_1 and α_2 will range from 0.0 to 0.8. The operators used to aggregate two neurons (i.e. $\oplus$ and $\odot$) are the average in the feature pooling process, and maximum in the label pooling. In addition, the value of the similarity threshold parameter used in Eq. (1) is fixed to 0.85, although other values are also possible.

In all experiments conducted in this section, we use 80% of the dataset to build the model and 20% for testing purposes, while the reported results are averaged over 10 trials to draw consistent conclusions.

4.1   Results and Discussion

Table 2 displays the results of our measure in the model proposed by [1]. These tables report the number of high-level features, the reduction percentage those high-level features represent (%Red-Features), the number of high-level labels, the reduction percentage in the number of labels (%Red-Labels), the accuracy obtained when using only the high-level features and labels, and the accuracy loss with respect the model using all features and labels (i.e. neural network model without performing the pooling operations).

Table 2. Results achieved by the GSE_1 and GSE_2 measures.

Dataset	HL-Features	%Red-Features	HL-Labels	%Red-Labels	Accuracy	Loss
D1	3	95.83%	3	50%	0.515	−0.308
D2	10	96.60%	3	50%	0.771	−0.144
D3	4	96.12%	4	71.43%	0.765	−0.036
D4	17	96.85%	22	87.43%	0.988	0
D5	19	96.75%	29	87.22%	0.99	0
D6	19	96.71%	50	87.5%	0.995	0
D7	20	96.85%	35	87.23%	0.991	0
D8	14	96.82%	4	50%	0.864	−0.054
D9	14	96.82%	4	0%	0.646	−0.22
D10	14	96.82%	3	50%	0.736	−0.058

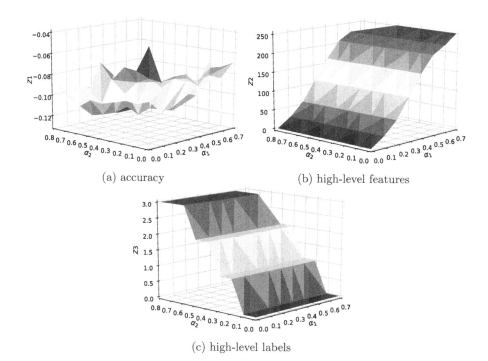

(a) accuracy   (b) high-level features

(c) high-level labels

Fig. 3. Average statistics over $G1$ using the GSE_1 and GSE_2 measures.

Our proposal (i.e. to compute the association between variables from GSE_1 and GSE_2) obtains a percentage of reduction in the features over 95%, and in the labels over 50% in most cases. On the other hand, the loss of accuracy is significant in those datasets that present a high correlation (e.g., $D1, D2, D9$), which means that this measure is not suitable in this datasets. It is remarkable the accuracy loss for $D1$, which is also the dataset with the lowest number of features in our study. However, the proposal reports a very small loss in those datasets having a lower correlation (e.g., $D4, D6, D7$).

Figures 3 and 4 show the comparison of our proposal against the one using Pearson's correlation (baseline). In these figures, we report the differences in accuracy between our method versus the baseline ($Z1$), the differences in the number of high-level features ($Z2$), and the number of high-level labels ($Z3$), when using different α_1 and α_2 values. Figure 3 summarizes the results for a first group of datasets $G1 = \{D1, D2, D3, D8, D9, D10\}$, while Fig. 4 shows the result of the second group $G2 = \{D4, D5, D6, D7\}$. The first group contains datasets having high correlation between their features and a middle correlation between their labels. Meanwhile, the second group consists of datasets having low correlation between their features and their labels.

For both groups, our proposal obtains a higher reduction rates when it comes to the number of high-level features and labels describing the problem.

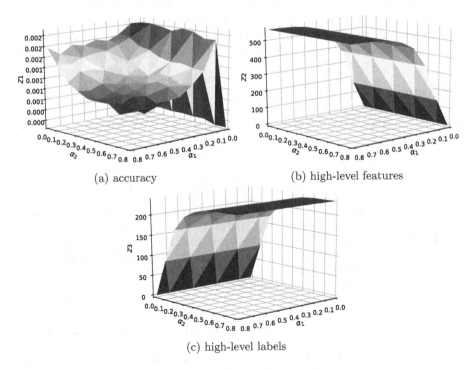

(a) accuracy

(b) high-level features

(c) high-level labels

Fig. 4. Average statistics over $G2$ using the GSE_1 and GSE_2 measures.

This difference is more significant when using high values for the association thresholds. In terms of accuracy, although the differences are not significant, the greatest differences are obtained in the $G1$ datasets, and that is, when high thresholds of association are used, while for the $G2$ datasets the opposite occurs. Our proposal achieves better results in datasets that have low correlation (i.e, those in $G2$), which confirms the hypothesis of our research.

It is worth mentioning that this model does not aim at increasing the prediction rates but to reduce of features and labels associated with the MLC problem. However, our results cry for the implementation of a convolutional operator to also increase networks' discriminatory power.

5 Concluding Remarks

In this paper, we have presented a method to quantify the association between problem variables (features and labels). This measure detects pairs of features (or labels) that are highly associable, and that will be used to perform an aggregation operation resulting in high-level features and labels. Unlike the pooling approach proposed in [1], our proposal does not require that either the features or labels have a certain degree of correlation with each other. Numerical results have shown that our proposal is able to significantly reduce the number of parameters in deep neural networks. When compared with the correlation-based variant, our model reported higher reduction values in datasets having low correlation values among their features and labels. As a result, we obtained simpler models without significantly affecting networks' discriminatory power.

References

1. Bello, M., Nápoles, G., Sánchez, R., Vanhoof, K., Bello, R.: Deep neural network to extract high-level features and labels in multi-label classification problems. Neurocomputing (Submitted)
2. Bello, R., Falcon, R., Pedrycz, W., Kacprzyk, J.: Granular Computing: At the Junction of Rough Sets and Fuzzy Sets. Springer, Heidelberg (2008). https://doi.org/10.1007/978-3-540-76973-6
3. Bonikowski, Z., Bryniarski, E., Wybraniec-Skardowska, U.: Extensions and intentions in the rough set theory. Inf. Sci. **107**(1–4), 149–167 (1998)
4. Charte, F., Charte, D., Rivera, A., del Jesus, M.J., Herrera, F.: R Ultimate multilabel dataset repository. In: Martínez-Álvarez, F., Troncoso, A., Quintián, H., Corchado, E. (eds.) HAIS 2016. LNCS (LNAI), vol. 9648, pp. 487–499. Springer, Cham (2016). https://doi.org/10.1007/978-3-319-32034-2_41
5. Choi, K., Fazekas, G., Sandler, M.: Automatic tagging using deep convolutional neural networks. arXiv preprint arXiv:1606.00298 (2016)
6. Goodfellow, I., Bengio, Y., Courville, A.: Deep Learning. The MIT Press, Cambridge (2016)
7. Herbert, J.P., Yao, J.: A granular computing framework for self-organizing maps. Neurocomputing **72**(13–15), 2865–2872 (2009)

8. Herrera, F., Charte, F., Rivera, A.J., del Jesus, M.J.: Multilabel Classification. Problem Analysis, Metrics and Techniques. Springer, Cham (2016). https://doi.org/10.1007/978-3-319-41111-8

9. Hinton, G.E., Salakhutdinov, R.R.: Reducing the dimensionality of data with neural networks. Science **313**(5786), 504–507 (2006)

10. LeCun, Y., Bengio, Y., Hinton, G.: Deep learning. Nature **521**(7553), 436–444 (2015)

11. Lee, C.Y., Gallagher, P.W., Tu, Z.: Generalizing pooling functions in convolutional neural networks: mixed, gated, and tree. In: Artificial Intelligence and Statistics, pp. 464–472 (2016)

12. Lin, M., Chen, Q., Yan, S.: Network in network. arXiv preprint arXiv:1312.4400 (2013)

13. Pawlak, Z.: Rough sets. Int. J. Comput. Inf. Sci. **11**(5), 341–356 (1982)

14. Pedrycz, W., Skowron, A., Kreinovich, V.: Handbook of Granular Computing. Wiley, New York (2008)

15. Rios, A., Kavuluru, R.: Convolutional neural networks for biomedical text classification: application in indexing biomedical articles. In: Proceedings of the 6th ACM Conference on Bioinformatics, Computational Biology and Health Informatics, pp. 258–267. ACM (2015)

16. Slowinski, R., Vanderpooten, D.: A generalized definition of rough approximations based on similarity. IEEE Trans. Knowl. Data Eng. **12**(2), 331–336 (2000)

17. Wei, Y., et al.: HCP: a flexible CNN framework for multi-label image classification. IEEE Trans. Pattern Anal. Mach. Intell. **38**(9), 1901–1907 (2015)

18. Yao, J.T., Yao, Y.Y.: Induction of classification rules by granular computing. In: Alpigini, J.J., Peters, J.F., Skowron, A., Zhong, N. (eds.) RSCTC 2002. LNCS (LNAI), vol. 2475, pp. 331–338. Springer, Heidelberg (2002). https://doi.org/10.1007/3-540-45813-1_43

19. Yao, Y.: Probabilistic approaches to rough sets. Expert Syst. **20**(5), 287–297 (2003)

20. Zhang, M.L., Zhou, Z.H.: A review on multi-label learning algorithms. IEEE Trans. Knowl. Data Eng. **26**(8), 1819–1837 (2013)

21. Zhu, J., Liao, S., Lei, Z., Li, S.Z.: Multi-label convolutional neural network based pedestrian attribute classification. Image Vis. Comput. **58**, 224–229 (2017)

Multi-granularity Complex Network Representation Learning

Peisen Li, Guoyin Wang$^{(\boxtimes)}$, Jun Hu, and Yun Li

Chongqing Key Laboratory of Computational Intelligence,
Chongqing University of Posts and Telecommunications,
Chongqing 400065, People's Republic of China
peisenli1@gmail.com, {wanggy,hujun,liyun}@cqupt.edu.cn

Abstract. Network representation learning aims to learn the low dimensional vector of the nodes in a network while maintaining the inherent properties of the original information. Existing algorithms focus on the single coarse-grained topology of nodes or text information alone, which cannot describe complex information networks. However, node structure and attribution are interdependent, indecomposable. Therefore, it is essential to learn the representation of node based on both the topological structure and node additional attributes. In this paper, we propose a multi-granularity complex network representation learning model (MNRL), which integrates topological structure and additional information at the same time, and presents these fused information learning into the same granularity semantic space that through fine-to-coarse to refine the complex network. Experiments show that our method can not only capture indecomposable multi-granularity information, but also retain various potential similarities of both topology and node attributes. It has achieved effective results in the downstream work of node classification and the link prediction on real-world datasets.

Keywords: Multi-granularity · Network representation learning · Information fuses

1 Introduction

Complex network is the description of the relationship between entities and the carrier of various information in the real world, which has become an indispensable form of existence, such as medical systems, judicial networks, social networks, financial networks. Mining Knowledge in networks has drown continuous attention in both academia and industry. How to accurately analyze and make decisions on these problems and tasks from different information networks is a vital research. e.g. in the field of sociology, a large number of interactive social platforms such as Weibo, WeChat, Facebook, and Twitter, create a lot of social networks including relationships between users and a sharp increase in interactive review text information. Studies have shown that these large, sparse

© Springer Nature Switzerland AG 2020
R. Bello et al. (Eds.): IJCRS 2020, LNAI 12179, pp. 236–250, 2020.
https://doi.org/10.1007/978-3-030-52705-1_18

new social networks at different levels of cognition will present the same small-world nature and community structure as the real world. Then, based on these interactive information networks for data analysis [1], such as the prediction of criminal associations and sensitive groups, we can directly apply it to the real world.

Network representation learning is an effective analysis method for the recognition and representation of complex networks at different granularity levels, while preserving the inherent properties, mapping high-dimensional and sparse data to a low-dimensional, dense vector space. Then apply vector-based machine learning techniques to handle tasks in different fields [2,3]. For example, link prediction [4], community discovery [5], node classification [6], recommendation system [7], etc.

In recent years, various advanced network representation learning methods based on topological structure have been proposed, such as Deepwalk [8], Node2vec [9], Line [10], which has become a classical algorithm for representation learning of complex networks, solves the problem of retaining the local topological structure. A series of deep learning-based network representation methods were then proposed to further solve the problems of global topological structure preservation and high-order nonlinearity of data, and increased efficiency. e.g., SDNE [13], GCN [14] and DANE [12]. However, the existing researches has focused on coarser levels of granularity, that is, a single topological structure, without comprehensive consideration of various granular information such as behaviors, attributes, and features. It is not interpretable, which makes many decision-making systems unusable.

In addition, the structure of the entity itself and its attributes or behavioral characteristics in a network are indecomposable [18]. Therefore, analyzing a single granularity of information alone will lose a lot of potential information. For example, in a job-related crime relationship network is show in Fig. 1, the anti-reconnaissance of criminal suspects leads to a sparse network than common social networks. The undiscovered edge does not really mean two nodes are not related like P2 and P3 or (P1 and P2), but in case detection, additional information of the suspect needs to be considered. The two without an explicit relationship were involved in the same criminal activity at a certain place (L1), they may have some potential connection. The suspect P4 and P7 are related by the attribute A4, the topology without attribute cannot recognize why the relation between them is generated. So these location attributes and activity information are inherently indecomposable and interdependence with the suspect, making the two nodes recognize at a finer granularity based on the additional information and relationship structure that the low-dimensional representation vectors learned have certain similarities. We can directly predict the hidden relationship between the two suspects based on these potential similarities. Therefore, it is necessary to consider the network topology and additional information of nodes.

The cognitive learning mode of information network is exactly in line with the multi-granularity thinking mechanism of human intelligence problem solving,

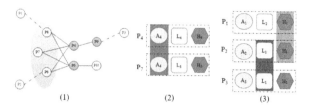

Fig. 1. The example of job-related crime relationship network

data is taken as knowledge expressed in the lowest granularity level of a multiple granularity space, while knowledge as the abstraction of data in coarse granularity levels [15]. Multi-granularity cognitive computing fuses data at different granularity levels to acquire knowledge [16]. Similarly, network representation learning can represent data into lower-dimensional granularity levels and preserve underlying properties and knowledge. To summarize, Complex network representation learning faces the following challenges:

Information Complementarity: The node topology and attributes are essentially two different types of granular information, and the integration of these granular information to enrich the semantic information of the network is a new perspective. But how to deal with the complementarity of its multiple levels and represent it in the same space is an arduous task.

Similarity Preservation: In complex networks, the similarity between entities depends not only on the topology structure, but also on the attribute information attached to the nodes. They are indecomposable and highly non-linear, so how to represent potential proximity is still worth studying.

In order to address the above challenges, this paper proposes a multi-granularity complex network learning representation method (MNRL) based on the idea of multi-granularity cognitive computing.

2   Related Works

Network representation learning can be traced back to the traditional graph embedding, which is regarded as a process of data from high-dimensional to low-dimensional. The main methods include principal component analysis (PCA) [19] and multidimensional scaling (MDS) [21]. All these methods can be understood as using an $n \times k$ matrix to represent the original $n \times m$ matrix, where $k \ll m$. Later, some researchers proposed IsoMap and LLE to maintain the overall structure of the nonlinear manifold [20]. In general, these methods have shown good performance on small networks. However, the time complexity is extremely high, which makes them unable to work on large-scale networks. Another popular class of dimensionality reduction techniques uses the spectral characteristics (e.g. feature vectors) of a matrix that can be derived from a graph to embed the nodes. Laplacian Eigenmaps [22] obtain low-dimensional vector representations of each node in the feature vector representation graph associated with its k smallest non-trivial feature values.

Recently, DeepWalk was inspired by Word2vec [24], a certain node was selected as the starting point, and the sequence of the nodes was obtained by random walk. Then the obtained sequence was regarded as a sentence and input to the Word2vec model to learn the low-dimensional representation vector. Deep-Walk can obtain the local context information of the nodes in the graph through random walks, so the learned representation vector reflects the local structure of the point in the network [8]. The more neighboring points that two nodes share in the network, the shorter the distance between the corresponding two vectors. Node2vec uses biased random walks to make a choose between breadth-first (BFS) and depth-first (DFS) graph search, resulting in a higher quality and more informative node representation than DeepWalk, which is more widely used in network representation learning. LINE [10] proposes first-order and second-order approximations for network representation learning from a new perspective. HARP [25] obtains a vector representation of the original network through graph coarsening aggregation and node hierarchy propagation. Recently, Graph convolutional network (GCN) [14] significantly improves the performance of network topological structure analysis, which aggregates each node and its neighbors in the network through a convolutional layer, and outputs the weighted average of the aggregation results instead of the original node's representation. Through the continuous stacking of convolutional layers, nodes can aggregate high-order neighbor information well. However, when the convolutional layers are super-imposed to a certain number, the new features learned will be over-smoothed, which will damage the network representation performance. Multi-GS [23] combines the concept of multi-granularity cognitive computing, divides the network structure according to people's cognitive habits, and then uses GCN to convolve different particle layers to obtain low-dimensional feature vector representations. SDNE [13] directly inputs the network adjacency matrix to the autoencoder [26] to solve the problem of preserving highly nonlinear first-order and second-order similarity.

The above network representation learning methods use only network structure information to learn low-dimensional node vectors. But nodes and edges in real-world networks are often associated with additional information, and these features are called attributes. For example, in social networking sites such as Weibo, text content posted by users (nodes) is available. Therefore, the node representation in the network also needs to learn from the rich content of node attributes and edge attributes. TADW studies the case where nodes are associated with text features. The author of TADW first proved that DeepWalk essentially decomposes the transition probability matrix into two low-dimensional matrices. Inspired by this result, TADW low-dimensionally represents the text feature matrix and node features through a matrix decomposition process [27]. CENE treats text content as a special type of node and uses node-node structure and node-content association for node representation [28]. More recently, DANE [12] and CAN [34] uses deep learning methods [11] to preserve potentially non-linear node topology and node attribute information. These two kinds of information provide different views for each node, but their heterogeneity is not considered. ANRL optimizes the network structure and attribute information separately, and uses the Skip-Gram model to skillfully handle the heterogeneity of the two different types of information [29]. Nevertheless, the consistent and complementary information in the topology and attributes is lost and the sensitivity to noise is increased, resulting in a lower robustness.

To process different types of information, Wang put forward the concepts of "from coarse to fine cognition" and "fine to coarse" fusion learning in the study of multi-granularity cognitive machine learning [30]. People usually do cognition at a coarser level first, for example, when we meet a person, we first recognize who the person is from the face, then refine the features to see the freckles on the face. While computers obtain semantic information that humans understand by fusing fine-grained data to coarse-grained levels. Refining the granularity of complex networks and the integration between different granular layers is still an area worthy of deepening research [17,31]. Inspired by this, divides complex networks into different levels of granularity: Single node and attribute data are microstructures, meso-structures are role similarity and community similarity, global network characteristics are extremely macro-structured. The larger the granularity, the wider the range of data covered, the smaller the granularity, the narrower the data covered. Our model learns the semantic information that humans can understand at above mentioned levels from the finest-grained attribute information and topological structure, finally saves it into low-dimensional vectors.

3   Multi-granular Network Representation Learning

3.1   Problem Definition

Let $G = (V, E, A)$ be a complex network, where V represents the set of n nodes and E represents the set of edges, and A represents the set of attributes. In detail, $A \in \Re^{n \times m}$ is a matrix that encodes all node additional attributes information, and $a_i \in A$ describes the attributes associated with node v_i, where $v_i \in V$. $e_{ij} = (v_i, v_j) \in E$ represents an edge between v_i and v_j. We formally define the multi-granularity network representation learning as follows:

Definition 1. Given a complex network $G = (V, E, A)$, we represent each node v_i and attribute a_i as a low-dimensional vector y_i by learning a function f_G : $V \rightarrow \Re^d$, where $d \ll |V|$ and y_i not only retains the topology of the nodes but also the node attribute information.

Definition 2. Given network $G = (V, E, A)$. Semantic similarity indicates that two nodes have similar attributes and neighbor structure, and the low-dimensional vector obtained by the network representation learning maintains the same similarity with the original network. E.g., if $v_i \sim v_j$ through the mapping function f_G to get the low-dimensional vectors $y_i = f_G(v_i)$, $y_j = f_G(v_j)$, y_i and y_j are still similar, $y_i \sim y_j$.

Definition 3. Complex networks are composed of node and attribute granules (elementary granules), which can no longer be decomposed. Learning these grains to get different levels of semantic information includes topological structure (micro), role acquaintance (meso) and global structure (macro). The complete low-dimensional representation of a complex network is the aggregation of these granular layers of information.

3.2 Multi-granularity Representation Model

In order to solve the problems mentioned above, inspired by multi-granularity cognitive computing, we propose a multi-granularity network representation learning method (MNRL), which refines the complex network representation learning from the topology level to the node's attribute characteristics and various attachments. The model not only fuses finer granular information but also preserves the node topology, which enriches the semantic information of the relational network to solve the problem of the indecomposable and interdependence of information. The algorithm framework is shown in Fig. 2.

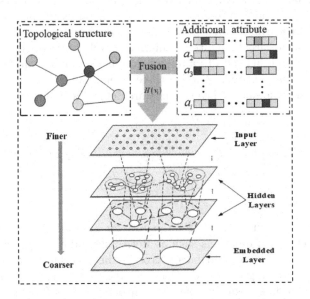

Fig. 2. The architecture of the proposed MNRL model.

Firstly, the topology and additional information are fused through the function H, then the variational encoder is used to learn network representation from fine to coarse. The output of the embedded layer are low-dimensional vectors, which combines the attribute information and the network topology.

Multi-granularity Information Fusion. To better characterize multiple granularity complex networks and solve the problem of nodes with potential associations that cannot be processed through the relationship structure alone, we refine the granularity to additional attributes, and designed an information fusion method, which are defined as follows:

$$x_i = H(v_i)$$

$$H(v_i) = a_i + \sum_{v_j \in N(v_i)} \frac{d(v_j)}{d(v_i)} w_{ij} a_j \tag{1}$$

Where $N(v_i)$ is the neighbors of node v_i in the network, a_i is the attributes associated with node v_i. $w_{ij} > 0$ for weighted networks and $w_{ij} = 1$ for unweighted networks. $d(v_j)$ is the degree of node v_j. x_i contains potential information of multiple granularity information, both the neighbor attribute information and the node itself.

Information Complementarity Capture. To capture complementarity of different granularity hierarchies and avoid the effects of various noises, our model in Fig. 1 is a variational auto-encoder, which is a powerful unsupervised deep model for feature learning. It has been widely used for multi-granularity cognitive computing applications. In multi-granularity complex networks, auto-encoders fuse different granularity data to a unified granularity space from fine to coarse. The variational auto-encoder contains three layers, namely, the input layer, the hidden layer, and the output layer, which are defined as follows:

$$
\begin{aligned}
y_i^1 &= \sigma\left(w^{(1)}x_i + b^{(1)}\right) \\
y_i^k &= \sigma\left(w^{(k)}y_i^{k-1} + b^{(k)}\right), k = 2, \ldots, K-2 \\
y_\mu &= w^{(K-1)}y_i^k + b^{(K-1)}, y_\sigma = w^{(K-1)}y_i^k + b^{(K-1)} \\
y_i^K &= \sigma(y_\mu + E * y_\sigma)
\end{aligned}
\tag{2}
$$

Here, K is the number of layers for the encoder and decoder. $\sigma(\cdot)$ represents the possible activation functions such as ReLU, sigmod or tanh. w^k and b^k are the transformation matrix and bias vector in the k-th layer, respectively. y_i^K is the unified vector representation that learning from model, which obeys the distribution function E, reducing the influence of noise. $E \sim (0, 1)$ is the standard normal distribution in this paper. In order to make the learned representation as similar as possible to the given distribution,it need to minimize the following loss function:

$$L_{KL} = y_\mu^2 + y_\sigma^2 - \log\left(y_\sigma^2\right) - 1 \tag{3}$$

To reduce potential information loss of original network, our goal is to minimize the following auto-encoder loss function:

$$L_{RE} = \sum_i^n \|\hat{x}_i - x_i\|_2^2 \tag{4}$$

where $\hat{x}_i$ is the reconstruction output of decoder and x_i incorporates prior knowledge into the model.

Semantic Similarity Preservation. To formulate the homogeneous network structure information, skip-gram model has been widely adopted in recent works and in the field of heterogeneous network research, Skip-grams suitable for different types of nodes processing have also been proposed [32]. In our model, the context of a node is the low-dimensional potential information. Given the node v_i and the associated reconstruction information y_i, we randomly walk $c \in C$ by maximizing the loss function:

$$L_{HS} = \arg\max \sum_{i=1}^{n} \sum_{c \in C} \sum_{j \leq |B|, j \neq 0} log p\left(v_{i+j} | y_i\right) \tag{5}$$

Where B is the size of the generation window and the conditional probability $p\left(v_{i+j} | y_i\right)$ is defined as the Softmax function:

$$p\left(v_{i+j} | y_i\right) = \frac{e^{v'^T_{i+j} y_i}}{\sum_{k=1}^{n} e^{v'^T_k y_i}} \tag{6}$$

In the above formula, v'_i is the node context representation of node v_i, and y_i is the result produced by the auto-encoder. Directly optimizing Eq. (6) is computationally expensive, which requires the summation over the entire set of nodes when computing the conditional probability of $p\left(v_{i+j} | y_i\right)$. We adopt the negative sampling approach proposed in Metapath2vec++ that samples multiple negative samples according to some noisy distributions:

$$\log \sigma\left(v'^T_{i+j} y_i\right) + \sum_{t=1}^{S} E_{v \sim P_n(v)} \left[\log \sigma\left(-v'^T_n y_i\right)\right] \tag{7}$$

Where $\sigma(\cdot) = 1/(1 + \exp(\cdot))$ is the sigmoid function and S is the number of negative samples. We set $P_n(v) \propto d_v^{\frac{3}{4}}$ as suggested in Wode2vec, where d_v is the degree of node v_i [24,32]. Through the above methods, the node's attribute information and the heterogeneity of the node's global structure are processed and the potential semantic similarity kept in a unified granularity space.

MNRL Model Joint Optimization. Multi-granularity complex network representation learning through the fusion of multiple kinds of granularity information, learning the basic granules through an autoencoder, and representing different levels of granularity in a unified low-dimensional vector solves the potential semantic similarity between nodes without direct edges. The model simultaneously optimizes the objective function of each module to make the final result robust and effective. The function is shown below:

$$L = \alpha L_{RE} + \beta L_{KL} + \psi L_{VAE} + \gamma\left(-L_{HS}\right) \tag{8}$$

In detail, L_{RE} is the auto-encoder loss function of Eq. (4), L_{KL} has been stated in formula (3), and L_{HS} is the loss function of the skip-gram model in Eq. (5). $\alpha, \beta, \psi, \gamma$ are the hyper parameters to balance each module. L_{VAE} is the parameter optimization function, the formula is as follows:

$$L_{VAE} = \frac{1}{2}\sum_{k=1}^{K}(\left\|w^k\right\|_F^2 + \left\|b^k\right\|_F^2 + \left\|\hat{w}^k\right\|_F^2 + \left\|\hat{b}^k\right\|_F^2) \qquad (9)$$

Where w^k, $\hat{w}^k$ are weight matrices for encoder and decoder respectively in the k-th layer, and b^k, $\hat{b}^k$ are bias matrix. The complete objective function is expressed as follows:

$$\begin{aligned}
L = &\ \alpha \sum_{i}^{n} \left\|\hat{x}_i - x_i\right\|_2^2 \\
&+ \beta \left(y_\mu^2 + y_\sigma^2 - log\left(y_\sigma^2\right) - 1\right) \\
&+ \gamma \sum_{i=1}^{n}\sum_{c \in C}\sum_{j \le |B|, j \ne 0} logp\left(v_{i+j}|y_i\right) \\
&+ \frac{\psi}{2}\sum_{k=1}^{K}(\left\|w^k\right\|_F^2 + \left\|b^k\right\|_F^2 + \left\|\hat{w}^k\right\|_F^2 + \left\|\hat{b}^k\right\|_F^2)
\end{aligned} \qquad (10)$$

MNRL preserves multiple types of granular information include node attributes, local network structure and global network structure information in a unified framework. The model solves the problems of highly nonlinearity and complementarity of various granularity information, and retained the underlying semantics of topology and additional information at the same time. Finally, we optimize the object function L in Eq. (10) through stochastic gradient descent. To ensure the robustness and validity of the results, we iteratively optimize all components at the same time until the model converges. The learning algorithm is summarized in Algorithm 1.

Algorithm 1. The Model of MNRL

Input: Graph $G = (V, E, A)$, Window size B, times of walk P, walk length U, hyper-parameter $\alpha, \beta, \psi, \gamma$, embedding size d.

Output: Node representations $y^k \in \Re^d$.

1: Generate node context starting P times with random walks with length U at each node.

2: Multiple granularity information fusion for each node by function $H(\cdot)$

3: Initialize all parameters

4: **While not converged do**

5:     Sample a mini-batch of nodes with its context

6:     Compute the gradient of ∇L

7:     Update auto-encoder and skip-gram module parameters

8: **End while**

9: Save representations $Y = y^K$

---

4   Experiment

4.1   Datasets and Baselines

Datasets: In our experiments, we employ four benchmark datasets: Facebook[1], Cora, Citeseer and PubMed[2]. These datasets contain edge relations and various attribute information, which can verify that the social relations of nodes and individual attributes have strong dependence and indecomposability, and jointly determine the properties of entities in the social environment. The first three datasets are paper citation networks, and these datasets are consist of bibliography publication data. The edge represents that each paper may cite or be cited by other papers. The publications are classified into one of the following six classes: Agents, AI, DB, IR, ML, HCI in Citeseer and one of the three classes (i.e., "Diabetes Mellitus Experimental", "Diabetes Mellitus Type 1", "Diabetes Mellitus Type 2") in Pubmed. The Cora dataset consists of Machine Learning papers which are classified into seven classes. Facebook dataset is a typical social network. Nodes represent users and edges represent friendship relations. We summarize the statistics of these benchmark datasets in Table 1.

Table 1. Statistics of the datasets. '-' indicates unknown labels.

Dataset	Nodes	Edges	Attributes	Labels
Citeseer	3312	4660	3703	6
PubMed	19717	44338	500	3
Cora	2708	5278	1433	7
Facebook	4039	88234	1238	-

Baselines: To evaluate the performance of our proposed MNRL, we compare it with 9 baseline methods, which can be divided into two groups. The former category of baselines leverage network structure information only and ignore the node attributes contains DeepWalk, Node2Vec, GraRep [33], LINE and SDNE. The other methods try to preserve node attribute and network structure proximity, which are competitive competitors. We consider TADW, GAE, VGAE, DANE as our compared algorithms. For all baselines, we used the implementation released by the original authors. The parameters for baselines are tuned to be optimal. For DeepWalk and Node2Vec, we set the window size as 10, the walk length as 80, the number of walks as 10. For GraRep, the maximum transition step is set to 5. For LINE, we concatenate the first-order and second-order result together as the final embedding result. For the rest baseline methods, their parameters are set following the original papers. At last, the dimension of the node representation is set as 128. For MNRL, the number of layers and dimensions for each dataset are shown in Table 2.

---

[1] https://snap.stanford.edu/data.

[2] https://linqs.soe.ucsc.edu/data.

Table 2. Detailed network layer structure information.

Dataset	Number of neurons in each layer
Citeseer	3703-1500-500-128-500-1500-3703
Pubmed	500-200-128-200-500
Cora	1433-500-128-500-1433
FaceBook	1238-500-128-500-1238

4.2   Node Classification

To show the performance of our proposed MNRL, we conduct node classification on the learned node representations. Specifically, we employ SVM as the classifier. To make a comprehensive evaluation, we randomly select 10%, 30%, 50% nodes as the training set and the rest as the testing set respectively. With these randomly chosen training sets, we use five-fold cross validation to train the classifier and then evaluate the classifier on the testing sets. To measure the classification result, we employ Micro-F1 (Mi-F1) and Macro-F1 (Ma-F1) as metrics. The classification results are shown in Table 3, 4, 5 respectively. From these four tables, we can find that our proposed MNRL achieves significant improvement compared with plain network embedding approaches, and beats other attributed network embedding approaches in most situations.

Table 3. Node classification result of Citeseer

Method	10%		30%		50%	
	Mi-F1	Ma-F1	Mi-F1	Ma-F1	Mi-F1	Ma-F1
DeepWalk	0.5138	0.4711	0.5658	0.5301	0.5961	0.5415
Node2Vec	0.5302	0.4786	0.6233	0.5745	0.6317	0.5929
GraRep	0.4796	0.4613	0.5477	0.5098	0.5662	0.5026
LINE	0.5178	0.4825	0.5679	0.5249	0.6167	0.5733
SDNE	0.5013	0.4896	0.5691	0.5283	0.5877	0.5447
TADW	0.5939	0.5218	0.6361	0.5707	0.6631	0.5660
GAE	0.5912	0.5441	0.6439	0.5802	0.6451	0.5767
VGAE	0.6201	0.5638	0.6413	0.5789	0.6311	0.5799
DANE	0.6217	0.5740	0.6889	**0.6495**	**0.7332**	0.6832
MRNL	**0.6833**	**0.6365**	**0.7176**	0.6451	0.7301	**0.6905**

Experimental results show that the representation results of each comparison algorithm perform well in node classification in downstream tasks. In general, a model that considers node attribute information and node structure information performs better than structure alone.

Table 4. Node classification result of Cora

Method	10%		30%		50%	
	Mi-F1	Ma-F1	Mi-F1	Ma-F1	Mi-F1	Ma-F1
DeepWalk	0.7567	0.7359	0.7947	0.7892	0.8234	0.8091
Node2Vec	0.7489	0.7311	0.8168	0.8103	0.8264	0.8135
GraRep	0.7456	0.7387	0.7991	0.7732	0.8011	0.7849
LINE	0.7212	0.7055	0.8193	0.8140	0.8429	0.8163
TADW	0.7400	0.7189	0.8127	0.7832	0.8413	0.8091
GAE	0.7713	0.7540	0.7985	0.7817	0.8101	0.7996
VGAE	0.7890	0.7667	0.8096	0.8001	0.8137	0.7996
DANE	0.7789	0.7703	0.8023	0.7905	0.8314	**0.8299**
MRNL	**0.8047**	**0.7736**	**0.8169**	**0.7974**	**0.8450**	0.8225

Table 5. Node classification result of PubMed

Method	10%		30%		50%	
	Mi-F1	Ma-F1	Mi-F1	Ma-F1	Mi-F1	Ma-F1
DeepWalk	0.7831	0.7698	0.8067	0.7891	0.8107	0.8012
Node2Vec	0.7984	0.7749	0.8146	0.7907	0.8103	0.7859
GraRep	0.8015	0.7771	0.8052	0.7861	0.8125	0.7958
LINE	0.8067	0.7889	0.8169	0.8012	0.8222	0.8011
TADW	0.8355	0.8304	0.8561	0.8413	0.8719	0.8636
GAE	0.8247	0.8191	0.8278	0.8201	0.8266	0.8217
VGAE	0.8346	0.8202	0.8331	0.8276	0.8355	0.8303
DANE	0.8501	**0.8483**	0.8538	0.8496	0.8645	**0.8643**
MRNL	**0.8532**	0.8411	**0.8597**	**0.8501**	**0.8677**	0.8605

From these three tables, we can find that our proposed MRNL achieves significant improvement compared with single granularity network embedding approaches. For joint representation, our model performs more effectively than most similar types of algorithms, especially in the case of sparse data, because our model input is the fusion information of multiple nodes with extra information. When comparing DANE, our experiments did not improve significantly but it achieved the expected results. DANE uses two auto-encoders to learn and express the network structure and attribute information separately, since the increase of parameters makes the optimal selection in the learning process, the performance will be better with the increase of training data, but the demand for computing resources will also increase and the interpretability of the algorithm is weak. While MRNL uses a variational auto-encoder to learn the structure and attribute information at the same time, the interdependence of information is preserved, which handles heterogeneous information well and reduces the impact of noise.

4.3 Link Prediction

In this subsection, we evaluate the ability of node representations in reconstructing the network structure via link prediction, aiming at predicting if there exists an edge between two nodes, is a typical task in networks analysis. Following other model works do, to evaluate the performance of our model, we randomly holds out 50% existing links as positive instances and sample an equal number of non-existing links. Then, we use the residual network to train the embedding models. Specifically, we rank both positive and negative instances according to the cosine similarity function. To judge the ranking quality, we employ the AUC to evaluate the ranking list and a higher value indicates a better performance. We perform link prediction task on Cora datasets and the results is shown in Fig. 3.

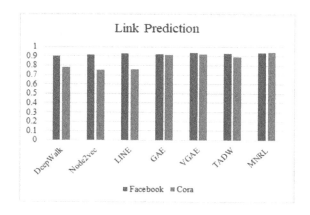

Fig. 3. Link prediction task on Cora and Facebook datasets

Compared with traditional algorithms that representation learning from a single granular structure information, the algorithms that both on structure and attribute information is more effective. TADW performs well, but the method based on matrix factorization has the disadvantage of high complexity in large networks. GAE and VGAE perform better in this experiment and are suitable for large networks. MNRL refines the input and retains potential semantic information. Link prediction relies on additional information, so it performs better than other algorithms in this experiment.

5   Conclusion

In this paper, we propose a multi-granularity complex network representation learning model (MNRL), which integrates topology structure and additional information, and presents these fused information learning into the same granularity semantic space that through fine-to-coarse to refine the complex network.

The effectiveness has been verified by extensive experiments, shows that the relation of nodes and additional attributes are indecomposable and complementarity, which together jointly determine the properties of entities in the network. In practice, it will have a good application prospect in large information network. Although the model saves a lot of calculation cost and well represents complex networks of various granularity, it needs to set different parameters in different application scenarios, which is troublesome and needs to be optimized in the future. The multi-granularity complex network representation learning also needs to consider the dynamic network and adapt to the changes of network nodes, so as to realize the real-time information network analysis.

Acknowledgments. This work is supported by the National Key Research and Development Program of China under Grant 2017YFC0804002, the National Natural Science Foundation of China (No. 61936001, No. 61772096).

References

1. Marsden, P.V., Lin, N. (eds.): Social Structure and Network Analysis, pp. 201–218. Sage, Beverly Hills (1982)
2. Zhang, D., Yin, J., Zhu, X., Zhang, C.: Network representation learning: a survey. IEEE Trans. Big Data. **6**(1), 3–28 (2020)
3. Fischer, A., Botero, J.F., Beck, M.T., De Meer, H., Hesselbach, X.: Virtual network embedding: a survey. IEEE Commun. Surv. Tutor. **15**(4), 1888–1906 (2013)
4. Liben-Nowell, D., Kleinberg, J.: The link-prediction problem for social networks. J. Am. Soc. Inform. Sci. Technol. **58**(7), 1019–1031 (2007)
5. Wang, F., Li, T., Wang, X., Zhu, S., Ding, C.: Community discovery using nonnegative matrix factorization. Data Min. Knowl. Disc. **22**(3), 493–521 (2011)
6. Bhagat, S., Cormode, G., Muthukrishnan, S.: Node classification in social networks. In: Aggarwal, C. (ed.) Social Network Data Analytics, pp. 115–148. Springer, Boston (2011). https://doi.org/10.1007/978-1-4419-8462-3_5
7. Resnick, P., Varian, H.R.: Recommender systems. Commun. ACM **40**(3), 56–58 (1997)
8. Perozzi, B., Al-Rfou, R., Skiena, S.: Deepwalk: online learning of social representations. In: Proceedings of the 20th ACM SIGKDD International Conference on Knowledge Discovery and Data Mining, pp. 701–710 (2014)
9. Grover, A., Leskovec, J.: node2vec: scalable feature learning for networks. In: Proceedings of the 22nd ACM SIGKDD International Conference on Knowledge Discovery and Data Mining, pp. 855–864 (2016)
10. Tang, J., Qu, M., Wang, M., Zhang, M., Yan, J., Mei, Q.: Line: large-scale information network embedding. In: Proceedings of the 24th International Conference on World Wide Web, pp. 1067–1077 (2015)
11. LeCun, Y., Bengio, Y., Hinton, G.: Deep learning. Nature **521**(7553), 436–444 (2015)
12. Gao, H., Huang, H.: Deep attributed network embedding. In: IJCAI 2018, pp. 3364–3370 (2018)
13. Wang, D., Cui, P., Zhu, W.: Structural deep network embedding. In: Proceedings of the 22nd ACM SIGKDD International Conference on Knowledge Discovery and Data Mining, pp. 1225–1234 (2016)

14. Kipf, T.N., Welling, M.: Semi-supervised classification with graph convolutional networks. arXiv preprint arXiv:1609.02907 (2016)

15. Wang, G.: DGCC: data-driven granular cognitive computing. Granular Comput. **2**(4), 343–355 (2017). https://doi.org/10.1007/s41066-017-0048-3

16. Bargiela, A., Pedrycz, W.: Granular computing. In: Handbook on Computational Intelligence: Volume 1: Fuzzy Logic, Systems, Artificial Neural Networks, and Learning Systems, pp. 43–66 (2016)

17. Lin, T.Y., Yao, Y.Y., Zadeh, L.A. (eds.) Data Mining, Rough Sets and Granular Computing, vol. 95. Physica (2013)

18. Tu, K., Cui, P., Wang, X., Wang, F., Zhu, W.: Structural deep embedding for hypernetworks. In: Thirty-Second AAAI Conference on Artificial Intelligence (2018)

19. Wold, S., Esbensen, K., Geladi, P.: Principal component analysis. Chemometr. Intell. Lab. Syst. **2**(1–3), 37–52 (1987)

20. Balasubramanian, M., Schwartz, E.L.: The isomap algorithm and topological stability. Science **295**(5552), 7 (2002)

21. Cox, T.F., Cox, M.A.: Multidimensional Scaling. Chapman and Hall/CRC, Boca Raton (2000)

22. Belkin, M., Niyogi, P.: Laplacian eigenmaps for dimensionality reduction and data representation. Neural Comput. **15**(6), 1373–1396 (2003)

23. Zhang, L., Qian, F., Zhao, S., et al.: Network representation learning based on multi-granularity structure. CAAI Trans. Intell. Syst. **14**(6), 1233–1242 (2019). https://doi.org/10.11992/tis.201905045

24. Goldberg, Y., Levy, O.: word2vec explained: deriving Mikolov et al'.s negative-sampling word-embedding method. arXiv preprint arXiv:1402.3722 (2014)

25. Chen, H., Perozzi, B., Hu, Y., Skiena, S.: Harp: hierarchical representation learning for networks. In: Thirty-Second AAAI Conference on Artificial Intelligence (2018)

26. Ng, A.: Sparse autoencoder. In: CS294A Lecture notes, vol. 72, pp. 1–19 (2011)

27. Yang, C., Liu, Z., Zhao, D., Sun, M., Chang, E.: Network representation learning with rich text information. In: Twenty-Fourth International Joint Conference on Artificial Intelligence (2015)

28. Sun, X., Guo, J., Ding, X., Liu, T.: A general framework for content-enhanced network representation learning. arXiv preprint arXiv:1610.02906 (2016)

29. Zhang, Z., et al.: ANRL: attributed network representation learning via deep neural networks. In: IJCAI, vol. 18, pp. 3155–3161 (2018)

30. Wang, G., Xu, J.: Granular computing with multiple granular layers for brain big data processing. Brain Inform. **1**(4), 1–10 (2014). https://doi.org/10.1007/s40708-014-0001-z

31. Chang, L.Y., Wang, G.Y., Wu, Y.: An approach for attribute reduction and rule generation based on rough set theory. J. Softw. **10**(11), 1206–1211 (1999)

32. Dong, Y., Chawla, N.V., Swami, A.: metapath2vec: scalable representation learning for heterogeneous networks. In: Proceedings of the 23rd ACM SIGKDD International Conference on Knowledge Discovery and Data Mining, pp. 135–144 (2017)

33. Cao, S., Lu, W., Xu, Q.: GraRep: learning graph representations with global structural information. In: Proceedings of the 24th ACM International on Conference on Information and Knowledge Management, pp. 891–900 (2015)

34. Meng, Z., Liang, S., Bao, H., Zhang, X.: Co-embedding attributed networks. In: Proceedings of the Twelfth ACM International Conference on Web Search and Data Mining, pp. 393–401 (2019)

Towards Student Centric Rough Concept Inventories

A. Mani[1,2]([ORCID])

[1] HBCSE, Tata Institute of Fundamental Research, Mumbai 400088, India
a.mani.cms@gmail.com, mani@hbcse.tifr.res.in
[2] Indian Statistical Institute, Kolkata 700108, India
https://www.logicamani.in

Abstract. In the context of education research, a concept inventory is an instrument (that consists of a number of multiple-choice questions) designed to test the understanding of concepts (and possibly the reasons for failure to understand) by learners. Subject to a few caveats they are known to be somewhat effective in non student centric learning environments. In this research the issue of adapting the subject/concept-specific instruments to make room for diverse response patterns (including vague ones) is explored in some detail by the present author. It is shown that higher granular operator spaces (or partial algebras) with additional temporal and key operators are well suited for handling them. An improved version of concept inventory called *rough concept inventory* that can handle vague subjective responses is also proposed in this research.

Keywords: Concept inventory · Student centric learning · Rough objects · Mereology · High granular operator partial algebras · Contamination problem · Education research · Force CI · Function CI

1 Introduction

A test that focuses on evaluating a student's competence in a specific skill is a criterion-referenced test. Usually a person's test scores are intended to suggest a general statement about their capabilities and behavior. Concept inventories (CIs) are criterion-referenced test designed to test a student's functional understanding of concepts. However they are mostly used by education researchers to assess the effectiveness of pedagogical methods.

The standard way to construct concept inventories is as follows:

- select a number of key concepts in a subject or topic;
- formulate multiple choice questions (MCQs) that aim to test key aspects of applications of the chosen concepts;
- each question is required to have at least one correct answer and a number of incorrect answers (distractors) based on student misconceptions or alternative conceptions. Individual steps may require plenty of additional work as can be seen in [1] also because the stakeholders views may not be clear in the first place.

© Springer Nature Switzerland AG 2020
R. Bello et al. (Eds.): IJCRS 2020, LNAI 12179, pp. 251–266, 2020.
https://doi.org/10.1007/978-3-030-52705-1_19

They have been used in studying the effectiveness of a number of pedagogical methods that target concept maturity. These involve pre and post tests that concern assessing students or participants before and after the implementation of the pedagogical procedure (see for example [2]). Like most other practices in education research, a number of concerns have been raised about the methodology and its supposed effectiveness in measuring all that it claims to measure (see Section 3 for some details). It is accepted that descriptive explanations by students for their choice of answers can significantly boost the quality of assessment offered by the concept inventory. Usually these are evaluated by instructors or researchers and the scores from MCQs are accordingly modified.

Apart from the sheer volume of responses generated in specific studies involving the use of concept inventories, in some learning environments it is often the case that instructors themselves may not be sufficiently competent in handling concepts [1]. Therefore automated methods are relevant for evaluation of CIs and for extending their use to provide relevant feedback to learners and instructors.

In student-centered learning students are put at the center of the learning process, and are encouraged to learn through active methods. Arguably, students become more responsible for their learning in such environments. In traditional teacher-centered classrooms, teachers have the role of instructors and are intended to function as the only source of knowledge. By contrast, teachers are typically intended to perform the role of facilitators in student-centered learning contexts. A number of best practices for teaching in such contexts [3] have evolved over time. These methodologies are naturally at odds with concept inventories.

Granular operator spaces and variants [4–7] are abstract frameworks for extending granularity and parthood in the context of general rough sets, and are also variants of rough Y-systems studied by the present author [8]. It has been recently shown by her that all types of granular operator spaces and variants can be transformed into partial algebras that satisfy additional conditions.

General rough sets are used in knowledge representation in a number of contexts [5,6,9–16]. But the problem of knowledge representation in the present context is more complicated. Because the concepts associated with concept inventories have some ontology associated, it is not a good idea to directly reduce them to information table format. Reduction of additional information from descriptive responses may be reasonable in supervised perspective, but it is far more easier to reduce them to higher granular operator perspectives or abstract operator perspectives. Learning contexts (especially constructive learning) adopt perspectives that are most compatible with axiomatic granular perspective because of the hierarchies imposed on any body of knowledge. For example, it is usually imposed that multiplication of natural numbers should be taught only after addition has been taught (and therefore this corresponds to an instance of context dependency). It can be shown that in most constructive teaching contexts, teachers stick to an approximately fixed hierarchy of concept dependence and that student centric activities are pursued (if at all) within a relatively looser variant of the same.

In this research, aspects of concept inventories are explained, a rough variant is introduced and it is shown that higher granular operator spaces with additional temporal operators are optimal for representing related knowledge. The paper is organized as follows: in the next section necessary background and recent results on partial groupoids are mentioned, essential aspects of concept inventories are explained and variants are proposed in the third section, higher granular operator spaces are explained and enhanced versions introduced in the next section, rough concept inventories are proposed in the fifth, and an example is explored in the sixth section.

2 Background, Recent Results

An *information table* $\mathcal{I}$, is a tuple of the form

$$\mathcal{I} = \langle S, \mathbb{A}, \{V_a : a \in \mathbb{A}\}, \{f_a : a \in \mathbb{A}\}\rangle$$

with S, $\mathbb{A}$ and V_a being sets of *objects, attributes* and *values* respectively. Information tables generate various types of relational or relator spaces which in turn relate to approximations of different types and form a substantial part of the problems encountered in general rough sets.

The rough domain corresponds to rough objects of specific type, while the classical and hybrid one correspond to all and mixed types of objects respectively [8]. Boolean algebra with approximation operators forms a classical rough semantics [9]. This fails to deal with the behavior of rough objects alone. The scenario remains true even when R in the approximation space is replaced by arbitrary binary relations. In general, $\wp(S)$ can be replaced by a set with a parthood relation and some approximation operators defined on it as in [8]. The associated semantic domain is the classical one for general Rough sets. The domain of discourse associated with roughly equivalent sets is a *rough semantic domain*. Hybrid domains can also be generated and have been used in the literature [6].

The problem of reducing confusion among concepts from one semantic domain in another is referred to as the contamination problem. Use of numeric functions like rough membership and inclusion maps based on cardinalities of subsets are also sources of contamination. The rationale can also be seen in the definition of operations like $\sqcup$ in pre-rough algebra (for example) that seek to define interaction between rough objects but use classical concepts that do not have any interpretation in the rough semantic domain. Details can be found in [17]. In machine learning practice, whenever inherent shortcomings in algorithmic framework being used are the source of noise then the frameworks may be said to be contaminated.

Key concepts used in the context of general rough sets (and also high granular operator spaces [4,6]) are mentioned next.

- A *crisp object* is one that has been designated as *crisp* or is an approximation of some other object.

- A *vague object* is one whose approximations do not coincide with itself or that which has been designated as a *vague* object.
- An object that is explicitly available for computations in a rough semantic domain (in a contamination avoidance perspective) is a *discernible object*.
- Many definitions and representations are associated with the idea of *rough objects*. From the representation point of view these are usually functions of definite or crisp or approximations of objects. Objects that are invariant relative to an approximation process are said to be *definite objects*. In rough perspectives of knowledge [5,9], algebraic combinations of definite objects (in some sense) or granules are assumed to correspond to crisp concepts, and knowledge to specific collections of crisp concepts. *It should be mentioned that non algebraic definitions are excluded in the present author's axiomatic approach* [4,6,8].

Definition 1. *A partial algebra (see [18])P is a tuple of the form*

$$\langle \underline{P}, f_1, f_2, \ldots, f_n, (r_1, \ldots, r_n) \rangle$$

with $\underline{P}$ being a set, f_i's being partial function symbols of arity r_i. The interpretation of f_i on the set $\underline{P}$ should be denoted by $f_i^{\underline{P}}$, but the superscript will be dropped in this paper as the application contexts are simple enough. If predicate symbols enter into the signature, then P is termed a partial algebraic system.

In this paragraph the terms are not interpreted. For two terms s, t, $s \stackrel{\omega}{=} t$ shall mean, if both sides are defined then the two terms are equal (the quantification is implicit). $\stackrel{\omega}{=}$ is the same as the existence equality (also written as $\stackrel{e}{=}$) in the present paper. $s \stackrel{\omega^*}{=} t$ shall mean if either side is defined, then the other is and the two sides are equal (the quantification is implicit). Note that the latter equality can be defined in terms of the former as

$$(s \stackrel{\omega}{=} s \longrightarrow s \stackrel{\omega}{=} t) \,\&\, (t \stackrel{\omega}{=} t \longrightarrow s \stackrel{\omega}{=} t)$$

In relational approach to general rough sets various granular, pointwise or abstract approximations are defined, and rough objects of various kinds are studied [6,8,12,19–21]. These approximations may be derived from information tables or may be abstracted from data relating to human (or machine) reasoning. A *general approximation space* is a pair of the form $S = \langle \underline{S}, R \rangle$ with $\underline{S}$ being a set and R being a binary relation (S and $\underline{S}$ *will be used interchangeably throughout this paper*). Approximations of subsets of $\underline{S}$ may be generated from these and studied at different levels of abstraction in theoretical approaches to rough sets. In relational approaches to rough sets a number of types of knowledge are representable starting from those by classical rough sets [9] to general rough sets as in [4,11,12,22]. However those based on relational approaches are not always applicable in evaluation and representation of academic data.

Mereology, the study of parts and wholes, has been studied from philosophical, logical, algebraic, topological and applied perspectives. In the literature on mereology [11,23,24], it is argued that most ideas of binary *part of* relations in human reasoning are at least antisymmetric and reflexive. *A major reason for not requiring transitivity of the parthood relation is because of the functional reasons that lead to its failure* (see [23]), and to accommodate *apparent parthood* [24]. The study of mereology in the context of rough sets can be approached in at least two essentially different ways. In the approach aimed at reducing contamination by the present author [4–6,8], the primary motivation is to avoid intrusion into the data by way of additional assumptions about the data and to capture rough reasoning at the level. In numeric function based approaches [25], the strategy is to base definitions of parthood on the degree of rough inclusion or membership – this differs substantially from the former approach. Rough Y-systems and granular operator spaces, introduced and studied extensively by the present author [4–6,8,24], are essentially higher order abstract approaches in general rough sets in which the primitives are ideas of approximations, parthood, and granularity.

2.1   Relations and Groupoids

Under certain conditions, partial or total groupoid operations can correspond to binary relations on a set. This subsection is repeated from a forthcoming paper for Sect. 4.

Definition 2. *In a general approximation space $S = \langle \underline{S}, R \rangle$ consider the following conditions:*

$$(\forall a, b)(\exists c) Rac \text{ \& } Rbc \qquad \text{(up-dir)}$$

$$(\forall a) Raa \qquad \text{(reflexivity)}$$

$$(\forall a, b)(Rab \text{ \& } Rba \longrightarrow a = b) \qquad \text{(anti-sym)}$$

If S satisfies up-dir, then it shall said to be a up-directed approximation space. If it satisfies the last two then it shall said to be a parthood space and a up-directed parthood space when it satisfies all three.

The condition up-dir is equivalent to the set $U_R(a,b) = \{x : Rax \text{ \& } Rbx\}$ being nonempty for every $a, b \in S$ and is also referred to as *directed* in the literature. It is avoided because it may cause confusion.

Definition 3. *If R is a binary relation on S, then a type-1 partial groupoid operation (1PGO) determined by R is defined as follows:*

$$(\forall a, b) \, a \circ b = \begin{cases} b & \text{if } Rab \\ c & c \in U_R(a,b) \text{ \& } \neg Rab \\ \text{undefined otherwise} \end{cases}$$

If R is up-directed, then the operation is total. In this case, the collection of groupoids satisfying the condition will be denoted by $\mathfrak{B}(S)$ and an arbitrary element of it will be denoted by $\mathsf{B}(S)$. The term 'a $\circ$ b' will be written as 'ab' for convenience.

Theorem 1. *The partial operation* $\circ$ *corresponds to a binary relation* R *if and only if*

$$(\forall a, b)(\exists z)(ab \neq b \ \& \ az = bz = z \rightarrow a(ab) = b(ab) = ab)$$
$$(\forall a, b, c)(ab = c \rightarrow c = b \ or \ (\exists z)az = bz = z)$$

The following results have been proved for relational systems in [26,27].

Theorem 2. *For a groupoid* A, *the following are equivalent*

- *A reflexive up-directed approximation space* S *corresponds to* A
- *A satisfies the equations*

$$aa = a \ \& \ a(ab) = b(ab) = ab$$

Definition 4. *If* A *is a groupoid, then two general approximation spaces corresponding to it are* $\Re(A) = \langle \underline{A}, R_A \rangle$ *and* $\Re^*(A) = \langle \underline{A}, R_A^* \rangle$ *with*

$$R_A = \{(a, b) : ab = b\}$$
$$R_A^* = \bigcup\{(a, ab), (b, ab)\}$$

Theorem 3.
- *If* A *is a groupoid then* $\Re^*(A)$ *is up-directed.*
- *If a groupoid* $A \models a(ab) = b(ab) = ab$ *then* $\Re(A) = \Re^*(A)$.
- *If* S *is an up-directed approximation space then* $\Re((B)(S)) = S$.

Theorem 4. *If* $S = \langle \underline{S}, R \rangle$ *is a up-directed approximation space, then*

- R *is reflexive* $\Leftrightarrow$ $\mathsf{B}(S) \models aa = a$.
- R *is symmetric* $\Leftrightarrow$ $\mathsf{B}(S) \models (ab)a = a$.
- R *is transitive* $\Leftrightarrow$ $\mathsf{B}(S) \models a((ab)c) = (ab)c$.
- *If* $\mathsf{B}(S) \models ab = ba$ *then* R *is antisymmetric.*
- *If* $\mathsf{B}(S) \models (ab)a = ab$ *then* R *is antisymmetric.*
- *If* $\mathsf{B}(S) \models (ab)c = a(bc)$ *then* R *is transitive.*

Morphisms between up-directed approximation spaces are preserved by corresponding groupoids in a nice way. This is an additional reason for investigating the algebraic perspective.

3   Ontology Matters

Concept inventories are expected to fulfill a number of requirements for assessment of the effect on learning. In particular, they are expected to

- be designed for measuring understanding as opposed to declarative knowledge,
- measure what they claim to measure (that is they should be valid),
- be standardized for use over diverse educational institutions at the level, and

- be longitudinal (that is they should be amenable for reuse at different points of time for evaluation with relatively less interference).

The well known *force concept inventory* (FCI)[28] and mechanics diagnostic test (MDT) are among the earliest concept inventories developed. They are used in the context of assessment of teaching procedures in physics and have played a significant role in influencing the development of concept inventories in other subjects. A number of concept inventories for specific subjects or topics in mathematics such as the calculus concept inventory [29] and function CI [30] are known. Some have claimed that FCI is a test of mastery of certain contexts and content relating to force and not a test of the force concept itself [31]. Others have tried to show that conceptual understanding is actually addressed in FCI [32]. Though people differ on their opinions about the thing that is actually being measured by FCI [33], FCI is known to measure something useful, and has been widely used.

The literature on concept inventories is large, but ontologies are not commonly used in their analysis or evaluation (though in principle much seems to be possible). Computer-based assessment software do use conceptual models such as labeled conceptual graphs and formal concept analysis. But related exercises require careful formalism to avoid misunderstanding and automatic evaluation is known to miss conceptual problems [34,35]. In the present author's view this is also because they try to avoid (rather than confront) vagueness inherent to the available knowledge.

Most authors agree (see [2]) that notions of misconceptions or alternative conceptions have a important role in determining measurements of conceptual understanding. In the present author's view alternative conceptions and apparent or real misconceptions have a dialectical relationship with conceptual understanding as a whole. This is corroborated by studies that show that students may or may not consistently apply their understanding of concepts (that they seem to have understood). The idea of consistency is a very relative notion that is typically associated with rigid goals in the teaching perspective. It is also very difficult to explore misconceptions with formal concept analysis and concept maps because of simply misreading the intended interpretation of students. Identification of student misconceptions depends on choice of domains and related specification of distractors that can actually relate to alternative conceptions. Studies [36] suggest that often they are not properly included in concept inventories.

The biggest deficiency of concept inventories that use questions in the MCQ format alone is its incompatibility with student centric approaches to evaluation of understanding. The MCQ in CIs (unlike those used in ordinary MCQs) are formulated after estimating possibilities on range and modalities of student responses. Further they are evaluated to ensure test reliability and validity. It is known that evaluators may not know the exact reasons for students choice of an incorrect answer and that understanding may not correlate with correct response [2,33]. A number of proposals have been put forward to address these deficiencies. The most popular have been ones that require students to add explanations for their choice (see for example [1]. MCQ scores obtained by students are adjusted based on the teacher's evaluations of the explanations. *This immediately suggests the problem of improving the methodology towards minimizing biased evaluation by teachers or evaluators.*

Competence levels (relating to a concept inventory) are typically constructed through abstract specification. For example in the function CI proposed in [30], six levels of understanding are identified. This can be enhanced to the following:

1. the ability to distinguish between functions and equations;
2. the ability to recognize and relate different representations of functions and use them interchangeably;
3. the ability to classify relationships as functions or not functions;
4. the ability to have a working familiarity with properties of functions such as $1 - 1$, many-one, increasing, decreasing, linearity;
5. the ability to have a working familiarity with properties of sets of functions such as composition and inverses;
6. the ability to use functions in context, modeling and interpreting;
7. the ability to use functions to preserve relationships across models;
8. the ability to engage with co-variational reasoning;
9. the ability to engage with algebraic reasoning.

Needless to say, the problem of representing data of this form is well beyond the capabilities of relational approach to rough sets. From an general rough set perspective, representing such ideas within the context of collection of relevant concepts subject to the granularities of constructivist ideas of knowledge and human learning are of much interest. While it is not hard to see that a number of abstract granular approximations are involved, it is necessary to identify and classify granules, represent the fine structure of concepts and the process of transformation of concepts by the pedagogical practice.

Not all concepts are constructed equal. Some are more relevant target concepts and can be regarded as *key* concepts. As can be seen in the list of abstract conceptual states (that may be read as key concepts) relating to the function concept, key concepts need not be simple from a representation point of view.

4   High Granular Operator Partial Algebras

Granular operator spaces and variants [4–7] are abstract frameworks for extending granularity and parthood in the context of general rough sets, and are also variants of rough Y-systems studied by the present author [8]. They are well suited for handling approximations of unclear aetiology (relative to construction from information systems) but subject to certain minimal conditions on granularity. In [37], it is shown by the present author that all types of granular operator spaces and variants can be transformed into partial algebras that satisfy additional conditions. Part of this is repeated for convenience in this section. It is also nontrivial because all covering approximation spaces cannot be transformed in the same way.

Definition 5. *A* High General Granular Operator Space *(GGS)* $\mathbb{S}$ *shall be a partial algebraic system of the form* $\mathbb{S} = \langle \underline{\mathbb{S}}, \gamma, l, u, \mathbf{P}, \leq, \vee, \wedge, \perp, \top \rangle$ *with* $\underline{\mathbb{S}}$ *being a set,* γ *being a unary predicate that determines* $\mathcal{G}$ *(by the condition* γx *if and only if* $x \in \mathcal{G}$*) an* admissible granulation*(defined below) for* $\mathbb{S}$ *and* l, u *being operators:* $\underline{\mathbb{S}} \longmapsto \underline{\mathbb{S}}$ *satisfying the following (* $\underline{\mathbb{S}}$ *is replaced with* $\mathbb{S}$ *if clear from the context.* $\vee$ *and* $\wedge$ *are idempotent partial operations and* $\mathbf{P}$ *is a binary predicate. Further* γx *will be replaced by* $x \in \mathcal{G}$ *for convenience.):*

$$(\forall x)\mathbf{P}xx \tag{PT1}$$

$$(\forall x, b)(\mathbf{P}xb \,\&\, \mathbf{P}bx \longrightarrow x = b) \tag{PT2}$$

$$(\forall a, b)a \vee b \overset{\omega}{=} b \vee a \,;\, (\forall a, b)a \wedge b \overset{\omega}{=} b \wedge a \tag{G1}$$

$$(\forall a, b)(a \vee b) \wedge a \overset{\omega}{=} a \,;\, (\forall a, b)(a \wedge b) \vee a \overset{\omega}{=} a \tag{G2}$$

$$(\forall a, b, c)(a \wedge b) \vee c \overset{\omega}{=} (a \vee c) \wedge (b \vee c) \tag{G3}$$

$$(\forall a, b, c)(a \vee b) \wedge c \overset{\omega}{=} (a \wedge c) \vee (b \wedge c) \tag{G4}$$

$$(\forall a, b)(a \leq b \leftrightarrow a \vee b = b \leftrightarrow a \wedge b = a) \tag{G5}$$

$$(\forall a \in \mathbb{S})\,\mathbf{P}a^l a \,\&\, a^{ll} = a^l \,\&\, \mathbf{P}a^u a^{uu} \tag{UL1}$$

$$(\forall a, b \in \mathbb{S})(\mathbf{P}ab \longrightarrow \mathbf{P}a^l b^l \,\&\, \mathbf{P}a^u b^u) \tag{UL2}$$

$$\perp^l = \perp \,\&\, \perp^u = \perp \,\&\, \mathbf{P}\top^l\top \,\&\, \mathbf{P}\top^u\top \tag{UL3}$$

$$(\forall a \in \mathbb{S})\,\mathbf{P}\perp a \,\&\, \mathbf{P}a\top \tag{TB}$$

Let $\mathbb{P}$ *stand for proper parthood, defined via* $\mathbb{P}ab$ *if and only if* $\mathbf{P}ab \,\&\, \neg\mathbf{P}ba$*). A granulation is said to be* admissible *if there exists a term operation* t *formed from the weak lattice operations such that the following three conditions hold:*

$$(\forall x \exists x_1, \ldots x_r \in \mathcal{G})\, t(x_1, x_2, \ldots x_r) = x^l$$
$$\text{and } (\forall x)\,(\exists x_1, \ldots x_r \in \mathcal{G})\, t(x_1, x_2, \ldots x_r) = x^u, \qquad \text{(Weak RA, WRA)}$$

$$(\forall a \in \mathcal{G})(\forall x \in \underline{\mathbb{S}})\,(\mathbf{P}ax \longrightarrow \mathbf{P}ax^l), \qquad \text{(Lower Stability, LS)}$$

$$(\forall x, a \in \mathcal{G})(\exists z \in \underline{\mathbb{S}})\,\mathbb{P}xz, \,\&\, \mathbb{P}az \,\&\, z^l = z^u = z, \qquad \text{(Full Underlap, FU)}$$

The conditions defining admissible granulations mean that every approximation is somehow representable by granules in a algebraic way, that every granule coincides with its lower approximation (granules are lower definite), and that all pairs of distinct granules are part of definite objects (those that coincide with their own lower and upper approximations). Special cases of the above are defined next.

Definition 6. • *In a GGS, if the parthood is defined by* **P***ab if and only if $a \leq b$ then the GGS is said to be a* high granular operator space *GS*.

• *A* higher granular operator space *(HGOS)* $\mathbb{S}$ *is a GS in which the lattice operations are total.*

• *In a higher granular operator space, if the lattice operations are set theoretic union and intersection, then the HGOS will be said to be a* set HGOS.

Theorem 5. *In the context of Definition 5, the binary predicates* **P** *can be replaced by partial two-place operations 1PGO $\odot$ and γ is replaceable by a total unary operation h defined as follows:*

$$
hx = \begin{cases} x & \text{if } \gamma x \\ \bot & \text{if } \neg\gamma x \end{cases} \tag{1}
$$

Consequently $\mathbb{S}^+ = \langle \underline{\mathbb{S}}, h, l, u, \odot, \vee, \wedge, \bot, \top \rangle$ *is a partial algebra that is semantically (and also in a category-theoretic sense) equivalent to the original GGS* $\mathbb{S}$.

Proof. Because of the restriction UL3 on $\bot$ and the redundancy of $\leq$ (because of G5), the result follows.   ‸

Definition 7. *The partial algebra formed in the above theorem will be referred to a* high granular operator partial algebra *(GGSo)*.

Problem 1. All covering approximation spaces considered in the rough set literature actually assume partial Boolean or partial lattice theoretical operations. Some authors (especially in modal logic perspectives) [12,20,38] presume that all Boolean operations are admissible – this view can be argued against. A natural question is *Are the modal logic semantics themselves only a possible interpretation of the actuality?* All this suggests the problem of finding minimal operations involved in the context.

Because all covering approximation spaces do not use granular approximations in the sense mentioned above, it follows that they do not form GGSo always.

5   Rough Concept Inventory and Its Model

A rough concept inventory is intended to be a concept inventory that can effectively handle vagueness inherent in relatively student centric perspectives through methodological improvements, and representations of approximate evaluations. *Relatively*, because the central process of concept inventories is not compatible beyond a point with student-centric approaches.

While it would be best if the methodology and the final analysis are all integrated together, it may be useful in practice to separate the two. The methodological aspect would be as follows:

1. select a number of key concepts in a subject or topic;
2. situate them relative to the concepts and granular concepts described in the model in the subsection below (or alternatively situate the concepts relative to a concept map in terms *was constructed from* and *is a part of*, and basic well-understood concepts);
3. formulate multiple choice questions that aim to test key aspects of applications of the chosen concepts;
4. each question is required to have at least one correct answer and a number of incorrect answers (distractors) based on student misconceptions or alternative conceptions;
5. require explanation from students for their choice;
6. evaluate explanations relative to model in terms of concept approximations (or alternatively evaluate explanations relative to concepts that are definitely understood and those that are possibly understood).

In the latter case, the methodology would follow the alternatives suggested in the second and the sixth step. The end result in this approach would also include a temporal extension of GGSo described below.

5.1   Temporal Extension of GGSo

A temporal extension of GGSo is introduced next to model rough concept inventories from a minimalist perspective. Essentially this is an extension of a GGSo with two unary temporal operations for specifying *before* and *after* states under few constraints and an additional operation for indicating *key* concepts. If desired a GGS can also be extended in the same way for simplicity. *This is intended to be used for the purpose of constructing a single model for the entire procedure of administering the inventory first, applying the pedagogical practice and then applying the concept inventory in the final stage.*

Definition 8. *In the context of Theorem 5, the partial algebra*

$$\mathbb{S}^* = \langle \underline{\mathbb{S}}, h, l, u, \mathfrak{B}, \mathfrak{A}, \odot, \vee, \wedge, \perp, \top \rangle$$

formed by adjoining three unary operations $\Bbbk$, $\mathfrak{A}$ *and* $\mathfrak{B}$ *to the GGSo* S^+ *will be said to be a* basic temporal high granular operator partial algebra *(TGGSo) provided the following properties are satisfied:*

$$(\forall x)\,\mathfrak{A}\mathfrak{A}x = \mathfrak{A}x \qquad\qquad \text{(idempotence-1)}$$

$$(\forall x)\,\mathfrak{B}\mathfrak{B}x = \mathfrak{B}x \qquad\qquad \text{(idempotence-2)}$$

$$(\forall x)\,\mathfrak{A}\mathfrak{B}x = \mathfrak{A}x \qquad\qquad \text{(supercedence-1)}$$

$$(\forall x)\,\mathfrak{B}\mathfrak{A}x = \mathfrak{B}x \qquad\qquad \text{(supercedence-2)}$$

$$(\forall a,b)\,(a \wedge b = a \longrightarrow \Bbbk a \wedge \Bbbk b = \Bbbk a) \qquad\qquad \text{(key-1)}$$

Compared with common usage of these temporal operators (see [39]) this may appear to be very minimalist. But the application context dictates that it would be a good idea to avoid imposing any connections with l, u, h, and $\odot$. Additional approximation operators may also be needed in practice. An element x that satisfies $\Bbbk x = x$ will be said to be a *key* concept.

The most direct interpretation of the different components of the model (or its equivalent formed from a GGS instead) from a practical perspective are as follows:

1. $\mathbb{S}$ can be read as the collection of relevant concepts tagged by real or dummy student/instructor names (including those that are not apparently part of the concept inventory);
2. h can be read as a partial function that helps in identifying granules in $\mathbb{S}$ (the relatively definite concepts from which other definite concepts are made up of in a simple way). The *simple ways* must be related to the definitions of other operations;
3. $\odot$ corresponds to parthood and a perspective of aggregation.
4. $a \vee b = b$ can be read as b was constructed from a.
5. $a \vee b$ can be read as that which is constructed out of an aggregation of a and b.

The easiest *simple way* can be by way of aggregation. When multiple approximations are used then the number of ways can be increased. Note that the model is not tied down to a single idea of concept evolution and has scope for handling the structures generated by the entire sample because of the very definition of $\mathbb{S}$.

6   Example Application

A real application requires datasets that include explicit student responses, and because of ethics concerns it is necessary to form synthetic versions of the same. Due to limited time, aspects of the proposed model are considered in relation to secondary information derived from a typical concept inventory (the temporal aspect is not used in the study).

In [1], the development and analysis of a concept inventory on rotational kinematics is considered. The questions and answers can be found in the appendix of the paper. The authors restrict themselves to questions probing angular velocity of a rigid body, trajectory of an arbitrary particle on a rotating rigid body, angular and linear velocities of particles on a rigid body, angular acceleration of a rigid body, validity of the equation $\tau = I\alpha$, dependence of angular velocity on the origin, relation between angular acceleration and tangential acceleration, relation between angular acceleration and centripetal acceleration, and finally components of linear acceleration. Thus the concept inventory is focused on a very specific set of *key concepts* (other key concepts may be latent). Apparently the questions have been optimized for testing the conception of a specific set of students through a number of steps. The authors mention that responses to questions were verified in the light of the explanations offered (if any).

The authors have this to say on the concept maps associated with angular velocity and angular acceleration:

> Consider the operational definition of the angular velocity of a rigid body as an illustrative example. Identifying the angle $\Delta\theta$ in $\omega = \frac{\Delta\theta}{\Delta t}$ would require the selection of an arbitrary particle on the body, not necessarily the center of mass, drawing a perpendicular line from the particle to the axis, noting the angle traced by this line as the rigid body rotates, etc. As another example, consider the case of α (angular acceleration), which may be nonzero even if the instantaneous angular velocity is zero. Operationally this would entail, among other things, identifying the angular velocities at two different instances and subtracting them. We noted similar intricacies that helped us probe pitfalls in student thinking.

This suggests that the authors have specific ideas of how the concepts being tested must evolve. While, this can be read as an idea of *standard suggested conception*, it is necessary to look into possible alternative conceptions that may be in the explanations offered by the students. In the methodology adopted, these aspects are to be discovered through pilot studies.

A specific question in the inventory is the following: A ceiling fan is rotating around a fixed axis. Consider the following statements for the particles not on the axis at a given instant.

Statement I: Every particle on the fan has the same linear velocity.

Statement II: Every particle on the fan has the same angular velocity.

The correct statement(s) is (are)

1. statement I only
2. statement II only
3. both statements I and II
4. neither statement I nor II.

The correct answer is the second option. Explanations offered by a student for this choice and others should be approximated by the evaluator. In the approach of [1], they are simply used for verifying the correctness of the response.

Statistical analysis is also used for evaluating the conceptual maturity of the set of participants studied.

In the approach suggested in the present paper, the entire dataset can be recoded as a TGGSo (with possibly multiple lower and upper approximation operators) and studied with minimum intrusion. Example granules can be *The angular velocity is computed by* $\omega = \frac{\Delta \theta}{\Delta t}$, and *angular velocities are computed relative to an axis*. Erroneous responses can also be seen as part of the data because objects are all labeled by student/instructor or evaluator names. Last but not in the least explanations can be read as approximation of correct or incorrect concepts and used to help in constructing a vivid characterization of the data set.

For using descriptive statistical methods on the relatively enlarged dataset, it would still be possible to permit additional categories based on the names of sub-concepts introduced or qualifiers on concept names. This would also lead to a descriptive statement on *learning* as opposed to grades or marks. Thus it leads to less contamination of the essence by numeric simplifications.

Remarks

In this research rough concept inventories are introduced, and their relation to existing approaches are discussed by present author. This is motivated by the need to make concept inventories more student centric, reduce contamination, and address a number of other known deficiencies. In particular, this can be a step towards answering the deeper question: *what does a concept inventory actually measure?*.

References

1. Mashood, K.K., Singh, V.: Rotational kinematics of a rigid body about a fixed axis: development and analysis of an inventory. Eur. J. Phys. **36**, 1–21 (2015)
2. Sands, D., Parker, M., Hedgeland, H., Jordan, S., Galloway, R.: Using concept inventories to measure understanding. Higher Educ. Pedagogies **3**(1), 173–182 (2018)
3. Jacobs, G.M., Renandya, W.A., Power, M.: Simple, Powerful Strategies for Student Centered Learning. SE. Springer, Cham (2016). https://doi.org/10.1007/978-3-319-25712-9
4. Mani, A.: High granular operator spaces and less-contaminated general rough mereologies. Forthcoming, pp. 1–77 (2019)
5. Mani, A.: Knowledge and Consequence in AC Semantics for General Rough Sets. In: Wang, G., Skowron, A., Yao, Y., Ślęzak, D., Polkowski, L. (eds.) Thriving Rough Sets. SCI, vol. 708, pp. 237–268. Springer, Cham (2017). https://doi.org/10.1007/978-3-319-54966-8_12
6. Mani, A.: Algebraic methods for granular rough sets. In: Mani, A., Düntsch, I., Cattaneo, G. (eds.) Algebraic Methods in General Rough Sets. Trends in Mathematics. Birkhauser Basel, pp. 157–336 (2018)
7. Mani, A.: Antichain based semantics for rough sets. In: Ciucci, D., Wang, G., Mitra, S., Wu, W.-Z. (eds.) RSKT 2015. LNCS (LNAI), vol. 9436, pp. 335–346. Springer, Cham (2015). https://doi.org/10.1007/978-3-319-25754-9_30

8. Mani, A.: Dialectics of counting and the mathematics of vagueness. In: Peters, J.F., Skowron, A. (eds.) Transactions on Rough Sets XV. LNCS, vol. 7255, pp. 122–180. Springer, Heidelberg (2012). https://doi.org/10.1007/978-3-642-31903-7_4

9. Pawlak, Z.: Rough Sets: Theoretical Aspects of Reasoning About Data. Kluwer Academic Publishers, Dodrecht (1991)

10. Mani, A.: Choice Inclusive General Rough Semantics. Inf. Sci. **181**(6), 1097–1115 (2011)

11. Mani, A.: Algebraic semantics of proto-transitive rough sets. In: Peters, J.F., Skowron, A. (eds.) Transactions on Rough Sets XX. LNCS, vol. 10020, pp. 51–108. Springer, Heidelberg (2016). https://doi.org/10.1007/978-3-662-53611-7_3

12. Pagliani, P., Chakraborty, M.: A Geometry of Approximation: Rough Set Theory: Logic, Algebra and Topology of Conceptual Patterns. Springer, Berlin (2008). https://doi.org/10.1007/978-1-4020-8622-9

13. Polkowski, L., Semeniuk–Polkowska, M.: Reasoning about concepts by rough mereological logics. In: Wang, G., Li, T., Grzymala-Busse, J.W., Miao, D., Skowron, A., Yao, Y. (eds.) RSKT 2008. LNCS (LNAI), vol. 5009, pp. 205–212. Springer, Heidelberg (2008). https://doi.org/10.1007/978-3-540-79721-0_31

14. Yao, Y.Y.: Rough-set concept analysis: interpreting rs-definable concepts based on ideas from formal concept analysis. Inf. Sci. **347**, 442–462 (2016)

15. Bazan, J., Son, N.H., Skowron, A., Szczuka, M.: A view on rough set concept approximations. In: Wang, G., Liu, Q., Yao, Y., Skowron, A. (eds.) RSFDGrC 2003. LNCS (LNAI), vol. 2639, pp. 181–188. Springer, Heidelberg (2003). https://doi.org/10.1007/3-540-39205-X_23

16. Skowron, A.: Rough sets and vague concepts. Fund. Inform. **64**(1–4), 417–431 (2005)

17. Mani, A.: Contamination-free measures and algebraic operations. In: 2013 IEEE International Conference on Fuzzy Systems (FUZZ), pp. 1–8. IEEE (2013)

18. Ljapin, E.S.: Partial Algebras and Their Applications. Academic, Kluwer (1996)

19. Cattaneo, G.: Algebraic methods for rough approximation spaces by lattice interior–closure operations. In: Mani, A., Cattaneo, G., Düntsch, I. (eds.) Algebraic Methods in General Rough Sets. TM, pp. 13–156. Springer, Cham (2018). https://doi.org/10.1007/978-3-030-01162-8_2

20. Pagliani, P.: Three lessons on the topological and algebraic hidden core of rough set theory. In: Mani, A., Cattaneo, G., Düntsch, I. (eds.) Algebraic Methods in General Rough Sets. TM, pp. 337–415. Springer, Cham (2018). https://doi.org/10.1007/978-3-030-01162-8_4

21. Cattaneo, G., Ciucci, D.: Algebraic methods for orthopairs and induced rough approximation spaces. In: Mani, A., Düntsch, I., Cattaneo, G. (eds.): Algebraic Methods in General Rough Sets, pp. 553–640. Birkhauser Basel (2018)

22. Mani, A.: Algebraic semantics of similarity-based bitten rough set theory. Fundamenta Informaticae **97**(1–2), 177–197 (2009)

23. Shafer, W.: Transitivity. In: Durlauf, S.N., Blume, L.E. (eds.) The New Palgrave: Dictionary of Economics. TM, pp. 6736–6738. Palgrave Macmillan UK, London (2008). https://doi.org/10.1007/978-1-349-58802-2_1731

24. Mani, A.: Dialectical rough sets, parthood and figures of opposition-I. In: Peters, J.F., Skowron, A. (eds.) Transactions on Rough Sets XXI. LNCS, vol. 10810, pp. 96–141. Springer, Heidelberg (2019). https://doi.org/10.1007/978-3-662-58768-3_4

25. Polkowski, L.: Approximate Reasoning by Parts. Springer, Heidelberg (2011). https://doi.org/10.1007/978-3-642-22279-5

26. Chajda, I., Langer, H., Sevcik, P.: An algebraic approach to binary relations. Asian European J. Math **8**(2), 1–13 (2015)

27. Chajda, I., Langer, H.: Groupoids assigned to relational systems. Math Bohemica **138**, 15–23 (2013)
28. Hestenes, D., Wells, M., Swackhamer, G.: Force concept inventory. Phys. Teacher **30**, 141–158 (1992)
29. Epstein, J.: The calculus concept inventory - measurement of the effect of teaching methodology in mathematics. Notices Amer. Math. Soc. **60**(8), 1018–1026 (2013)
30. O'Shea, A., Breen, S., Jaworski, B.: The development of a function concept inventory. Int. J. Res. Undergraduate Math. Educ. **2**(3), 279–296 (2016). https://doi.org/10.1007/s40753-016-0030-5
31. Huffman, D., Heller, P.: What does the force concept inventory actually measure? Phys. Teacher **33**, 138–143 (1995)
32. Hestenes, D., Halloum, I.: Interpreting the force concept inventory: a response to, Critique by Huffman and Heller. Phys. Teacher **33**(1995), 502–506 (1995)
33. Wang, J., Bao, L.: Analyzing force concept inventory with item response theory. Am. J. Phys. **78**(10), 1064–1070 (2010)
34. Priss, U., Reigler, U., Jensen, N.: Using FCA for modeling conceptual difficulties in learning processes. In: Domenach, F., et al. (eds.): ICFCA 2012. LNCS 7278, pp. 161–173. Springer, Heidelberg (2012)
35. Priss, U., Jensen, N., Rod, O.: Using conceptual structures in the design of computer-based assessment software. In: Pfeiffer, H.D., Ignatov, D.I., Poelmans, J., Gadiraju, N. (eds.) ICCS-ConceptStruct 2013. LNCS (LNAI), vol. 7735, pp. 121–134. Springer, Heidelberg (2013). https://doi.org/10.1007/978-3-642-35786-2_10
36. Lindell, R.S., Peak, E., Foster, T.M.: Are they all created equal? - a comparison of different concept inventory development methodologies. In: AIP Conference Proceedings 883, New York, Syracuse, pp. 14–17 (2007)
37. Mani, A.: Functional extensions of knowledge representation in general rough sets. In: Bello, R., et al. (eds.) IJCRS 2020. LNAI, pp. 1–15. Springer, Heidelberg (2020)
38. Samanta, P., Chakraborty, M.K.: Interface of rough set systems and modal logics: a survey. In: Peters, J.F., Skowron, A., Ślęzak, D., Nguyen, H.S., Bazan, J.G. (eds.) Transactions on Rough Sets XIX. LNCS, vol. 8988, pp. 114–137. Springer, Heidelberg (2015). https://doi.org/10.1007/978-3-662-47815-8_8
39. Goranko, V., Rumberg, A.: Temporal Logic. In: Zalta, E.N., (ed.) The Stanford Encyclopedia of Philosophy. Spring 2020 edn. (2020)

HGAR: Hybrid Granular Algorithm for Rating Recommendation

Fulan Qian[1]([✉]), Yafan Huang[1], Jianhong Li[1], Shu Zhao[1], Jie Chen[1], Xiangyang Wang[2], and Yanping Zhang[1]

[1] School of Computer Science and Technology, Anhui University, Hefei 230601, Anhui, People's Republic of China
qianfulan@hotmail.com
[2] Anhui Electrical Engineering Professional Technique College, Hefei 230051, Anhui, People's Republic of China

Abstract. Recommendation algorithms based on collaborative filtering show products which people might like and play an important role in personalized service. Nevertheless, the most of them just adopt explicit information feedback and achieve low recommendation accuracy. In recent years, deep learning methods utilize non-linear network framework to receive feature representation of massive data, which can obtain implicit information feedback. Therefore, many algorithms are designed based on deep learning to improve recommendation effects. Even so, the results are unsatisfactory. The reason is that they never consider explicit information feedback. In this paper, we propose a Hybrid Granular Algorithm for Rating Recommendation (HGAR), which is based on granulation computing. The core idea is to explore the multi-granularity of interaction information for both explicit and implicit feedback to predict the users ratings. Thus, we used Singular Value Decomposition model to get explicit information and implicit information can be received by multi-layer perception of deep learning. In addition, we fused the two part information when the two models are jointly trained. Therefore, HGAR can explore the multi-granularity of interaction information which learned explicit interaction information and mined implicit information in different information granular level. Experiment results show that HGAR significantly improved recommendation accuracy compared with different recommendation models including collaborative filtering and deep learning methods.

Keywords: Information granular · Information feedback · Rating recommendation

1 Introduction

As a tool to help users find useful information quickly, the recommendation algorithm solves information overload and implements personalized recommendation, so it has many application scenarios and commercial values. However, with the

R. Bello et al. (Eds.): IJCRS 2020, LNAI 12179, pp. 267–279, 2020.
https://doi.org/10.1007/978-3-030-52705-1_20

rapidly growth of the amount of data, methods based on collaborative filtering encountered some problems. For example, users' preferences cannot be easily obtained, so it is impossible to achieve good recommendation accuracy. Many researchers try to find useful information to improve recommendation accuracy in big data. This kind of demand promotes the application and development of particle computing theory. As shown in Fig. 1, the interactive information can be divided into explicit information feedback and implicit information feedback in the user-item bipartite graph. Explicit information includes ratings, purchases, friends, follow-ups, and other information that actually happens. While implicit information feedback is the relationships and information hidden behind the actual data, such as browsing, clicking, adding to the shopping cart, etc. In general, explicit information can more directly reflect user preferences. However, explicit information is difficult to obtain and data volume is small. The amount of implicit feedback information is large, easy to obtain, but also can tap into the user's more interests. According to information granulation, we can think of information as consisting of explicit information and implicit information. Therefore, from the perspective of information granulation, by effectively mining the explicit information granule and implicit information granule, the recommendation effect can be better improved.

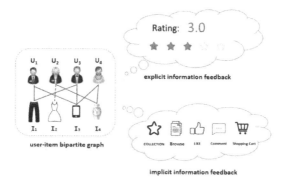

Fig. 1. An example of our proposed recommender system based on explicit and implicit information feedback

Since the information can split different granularity, we can use the idea of granulation to solve the problems in the recommendation system. For example, item-based and user-based recommendation algorithms actually granulate the user set or the item set in the form of targeted user's nearest neighbor. The granulation method is introduced in three-way decision, which uses the explicit information feedback to reflect the information granular. At present, many scholars try to solve the recommendation problem with three-way decision. Huang et al. [1] presented a three-way decision method for recommendation which considers the variable cost as a function of project popularity. Zhang et al. [2]

proposed a regression-based three-way recommender system that aims to minimize the average cost by adjusting the thresholds for different behaviors. Xu et al. [3] designed a model that adds a set of items that may be recommended to users. Zhang et al. [4] created a framework that integrates three-way decision and random forests to build recommender systems. Qian et al. [5] proposed a three-way decision collaborative recommendation algorithm based on user reputation by giving each user a corresponding reputation coefficient. These methods make rating prediction, but the accuracy is not good. Therefore, only relying on explicit information feedback is not a good solution.

Due to the powerful capacity of mining implicit information, deep learning techniques have gained much success in many domains. Therefore, much effort has been made to introduce deep learning techniques to rating recommendations. Cheng et al. [6] jointly trained wide linear models and deep neural networks to combine the benefits of memorization and generalization for recommender systems. Guo et al. [7] combines Factorization Machine (FM) with Deep Neural Networks (DNN) to improve the model ability of learning feature interactions. Covington et al. [8] proposed deep neural network to learn both user and item's embedding, which is generated from their corresponding features separately. However, the above method of deep learning uses implicit information feedback and does not consider explicit information feedback. Therefore, the recommended results could not receive superior accuracy.

To address the challenges we mentioned, in this paper, we propose a hybrid granular algorithm for rating recommendation (HGAR), by combining the advantages of explicit and implicit information feedback to achieve the effect of combinatorial optimization. Explicit information feedback is obtained by user ratings while implicit information is trained by deep learning framework. We can further get new granular by fusing these two information granularity. For a large number of data, HGAR reduced irrelevant information of data and extracted the most accurate user preferences to acquire better recommendation effect. Experiments demonstrate that our model outperforms the compared methods for rating recommendation.

The following sections of this paper are organized as follows: Sect. 2 introduces the problem formulations for quotient space attribute sets; Sect. 3 describes hybrid granular algorithm for rating recommendation in detail; Sect. 4 presents the experimental results and analysis; Sect. 5 is the conclusion of the full paper.

2 Problem Formulation

According to the idea of granulation, we turn the interactive information granulation into explicit information feedback and implicit information feedback. The granular computing theory abstracts the problems into triples to describe them, and then solving them from different granular. Then discussing the representation of different domain attribute in different granularity, and exploring the interdependence and transformation of these representations. In this paper, we define information granular notations of data.

Let $\{x_1, x_2, x_3, ..., x_n\}$ denotes interactive information attribute, n is referred to as the number of attributes. For a recommendation system, X contains explicit information and implicit information based on previous discussions. So we can formulate the equation $X = X_1 + X_2$, X_1 is explicit information granule and X_2 is implicit information granule. Thus, we define $x_i \in X_j$ as interactive information attribute is classified into explicit and implicit, in which $i \in \{1, 2, 3, ..., n\}, j \in \{1, 2\}$. And Y denotes the domain of the rating values. The domain of ratings is made on a 5-star scale (whole-star ratings only). Besides $f : X \rightarrow Y$ is a property function, and if f is a single value, then f can be used to define the partition. Generally speaking, we can easily figure out the structure of Y. For example, if Y is a set of real numbers or Euclidean space, we can define the corresponding classification in Y by using the information feedback of X (i.e. taking different information granularity for rating).

The method is as follows: define $X_j = \{x_i | f(x_i) \in Y\}, i \in \{1, 2, 3, ..., n\}, j \in \{1, 2\}$. So $\{X_j\}$ is a partition of X. Specifically, the notion of explicit information granule can be defined as: $X_1 = f_{explicit}(x_i)$, and the corresponding method is described by the information particle as $Y = f(X_1)$. Similarly, the notion of implicit information granule is $X_2 = f_{implicit}(x_i)$. And $Y = f(X_2)$ is the method described by explicit information particles. To sum up, the final output Y is defined as: $Y = f(x_1) + f(x_2)$, the framework is shown in Fig. 2. In the following sections, we will introduce the detail operation of this algorithm framework.

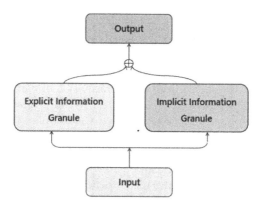

Fig. 2. The basic framework of HGAR

3   Hybrid Granular Algorithm for Rating Recommendation

In this section, we use Singular Value Decomposition (SVD) to represent the explicit information granule and Multi-layer Perceptron (MLP) to represent the implicit information granule. We first present how SVD and MLP worked separately and explain how they serve as a rating recommendation framework.

Figure 5 depicts the architecture of the proposed hybrid granular model. Then, we fuse these modules to predict ratings through the HGAR model which has been trained.

Embedding Layer. We adopt an embedding layer to present user and item. The user-id and the item-id are input information that needs to be preprocessed before entering the model. This is done by mapping the input information to a dense vector. In this way, we can obtain *uemb* as a set of feature vector from user, and *iemb* as a set of feature vector from item. The processing of the embedded layer is represented as follows:

$$uemb = embedding\_lookup(userid) \tag{1}$$

$$iemb = embedding\_lookup(itemid) \tag{2}$$

Where *embedding_lookup* represents the embedding operation, *userid* and *itemid* are the input of embedding layer, *uemb* and *iemb* are the output vectors.

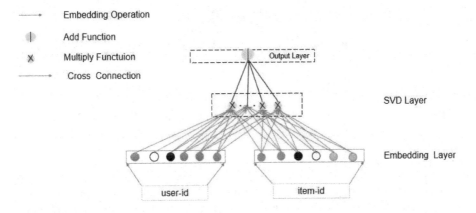

Fig. 3. The architectures of SVD layer for explicit information feedback

3.1   SVD Layer

In this layer, we take advantage of explicit feedback from user and item to implement rating prediction. The model of SVD layer is shown in Fig. 3. SVD is a matrix factorization method. The high dimensional user-item rating matrix is converted into two low dimensional user factor matrices and item factor matrices. In order to obtain feedback information to obtain the user's rating of the item. The formula is shown in:

$$X_1 = f_{explicit}(uemb, iemb) \tag{3}$$

where $f_{explicit}$ is · operation.

The rating consists of four components: global average, user bias, item bias and user-item interaction. The following equation shows the calculation process:

$$\widehat{r_{ui}} = \mu + b_i + b_u + X_1 \tag{4}$$

Where the rating $\widehat{r_{ui}}$ is the output of the SVD layer, μ denotes the overall average rating, b_i and b_u respectively indicate the observed deviations of user u and item i. Obviously, SVD directly adopts explicit information feedback (rating information) to adjust model prediction errors and to get better recommendation accuracy.

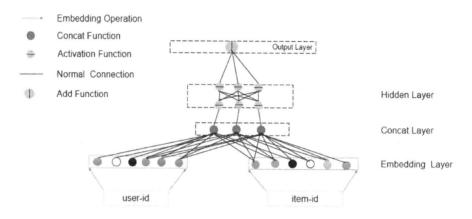

Fig. 4. The architectures of deep component for implicit information feedback

3.2   Deep Component

Contact Layer. Before mining the implicit information, we have preprocessed the embedding vector. After that, we need to adopt a contact layer to concatenate $uemb$ and $iemb$ into one vector. Mapping the two vectors to a vector space and reducing data dimension. The formulation is shown by:

$$\alpha = uemb \oplus iemb \tag{5}$$

where $\oplus$ represents the concat operation, α is the output of contact layer.

Hidden Layer. The MLP model is designed to learn implicit information from hidden layer, as shown in Fig. 4. It consists of an input layer, an output layer and a number of hidden layers. In the process of model training, the embedded vector is randomly initialized firstly, and then the value of the embedded vector is trained to minimize the loss function. These low-dimensional dense embedding vectors are fed into the hidden layer of the neural network in the forward channel. MLP can enhance the expressiveness of the model through multiple hidden layers, but it also increases the complexity of the model. High-dimensional features can be converted into a low-dimensional but dense valuable features by multi-layer. According to the definition of implicit information particles, the hidden layer denotes as:

$$X_2 = f_{implicit}\left(W^{(l+1)}\alpha^l + b^l\right) \qquad (6)$$

$$\alpha^{l+1} = f(X_2) \qquad (7)$$

Where $f_{implicit}$ denotes non-linear activation, l is the number of layer, W^l, b^l, α^l are the l-th weight, the l-th bias, the l-th input. f shows the linear activation function.

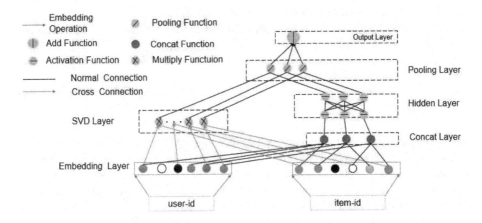

Fig. 5. The model of hybrid granular algorithm for rating recommendation

3.3   Joint Training of HGAR Model

Pooling Layer. Through the previous operation, we obtained the explicit rating and the implicit rating respectively. Now, we need to convert ratings with sum pooling to descend to 1-dimension. The operation is defined as follows:

$$m = \sum_{i}^{n} e_i, \forall i = 2, 3, ..., n \qquad (8)$$

where e_i represents the i-dimension vector of input, and m is the 1-dimension output.

Output Layer. Finally, we combine both explicit and implicit rating into a single vector representation to predict the final rating. The output after fusion is formulated as:

$$\widehat{R_{u,i}} = f\left(m_{ui}^{SVD}, m_{ui}^{MLP}\right) \tag{9}$$

Where $\widehat{R_{u,i}}$ denotes the user rating for a specific item, m_{ui}^{SVD} is the pooling result of the SVD model, m_{ui}^{MLP} is the pooling result of the deep component.

4   Experimental Analysis

In this section, we present our experimental setup and empirical evaluation. We aim to answer the following questions in our experiments:

Q1: How does HGAR perform in terms of efficiency and effectiveness, compared to other state-of-the-art methods based on explicit feedback?

Q2: How does HGAR perform as compared to the state-of-the-art deep learning methods based on implicit feedback?

Q3: How do Singular Value Decomposition (SVD) and Multi-layer Perceptron (MLP) affect the performance of HGAR?

4.1   Data Description

We perform experiments on two well-known and widely used datasets in recommendation: Movielens-100k and Movielens-1M. In Movielens-100k dataset, it contains nearly 100,000 rating records of 943 users on 1,682 movies. As Movielens-1M dataset contains UserIDs which ranged between 1 and 6040 and MovieIDs which ranged between 1 and 3952. Ratings are made on a 5-star scale (whole-star ratings only). Each user has at least 20 ratings. We divide the dataset into training and test set as 8:2, and we use 5-fold cross-validation to get the average results. The basic statistical information of two datasets are illustrated by Table 1.

Table 1. Statistics of the MovieLens datasets

	MovieLens 100k	MovieLens 1M
Users	943	6,040
Items	1,682	3,952
Ratings	100,000	1,000,209
Ratings of per user	106.4	165.6
Rating of per item	59.5	253.1
Rating Sparsity	93.7%	95.8%

4.2 Evaluation Metrics

We use Mean Absolute Error (MAE) and Root Mean Square Error (RMSE) to evaluate the prediction performance of all algorithms.

$$MAE = \frac{\sum_{(u,i) \in N} \left| R_{u,i} - \widehat{R_{u,i}} \right|}{|N|} \tag{10}$$

$$RMSE = \sqrt{\frac{\sum_{(u,i) \in N} \left(R_{u,i} - \widehat{R_{u,i}} \right)^2}{|N|}} \tag{11}$$

where N denotes the whole number of ratings, $R_{u,i}$ denotes the rating user u gives to item i, and $\widehat{R_{u,i}}$ denotes the rating user u gives to item i as prediction. The smaller values of MAE and RMSE indicate the better performance.

4.3 Baselines

We compared our method with the following baseline methods, including the state-of-the-art recommendation methods and the proposed model with its two parts (SVD and MLP). Below we provide the names of algorithms as well as its brief introduction that will used in the following experiments.

- **SVD**: A classical SVD algorithm based on user and item bias.
- **MLP**: A traditional neural network to solve the nonlinear problem that is trained by error backpropagation.
- **PMF** [9]: Probabilistic matrix factorization model, which is a widely used matrix factorization model.
- **BPMF** [10]: Bayesian probabilistic matrix factorization for recommendation.
- **RLMC** [11]: A new robust local matrix completion algorithm that characterize the bias and variance of the estimator in a finite sample setting.
- **RegSVD** [12]: A rating prediction algorithm based on SVD.
- **PRMF** [13]: A novel recommendation method that can automatically learn the dependencies between users to improve recommendation accuracy.
- **TWDA** [5]: A three-way decision methods to process the boundary region and divided all ratings in boundary region into positive region or negative region reasonably.
- **PRA** [14]: Probabilistic rating auto-encoder that uses autoencoder to generate latent user feature profiles.
- **CDAE** [15]: A novel method called collaborative denoising auto-encoder for top-N recommendation that utilizes the idea of denoising auto-encoders.
- **SR**^{imp} [16]: Exploiting users implicit social relationships for recommendation.
- **SVD++** [17]: Merging the latent factor model and neighborhood model for recommendation.
- **Wide and Deep** [6]: Jointly trained wide linear models and deep neural networks to combine the benefits of memorization and generalization for recommender systems.

- **Hybird IC-CRBMF** [18]: An improved item category aware conditional restricted Boltzmann machine frame model for recommendation by integrating item category information as the conditional layer.
- **HACF** [19]: A fundamentally new architecture of hierarchical autoencoder where each layer reconstructs and provides complimentary information.
- **HGAR**: Our proposed method combines SVD and MLP to obtain explicit and implicit information simultaneously, which further improves recommendation accuracy.

4.4   Comparison of Performance with Other State-of-the-art Methods Based on Explicit Feedback (Q1)

The Table 2 represents all MAE results of two data sets based on explicit information feedback. From the results, we can see clearly that: Results for MAE, HGAR outperforms all other methods based on explicit information feedback. To be specific, HGAR is equal to TWDA on Movielens-100k, but HGAR shows an improvement of 2% compared to TWDA on Movielens-1M. This shows that our model is better at large data sets. The results reveal that other methods only based on explicit feedback cannot obtain higher precision. Thus, our method of hybrid granular which combines explicit and implicit features has better performance on MAE.

Table 2. Experimental performance MAE metrics of HGAR compared to explicit feedback baselines on the MovieLens datasets.

	MovieLens 100k	MovieLens 1M
PMF	0.782	0.690
BPMF	0.881	0.680
RLMC	0.760	0.736
RegSVD	0.733	0.698
PRMF	0.721	0.673
TWDA	0.717	0.670
HGAR	0.717	0.668

4.5   Comparison of Performance with Other State-of-the-art Methods Based on Implicit Feedback (Q2)

Table 3 shows the performance of HGAR compared with other algorithms for implicit feedback. The benchmark algorithms, for example, SP, SVD++, Wide and Deep, they all take advantage of implicit information feedback for rating recommendation. We compared HGAR with them and obtained better experimental results. In particular, the result of HACF on Movielens-100k is the same as ours, but on the 1M dataset, our result is better. Similarly, on Movielens-1M, SVD++ is equal to us, but in the 100k dataset, we show an advantage.

Table 3. Experimental performance MAE metrics of HGAR compared to implicit feedback baselines on the MovieLens datasets.

	MovieLens 100k	MovieLens 1M
PRA	0.759	0.714
CDAE	0.735	0.691
SR^{imp}	0.729	0.674
SVD++	0.726	0.668
Wide and Deep	0.723	0.671
Hybrid IC-CRBMF	0.719	0.681
HACF	0.717	0.681
HGAR	0.717	0.668

Given all above analysis, our approach makes a good result on two public real-world datasets, which could explain that the granulation of explicit and implicit information plays an important role and brings a significant improvement.

4.6   The Impact of SVD and MLP (Q3)

SVD and MLP are two parts of our model, thus we experiment these two separate algorithms to make sure whether combination is better. From Table 4, we can see that HGAR makes significant improvements compared to the MLP, whatever MAE or RMSE on Movielens-100k or 1M. Meanwhile, as shown in Table 4 compared to SVD, the MAE value of HGAR is better with 0.1% in Movielens-1M and poorer with 0.5% in Movielens-100k. In addition, the RMSE and MAE values of HGAR show good results in Movielens 1M. Thus, we find that SVD only gets explicit feedback as well as MLP merely obtained implicit feedback. They all perform badly because merely from a single attribute perspective is not as good as from the idea of multi-granularity decomposition to recommend.

Table 4. Experimental performance of SVD and MLP on the MovieLens datasets.

Dataset	Movielens 100k		Movielens 1M	
	MAE	RMSE	MAE	RMSE
SVD	0.718	0.916	0.673	0.861
MLP	0.741	0.947	0.716	0.910
HGAR	0.717	0.921	0.668	0.856

5   Conclusion

In this paper, we proposed Hybrid Granular Algorithm for Rating Recommendation. Considering the large amount of data in the recommendation system, we put the problem on the space of different granularity for analysis and research. To make full use of information granularity, we study the attributes of interactive information and conclude that it can be divided into explicit information and implicit information. In this way, the fine-grained and precise user preferences can be captured. Results on two public datasets show that the proposed model produces comparative performance compared to state-of-the-art methods based on explicit or implicit information feedback.

Acknowledgements. This work was partially supported by the National Natural Science Foundation of China (Grants #61673020 #61702003 #61876001), the Provincial Natural Science Foundation of Anhui Province (Grants #1808085MF175). The authors would like to thank the anonymous reviewers for their valuable comments.

References

1. Huang, J., Wang, J., Yao, Y., Zhong, N.: Cost-sensitive three-way recommendations by learning pair-wise preferences. Int. J. Approx. Reason. **86**, 28–40 (2017)
2. Zhang, H.R., Min, F., Shi, B.: Regression-based three-way recommendation. Inf. Sci. **378**, 444–461 (2016)
3. Xu, Y.-Y., Zhang, H.-R., Min, F.: A three-way recommender system for popularity-based costs. In: Polkowski, L., et al. (eds.) IJCRS 2017. LNCS (LNAI), vol. 10314, pp. 278–289. Springer, Cham (2017). https://doi.org/10.1007/978-3-319-60840-2_20
4. Zhang, H.R., Min, F.: Three-way recommender systems based on random forests. Knowl. Based Syst. **91**, 275–286 (2016)
5. Qian, F., Min, Q., Zhao, S., Chen, J., Wang, X., Zhang, Y.: Three-way decision collaborative recommendation algorithm based on user reputation. In: Mihálydeák, T., et al. (eds.) IJCRS 2019. LNCS (LNAI), vol. 11499, pp. 424–438. Springer, Cham (2019). https://doi.org/10.1007/978-3-030-22815-6_33
6. Cheng, H.T., Koc, L., Harmsen, J., Shaked, T., Chandra, T., Aradhye, H., et al.: Wide & deep learning for recommender systems. In: Proceedings of the 1st Workshop on Deep Learning for Recommender Systems, pp. 561–568. ACM(2009)
7. Guo, H., Tang, R., Ye, Y., Li, Z., He, X.: Deepfm: a factorization-machine based neural network for CTR prediction. arXiv preprint arXiv:1703.04247 (2017)
8. Covington, P., Adams, J., Sargin, E.: Deep neural networks for youtube recommendations. In: 2016 ACM Conference on Recommender Systems, pp. 191–198. ACM (2016)
9. Salakhutdinov, R.: Probabilistic matrix factorization. In: 2008 Advances in neural information processing systems, pp. 1257–1264. ACM (2008)
10. Salakhutdinov, R., Mnih, A.: Bayesian probabilistic matrix factorization using Markov chain Monte Carlo. In: 2008 International Conference on Machine Learning, pp. 880–887(2008)
11. Sabetsarvestani, Z., Kiraly, F., Miguel, R., Rodrigues, D.: Entry-wise matrix completion from noisy entries. In: 2018 European Signal Processing Conference (EUSIPCO), pp. 2603–2607. IEEE(2018)
12. Paterek, A.: Improving regularized singular value decomposition for collaborative filtering. In: Proceedings of KDD cup and workshop, vol. 2007, pp. 5–8. ACM (2007)

13. Liu, Y., Zhao, P., Liu, X., Wu, M., Li, X. L.: Learning user dependencies for recommendation. In: 2016 International Joint Conference on Artificial Intelligence, pp. 2379–2385 (2017)
14. Liang, H., Baldwin, T.: A probabilistic rating auto-encoder for personalized recommender systems. In: 2015 ACM International on Conference on Information and Knowledge Management, pp. 1863–1866. ACM (2015)
15. Wu, Y., DuBois, C., Zheng, A. X., Ester, M.: Collaborative denoising auto-encoders for top-n recommender systems. In: 2016 ACM International Conference on Web Search and Data Mining, pp. 153–162. ACM (2016)
16. Ma, H.: An experimental study on implicit social recommendation. In: 2013 ACM SIGIR conference on Research and development in information retrieval, pp. 73–82. ACM (2013)
17. Koren, Y.: Factorization meets the neighborhood: a multifaceted collaborative filtering model. In: 2008 ACM SIGKDD International Conference on Knowledge Discovery and Data Mining, pp. 426–434. ACM (2008)
18. Liu, X., Ouyang, Y., Rong, W., Xiong, Z.: Item category aware conditional restricted boltzmann machine based recommendation. In: Arik, S., Huang, T., Lai, W.K., Liu, Q. (eds.) ICONIP 2015. LNCS, vol. 9490, pp. 609–616. Springer, Cham (2015). https://doi.org/10.1007/978-3-319-26535-3_69
19. Maheshwari, S., Majumdar, A.: Hierarchical autoencoder for collaborative filtering. In: 2018 International Joint Conference on Neural Networks (IJCNN), pp. 1–7. IEEE (2018)

Formal Concept Analysis

Concept Analysis Using Quantitative Structured Three-Way Rough Set Approximations

Mengjun Hu[✉][iD]

Department of Computer Science, University of Regina,
Regina, SK S4S 0A2, Canada
mengjun.hu@uregina.ca

Abstract. One important topic of concept analysis is to learn an intension of a concept through a given extension. In the case where an exact intension cannot be formulated due to limited information, rough set theory introduces approximations to roughly learn the intension. Pawlak originally proposes a qualitative formulation of approximations which allows no error in the learned intension. Various quantitative formulations have been studied as generalizations, most of which use probabilistic measures. In contrast, non-probabilistic formulations have not been fully investigated. On the other hand, three-way approximations and structured approximations have been proposed to emphasize the semantics of approximations for the purpose of learning and interpreting intension. To combine the benefits of these two directions of generalizations, this paper investigates quantitative structured three-way approximations based on both probabilistic and non-probabilistic measures in the context of both complete and incomplete information.

Keywords: Concept analysis · Three-way · Rough set · Incomplete information · Subsethood measure

1 Introduction

Concept analysis is one common application of rough set theory [18, 19]. A concept can be formally represented by a pair of intension and extension [3] where the intension describes the definition and the extension lists all instances. Concept analysis is usually based on a dataset represented in a tabular form with rows as objects and columns as attributes [7, 20, 21, 24, 31, 32]. The attributes are used to describe the properties of objects as well as to formulate intensions. While learning extension from a given intension is not difficult, the opposite task may be complicated. In particular, we may not be able to find an exact intension of a given extension due to insufficient, incomplete, or limited information.

To solve the above issue, rough set theory introduces the concept of approximations to roughly approximate the true intension. As illustrated by Fig. 1(a) [12], the set of all objects in a given table, represented by the biggest

© Springer Nature Switzerland AG 2020
R. Bello et al. (Eds.): IJCRS 2020, LNAI 12179, pp. 283–297, 2020.
https://doi.org/10.1007/978-3-030-52705-1_21

rectangle, is divided into pieces called definable sets. Each definable set can be precisely described by an intension and is used to approximate a given extension. Pawlak [18,19] proposes a pair of lower and upper approximations, where the lower approximation corresponds to the positive region in Fig. 1(a) and the upper approximation corresponds to the union of the positive and boundary regions. A set of classification rules is derived from an approximation by using the intension of a definable set as the premise of one rule. All such rules together approximate the true intension and can be used to classify instances of the concept. By interpreting classification rules through associated actions, Yao [26] further considers the negative region, which leads to a three-way approximation consisting of the positive, boundary, and negative regions. Semantically, the positive and negative regions are associated with actions of accepting and rejecting instances of the concept, respectively. The boundary region is associated with a non-commitment action, which reflects the limitation of our knowledge.

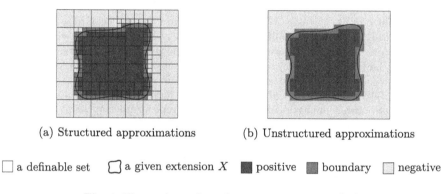

(a) Structured approximations   (b) Unstructured approximations

□ a definable set   ⬭ a given extension X ■ positive   ■ boundary   □ negative

Fig. 1. Illustrations of rough set approximations [12]

There are at least two directions in which the above approximations can be improved. The first direction is to allow a certain rate of misclassification in order to enlarge the positive and negative regions and shrink the boundary region. Instead of the qualitative set-inclusion used in Fig. 1(a), a consideration of quantitative measures results in various quantitative rough set models. Most related research uses probabilistic measures [9,23,29,33] and a few considers non-probabilistic [10,28]. Yao and Deng [28] propose a general framework of formulating both probabilistic and non-probabilistic approximations based on subsethood measures whose properties are further studied in [11]. A few related works regarding concept-based non-probabilistic classifiers are investigated in [17], which may inspire the research on non-probabilistic quantitative rough set models.

The second direction is to build explanation-oriented approximations that emphasize on explaining and understanding the semantics. Most formulations of approximations focus on which objects should be included without due attention to their descriptions which are necessary to formulate rules. The lower and upper approximations proposed by Pawlak in [18,19] and the three-way approximations proposed by Yao in [26] are defined as sets of objects, which are better illustrated by Fig. 1(b). A lack of internal structure leads to certain difficulties in deriving and interpreting classification rules. In contrast, a structured approximation [2,12] is defined as a set of definable sets instead of a set of objects. With clearer semantics, structured approximations can be conveniently and meaningfully applied to learn concepts with incomplete information, where most existing rough set models face a common challenge of interpreting their approximations in order to formulate classification rules [12,15].

This paper studies quantitative structured approximations as improvements in both directions. More specifically, we investigate quantitative structured approximations based on both probabilistic and non-probabilistic measures with both complete and incomplete information. This work focuses on exploring meaningful approaches to building explanation-oriented quantitative structured approximations. Accordingly, we present conceptual formulations of approximations that emphasize on the semantics, rather than computational formulations that emphasize efficient computations in practice. Further discussions on conceptual and computational formulations can be found in [4,12,16,25].

In the remainder of this paper, Sect. 2 reviews qualitative structured approximations with both complete (Sect. 2.1) and incomplete (Sect. 2.2) information. Section 3 explores the generalizations into quantitative structured approximations, including both complete (Sect. 3.1) and incomplete (Sects. 3.2 and 3.3) information. Conclusion and future work are discussed in Sect. 4.

2 Concept Analysis Using Qualitative Structured Three-Way Approximations

This section reviews the main results of qualitative structured approximations proposed in [12] with both complete and incomplete information.

2.1 Learning Intension with Complete Information

An information table is formally used in rough sets to represent a given dataset. In the case of complete information, a complete information table is formulated as the following tuple:

$$T = (OB, AT, \{V_a \mid a \in AT\}, \{I_a : OB \to V_a \mid a \in AT\}), \tag{1}$$

where OB is a finite nonempty set of objects, AT is a finite nonempty set of attributes, V_a is the domain of an attribute a, and I_a is an information function which maps each object to a unique value in V_a. This unique value reflects the complete information or our complete knowledge. Logic formulas regarding attributes and their values are used as formal descriptions of objects. By arguing that a consideration of logic conjunction is sufficient for the rule-learning purpose, Hu and Yao [12] use a conjunctive description language DL_c consisting of formulas defined as follows:

(1) Atomic formulas: $(a = v) \in \mathrm{DL}_c$, where $a \in AT$ and $v \in V_a$;
(2) Composite formulas: if $p, q \in \mathrm{DL}_c$, and p and q do not share any attribute, then $p \wedge q \in \mathrm{DL}_c$.

The satisfiability of a formula by an object, denoted by $\models$, is defined as:

$$\begin{aligned}
(1) \quad & o \models (a = v) \Longleftrightarrow I_a(o) = v, \\
(2) \quad & o \models (p \wedge q) \Longleftrightarrow o \models p \wedge o \models q,
\end{aligned} \tag{2}$$

where $o \in OB$, $a \in AT$, $v \in V_a$, and $p, q, p \wedge q \in \mathrm{DL}_c$. Accordingly, a formula is associated with a set of objects exhibiting its meaning.

Definition 1. *For a formula $p \in \mathrm{DL}_c$, the set of objects:*

$$m(p) = \{o \in OB \mid o \models p\}, \tag{3}$$

is called the meaning set of p.

On the other hand, objects in $m(p)$ can be uniformly described by p and thus, is considered to be definable. By using a formula as intension and its meaning set as extension, one can form a definable concept.

Definition 2. *A set of objects $O \subseteq OB$ is a conjunctively definable set if there exists a formula $p \in \mathrm{DL}_c$ such that $O = m(p)$. The pair $(p, m(p))$ is a conjunctively definable concept.*

$\mathrm{DEF}(T)$ is widely used to represent the family of definable sets in recent works [4, 22, 25]. Accordingly, we use $\mathrm{CDEF}(T)$ to represent the family of conjunctively definable concepts which is used to construct structured approximations.

Definition 3. *For a set of objects $X \subseteq OB$, its structured positive and negative regions [12] are defined as:*

$$\begin{aligned}
\mathrm{SPOS}(X) &= \{(p, m(p)) \in \mathrm{CDEF}(T) \mid m(p) \neq \emptyset, m(p) \subseteq X\}, \\
\mathrm{SNEG}(X) &= \{(p, m(p)) \in \mathrm{CDEF}(T) \mid m(p) \neq \emptyset, m(p) \subseteq X^c\},
\end{aligned} \tag{4}$$

where X^c is the complement of X.

The boundary region is commonly defined through the positive and negative regions. With respect to Definition 3, the structured boundary region of X can be defined as:

$$\mathrm{SBND}(X) = \{(p, m(p)) \in \mathrm{CDEF}(T) \mid \neg(m(p) \subseteq X) \wedge \neg(m(p) \subseteq X^c)\}. \tag{5}$$

From the view of learning intension, we are not interested in the boundary region since it doesn't lead to classification rules for recognizing either positive or negative instances of the concept.

Most research in the literature applies unstructured approximations [18, 19] which can be expressed as [12]:

$$\begin{aligned}
\mathrm{POS}(X) &= \bigcup \{m(p) \mid (p, m(p)) \in \mathrm{SPOS}(X)\}, \\
\mathrm{NEG}(X) &= \bigcup \{m(p) \mid (p, m(p)) \in \mathrm{SNEG}(X)\}.
\end{aligned} \tag{6}$$

As argued and illustrated in [12], the structured approximations benefit rule learning with clear semantics obtained through preserving the internal structure as well as introducing intensions which are left-hand-sides of rules.

2.2 Learning Intension with Incomplete Information

Although an object actually takes exactly one value on an attribute, due to our limited or incomplete information, we may not be able to know this actual value. In such a case, an incomplete information table is used, which can be formally represented as the following tuple:

$$\widetilde{T} = (OB, AT, \{V_a \mid a \in AT\}, \{\widetilde{I}_a : OB \to 2^{V_a} - \{\emptyset\} \mid a \in AT\}), \tag{7}$$

where OB, AT, and V_a have the same meanings as in a complete table, and $\widetilde{I}_a$ maps each object to a nonempty subset of V_a. Every value in $\widetilde{I}_a(x)$ may be the actual value of an object $x \in OB$ on an attribute $a \in AT$, but exactly one value is indeed the actual one which we do not know due to incomplete information.

Lipski [14] equivalently interprets an incomplete table as a family of complete tables. A complete table $(OB, AT, \{V_a \mid a \in AT\}, \{I_a : OB \to V_a \mid a \in AT\})$ is called a completion of $\widetilde{T}$ if:

$$\forall x \in OB \, \forall a \in AT (I_a(x) \in \widetilde{I}_a(x)). \tag{8}$$

One gets a completion of $\widetilde{T}$ by picking up exactly one value for each object on each attribute. Since each value in $\widetilde{I}_a(x)$ represents one possibility of the actual value, a completion is a possibility of the actual table and called a possible world. Accordingly, Lipski's interpretation is called the possible-world semantics of an incomplete table. The family of all completions of $\widetilde{T}$ is denoted as $\mathrm{COMP}(\widetilde{T})$.

The meaning set of a formula p in a completion T, denoted by $m(p|T)$, is a possibility of its actual meaning set. The collection of p's meaning sets in all completions covers all possibilities of p's actual meaning set and can be used to interpret p.

Definition 4. *The meaning set of a formula $p \in \mathrm{DL}_c$ in an incomplete table $\widetilde{T}$ is defined as:*

$$\widetilde{m}(p) = \{m(p|T) \mid T \in \mathrm{COMP}(\widetilde{T})\}, \tag{9}$$

It is verified that $\widetilde{m}(p)$ is actually an interval set defined as [12]:

$$\widetilde{m}(p) = [m_*(p), m^*(p)] = \{S \subseteq OB \mid m_*(p) \subseteq S \subseteq m^*(p)\}. \tag{10}$$

The sets $m_*(p)$ and $m^*(p)$ are the lower and upper bounds of $\widetilde{m}(p)$, respectively. The interval set $[m_*(p), m^*(p)]$ contains all sets in-between these two bounds (inclusive). Moreover, the two bounds can be computed as:

$$m_*(p) = \bigcap_{T \in \mathrm{COMP}(\widetilde{T})} m(p|T) = \{x \in OB \mid \forall T \in \mathrm{COMP}(\widetilde{T}), x \in m(p|T)\},$$

$$m^*(p) = \bigcup_{T \in \mathrm{COMP}(\widetilde{T})} m(p|T) = \{x \in OB \mid \exists T \in \mathrm{COMP}(\widetilde{T}), x \in m(p|T)\}. \tag{11}$$

The lower bound $m_*(p)$ contains objects satisfying p in every possible world, that is, they must satisfy p in the actual table and be included in p's actual meaning set. Similarly, the upper bound $m^*(p)$ contains objects satisfying p in at least one possible world, that is, they possibly satisfy p in the actual table and may be included in p's actual meaning set. By means of $\widetilde{m}(p)$, the definability can be generalized with respect to an incomplete table.

Definition 5. *An interval set $\mathcal{O}$ on OB is conjunctively definable if there exists a conjunctive formula $p \in \mathrm{DL}_c$ such that $\mathcal{O} = \widetilde{m}(p)$. The pair $(p, \widetilde{m}(p))$ is a conjunctively definable interval concept.*

The family of conjunctively definable interval concepts $\mathrm{CDEFI}(\widetilde{T})$ is used to construct the structured approximations in an incomplete table.

Definition 6. *Given a set of objects $X \subseteq OB$ in an incomplete table $\widetilde{T}$, two pairs of structured regions are constructed as [12]:*

$$(1) \quad \mathrm{SPOS}_*(X) = \{(p, \widetilde{m}(p)) \in \mathrm{CDEFI}(\widetilde{T}) \mid \forall S \in \widetilde{m}(p), S \neq \emptyset, S \subseteq X\},$$
$$\mathrm{SNEG}_*(X) = \{(p, \widetilde{m}(p)) \in \mathrm{CDEFI}(\widetilde{T}) \mid \forall S \in \widetilde{m}(p), S \neq \emptyset, S \subseteq X^c\};$$
$$(2) \quad \mathrm{SPOS}^*(X) = \{(p, \widetilde{m}(p)) \in \mathrm{CDEFI}(\widetilde{T}) \mid \exists S \in \widetilde{m}(p), S \neq \emptyset, S \subseteq X\},$$
$$\mathrm{SNEG}^*(X) = \{(p, \widetilde{m}(p)) \in \mathrm{CDEFI}(\widetilde{T}) \mid \exists S \in \widetilde{m}(p), S \neq \emptyset, S \subseteq X^c\}. \quad (12)$$

$\mathrm{SPOS}_*(X)$ *and* $\mathrm{SNEG}_*(X)$ *are called lower structured regions, and* $\mathrm{SPOS}^*(X)$ *and* $\mathrm{SNEG}^*(X)$ *are upper structured regions.*

A lower structured region requires an exhaustivity of the set-inclusion relationship between a set in $\widetilde{m}(p)$ and X (or X^c), and an upper structured region requires an existence of such a relationship. The two lower structured regions give the lower bounds of the actual structured positive and negative regions, respectively, and the upper structured regions give the upper bounds [12].

3   Concept Analysis Using Quantitative Structured Three-Way Approximations

In this section, we generalize the qualitative structured regions into quantitative structured regions, in both complete and incomplete tables, based on two types of quantitative measures, namely, probabilities and subsethood measures. The generalization with respect to a complete table is straightforward based on existing research on quantitative unstructured approximations. In contrast, the generalization with respect to an incomplete table needs further investigation.

3.1   Probabilistic and Non-probabilistic Structured Approximations in a Complete Table

A probabilistic rough set model [27] replaces the qualitative set-inclusion with quantitative probabilities in defining unstructured approximations. By the same idea, if an object described by $p \in DL_c$ has a high probability of being a positive instance of X, then the concept $(p, m(p))$ is included in the structured positive region. This leads to the following probabilistic structured regions.

Definition 7. *For a set of objects $X \subseteq OB$, its probabilistic structured positive and negative regions are defined as:*

$$\mathrm{SPOS}^{pr}_{(\alpha,\cdot)}(X) = \{(p, m(p)) \in \mathrm{CDEF}(T) \mid Pr(X|m(p)) \geq \alpha\},$$
$$\mathrm{SNEG}^{pr}_{(\cdot,\gamma)}(X) = \{(p, m(p)) \in \mathrm{CDEF}(T) \mid Pr(X^c|m(p)) \geq \gamma\}, \qquad (13)$$

where $0 \leq \alpha, \gamma \leq 1$ are two thresholds, a dot represents a non-relevant threshold, and the probabilities are computed as:

$$Pr(X|m(p)) = \frac{|X \cap m(p)|}{|m(p)|}, \quad Pr(X^c|m(p)) = \frac{|X^c \cap m(p)|}{|m(p)|}. \qquad (14)$$

The qualitative structured regions can be viewed as a special case of the probabilistic structured regions with $\alpha = \gamma = 1$.

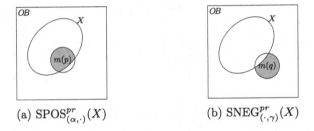

(a) $\mathrm{SPOS}^{pr}_{(\alpha,\cdot)}(X)$ (b) $\mathrm{SNEG}^{pr}_{(\cdot,\gamma)}(X)$

Fig. 2. Relationships between X and sets in $\mathrm{SPOS}^{pr}_{(\alpha,\cdot)}(X)$ and $\mathrm{SNEG}^{pr}_{(\cdot,\gamma)}(X)$

As shown in Fig. 2(a), $(p, m(p))$ is included in $\mathrm{SPOS}^{pr}_{(\alpha,\cdot)}(X)$ if the intersection between $m(p)$ and X occupies a large portion of $m(p)$. Accordingly, p is used to classify positive instances of X with an error rate:

$$\mathrm{IAE}((p, m(p))) = \frac{|X^c \cap m(p)|}{|m(p)|} = 1 - \frac{|X \cap m(p)|}{|m(p)|} \leq 1 - \alpha, \qquad (15)$$

which is called the rate of incorrect acceptance error (IAE) [5]. Similarly, a concept $(q, m(q)) \in \mathrm{SNEG}^{pr}_{(\cdot,\gamma)}(X)$ is associated with the following rate of incorrect rejection error (IRE):

$$\mathrm{IRE}((q, m(q))) = \frac{|X \cap m(q)|}{|m(q)|} = 1 - \frac{|X^c \cap m(q)|}{|m(q)|} \leq 1 - \gamma. \qquad (16)$$

At the expenses of IAE and IRE, we are able to approximate a larger part of X compared with the qualitative regions.

Most existing research on quantitative rough sets is based on the probabilistic (unstructured) approximations, such as decision-theoretic rough sets [29], game-theoretic rough sets [10], information-theoretic rough sets [6], naive Bayesian rough sets [30], confirmation-theoretic rough sets [9], and Bayesian rough sets [23]. Variable precision rough sets [33] can be viewed as both probabilistic and non-probabilistic in the sense that approximations are defined in terms of precisions which are estimated through probabilities.

In contrast, there is limited research [8,13,28] on non-probabilistic rough set models, which mainly uses subsethood measures and similarity measures instead of probabilities. Subsethood measure is a quantitative generalization of the qualitative set-inclusion. Given a universe OB, a normalized subsethood measure is defined as a mapping $sh : 2^{OB} \times 2^{OB} \to [0,1]$ where 2^{OB} is the power set of OB. For two sets $A, B \subseteq OB$, $sh(A \sqsubseteq B)$ represents the degree to which A is a subset of B. Yao and Deng [28] formulate quantitative unstructured approximations through subsethood measures, which unifies both probabilistic and non-probabilistic models. Following their formulation, we present the following quantitative structured regions based on subsethood measures.

Definition 8. *Suppose $sh : 2^{OB} \times 2^{OB} \to [0,1]$ is a normalized subsethood measure. For a given set of objects $X \subseteq OB$, its quantitative structured positive and negative regions can be defined as:*

$$\text{SPOS}^{sh}_{(\alpha,\cdot)}(X) = \{(p, m(p)) \in \text{CDEF}(T) \mid m(p) \neq \emptyset, sh(m(p) \sqsubseteq X) \geq \alpha\},$$

$$\text{SNEG}^{sh}_{(\cdot,\gamma)}(X) = \{(p, m(p)) \in \text{CDEF}(T) \mid m(p) \neq \emptyset, sh(m(p) \sqsubseteq X^c) \geq \gamma\}. \quad (17)$$

By using a subsethood measure $sh(A \sqsubseteq B) = \frac{|A \cap B|}{|A|}$, we have:

$$sh(m(p) \sqsubseteq X) = \frac{|m(p) \cap X|}{|m(p)|} = Pr(X|m(p)), \quad (18)$$

and similarly, $sh(m(p) \sqsubseteq X^c) = Pr(X^c|m(p))$. Thus, the two probabilistic regions in Definition 7 are special cases of the above two regions defined through subsethood measures. One may also consider many other meaningful cardinality-based subsethood measures [1] to formulate non-probabilistic approximations. For example, by using a measure $sh_{R_5^c}$ [11]:

$$sh_{R_5^c}(A \sqsubseteq B) = \begin{cases} \frac{|A^c|}{|(A \cap B)^c|}, & \neg(A = B = OB), \\ 1, & A = B = OB. \end{cases} \quad (19)$$

we can formulate a pair of non-probabilistic structured regions of $X \subseteq OB$ as:

$$\text{SPOS}^{sh_{R_5^c}}_{(\alpha,\cdot)}(X) = \{(p, m(p)) \in \text{CDEF}(T) \mid m(p) \neq \emptyset, \frac{|m(p)^c|}{|(m(p) \cap X)^c|} \geq \alpha\},$$

$$\text{SNEG}^{sh_{R_5^c}}_{(\cdot,\gamma)}(X) = \{(p, m(p)) \in \text{CDEF}(T) \mid m(p) \neq \emptyset, \frac{|m(p)^c|}{|(m(p) \cap X^c)^c|} \geq \gamma\}. \quad (20)$$

3.2 Probabilistic Structured Approximations in an Incomplete Table

One may also consider both probabilities and subsethood measures in defining quantitative regions in an incomplete table. In this subsection, we present two ways to define probabilistic structured regions. An intuitive way is to simply replace the set-inclusion in Definition 6 with probabilities, which leads to the following definition.

Definition 9. *Given a set of objects $X \subseteq OB$ in an incomplete table $\widetilde{T}$, one can construct the following probabilistic lower and upper structured regions:*

(1) $\mathrm{SPOS}^{pr}_{*(\alpha,\cdot)}(X) = \{(p, \widetilde{m}(p)) \mid \forall S \in \widetilde{m}(p), Pr(X|S) \geq \alpha\}$,

$\qquad \mathrm{SNEG}^{pr}_{*(\cdot,\gamma)}(X) = \{(p, \widetilde{m}(p)) \mid \forall S \in \widetilde{m}(p), Pr(X^c|S) \geq \gamma\}$;

(2) $\mathrm{SPOS}^{*pr}_{(\alpha,\cdot)}(X) = \{(p, \widetilde{m}(p)) \mid \exists S \in \widetilde{m}(p), Pr(X|S) \geq \alpha\}$,

$\qquad \mathrm{SNEG}^{*pr}_{(\cdot,\gamma)}(X) = \{(p, \widetilde{m}(p)) \mid \exists S \in \widetilde{m}(p), Pr(X^c|S) \geq \gamma\}$,   (21)

where the probabilities are computed as:

$$Pr(X|S) = \frac{|X \cap S|}{|S|}, \quad Pr(X^c|S) = \frac{|X^c \cap S|}{|S|}. \qquad (22)$$

The condition $(p, \widetilde{m}(p)) \in \mathrm{CDEFI}(\widetilde{T})$ is omitted in Definition 9 and the following definitions where this doesn't cause misunderstanding. Since $\widetilde{m}(p)$ is a collection of p's meaning sets in all completions, the above regions can be equivalently expressed through the family $\mathrm{COMP}(\widetilde{T})$.

Proposition 1. *Given a set of objects $X \subseteq OB$, one may construct the probabilistic lower and upper structured regions of X as follows:*

(1) $\mathrm{SPOS}^{pr}_{*(\alpha,\cdot)}(X) = \{(p, \widetilde{m}(p)) \mid \forall T \in \mathrm{COMP}(\widetilde{T}), Pr(X|m(p|T)) \geq \alpha\}$,

$\qquad \mathrm{SNEG}^{pr}_{*(\cdot,\gamma)}(X) = \{(p, \widetilde{m}(p)) \mid \forall T \in \mathrm{COMP}(\widetilde{T}), Pr(X^c|m(p|T)) \geq \gamma\}$;

(2) $\mathrm{SPOS}^{*pr}_{(\alpha,\cdot)}(X) = \{(p, \widetilde{m}(p)) \mid \exists T \in \mathrm{COMP}(\widetilde{T}), Pr(X|m(p|T)) \geq \alpha\}$,

$\qquad \mathrm{SNEG}^{*pr}_{(\cdot,\gamma)}(X) = \{(p, \widetilde{m}(p)) \mid \exists T \in \mathrm{COMP}(\widetilde{T}), Pr(X^c|m(p|T)) \geq \gamma\}$. (23)

The possible-world semantics also connects the above probabilistic regions in an incomplete table to those in a complete table (i.e., Definition 7). The condition $Pr(X|m(p|T)) \geq \alpha$ implies that the concept $(p, m(p|T))$ is included in $\mathrm{SPOS}^{pr}_{(\alpha,\cdot)}(X|T)$. Accordingly, we get the following theorem.

Theorem 1. *Given a set of objects $X \subseteq OB$, we have:*

(1) $(p, \widetilde{m}(p)) \in \mathrm{SPOS}^{pr}_{*(\alpha,\cdot)}(X) \Leftrightarrow \forall T \in \mathrm{COMP}(\widetilde{T}), (p, m(p|T)) \in \mathrm{SPOS}^{pr}_{(\alpha,\cdot)}(X|T)$,

$\qquad (p, \widetilde{m}(p)) \in \mathrm{SNEG}^{pr}_{*(\cdot,\gamma)}(X) \Leftrightarrow \forall T \in \mathrm{COMP}(\widetilde{T}), (p, m(p|T)) \in \mathrm{SNEG}^{pr}_{(\cdot,\gamma)}(X|T)$;

(2) $(p, \widetilde{m}(p)) \in \mathrm{SPOS}^{*pr}_{(\alpha,\cdot)}(X) \Leftrightarrow \exists T \in \mathrm{COMP}(\widetilde{T}), (p, m(p|T)) \in \mathrm{SPOS}^{pr}_{(\alpha,\cdot)}(X|T)$,

$\qquad (p, \widetilde{m}(p)) \in \mathrm{SNEG}^{*pr}_{(\cdot,\gamma)}(X) \Leftrightarrow \exists T \in \mathrm{COMP}(\widetilde{T}), (p, m(p|T)) \in \mathrm{SNEG}^{pr}_{(\cdot,\gamma)}(X|T)$.

(24)

Definition 9 and Proposition 1 provide two mathematically equivalent but semantically different formulations. Definition 9 is a straightforward generalization of the qualitative regions. Proposition 1 provides an equivalent version through the family $\mathrm{COMP}(\widetilde{T})$, which offers a clearer semantics. This clear semantics enables us to explore the relationships stated in Theorem 1.

Instead of considering $Pr(X|S)$ for every set $S \in \widetilde{m}(p)$, we may generalize $Pr(X|S)$ into a probability $Pr(X|\widetilde{m}(p))$ regarding a set X and an interval set $\widetilde{m}(p)$ which has not been well studied. Different interpretations of $Pr(X|\widetilde{m}(p))$ may lead to different formulas. In our work, we interpret $Pr(X|\widetilde{m}(p))$ as the probability of a set in $\widetilde{m}(p)$ being included in X. Accordingly, we define the following probabilistic structured regions.

Definition 10. *Given a set of objects $X \subseteq OB$, one may define the following pair of probabilistic structured positive and negative regions:*

$$\widetilde{\mathrm{SPOS}}^{pr}_{(\alpha,\cdot)}(X) = \{(p, \widetilde{m}(p)) \in \mathrm{CDEFI}(\widetilde{T}) \mid Pr(X|\widetilde{m}(p)) \geq \alpha\},$$

$$\widetilde{\mathrm{SNEG}}^{pr}_{(\cdot,\gamma)}(X) = \{(p, \widetilde{m}(p)) \in \mathrm{CDEFI}(\widetilde{T}) \mid Pr(X^c|\widetilde{m}(p)) \geq \gamma\}, \qquad (25)$$

where the probabilities are computed as:

$$Pr(X|\widetilde{m}(p)) = \frac{|\{S \in \widetilde{m}(p)|\emptyset \neq S \subseteq X\}|}{|\widetilde{m}(p)|},$$

$$Pr(X^c|\widetilde{m}(p)) = \frac{|\{S \in \widetilde{m}(p)|\emptyset \neq S \subseteq X^c\}|}{|\widetilde{m}(p)|}. \qquad (26)$$

Since each set in $\widetilde{m}(p)$ represents a possibility of p's actual meaning set, a high probability $Pr(X|\widetilde{m}(p))$ means that, in a large portion of all possible worlds $\mathrm{COMP}(\widetilde{T})$, the meaning set of p is included in X, or equivalently, p appears in the qualitative structured positive region of X. Thus, it is with high probability that p appears in the actual qualitative structured positive region of X.

Proposition 2. *Given a set of objects $X \subseteq OB$, we have:*

$$(p, \widetilde{m}(p)) \in \widetilde{\mathrm{SPOS}}^{pr}_{(\alpha,\cdot)}(X) \Longrightarrow Pr\big((p, m(p|T_0)) \in \mathrm{SPOS}(X|T_0)\big) \geq \alpha,$$

$$(p, \widetilde{m}(p)) \in \widetilde{\mathrm{SNEG}}^{pr}_{(\cdot,\gamma)}(X) \Longrightarrow Pr\big((p, m(p|T_0)) \in \mathrm{SNEG}(X|T_0)\big) \geq \gamma, \quad (27)$$

where $T_0 \in \mathrm{COMP}(\widetilde{T})$ is the actual table.

The two probabilities $Pr(X|\widetilde{m}(p))$ and $Pr(X^c|\widetilde{m}(p))$ can be efficiently computed through the two bounds of $\widetilde{m}(p)$. For $Pr(X|\widetilde{m}(p))$, if $m_*(p) \not\subseteq X$, then no set in $\widetilde{m}(p)$ is included X, that is, $Pr(X|\widetilde{m}(p)) = 0$. Otherwise, we have:

$$Pr(X|\widetilde{m}(p)) = \frac{|\{S \subseteq OB \mid m_*(p) \subseteq S \subseteq m^*(p) \wedge S \subseteq X\}|}{|\{S \subseteq OB \mid m_*(p) \subseteq S \subseteq m^*(p)\}|}$$

$$= \frac{|\{S \subseteq OB \mid m_*(p) \subseteq S \subseteq m^*(p) \cap X\}|}{|\{S \subseteq OB \mid m_*(p) \subseteq S \subseteq m^*(p)\}|} = \frac{2^{|m^*(p) \cap X - m_*(p)|}}{2^{|m^*(p) - m_*(p)|}}$$

$$= \frac{2^{|m^*(p) \cap X| - |m_*(p)|}}{2^{|m^*(p)| - |m_*(p)|}} = 2^{|m^*(p) \cap X| - |m^*(p)|}. \qquad (28)$$

The probability $Pr(X^c|\widetilde{m}(p))$ can be similarly computed and the following computational formulation of the structured regions can be accordingly obtained.

Theorem 2. *Given a set of objects $X \subseteq OB$, one may construct a pair of probabilistic structured regions of X as:*

$$\widetilde{\mathrm{SPOS}}^{pr}_{(\alpha,\cdot)}(X) = \{(p, \widetilde{m}(p)) \in \mathrm{CDEFI}(\widetilde{T}) \mid \mu \cdot 2^{|m^*(p) \cap X| - |m^*(p)|} \geq \alpha\},$$

$$\widetilde{\mathrm{SNEG}}^{pr}_{(\cdot,\gamma)}(X) = \{(p, \widetilde{m}(p)) \in \mathrm{CDEFI}(\widetilde{T}) \mid \mu_c \cdot 2^{|m^*(p) \cap X^c| - |m^*(p)|} \geq \gamma\}, \quad (29)$$

where μ and μ_c are two numbers defined as:

$$\mu = \begin{cases} 1, & \text{if } m_*(p) \subseteq X, \\ 0, & \text{otherwise.} \end{cases} \qquad \mu_c = \begin{cases} 1, & \text{if } m_*(p) \subseteq X^c, \\ 0, & \text{otherwise.} \end{cases} \quad (30)$$

While Definition 10 provides a conceptual understanding of the structured regions which requires an exhaustive scan of $\widetilde{m}(p)$ to compute the probabilities, Theorem 2 gives an equivalent computational formulation where the probabilities can be efficiently computed through the two bounds of $\widetilde{m}(p)$.

Example 1. We illustrate the above probabilistic structured regions in Definitions 9 and 10 with an incomplete table given by Table 1 [12]. The family $\mathrm{CDEFI}(\widetilde{T})$ is given by Table 2.

Table 1. An incomplete table $\widetilde{T}$ [12]

	a_1	a_2	a_3
o_1	{1}	{5}	{6}
o_2	{2}	{4}	{6}
o_3	{1}	{3}	{6}
o_4	{1}	{3,4}	{7}
o_5	{1,2}	{5}	{6}
o_6	{1}	{4}	{6}
o_7	{1,2}	{4}	{6}

Given a set of objects $X = \{o_2, o_5, o_6, o_7\}$ ($X^c = \{o_1, o_3, o_4\}$) and thresholds $\alpha = \gamma = 0.7$, the two pairs of lower and upper regions defined in Definition 9 are:

$$\begin{aligned}
\mathrm{SPOS}^{pr}_{*(\alpha,\cdot)}(X) = & \ \{IC_2, IC_4, IC_{12}, IC_{16}, IC_{20}, IC_{26}, IC_{32}\} \\
\mathrm{SNEG}^{pr}_{*(\cdot,\gamma)}(X) = & \ \{IC_3, IC_7, IC_8, IC_{15}, IC_{18}, IC_{24}\} \\
\mathrm{SPOS}^{*pr}_{(\alpha,\cdot)}(X) = & \ \{IC_2, IC_4, IC_9, IC_{12}, IC_{13}, IC_{16}, IC_{20}, IC_{26}, IC_{32}, IC_{34}\} \\
\mathrm{SNEG}^{*pr}_{(\cdot,\gamma)}(X) = & \ \{IC_1, IC_3, IC_7, IC_8, IC_{10}, IC_{15}, IC_{18}, IC_{19}, IC_{21}, IC_{24}, IC_{25}, \\
& \ \ IC_{27}, IC_{28}\} \, .
\end{aligned} \quad (31)$$

Table 2. The family CDEFI($\widetilde{T}$) for Table 1

Label	Intension	Extension	Label	Intension	Extension
IC_1	$a_1 = 1$	$[\{o_1, o_3, o_4, o_6\},$ $\{o_1, o_3, o_4, o_5, o_6, o_7\}]$	IC_{19}	$a_2 = 3 \wedge a_3 = 7$	$[\emptyset, \{o_4\}]$
IC_2	$a_1 = 2$	$[\{o_2\}, \{o_2, o_5, o_7\}]$	IC_{20}	$a_2 = 4 \wedge a_3 = 6$	$[\{o_2, o_6, o_7\}, \{o_2, o_6, o_7\}]$
IC_3	$a_2 = 3$	$[\{o_3\}, \{o_3, o_4\}]$	IC_{21}	$a_2 = 4 \wedge a_3 = 7$	$[\emptyset, \{o_4\}]$
IC_4	$a_2 = 4$	$[\{o_2, o_6, o_7\},$ $\{o_2, o_4, o_6, o_7\}]$	IC_{22}	$a_2 = 5 \wedge a_3 = 6$	$[\{o_1, o_5\}, \{o_1, o_5\}]$
IC_5	$a_2 = 5$	$[\{o_1, o_5\}, \{o_1, o_5\}]$	IC_{23}	$a_2 = 5 \wedge a_3 = 7$	$[\emptyset, \emptyset]$
IC_6	$a_3 = 6$	$[\{o_1, o_2, o_3, o_5, o_6, o_7\},$ $\{o_1, o_2, o_3, o_5, o_6, o_7\}]$	IC_{24}	$a_1 = 1 \wedge a_2 = 3 \wedge a_3 = 6$	$[\{o_3\}, \{o_3\}]$
IC_7	$a_3 = 7$	$[\{o_4\}, \{o_4\}]$	IC_{25}	$a_1 = 1 \wedge a_2 = 3 \wedge a_3 = 7$	$[\emptyset, \{o_4\}]$
IC_8	$a_1 = 1 \wedge a_2 = 3$	$[\{o_3\}, \{o_3, o_4\}]$	IC_{26}	$a_1 = 1 \wedge a_2 = 4 \wedge a_3 = 6$	$[\{o_6\}, \{o_6, o_7\}]$
IC_9	$a_1 = 1 \wedge a_2 = 4$	$[\{o_6\}, \{o_4, o_6, o_7\}]$	IC_{27}	$a_1 = 1 \wedge a_2 = 4 \wedge a_3 = 7$	$[\emptyset, \{o_4\}]$
IC_{10}	$a_1 = 1 \wedge a_2 = 5$	$[\{o_1\}, \{o_1, o_5\}]$	IC_{28}	$a_1 = 1 \wedge a_2 = 5 \wedge a_3 = 6$	$[\{o_1\}, \{o_1, o_5\}]$
IC_{11}	$a_1 = 2 \wedge a_2 = 3$	$[\emptyset, \emptyset]$	IC_{29}	$a_1 = 1 \wedge a_2 = 5 \wedge a_3 = 7$	$[\emptyset, \emptyset]$
IC_{12}	$a_1 = 2 \wedge a_2 = 4$	$[\{o_2\}, \{o_2, o_7\}]$	IC_{30}	$a_1 = 2 \wedge a_2 = 3 \wedge a_3 = 6$	$[\emptyset, \emptyset]$
IC_{13}	$a_1 = 2 \wedge a_2 = 5$	$[\emptyset, \{o_5\}]$	IC_{31}	$a_1 = 2 \wedge a_2 = 3 \wedge a_3 = 7$	$[\emptyset, \emptyset]$
IC_{14}	$a_1 = 1 \wedge a_3 = 6$	$[\{o_1, o_3, o_6\},$ $\{o_1, o_3, o_5, o_6, o_7\}]$	IC_{32}	$a_1 = 2 \wedge a_2 = 4 \wedge a_3 = 6$	$[\{o_2\}, \{o_2, o_7\}]$
IC_{15}	$a_1 = 1 \wedge a_3 = 7$	$[\{o_4\}, \{o_4\}]$	IC_{33}	$a_1 = 2 \wedge a_2 = 4 \wedge a_3 = 7$	$[\emptyset, \emptyset]$
IC_{16}	$a_1 = 2 \wedge a_3 = 6$	$[\{o_2\}, \{o_2, o_5, o_7\}]$	IC_{34}	$a_1 = 2 \wedge a_2 = 5 \wedge a_3 = 6$	$[\emptyset, \{o_5\}]$
IC_{17}	$a_1 = 2 \wedge a_3 = 7$	$[\emptyset, \emptyset]$	IC_{35}	$a_1 = 2 \wedge a_2 = 5 \wedge a_3 = 7$	$[\emptyset, \emptyset]$
IC_{18}	$a_2 = 3 \wedge a_3 = 6$	$[\{o_3\}, \{o_3\}]$			

With the same set X and the same thresholds, the pair of regions defined in Definition 10 are:

$$\widetilde{\text{SPOS}}_{(\alpha,\cdot)}^{pr}(X) = \{IC_2, IC_{12}, IC_{16}, IC_{20}, IC_{26}, IC_{32}\}$$

$$\widetilde{\text{SNEG}}_{(\cdot,\gamma)}^{pr}(X) = \{IC_3, IC_7, IC_8, IC_{15}, IC_{18}, IC_{24}\}. \tag{32}$$

3.3 Non-probabilistic Structured Approximations in an Incomplete Table Based on Subsethood Measures

We present a more general formulation of quantitative structured regions by using a subsethood measure instead of the probabilities in Definition 9.

Definition 11. *Suppose* $sh : 2^{OB} \times 2^{OB} \rightarrow [0,1]$ *is a normalized subsethood measure. Given a set* $X \subseteq OB$, *one can define the following structured regions:*

(1) $\text{SPOS}_{*(\alpha,\cdot)}^{sh}(X) = \{(p, \widetilde{m}(p)) \mid \forall S \in \widetilde{m}(p), S \neq \emptyset, sh(S \sqsubseteq X) \geq \alpha\}$,

   $\text{SNEG}_{*(\cdot,\gamma)}^{sh}(X) = \{(p, \widetilde{m}(p)) \mid \forall S \in \widetilde{m}(p), S \neq \emptyset, sh(S \sqsubseteq X^c) \geq \gamma\}$;

(2) $\text{SPOS}_{(\alpha,\cdot)}^{*sh}(X) = \{(p, \widetilde{m}(p)) \mid \exists S \in \widetilde{m}(p), S \neq \emptyset, sh(S \sqsubseteq X) \geq \alpha\}$,

   $\text{SNEG}_{(\cdot,\gamma)}^{*sh}(X) = \{(p, \widetilde{m}(p)) \mid \exists S \in \widetilde{m}(p), S \neq \emptyset, sh(S \sqsubseteq X^c) \geq \gamma\}. \tag{33}$

By using a subsethood measure $sh(A \sqsubseteq B) = \frac{|A \cap B|}{|A|}$, the probabilistic regions in Definition 9 become special cases of the above regions. Non-probabilistic structured regions can be constructed by applying non-probabilistic subsethood measures such as $sh_{R_5^c}$ given in Eq. (19). Since each set $S \in \widetilde{m}(p)$ is a meaning set of p in a completion, one can equivalently express the above regions through the family $\text{COMP}(\widetilde{T})$, for example:

$$\text{SPOS}^{sh}_{*(\alpha,\cdot)}(X) = \{(p, \widetilde{m}(p)) \mid \forall T \in \text{COMP}(\widetilde{T}), m(p|T) \neq \emptyset, sh(m(p|T) \sqsubseteq X) \geq \alpha\}. \quad (34)$$

Similar as in Theorem 1, one may also establish relationships between the above subsethood-based regions and those in the completions, for example:

$$(p, \widetilde{m}(p)) \in \text{SPOS}^{sh}_{*(\alpha,\cdot)}(X) \iff \forall T \in \text{COMP}(\widetilde{T}), (p, m(p|T)) \in \text{SPOS}^{sh}_{(\alpha,\cdot)}(X|T). \quad (35)$$

Alternatively, one may generalize subsethood measures for two sets into those for an interval set and a set to construct quantitative regions. Such subsethood measures evaluate the degree to which an interval set is included in a set.

Definition 12. *Suppose* $Sh : \mathcal{I}(OB) \times 2^{OB} \rightarrow [0,1]$ *is a normalized subsethood measure where* $\mathcal{I}(OB)$ *is the family of interval sets on OB. Given a set of objects* $X \subseteq OB$, *one may define the following pair of structured regions:*

$$\widetilde{\text{SPOS}}^{Sh}_{(\alpha,\cdot)}(X) = \{(p, \widetilde{m}(p)) \in \text{CDEFI}(\widetilde{T}) \mid Sh(\widetilde{m}(p) \sqsubseteq X) \geq \alpha\},$$

$$\widetilde{\text{SNEG}}^{Sh}_{(\cdot,\gamma)}(X) = \{(p, \widetilde{m}(p)) \in \text{CDEFI}(\widetilde{T}) \mid Sh(\widetilde{m}(p) \sqsubseteq X^c) \geq \gamma\}. \quad (36)$$

This definition depends on the specific definition of Sh which has not been well studied. One may define $Sh(\mathcal{A} \sqsubseteq B)$ through $sh(A \sqsubseteq B)$ where $A \in \mathcal{A}$, such as taking the average:

$$Sh(\mathcal{A} \sqsubseteq B) = \frac{\sum\limits_{A \in \mathcal{A}} sh(A \sqsubseteq B)}{|\mathcal{A}|}. \quad (37)$$

One may also define Sh through the qualitative set-inclusion such as using the proportion of subsets of B in $\mathcal{A}$:

$$Sh(\mathcal{A} \sqsubseteq B) = \frac{|\{A \in \mathcal{A} \mid \emptyset \neq A \subseteq B\}|}{|\mathcal{A}|}. \quad (38)$$

With the latter, the probabilistic regions in Definition 10 become special cases of the above regions in Definition 12. One may also construct non-probabilistic quantitative regions through Definition 12 by applying a subsethood measure Sh that cannot be explained through probabilities.

4   Conclusion and Future Work

To combine the advantages of both quantitative and structured approximations, this paper investigates quantitative formulations of structured approximations in both complete and incomplete tables. We consider both probabilistic formulations which are widely studied in the literature regarding unstructured approximations and non-probabilistic formulations which have not received due attention.

Our work brings up several interesting topics to work on. A first topic is the interpretation and determination of thresholds in various quantitative regions. While there are lots of existing related studies with respect to the probabilistic unstructured approximations in a complete table, the thresholds in subsethood-based quantitative regions and those regions in incomplete tables need further investigation. Solutions to this topic will help construct efficient computational formulations of approximations, which is a second topic for future work. A third topic is to investigate the relationships between this work and other concept analysis approaches such as lattice theory, formal concept analysis, and pattern structures. A fourth topic is the generalization of subsethood measures $sh(A \sqsubseteq B)$ regarding two sets into those regarding interval sets, including $\mathcal{S}h(\mathcal{A} \sqsubseteq \mathcal{B})$, $s\mathcal{H}(\mathcal{A} \sqsubseteq \mathcal{B})$, and $\mathcal{S}\mathcal{H}(\mathcal{A} \sqsubseteq \mathcal{B})$ where $\mathcal{A}, \mathcal{B}$ are interval sets and A, B are sets. The research on this topic will shed new light on defining meaningful quantitative approximations in an incomplete table.

Acknowledgement. The author thanks reviewers for their valuable comments and suggestions.

References

1. De Baets, B., De Meyer, H., Naessens, H.: On rational cardinality-based inclusion measures. Fuzzy Sets Syst. **128**, 169–183 (2002)
2. Bryniarski, E.: A calculus of rough sets of the first order. Bull. Polish Acad. Sci. Math. **37**, 71–78 (1989)
3. Buroker, J.: Port Royal Logic. Stanford Encyclopedia of Philosophy. http://plato.stanford.edu/entries/port-royal-logic/. Accessed January 2020
4. D'eer, L., Cornelis, C., Yao, Y.: A semantically sound approach to Pawlak rough sets and covering-based rough sets. Int. J. Approximate Reasoning **78**, 62–72 (2016)
5. Deng, X.: Three-way classification models. PhD thesis, University of Regina (2015)
6. Deng, X., Yao, Y.: A multifaceted analysis of probabilistic three-way decisions. Fundamenta Informaticae **132**, 291–313 (2014)
7. Ganter, B., Stumme, G., Wille, R. (eds.): Formal Concept Analysis: Foundations and Applications. LNCS (LNAI), vol. 3626. Springer, Heidelberg (2005). https://doi.org/10.1007/978-3-540-31881-1
8. Gomolińska, A.: Rough approximation based on weak q-RIFs. In: Peters, J.F., Skowron, A., Wolski, M., Chakraborty, M.K., Wu, W.-Z. (eds.) Transactions on Rough Sets X. LNCS, vol. 5656, pp. 117–135. Springer, Heidelberg (2009). https://doi.org/10.1007/978-3-642-03281-3_4
9. Greco, S., Matarazzo, B., Slowinski, R.: Parameterized rough set model using rough membership and Bayesian confirmation measures. Int. J. Approximate Reasoning **49**, 285–300 (2008)
10. Herbert, J.P., Yao, J.T.: Game-theoretic rough sets. Fundamenta Informaticae **108**, 267–286 (2011)
11. Hu, M., Deng, X., Yao, Y.: On the properties of subsethood measures. Inf. Sci. **494**, 208–232 (2019)

12. Hu, M., Yao, Y.: Structured approximations as a basis for three-way decisions with rough sets. Knowl.-Based Syst. **165**, 92–109 (2019)
13. Janicki, R., Lenarcic, A.: Optimal approximations with rough sets and similarities in measure spaces. Int. J. Approximate Reasoning **71**, 1–14 (2016)
14. Lipski, W.: On semantics issues connected with incomplete information table. ACM Trans. Database Syst. **4**, 262–296 (1979)
15. Luo, J., Fujita, H., Yao, Y., Qin, K.: On modeling similarity and three-way decision under incomplete information in rough set theory. Knowl.-Based Syst. **191**, 105251 (2020)
16. Luo, J., Hu, M., Qin, K.: Three-way decision with incomplete information based on similarity and satisfiability. Int. J. Approximate Reasoning **120**, 151–183 (2020)
17. Naidenova, X., Buzmakov, A., Parkhomenko, V., Schukin, A.: Notes on relation between symbolic classifiers. In: Watson, B.W., Kuznetsov, S.O. (eds.) CEUR Workshop Proceedings, CEUR-WS, vol. 1921, pp. 88–103 (2017)
18. Pawlak, Z.: Rough Sets: Theoretical Aspects of Reasoning about Data. Kluwer Academic, Dordrecht (1991)
19. Pawlak, Z.: Rough sets. Int. J. Comput. Inf. Sci. **11**, 341–356 (1982)
20. Qi, J., Qian, T., Wei, L.: The connections between three-way and classical concept lattices. Knowl.-Based Syst. **91**, 143–151 (2016)
21. Ren, R., Wei, L., Yao, Y.: An analysis of three types of partially-known formal concepts. Int. J. Mach. Learn. Cybernet. **9**(11), 1767–1783 (2017). https://doi.org/10.1007/s13042-017-0743-z
22. Sang, B., Yang, L., Chen, H., Xu, W., Guo, Y., Yuan, Z.: Generalized multigranulation double-quantitative decision-theoretic rough set of multi-source information system. Int. J. Approximate Reasoning **115**, 157–179 (2019)
23. Slezak, D., Ziarko, W.: Bayesian rough set model. In: Proceedings of the International Workshop on Foundation of Data Mining, pp. 131–135 (2002)
24. Yao, Y.: Three-way granular computing, rough sets, and formal concept analysis. Int. J. Approximate Reasoning **116**, 106–125 (2020)
25. Yao, Y.: The two sides of the theory of rough sets. Knowl.-Based Syst. **80**, 67–77 (2015)
26. Yao, Y.: Three-way decision: an interpretation of rules in rough set theory. In: Wen, P., Li, Y., Polkowski, L., Yao, Y., Tsumoto, S., Wang, G. (eds.) RSKT 2009. LNCS (LNAI), vol. 5589, pp. 642–649. Springer, Heidelberg (2009). https://doi.org/10.1007/978-3-642-02962-2_81
27. Yao, Y.: Probabilistic rough set approximations. Int. J. Approximate Reasoning **49**, 255–271 (2008)
28. Yao, Y., Deng, X.: Quantitative rough sets based on subsethood measures. Inf. Sci. **267**, 306–322 (2014)
29. Yao, Y., Wong, S.K.M., Lingras, P.J.: A decision-theoretic rough set model. In: Proceedings of the 5th International Symposium on Methodologies for Intelligent Systems, vol. 5, pp. 17–24 (1990)
30. Yao, Y., Zhou, B.: Naive Bayesian rough sets. In: Yu, J., Greco, S., Lingras, P., Wang, G., Skowron, A. (eds.) RSKT 2010. LNCS (LNAI), vol. 6401, pp. 719–726. Springer, Heidelberg (2010). https://doi.org/10.1007/978-3-642-16248-0_97
31. Zhang, T., Li, H., Liu, M., Rong, M.: Incremental concept-cognitive learning based on attribute topology. Int. J. Approximate Reasoning **118**, 173–189 (2020)
32. Zhi, H., Qi, J., Qian, T., Wei, L.: Three-way dual concept analysis. Int. J. Approximate Reasoning **114**, 151–165 (2020)
33. Ziarko, W.: Variable precision rough set model. J. Comput. Syst. Sci. **46**, 39–59 (1993)

On the Hierarchy of Equivalence Classes Provided by Local Congruences

Roberto G. Aragón, Jesús Medina, and Eloísa Ramírez-Poussa$^{(\boxtimes)}$

Department of Mathematics, University of Cádiz, Cádiz, Spain
{roberto.aragon,jesus.medina,eloisa.ramirez}@uca.es

Abstract. In this work, we consider a special kind of equivalence relations, which are called local congruences. Specifically, local congruences are equivalence relations defined on lattices, whose equivalence classes are convex sublattices of the original lattices. In the present paper, we introduce an initial study about how the set of equivalence classes provided by a local congruence can be ordered.

Keywords: Congruence · Local congruence · Concept lattice · Ordering relation

1 Introduction

The notion of local congruence arose in an attempt to weaken the conditions imposed in the definition of a congruence relation on a lattice, with the goal of taking advantage of different properties of these relations with respect to attribute reduction in formal concept analysis [11,17,21].

Formal concept analysis (FCA) is a theory of data analysis that organizes the information collected in a considered dataset, by means of the algebraic structure of a complete lattice. Moreover, this theory also offers diverse mechanisms for obtaining, handling and relating (by attribute implications) information from datasets. One of the most interesting mechanisms is attribute reduction. Its main goal is the selection of the main attributes of the given dataset and detecting the unnecessary ones to preserve the estructure of the complete lattice.

In [4,5], the authors remarked that when a reduction of the set of attributes in the dataset is carried out, an equivalence relation is induced. This induced equivalence relation satisfies that the generated equivalence classes have the structure of a join-semilattice. Inspired by this fact, the original idea given in [1] was to complement these studies by proposing the use of equivalence relations

Partially supported by the 2014-2020 ERDF Operational Programme in collaboration with the State Research Agency (AEI) in projects TIN2016-76653-P and PID2019-108991GB-I00, and with the Department of Economy, Knowledge, Business and University of the Regional Government of Andalusia in project FEDER-UCA18-108612, and by the European Cooperation in Science & Technology (COST) Action CA17124

R. Bello et al. (Eds.): IJCRS 2020, LNAI 12179, pp. 298–307, 2020.
https://doi.org/10.1007/978-3-030-52705-1_22

containing the induced equivalence relation and satisfying that the generated equivalence classes be convex sublattices of the original lattice.

For example, congruence relations [6, 10, 12, 13] hold the previously exposed requirements. In addition, congruence relations have already been applied to the framework of FCA [11, 15, 18–20]. Nevertheless, in [2] was proved that congruence relations are not suitable to complement the reductions in FCA, since the constraints imposed by this kind of equivalence relation entail a great loss of information. This reason is the main justification to weaken the notion of congruence relation, appearing the definition of local congruence. These new equivalence relations are also defined on lattices and only require that the equivalence classes be convex sublattices of the original lattice. The use of local congruences considerably reduces the problem of the loss of information.

However, the appearance of local congruences uncovers new open problems that require answers. One of these open problems is to provide an ordering relation on the set of equivalence classes, that is, on the quotient set associated with the local congruence. This is the main issue addressed in this paper. First of all, we will show that the usually considered ordering relations on the set of equivalence class of a congruence relation, cannot be used for local congruences. Then, we will define a new binary relation on lattices which turns out to be a pre-order when it is used to establish a hierarchy on the equivalence classes provided by a local congruence. Finally, we will also state under what conditions this pre-order is a partial order.

The paper is organized as follows: Sect. 2 recalls some preliminary notions used throughout of the paper. Section 3 presents the study of the hierarchy among the equivalence classes provided by local congruences. The paper finishes with some conclusions and prospects for future works, which are included in Sect. 4.

2   Preliminaries

In this section, we recall basic notions used in this paper. The first notion is related to a special kind of equivalence relation on lattices, which are called congruence relations.

Definition 1 ([10]). *Given a lattice $(L, \preceq)$, we say that an equivalence relation θ on L is a congruence if, for all $a_0, a_1, b_0, b_1 \in L$,*

$$(a_0, b_0) \in \theta, \ (a_1, b_1) \in \theta \ \text{imply that} \ (a_0 \vee a_1, b_0 \vee b_1) \in \theta, \ (a_0 \wedge a_1, b_0 \wedge b_1) \in \theta.$$

where $\wedge$ and $\vee$ are the infimum and the supremum operators.

Now, we recall the notion of quotient lattice from a congruence, based on the operations of the original lattice.

Definition 2 ([10]). *Given an equivalence relation θ on a lattice $(L, \preceq)$, the operators infimum and supremum, $\vee_\theta$ and $\wedge_\theta$, can be defined on the set of equivalence classes $L/\theta = \{[a]_\theta \mid a \in L\}$ for all $a, b \in L$, as follows:*

$$[a]_\theta \vee_\theta [b]_\theta = [a \vee b]_\theta \ \text{and} \ [a]_\theta \wedge_\theta [b]_\theta = [a \wedge b]_\theta.$$

$\vee_\theta$ and $\wedge_\theta$ are well defined on L/θ if and only if θ is a congruence.

When θ is a congruence on L, the tuple $(L/\theta, \vee_\theta, \wedge_\theta)$ is called quotient lattice of L modulo θ.

Now, let us suppose that $\{a, b, c, d\}$ is a subset of a given lattice $(L, \preceq)$. Then, the pairs a, b and c, d are said to be *opposite sides* of the quadrilateral $(a, b; c, d)$ if $a < b$, $c < d$ and either:

$$(a \vee d = b \text{ and } a \wedge d = c) \quad \text{or} \quad (b \vee c = d \text{ and } b \wedge c = a).$$

In addition, we say that the equivalence classes provided by an equivalence relation are *quadrilateral-closed* if whenever given two opposite sides of a quadrilateral $(a, b; c, d)$, such that $a, b \in [x]_\theta$, with $x \in L$ then there exists $y \in L$ such that $c, d \in [y]_\theta$. This notion leads us to the following result which is a characterization of the congruence notion in terms of their equivalence classes and plays a key role in the definition of local congruences as we will show later (more detailed information on the characterization and the notions involved in this result can be found in [10]).

Theorem 1 ([10]). *Let $(L, \preceq)$ be a lattice and θ an equivalence relation on L. Then, θ is a congruence if and only if*

(i) each equivalence class of θ is a sublattice of L,
(ii) each equivalence class of θ is convex,
(iii) the equivalence classes of θ are quadrilateral-closed.

With the goal of obtaining a less-constraining equivalence relations than congruences, but preserving some interesting properties satisfied by this kind of equivalence relations, the notion of local congruence arose [2] in the framework of attribute reduction in FCA [7–9,11,16], focused on providing an optimal reduction on FCA from the application of Rough Set techniques [4,5,14]. This notion is recalled in the following definition and mainly consist in the elimination of a restriction (last item) in the previous theorem.

Definition 3. *Given a lattice $(L, \preceq)$, we say that an equivalence relation δ on L is a* local congruence *if the following properties hold:*

(i) each equivalence class of δ is a sublattice of L,
(ii) each equivalence class of δ is convex.

Next section studies how we can define an ordering relation between the equivalence classes obtained from a local congruence.

3   Ordering Classes of Local Congruences

In this section, we are interested in studying ordering relations for local congruences. This fact is fundamental for establishing a proper hierarchy among the classes of concepts obtained after the reduction in FCA [3–5].

The set of equivalence classes of a congruence on a lattice L can be ordered by a partial order $\preceq_\theta$ which is defined, for all $a, b \in L$, by means of the operators $\vee_\theta$ and $\wedge_\theta$ presented in Definition 2, as follows:

$$[a]_\theta \preceq_\theta [b]_\theta \quad \text{if} \quad [a]_\theta = [a]_\theta \wedge_\theta [b]_\theta \quad \text{or} \quad [b]_\theta = [a]_\theta \vee_\theta [b]_\theta \tag{1}$$

This ordering relation cannot be used for local congruences since local congruences are not compatible with either supremum or infimum, that is, the operators $\vee_\theta$ and $\wedge_\theta$ could not be well defined when the considered relation is a local congruence due to they do not satisfy the quadrilateral-closed property unlike congruences. In the next example, we illustrate this fact.

Example 1. Let us consider the lattice $(L, \preceq)$ shown in the left side of Fig. 1, and the local congruence δ, highlighted by means of a Venn diagram, given in the right side of Fig. 1.

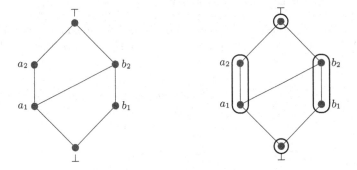

Fig. 1. Lattice (left) and local congruence (right) of Example 1.

It is easy to see that the considered local congruence δ provides four different equivalence classes which are listed below:

$$[\top]_\delta = \{\top\}$$
$$[a_1]_\delta = [a_2]_\delta = \{a_1, a_2\}$$
$$[b_1]_\delta = [b_2]_\delta = \{b_1, b_2\}$$
$$[\bot]_\delta = \{\bot\}$$

We can observe that $a_1, \bot$ and b_1, b_2 are opposite sides, but a_1 and $\bot$ are not in the same equivalence class, which means that the equivalence classes of δ are not quadrilateral-closed. As a consequence, the infimum and supremum operators described in Expression (1) are not well defined. For example, we have that

$$[a_2]_\delta \wedge_\delta [b_1]_\delta = [a_2 \wedge b_1]_\delta = [\bot]_\delta$$
$$[a_2]_\delta \wedge_\delta [b_1]_\delta = [a_2 \wedge b_2]_\delta = [a_1]_\delta$$

and clearly $[\bot]_\delta \neq [a_1]_\delta$. Therefore, the ordering $\preceq_\delta$ cannot be defined on local congruences.   □

A property of the ordering relation, shown in Expression (1), was shown in [10], which provides another possibility of defining an ordering on the set of local congruences.

Proposition 1 ([10]). *Let θ be a congruence on a lattice $(L, \preceq)$ and let $[a]_\theta$ and $[b]_\theta$ be equivalence classes of L/θ. Then, the binary relation $\leq$ defined on L/θ as: $[a]_\theta \leq [b]_\theta$, if there exist $a' \in [a]_\theta$ and $b' \in [b]_\theta$, for all $a' \preceq b'$, is an ordering relation.*

Clearly, the relation $\leq$ is the associated ordering relation with the algebraic lattice $(L/\theta, \vee_\theta, \wedge_\theta)$. Consequently, we cannot use either this alternative definition in the equivalence classes of a local congruence. In the following example, we show a case where the application of this ordering relation for a local congruence does not satisfies the transitivity property.

Example 2. We will consider the lattice $(L, \preceq)$ and the local congruence δ both given in Fig. 2. As we can observe, the local congruence provides five different equivalence classes:

$$[\top]_\delta = \{\top\}$$
$$[a_1]_\delta = [a_2]_\delta = \{a_1, a_2\}$$
$$[b_1]_\delta = [b_2]_\delta = \{b_1, b_2\}$$
$$[c_1]_\delta = [c_2]_\delta = \{c_1, c_2\}$$
$$[\bot]_\delta = \{\bot\}$$

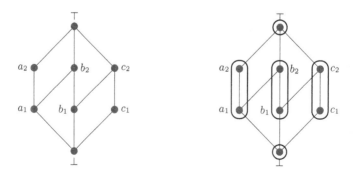

Fig. 2. Lattice (left) and local congruence (right) of Example 2.

If we try to order the equivalence classes of δ using the ordering relation described in Proposition 1, we obtain that $[a_1]_\delta \leq [b_1]_\delta$, since $a_1 \preceq b_2$, and

$[b_1]_\delta \le [c_1]_\delta$ because $b_1 \preceq c_2$. Nevertheless, we can see that $[a_1]_\delta$ is not lesser than $[c_1]_\delta$ because neither a_1 nor a_2 are lesser than c_1 or c_2 in L. Therefore, the ordering relation defined in Proposition 1 is not transitive for local congruences in general and thus, it is not a partial order for local congruences.   □

As we have seen in the previous example, the ordering relation defined in Proposition 1 cannot be used either to order the equivalence classes obtained from local congruences. However, the underlying idea of the ordering relation of Proposition 1 can be considered to define a more suitable ordering relation for being applied on local congruences. In order to achieve this goal, we formalize some notions presented in [6], which are related to the ordering of elements in the quotient set provided from equivalence relations defined on posets. The following notion is related to the way in which two elements of the original lattice can be connected via the local congruence.

Definition 4. *Let* $(L, \preceq)$ *be a lattice and a local congruence* δ *on* L.

(i) *A sequence of elements of* L, $(p_0, p_1, \ldots, p_n)$ *with* $n \ge 1$, *is called a* δ-*sequence, denoted as* $(p_0, p_n)_\delta$, *if for each* $i \in \{1, \ldots, n\}$ *either* $(p_{i-1}, p_i) \in \delta$ *or* $p_{i-1} \preceq p_i$ *holds.*

(ii) *If a* δ-*sequence* $(p_0, p_n)_\delta$ *satisfies that* $p_0 = p_n$, *then it is called a* δ-*cycle. In addition, if the* δ-*cycle satisfies that* $[p_0]_\delta = [p_1]_\delta = \cdots = [p_n]_\delta$, *then we say that the* δ-*cycle is* closed.

With the notions of Definition 4, we present a new binary relation on local congruences in the following definition.

Definition 5. *Given a lattice* $(L, \preceq)$ *and a local congruence* δ *on* L, *we define a binary relation* $\preceq_\delta$ *on* L/δ *as follows:*

$$[x]_\delta \preceq_\delta [y]_\delta \quad \text{if there exists a } \delta\text{-sequence } (x', y')_\delta$$

for some $x' \in [x]_\delta$ *and* $y' \in [y]_\delta$.

Now, we go back to Example 2 in order to illustrate this relation.

Example 3. Returning to Example 2, we want to establish a hierarchy among the equivalence classes depicted in Fig. 2 by means of the relation given in Definition 5. By considering this definition, it is clear that $[a_1]_\delta \preceq_\delta [b_1]_\delta$ and $[b_1]_\delta \preceq_\delta [c_1]_\delta$. In addition, we can observe that, in this case, we also have that $[a_1]_\delta \preceq_\delta [c_1]_\delta$ since there exists a δ-sequence that connects one element of the class $[a_1]_\delta$ with another element of the class $[c_1]_\delta$, this δ-sequence is shown below:

$$(a_1, c_2)_\delta = (a_1, b_2, b_1, c_2), \quad \text{since} \quad a_1 \preceq b_2, \quad (b_2, b_1) \in \delta \quad \text{and} \quad b_1 \preceq c_2$$

Therefore, the relationship among the elements in the quotient set L/δ given by $\preceq_\delta$ are shown in Fig. 3.   □

Fig. 3. Hasse diagram of the relation among the elements in L/δ of Example 3.

Observe that the binary relation $\preceq_\delta$ given in Definition 5 is a pre-order. Evidently, by definition, $\preceq_\delta$ is reflexive and transitive. Now, we present an example in which the previously defined relation $\preceq_\delta$ does not hold the antisymmetry property and, consequently, it cannot be used to establish an ordering among the equivalent classes obtained from a local congruence.

Example 4. Let us consider the lattice $(L, \preceq)$ and the local congruence δ given in Fig. 4. The equivalence classes provided by δ are:

$$[\top]_\delta = \{\top\}$$
$$[a_1]_\delta = [a_2]_\delta = [a_3]_\delta = [a_4]_\delta = \{a_1, a_2, a_3, a_4\}$$
$$[b_1]_\delta = [b_2]_\delta = \{b_1, b_2\}$$
$$[c_1]_\delta = [c_2]_\delta = \{c_1, c_2\}$$
$$[\bot]_\delta = \{\bot\}$$

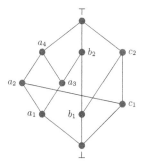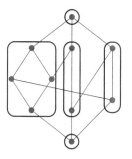

Fig. 4. Lattice and local congruence of Example 4.

If we try to establish a hierarchy among the equivalence classes using the binary relation given in Definition 5, we obtain that $[c_1]_\delta \preceq_\delta [a_1]_\delta$ since there exists another δ-sequence that connects c_1 with a_2:

$$(c_1, a_2)_\delta = (c_1, a_2) \quad \text{since} \quad c_1 \preceq a_2$$

In addition, we also have that $[a_1]_\delta \preceq_\delta [c_1]_\delta$, because there exists a δ-sequence that connects the elements a_3 and c_2:

$$(a_3, c_2)_\delta = (a_3, b_2, b_1, c_2) \quad \text{since} \quad a_3 \preceq b_2, (b_2, b_1) \in \delta \quad \text{and} \quad b_1 \preceq c_2$$

Therefore, we have that $[a_1]_\delta \preceq_\delta [c_1]_\delta$ and $[c_1]_\delta \preceq_\delta [a_1]_\delta$, but these classes are not equal. Thus, the antisymmetry property does not hold and, as a consequence, the obtained equivalent classes from the considered local congruence cannot be ordered by means of the considered binary relation. □

As we have seen in the previous example, the preorder $\preceq_\delta$ is not a partial order since the antisymmetry property is not satisfied for any local congruence, in general. Therefore, it is important to study sufficient conditions to ensure that $(L/\delta, \preceq_\delta)$ is a poset. The following result states a condition under which the binary relation of Definition 5 is a partial order on local congruences.

Theorem 2. *Given a lattice $(L, \preceq)$ and a local congruence δ on L, the preorder $\preceq_\delta$ given in Definition 5 is a partial order if and only if either no δ-cycle exists or every δ-cycle of elements in L is closed.*

Since no δ-cycle of elements in L exists with respect to the local congruences in Examples 1 and 2, we can ensure that the obtained quotient sets, together with the binary relation $\preceq_\delta$, are posets in both examples. The following example shows a local congruence on the lattice L given in Example 4, such as $(L/\delta, \preceq_\delta)$ is a poset.

Example 5. On the lattice $(L, \preceq)$ of Example 4, the quotient set L/δ_1 given by local congruence δ_1 depicted in the right side of Fig. 5, together with the binary relation defined in Definition 5, is a poset.

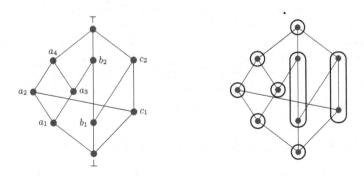

Fig. 5. Local congruence δ_1 (right) of Example 5 on the lattice of Example 4 (left).

We can ensure that because no δ_1-cycle exists. The right side of Fig. 6 shows an equivalence relation that contains the δ-cycle of Example 4 in one equivalence class. Therefore, the least local congruence, called δ_2, is the one that groups all the elements in a single class and, as a consequence, the δ-cycle is closed. Therefore, by Theorem 2, the pair $(L/\delta_2, \preceq_{\delta_2})$ it is also a poset. $\square$

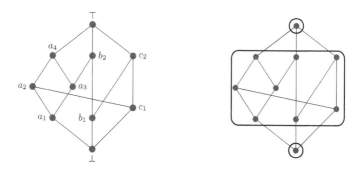

Fig. 6. Equivalence relation (right) of Example 5 on the lattice of Example 4 (left).

4   Conclusions and Future Work

In this paper, we have introduced an initial study about different ways of establish a hierarchy among the equivalence classes provided by local congruences. We have analyzed the results of applying the usually considered ordering relations on the quotient set of congruences, obtaining that these ordering relations are not suitable to be used on local congruences. Based on the underlying philosophy of one characterization of the ordering relation used for congruences, we have defined a new binary relation on the equivalence classes obtained from a local congruences. We have also proven that this binary relation is a preorder. Moreover, we have stated a sufficient condition on the lattice in which the local congruence is defined, in order to guarantee that this preorder is actually a partial order. All the ideas presented throughout this study have been illustrated by means of diverse examples.

As future work, we are interested in continuing this study and defining another binary relation, which will be a partial order on the equivalence classes of any local congruence. Furthermore, we will apply this type of equivalence relations in practical problems, such as in tasks related to the reduction of concept lattices in the framework of formal concept analysis.

References

1. Aragón, R.G., Medina, J., Ramírez-Poussa, E.: Weaken the congruence notion to reduce concept lattices. Studies in Computational Intelligence, pp. 1–7 (2019, in press)

2. Aragón, R.G., Medina, J., Ramírez-Poussa, E.: Weaken the congruence notion to reduce concept lattices. In: European Symposium on Computational Intelligence and Mathematics (ESCIM 2019), pp. 45–46 (2019)
3. Benítez-Caballero, M.J., Medina-Moreno, J., Ramírez-Poussa, E.: Bireducts in formal concept analysis. In: Kóczy, L.T., Medina-Moreno, J., Ramírez-Poussa, E., Šostak, A. (eds.) Computational Intelligence and Mathematics for Tackling Complex Problems. SCI, vol. 819, pp. 191–198. Springer, Cham (2020). https://doi.org/10.1007/978-3-030-16024-1_24
4. Benítez-Caballero, M.J., Medina, J., Ramírez-Poussa, E., Ślęzak, D.: A computational procedure for variable selection preserving different initial conditions. Int. J. Comput. Math. **97**, 387–404 (2020)
5. Benítez-Caballero, M.J., Medina, J., Ramírez-Poussa, E., Ślęzak, D.: Rough-set-driven approach for attribute reduction in fuzzy formal concept analysis. Fuzzy Sets Syst. (2020, in press)
6. Blyth, T.S.: Lattices and Ordered Algebraic Structures. Springer, London (2005). https://doi.org/10.1007/b139095
7. Cornejo, M.E., Medina, J., Ramírez-Poussa, E.: Attribute reduction in multi-adjoint concept lattices. Inf. Sci. **294**, 41–56 (2015)
8. Cornejo, M.E., Medina, J., Ramírez-Poussa, E.: On the use of irreducible elements for reducing multi-adjoint concept lattices. Knowl.-Based Syst. **89**, 192–202 (2015)
9. Cornejo, M.E., Medina, J., Ramírez-Poussa, E.: Characterizing reducts in multi-adjoint concept lattices. Inf. Sci. **422**, 364–376 (2018)
10. Davey, B., Priestley, H.: Introduction to Lattices and Order, 2nd edn. Cambridge University Press, Cambridge (2002)
11. Ganter, B., Wille, R.: Formal Concept Analysis: Mathematical Foundation. Springer, Heidelberg (1999). https://doi.org/10.1007/978-3-642-59830-2
12. Grätzer, G.: General Lattice Theory, 2nd edn. Birkhäuser, Basel (2007)
13. Grätzer, G.: Universal Algebra, 2nd edn. Springer, New York (2008)
14. Jiang, Z., Liu, K., Yang, X., Yu, H., Fujita, H., Qian, Y.: Accelerator for supervised neighborhood based attribute reduction. Int. J. Approx. Reason. **119**, 122–150 (2020)
15. Li, J.-Y., Wang, X., Wu, W.-Z., Xu, Y.-H.: Attribute reduction in inconsistent formal decision contexts based on congruence relations. Int. J. Mach. Learn. Cybernet. **8**(1), 81–94 (2016). https://doi.org/10.1007/s13042-016-0586-z
16. Medina, J.: Relating attribute reduction in formal, object-oriented and property-oriented concept lattices. Comput. Math. Appl. **64**(6), 1992–2002 (2012)
17. Medina, J., Ojeda-Aciego, M., Ruiz-Calviño, J.: Formal concept analysis via multi-adjoint concept lattices. Fuzzy Sets Syst. **160**(2), 130–144 (2009)
18. Viaud, J., Bertet, K., Demko, C., Missaoui, R.: The reverse doubling construction. In: 2015 7th International Joint Conference on Knowledge Discovery, Knowledge Engineering and Knowledge Management (IC3K), vol. 1, pp. 350–357, November 2015
19. Viaud, J.-F., Bertet, K., Demko, C., Missaoui, R.: Subdirect decomposition of contexts into subdirectly irreducible factors. In: International Conference on Formal Concept Analysis ICFCA2015, Nerja, Spain, June 2015
20. Viaud, J.-F., Bertet, K., Missaoui, R., Demko, C.: Using congruence relations to extract knowledge from concept lattices. Discrete Appl. Math. **249**, 135–150 (2018)
21. Wille, R.: Restructuring lattice theory: an approach based on hierarchies of concepts. In: Rival, I. (ed.) Ordered Sets, pp. 445–470. Reidel, Dordrecht (1982). https://doi.org/10.1007/978-94-009-7798-3_15

Object Oriented Protoconcepts and Logics for Double and Pure Double Boolean Algebras

Prosenjit Howlader and Mohua Banerjee[(⊠)]

Department of Mathematics and Statistics, Indian Institute of Technology,
Kanpur 208016, India
{prosen,mohua}@iitk.ac.in

Abstract. The notion of a protoconcept in the framework of Yao's object oriented concepts is proposed. Approximations by object oriented concepts are defined and these 'object oriented protoconcepts' are characterized using them. It is shown that the object oriented protoconcepts form a double Boolean algebra, and any double Boolean algebra is quasi-embedded in an algebra of protoconcepts. A logic **DBL** for the class of double Boolean algebras is proposed along with an extension **PDBL** for the class of pure double Boolean algebras. Utilizing the representation result for (pure) double Boolean algebras, it is established that **DBL** (**PDBL**) is sound and complete with respect to a semantics based on object oriented protoconcepts (semiconcepts).

1 Introduction

Formal concept analysis (FCA) [5] has seen wide applications since its inception. In order to study a conceptual knowledge system [10], Wille introduced the negation of a concept, leading to the notions of semiconcepts and protoconcepts. Algebraic studies of these notions led to the definition of double Boolean algebras and pure double Boolean algebras [11]. These structures have been investigated by many authors [1,7–9].

Over the years, there has been a lot of work on different kinds of intersections of the theories of rough sets and FCA. Two seminal works in this regard are by Düntsch and Gediga, who introduced property oriented concepts [4], and by Yao, who proposed object oriented concepts [14]. Algebraic studies of these concepts have been conducted by many [6,12,15]. Our interest lies in introducing negation in the study of these concepts, in the line of the study by Wille as mentioned above. In [6], object oriented semiconcepts were introduced. In this work, we define object oriented protoconcepts (Sect. 4), and characterize them through a notion of approximation by object oriented concepts. The algebraic structure formed by object oriented protoconcepts is studied giving a representation result

This work is supported by the *Council of Scientific and Industrial Research* (CSIR) India - Research Grant No. 09/092(0950)/2016-EMR-I.

R. Bello et al. (Eds.): IJCRS 2020, LNAI 12179, pp. 308–323, 2020.
https://doi.org/10.1007/978-3-030-52705-1_23

for double Boolean algebras. In Sect. 5, we propose a sequent calculus **DBL**, which is shown to be sound and complete with respect to the class of double Boolean algebras. **DBL** is extended to **PDBL** to give a logic for the class of pure double Boolean algebras. Utilizing the representation results for the algebras, it is next shown (Sect. 6) that these logics are also sound and complete for semantics based on object oriented protoconcepts and semiconcepts respectively.

Preliminaries required for the work are presented in the next section. Section 3 revisits the notions of concepts and semiconcepts related to rough set theory, illustrated through a running example. Section 7 concludes the paper.

2   FCA and Double Boolean Algebras

Definition 1 [5]. A *context* is a triple $\mathbb{K} := (G, M, R)$, where G is the set of *objects,* M is the set of *properties,* and $R \subseteq G \times M$.

For a context $\mathbb{K} := (G, M, R)$, its *complement* is the context $\mathbb{K}^c := (G, M, -R)$ where $-R := (G \times M) \setminus R$.

For any $A \subseteq G, B \subseteq M$, the following sets are defined:
$A' := \{m \in M : \forall g \in G(g \in A \Longrightarrow gRm)\}$, and
$B' := \{g \in G : \forall m \in M(m \in B \Longrightarrow gRm)\}$.
Then (A, B) is a *concept* of $\mathbb{K}$, provided $A' = B$ and $B' = A$.

An order relation $\leq$ is defined on the set of all concepts as follows. For concepts (A_1, B_1) and (A_2, B_2), $(A_1, B_1) \leq (A_2, B_2)$ if and only if $A_1 \subseteq A_2$ (equivalently $B_2 \subseteq B_1$).

Notation 1. We denote the class of all contexts by $\mathcal{K}$.

For a relation $R \subseteq G \times M$, R^{-1} is the converse of R, that is $R^{-1} \subseteq M \times G$ and $yR^{-1}x$ if and only if xRy. For any $x \in G, y \in M$, $R(x) := \{y \in M : xRy\}$, and $R^{-1}(y) := \{x \in G : xRy\}$.

The complement of a subset X of G (or M) will be denoted simply by X^c.

The set of all concepts of the context $\mathbb{K}$ is denoted by $\mathcal{B}(\mathbb{K})$. For a concept (A, B), $A := ext((A, B))$ is its *extent*, while $B := int((A, B))$ is its *intent*.

Attempting to introduce the negation of a formal concept, it was noticed that there is a problem of closure if set-complement is used to define it. So the notion of concept was generalized to that of a semiconcept, and also to a protoconcept [11].

Definition 2. Let $\mathbb{K} := (G, M, R)$ be a context and $A \subseteq G, B \subseteq M$. The pair (A, B) is a *semiconcept* of $\mathbb{K}$ if and only if $A' = B$ or $B' = A$, while it is a *protoconcept* of $\mathbb{K}$ if and only if $A'' = B'$ (equivalently $A' = B''$).

Notation 2. The set of all semiconcepts is denoted by $\mathfrak{H}(\mathbb{K})$ and the set of all protoconcepts is denoted by $\mathfrak{P}(\mathbb{K})$.

Observation 1. For a context $\mathbb{K} := (G, M, R)$, $\mathfrak{H}(\mathbb{K}) \subseteq \mathfrak{P}(\mathbb{K})$.

The following operations are defined in $\mathfrak{P}(\mathbb{K})$. Let $(A_1, B_1), (A_2, B_2), (A, B)$ be any protoconcepts.

$$(A_1, B_1) \sqcap (A_2, B_2) := (A_1 \cap A_2, (A_1 \cap A_2)'),$$

$$(A_1, B_1) \sqcup (A_2, B_2) := ((B_1 \cap B_2)', B_1 \cap B_2),$$

$$\neg(A, B) := (A', A^{c'}), \quad \lrcorner(A, B) := (B^{c'}, B^c), \quad \top := (G, \phi), \text{ and } \bot := (\phi, M).$$

Notation 3. $\mathfrak{P}(\mathbb{K})$ forms a abstract algebra of type $(2, 2, 1, 1, 0, 0)$ with respect to the above operations and is called the *protoconcept algebra* of a context $\mathbb{K}$. It is denoted by $\underline{\mathfrak{P}}(\mathbb{K}) := (\mathfrak{P}(\mathbb{K}), \sqcup, \sqcap, \neg, \lrcorner, \top, \bot)$.

$\mathfrak{H}(\mathbb{K})$ is closed under the above operations, forming a subalgebra of the protoconcept algebra. This algebra is called the *semiconcept algebra* of a context $\mathbb{K}$, and is denoted by $\underline{\mathfrak{H}}(\mathbb{K}) := (\mathfrak{H}(\mathbb{K}), \sqcup, \sqcap, \neg, \lrcorner, \top, \bot)$.

On abstraction of properties of the protoconcept algebra of a context, Wille defined the double Boolean algebra [11]. The semiconcept algebra leads to the notion of a pure double Boolean algebra.

Definition 3 [11]. A *double Boolean algebra* (dBa) is an abstract algebra $\boldsymbol{D} := (D, \sqcup, \sqcap, \neg, \lrcorner, \top, \bot)$ which satisfies the following properties, for any $x, y, z \in D$.

(1a) $(x \sqcap x) \sqcap y = x \sqcap y$ (1b) $(x \sqcup x) \sqcup y = x \sqcup y$

(2a) $x \sqcap y = y \sqcap x$ (2b) $x \sqcup y = y \sqcup x$

(3a) $x \sqcap (y \sqcap z) = (x \sqcap y) \sqcap z$ (3b) $x \sqcup (y \sqcup z) = (x \sqcup y) \sqcup z$

(4a) $\neg(x \sqcap x) = \neg x$ (4b) $\lrcorner(x \sqcup x) = \lrcorner x$

(5a) $x \sqcap (x \sqcup y) = x \sqcap x$ (5b) $x \sqcup (x \sqcap y) = x \sqcup x$

(6a) $x \sqcap (y \vee z) = (x \sqcap y) \vee (x \sqcap z)$ (6b) $x \sqcup (y \wedge z) = (x \sqcup y) \wedge (x \sqcup z)$

(7a) $x \sqcap (x \vee y) = x \sqcap x$ (7b) $x \sqcup (x \wedge y) = x \sqcup x$

(8a) $\neg\neg(x \sqcap y) = x \sqcap y$ (8b) $\lrcorner\lrcorner(x \sqcup y) = x \sqcup y$

(9a) $x \sqcap \neg x = \bot$ (9b) $x \sqcup \lrcorner x = \top$

(10a) $\neg\bot = \top \sqcap \top$ (10b) $\lrcorner\top = \bot \sqcup \bot$

(11a) $\neg\top = \bot$ (11b) $\lrcorner\bot = \top$

12 $(x \sqcap x) \sqcup (x \sqcap x) = (x \sqcup x) \sqcap (x \sqcup x),$

where $x \vee y := \neg(\neg x \sqcap \neg y)$ and $x \wedge y := \lrcorner(\lrcorner x \sqcup \lrcorner y)$.

On D, a quasi-order $\sqsubseteq$ is given by $x \sqsubseteq y \Longleftrightarrow x \sqcap y = x \sqcap x$ and $x \sqcup y = y \sqcup y$, for any $x, y \in D$.

A dBa $\boldsymbol{D}$ is called *pure*, if for all $x \in D$ either $x \sqcap x = x$ or $x \sqcup x = x$.

Theorem 1 [11]. $\underline{\mathfrak{P}}(\mathbb{K})$ forms a dBa and $\underline{\mathfrak{H}}(\mathbb{K})$ forms a pure dBa.

Notation 4. For any dBa $\boldsymbol{D} := (D, \sqcup, \sqcap, \neg, \lrcorner, \top, \bot)$, $D_\sqcap := \{x \in D \mid x \sqcap x = x\}$, $D_\sqcup := \{x \in D \mid x \sqcup x = x\}$. For $x \in D$, $x_\sqcap := x \sqcap x$ and $x_\sqcup := x \sqcup x$.

Let $\boldsymbol{D} := (D, \sqcup, \sqcap, \neg, \lrcorner, \top, \bot)$ be a dBa. Then we have

Proposition 1 [8].

    (i) $D_\sqcap := (D_\sqcap, \sqcap, \vee, \neg, \bot, \neg\bot)$ is a Boolean algebra, whose order relation is the restriction of $\sqsubseteq$ to $D_\sqcap$ and is denoted by $\sqsubseteq_\sqcap$.

    (ii) $D_\sqcup := (D_\sqcup, \sqcup, \wedge, \lrcorner, \top, \lrcorner\top)$ is a Boolean algebra, whose order relation is the restriction of $\sqsubseteq$ to $D_\sqcup$ and is denoted by $\sqsubseteq_\sqcup$.

    (iii) $x \sqsubseteq y$ if and only if $x \sqcap x \sqsubseteq y \sqcap y$ and $x \sqcup x \sqsubseteq y \sqcup y$ for $x, y \in D$, that is $x_\sqcap \sqsubseteq_\sqcap y_\sqcap$ and $x_\sqcup \sqsubseteq y_\sqcup$, for any $x, y \in D$.

The following proposition giving further properties of dBas is useful. Part (a) is proved in [7]; we prove (b) below.

Proposition 2.

    (a) [7] For any $x, y, a \in D$:

1. $x \sqcap \bot = \bot$ and $x \sqcup \bot = x \sqcup x$, that is $\bot \sqsubseteq x$.
2. $x \sqcup \top = \top$ and $x \sqcap \top = x \sqcap x$, that is $x \sqsubseteq \top$.
3. $x = y$ implies that $x \sqsubseteq y$ and $y \sqsubseteq x$.
4. $x \sqsubseteq y$ and $y \sqsubseteq x$ if and only if $x \sqcap x = y \sqcap y$ and $x \sqcup x = y \sqcup y$.
5. $x \sqcap y \sqsubseteq x, y \sqsubseteq x \sqcup y, x \sqcap y \sqsubseteq y, x \sqsubseteq x \sqcup y$.
6. $x \sqsubseteq y$ implies $x \sqcap a \sqsubseteq y \sqcap a$ and $x \sqcup a \sqsubseteq y \sqcup a$.

    (b) For any $x, y \in D$:

1. $\neg x = (\neg x)_\sqcap \in D_\sqcap$ and $\lrcorner x = (\lrcorner x)_\sqcup \in D_\sqcup$.
2. $x \sqsubseteq y$ if and only if $\neg y \sqsubseteq \neg x$ and $\lrcorner y \sqsubseteq \lrcorner x$.

Proof. (b) 1. Let $x \in D$. By axiom (1a), $x \sqcap x \in D_\sqcap$. Using axiom (4a) and Proposition 1(i) we have $\neg x = \neg(x \sqcap x) \in D_\sqcap$. The other is proved dually.

2. Let $x, y \in D$. Proposition 1(iii) gives $x \sqsubseteq y$ if and only if $x_\sqcap \sqsubseteq_\sqcap y_\sqcap$ and $x_\sqcup \sqsubseteq_\sqcup y_\sqcup$, which is if and only if $\neg y_\sqcap \sqsubseteq_\sqcap \neg x_\sqcap$ and $\lrcorner y_\sqcup \sqsubseteq_\sqcup \lrcorner x_\sqcup$ by Proposition 1(i)–(ii). The latter is if and only if $\neg y \sqsubseteq_\sqcap \neg x$ and $\lrcorner y \sqsubseteq_\sqcup \lrcorner x$ (using axioms (4a) and (4b)), which is if and only if $\neg y \sqsubseteq \neg x$ and $\lrcorner y \sqsubseteq \lrcorner x$, by Proposition 1(i)–(ii) and (b)(1) of this Proposition. $\square$

Definition 4. Let **D** and **M** be two dBas. A map $h : M \to D$ is called a *homomorphism* if h preserves the operations in the algebras.

    h is called *quasi-injective*, when $x \sqsubseteq y$ if and only if $h(x) \sqsubseteq h(y)$, for all $x, y \in M$. A quasi-injective and surjective homomorphism is called a *quasi-isomorphism* and a bijective homomorphism is called an *isomorphism*.

    In a dBa $\mathbf{D} := (D, \sqcup, \sqcap, \neg, \lrcorner, \top, \bot)$ a subset F of D is called a *filter*, if it is an upset and closed under $\sqcap$. Dually, a subset I of D is called an *ideal* if it is a downset and closed under $\sqcup$.

    $F_0(\subseteq D)$ is called a *base* for the filter F if $F = \{y \in D : x \sqsubseteq y \text{ for some } x \in F_0\}$. Base for an ideal is dually defined.

Notation 5. Let $\mathbf{D} := (D, \sqcup, \sqcap, \neg, \lrcorner, \top, \bot)$ be a dBa.

    $\mathcal{F}_p(\mathbf{D}) := \{F \subseteq D | F \text{ is a filter of } \mathbf{D} \text{ and } F \cap D_\sqcap \text{ is a prime filter in } \mathbf{D}_\sqcap\}$, and $\mathcal{I}_p(\mathbf{D}) := \{I \subseteq D | I \text{ is an ideal of } \mathbf{D} \text{ and } I \cap D_\sqcup \text{ is a prime ideal in } \mathbf{D}_\sqcup\}$.

    For any $x \in D$, $F_x := \{F \in \mathcal{F}_p(\mathbf{D}) \mid x \in F\}$ and $I_x := \{I \in \mathcal{I}_p(\mathbf{D}) \mid x \in I\}$.

    Define the context $\mathbb{K}(\mathbf{D}) := (\mathcal{F}_p(\mathbf{D}), \mathcal{I}_p(\mathbf{D}), \Delta)$, where for any $F \in \mathcal{F}_p(\mathbf{D})$, $I \in \mathcal{I}_p(\mathbf{D})$, $F \Delta I$ if and only if $F \cap I \neq \emptyset$.

Lemma 1 [11]. Let F be a filter of a dBa $\mathbf{D}$.

1. $F \cap D_\sqcap$ and $F \cap D_\sqcup$ are filters of the Boolean algebras $\mathbf{D}_\sqcap$, $\mathbf{D}_\sqcup$ respectively.
2. Each filter F_0 of the Boolean algebra $\mathbf{D}_\sqcap$ is the base of some filter F of $\mathbf{D}$ such that $F_0 = F \cap D_\sqcap$. Moreover if F_0 is a prime filter of $\mathbf{D}_\sqcap$, $F \in \mathcal{F}_p(\mathbf{D})$.

A similar result can be proved for ideals in a dBa.

Lemma 2. $(F_x)^c = F_{\neg x}$ and $(I_x)^c = I_{\lrcorner x}$, for any $x \in D$.

Theorem 2 [11]. The map $h : D \to \mathfrak{P}(\mathbb{K}(\mathbf{D}))$ defined by $h(x) := (F_x, I_x)$ for all $x \in D$, is a quasi-injective homomorphism.

Theorem 3 [1]. If $\mathbf{D} := (D, \sqcup, \sqcap, \neg, \lrcorner, \top, \bot)$ is a pure dBa, the map $h : D \to \mathfrak{H}(\mathbb{K}(\mathbf{D}))$ defined by $h(x) := (F_x, I_x)$ for all $x \in D$, is an injective homomorphism.

3   Concepts and Semiconcepts Based on Rough Set Theory

In [4], Düntsch and Gediga pointed out limitations of FCA as a tool for qualitive data analysis. They gave the example of a context $\mathbb{K} := (G, M, \Gamma)$, where G is a set of problems and M is a set of skills, and the relation $\Gamma \subseteq G \times M$ may be interpreted in two different ways:

I. Skill s is necessary to solve q and $\Gamma(q)$ is minimally sufficient to solve q.
II. It is possible to solve problem q with skill s.

In this section, we work with an instance of such a context given below.

Example 1 Let $G := \{q_1, q_2, q_3, q_4, q_5, q_6\}$ be a set of problems and consider a set of skills $S := \{s_1, s_2, s_3, s_4, s_5, s_6, s_7, s_8, s_9, s_{10}, s_{11}\}$. The context $\mathbb{K} := (G, M, \Gamma)$ is represented by the table below. A cross in the $i-j$-th cell of the table indicates that the relation $q_i \Gamma s_j$ holds.

Table 1. Context

	s_1	s_2	s_3	s_4	s_5	s_6	s_7	s_8	s_9	s_{10}	s_{11}
q_1	×		×		×		×		×		
q_2		×	×	×		×		×			
q_3	×	×		×		×			×		
q_4	×		×		×						
q_5				×			×	×			×
q_6											×

Case I: Let $q_i \Gamma s_j$ be interpreted as: skill s_j is necessary to solve q_i, and $\Gamma(q_i)$ is minimally sufficient to solve q_i. Let a student be asked to solve a set of problems $A := \{q_1, q_2\}$ in a test to check his skills (from the set S). In FCA, $A' = \{s_3\}$ gives the collection of the skills necessary to solve all the problems of A. However, different problems in A may require different sets of skills to solve them. For instance, q_1 requires s_1 while q_2 does not. So $\{s_1, s_2, s_3, s_4, s_5, s_6, s_7, s_8, s_{10}\}$ would more adequately represent the skills necessary to solve A.

Case II: Let us interpret $q_i \Gamma s_j$ as: it is possible to solve q_i with skill s_j. Considering the problem set $A := \{q_1, q_2\}$, again $A' = \{s_3\}$ does not give all the possible skills that could be used to solve the problems in A – for instance, it is possible to solve them with the skills s_1 and s_2.

Now the question is, how do we assign a skill set to the problem set A? Düntsch and Gediga address this question in [4], using modal style operators and introduce property oriented concepts. Let $\mathbb{K} := (G, M, R)$ be a context, $A \subseteq G$ and $B \subseteq M$.

$$B_R^\Diamond := \{x \in G | R(x) \cap B \neq \emptyset\} \text{ and } B_R^\Box := \{x \in G | R(x) \subseteq B\}$$

$$A_{R^{-1}}^\Diamond := \{y \in M | R^{-1}(y) \cap A \neq \emptyset\} \text{ and } A_{R^{-1}}^\Box := \{y \in M | R^{-1}(y) \subseteq A\}$$

If there is no confusion about the relation involved, we shall omit the subscript and denote $B_R^\Diamond$ by $B^\Diamond$, $B_R^\Box$ by $B^\Box$, and similarly for the case of A.

Definition 5 [3]. A *closure operator* on a set X is an operator C on the power set $\mathcal{P}(X)$ of X such that for all $A, B \in \mathcal{P}(X)$,

C1 $A \subseteq C(A)$,
C2 $A \subseteq B$ *implies* $C(A) \subseteq C(B)$,
C3 $C(C(A)) = C(A)$.

$A \in \mathcal{P}(X)$ is called *closed* if and only if $C(A) = A$.

An *interior operator* I on the set X is defined dually, and $A \in \mathcal{P}(X)$ is called *open* if and only if $I(A) = A$.

Some properties of the operators $\Box, \Diamond$ are as follows.

Theorem 4 [14]. *Let* $A, A_1, A_2 \subseteq G$ *and* $B, B_1, B_2 \subseteq M$.

1. $A_1 \subseteq A_2$ *implies that* $A_1^\Box \subseteq A_2^\Box$ *and* $A_1^\Diamond \subseteq A_2^\Diamond$.
2. $B_1 \subseteq B_2$ *implies that* $B_1^\Box \subseteq B_2^\Box$ *and* $B_1^\Diamond \subseteq B_2^\Diamond$.
3. $A_R^\Box = A_{-R}^{c' c}; B_R^\Box = B_{-R}^{c' c}$ *and* $A_R^\Diamond = A_{-R}^{'c}; B_R^\Diamond = B_{-R}^{'c}$.
4. $A^{\Box\Diamond\Box} = A^\Box$ *and* $B^{\Box\Diamond\Box} = B^\Box$.
5. $A^{\Diamond\Box\Diamond} = A^\Diamond$ *and* $B^{\Diamond\Box\Diamond} = B^\Diamond$.
6. $\Box\Diamond$ *is interior operator on* G *and* $\Diamond\Box$ *is closure operator on* M.

Let $\mathbb{K} := (G, M, R)$ be a context, $A \subseteq G, B \subseteq M$.

Definition 6 [4]. (A, B) is a *property oriented concept* of $\mathbb{K}$ if it satisfies the conditions $A^\diamond = B$ and $B^\square = A$.

Refer to Example 1, Case I. For any $A \subseteq G$, $A^\diamond$ can be interpreted as the set of all skills m that are necessary to solve problems in A $(\Gamma^{-1}(m) \cap A \neq \emptyset)$. For any $B \subseteq M$, $B^\square$ can be interpreted as the set of problems x such that B contains all skills sufficient to solve x $(\Gamma(x) \subseteq B)$. So the intent B of a property oriented concept (A, B) of the context given in Example 1, represents the set of skills that are necessary and sufficient to solve problems in the extent A. For instance, $\{s_1, s_2, s_3, s_4, s_5, s_6, s_7, s_8, s_{10}\}$ gives all the skills that are necessary and sufficient to solve the problems in $\{q_1, q_2, q_4\}$; hence $(\{q_1, q_2, q_4\}, \{s_1, s_2, s_3, s_4, s_5, s_6, s_7, s_8, s_{10}\})$ is a property-oriented concept.

Definition 7 [13]. (A, B) is an *object oriented concept* of the context $\mathbb{K}$ if it satisfies the condition $A^\square = B$ and $B^\diamond = A$.

An order relation $\leq$ can be defined on the set of such pairs. For any object oriented concepts $(A_1, B_1), (A_2, B_2)$, $(A_1, B_1) \leq (A_2, B_2)$ if and only if $A_1 \subseteq A_2$ (equivalently, $B_1 \subseteq B_2$).

Refer to Example 1, Case II. For $A \subseteq G$, $A^\square$ can be interpreted as the set of skills s such that A contains all the problems that are possible to solve with s $(\Gamma^{-1}(s) \subseteq A)$. For $B \subseteq M$, $B^\diamond$ is the set of problems that are possible to solve with some skill in B $(\Gamma(x) \cap B \neq \emptyset)$. So the extent A of an object oriented concept (A, B) in this context is the set of problems which are possible to solve with skills in B, while the intent B is the set of skills by which only problems in A can be solved. $(\{q_3, q_5\}, \{s_9\})$ then forms an object-oriented concept.

A comparative study of Wille's concepts, property and object oriented concepts and concept lattices is done extensively by Yao in [13].

We next turn to object-oriented semiconcepts, introduced in [6] to bring in the notion of negation.

Definition 8 [6]. (A, B) is an *object oriented semiconcept* of $\mathbb{K}$ if $A^\square = B$ or $B^\diamond = A$.

Notation 6. The set of all object oriented concepts is denoted by **RO-L**$(\mathbb{K})$, the set of all object oriented semiconcepts is denoted by $\mathfrak{S}(\mathbb{K})$.

The following are observed in [6].

Observation 2.

1. $(A, B) \in \mathfrak{S}(\mathbb{K})$ if and only if either $(A, B) = (A, A^\square)$ or $(A, B) = (B^\diamond, B)$.
2. $RO - L(\mathbb{K}) \subseteq \mathfrak{S}(\mathbb{K})$.
3. (A, B) is a semiconcept of $\mathbb{K}$ if and only if (A^c, B) is an object oriented semiconcept of the context $\mathbb{K}^c$.

The following operations are defined in $\mathfrak{S}(\mathbb{K})$. Let $(A_1, B_1), (A_2, B_2), (A, B)$ be any object oriented semiconcepts.

$$(A_1, B_1) \sqcap (A_2, B_2) := ((B_1 \cap B_2)^\Diamond, B_1 \cap B_2),$$
$$(A_1, B_1) \sqcup (A_2, B_2) := (A_1 \cup A_2, (A_1 \cup A_2)^\Box),$$
$$\neg(A, B) := (A^c, A^{c\Box}), \lrcorner(A, B) := (B^{c\Diamond}, B^c), \top := (G, M), \text{ and } \bot := (\phi, \phi).$$

Refer again to Example 1, Case II. It is clear that it is possible to solve the problems q_1, q_2, q_4 with skills other than s_9. We are able to express this observation in the framework of object oriented semiconcepts. Note that $a := (\{q_3, q_5\}, \{s_9\})$ is an object oriented semiconcept of the context. The extent of a gives exactly the problems that are possible to solve with skill s_9. The negation of a, $\lrcorner a$, is the object oriented semiconcept $(\{q_1, q_2, q_3, q_4, q_5, q_6\}, \{s_9\}^c)$ whose extent gives all the problems that are possible to solve with skills other than s_9, and includes q_1, q_2, q_4.

In this work, for the sake of simplicity in expressions of results, we consider operations on object oriented semiconcepts that are dual to the ones mentioned above. In other words, for $(A, B), (C, D) \in \mathfrak{S}(\mathbb{K})$, we consider

$$(A, B) \sqcap (C, D) := (A \cup C, (A \cup C)^\Box),$$
$$(A, B) \sqcup (C, D) := ((B \cap D)^\Diamond, B \cap D),$$
$$\lrcorner(A, B) := (B^{c\Diamond}, B^c), \ \neg(A, B) := (A^c, A^{c\Box}), \ \top := (\emptyset, \emptyset), \ \bot := (G, M).$$

From the results established in [6], we obtain

Theorem 5.

1. $\underline{S}(\mathbb{K}) := (\mathfrak{S}(\mathbb{K}), \sqcup, \sqcap, \neg, \lrcorner, \top, \bot)$ is a pure dBa.
2. $\mathfrak{H}(\mathbb{K})$ is isomorphic to $\underline{S}(\mathbb{K}^c)$.

The map in (2) of the above theorem is due to Observation 2(3). From Lemma 2, Theorem 5(2) and Theorem 3 we have the following.

Theorem 6. For a pure dBa $\mathbf{D}$ the map $h : D \to \mathfrak{S}(\mathbb{K}^c(\mathbf{D}))$ defined by $h(x) := (F_{\neg x}, I_x)$ for all $x \in D$, is an injective dBa homomorphism.

4 Object Oriented Protoconcepts

We now define and give some properties of object oriented protoconcepts. As before, $\mathbb{K} := (G, M, R)$ is a context, $A \subseteq G, B \subseteq M$.

Definition 9. (A, B) is an *object oriented protoconcept* of $\mathbb{K}$ if $A^{\Box\Diamond} = B^\Diamond$.

Notation 7. The set of all object oriented protoconcepts is denoted by $\mathfrak{R}(\mathbb{K})$.

Proposition 3.

1. $A^{\Box\Diamond} = B^\Diamond$ if and only if $A^\Box = B^{\Diamond\Box}$.

2. $\mathfrak{S}(\mathbb{K}) \subseteq \mathfrak{R}(\mathbb{K})$.

Proof. (1) Let $A^{\square\lozenge} = B^{\lozenge}$. Then $A^{\square\lozenge\square} = B^{\lozenge\square}$. Now from 4 of Theorem 4 we know that $A^{\square\lozenge\square} = A^{\square}$. Therefore $A^{\square} = B^{\lozenge\square}$. Let $A^{\square} = B^{\lozenge\square}$ then $A^{\square\lozenge} = B^{\lozenge\square\lozenge}$. Now from 5 of the Theorem 4 we know that $B^{\lozenge\square\lozenge} = B^{\lozenge}$. Therefore $A^{\square\lozenge} = B^{\lozenge}$.

(2) Let $(A, B) \in \mathfrak{S}(\mathbb{K})$ then either $A^{\square} = B$ or $B^{\lozenge} = A$. Now if $A^{\square} = B$ then we gate $A^{\square\lozenge} = B^{\lozenge}$ and hence $(A, B) \in \mathfrak{R}(\mathbb{K})$. Now if $B^{\lozenge} = A$ then $A^{\square} = B^{\lozenge\square}$. Therefore $(A, B) \in \mathfrak{R}(\mathbb{K})$ by (1) of this Proposition. Hence $\mathfrak{S}(\mathbb{K}) \subseteq \mathfrak{R}(\mathbb{K})$. $\square$

Recall Example 1 and Case II. Consider $A_1 := \{q_1, q_2, q_4, q_6\}$ and $B_1 := \{s_3\}$. Then $A_1^{\square\lozenge} = B_1^{\lozenge}$, so that (A_1, B_1) is an object oriented protoconcept of the context. Now observe that $B_1^{\lozenge} = \{q_1, q_2, q_4\}$ and so $B_1^{\lozenge} \neq A_1$; $A_1^{\square} = \{s_3, s_7, s_{10}\}$, so $A_1^{\square} \neq B_1$. This means $(A_1, B_1) \in \mathfrak{R}(\mathbb{K})$ but $(A_1, B_1) \notin \mathfrak{S}(\mathbb{K})$, indicating that the converse of (2) in Proposition 3 is not true.

We next characterize the object oriented protoconcepts of $\mathbb{K}$ (Theorem 7 below) using a notion of 'approximation' by object oriented concepts. Based on the facts that $\square\lozenge$ is an interior operator on G and $\lozenge\square$ is a closure operator on M, the discussion in [15] of 'definable' object sets gives the definition below.

Definition 10.

1. A is said to be *definable in* $\mathcal{P}(G)$ if and only if $A^{\square\lozenge} = A$.
2. B is said to be *definable in* $\mathcal{P}(M)$ if and only if $B^{\lozenge\square} = B$.

It is then easy to see the following.

Proposition 4. A is definable if and only if it is the extent of some object oriented concept of $\mathbb{K}$, and B is definable if and only if it is the intent of some object-oriented concept of $\mathbb{K}$.

Proof. From Definition 7 it follows that for any object oriented concept (A, B), A, B are definable. To see the converse assertions, for definable A and B, consider respectively the object oriented concepts $(A, A^{\square})$ and $(B^{\lozenge}, B)$. $\square$

So an object oriented concept (A, B) of the context $\mathbb{K}$ can be thought of as a pair of definable sets such that one can be determined by the other as $A = B^{\lozenge}$ and $A^{\square} = B$. However, for pairs $(E, F) \in \mathcal{P}(G) \times \mathcal{P}(M)$ which are not object oriented concepts, it may be worthwhile to determine the largest definable set C contained in E and determined by F as $C = F^{\lozenge}$, and also the smallest definable set D containing F and determined by E as $E^{\square} = D$. Such a pair (C, D) would be an object oriented concept, and unique in the above respect. For instance, referring to Example 1, Case II: for a set E of problems and set F of skills, C would be the largest definable set of problems inside E that are possible to be solved with the skills in F $(C = F^{\lozenge})$. We call the pair (C, D) an approximation of the pair (E, F).

Definition 11. An object oriented concept (C, D) is called an *approximation* of (A, B) if and only if C is the largest definable set in $\mathcal{P}(G)$ contained in A, and D is the smallest definable set in $\mathcal{P}(M)$ containing B.

Observation 3. If (A, B) has an approximation, it is unique.

Theorem 7. (A, B) is an object oriented protoconcept of $\mathbb{K}$ if and only if it has an approximation.

Proof. Let (A, B) be an object oriented protoconcept of $\mathbb{K}$. Then $A^{\square\Diamond} = B^{\Diamond}$, which is equivalent to $A^{\square} = B^{\Diamond\square}$. Now we set $C := A^{\square\Diamond}$ and $D := A^{\square} = B^{\Diamond\square}$. Then the pair $(C, D) = (A^{\square\Diamond}, A^{\square})$ is an object oriented concept. Since $\square\Diamond$ is an interior operator on G, $C \subseteq A$; since $\Diamond\square$ is a closure operator on M, $B \subseteq D$. As C is the extent of the object oriented concept (C, D), C is definable in $\mathcal{P}(G)$ by Proposition 4. Now let E be a definable set in $\mathcal{P}(G)$ such that $E \subseteq A$. By 1 of Theorem 4 we have $E = E^{\square\Diamond} \subseteq A^{\square\Diamond} = C$, making C the largest definable set contained in A. Similarly we get that D is the smallest definable set containing B. So (C, D) is an approximation of (A, B).

For the converse, let us assume that (X, Y) is an approximation of (A, B). Then (X, Y) is an object oriented concept and $X \subseteq A$ and $B \subseteq Y$. Using 1 and 2 of Theorem 4 we have $X = X^{\square\Diamond} \subseteq A^{\square\Diamond}$ and $B^{\Diamond\square} \subseteq Y^{\Diamond\square} = Y$. By 4 and 6 of Theorem 4, $A^{\square\Diamond\square\Diamond} = A^{\square\Diamond} \subseteq A$, making $A^{\square\Diamond}$ a definable set contained in A. Since X is the largest definable set contained in A, $A^{\square\Diamond} \subseteq X$. Dually we can show that $Y \subseteq B^{\Diamond\square}$. So $X = A^{\square\Diamond}$ and $Y = B^{\Diamond\square}$. As (X, Y) is an object oriented concept then $X^{\square} = Y$ and $X = Y^{\Diamond}$. So $A^{\square} = A^{\square\Diamond\square} = B^{\Diamond\square}$ and hence (A, B) is a object oriented protoconcept of $\mathbb{K}$. $\square$

In Example 1, Case II: as shown earlier, the pair $(\{q_1, q_2, q_4, q_6\}, \{s_3\})$ is an object oriented protoconcept but not an object oriented concept. This pair has the object oriented concept $(\{q_1, q_2, q_4\}, \{s_3, s_7, s_{10}\})$ as its unique approximation.

Proposition 5. (A, B) is an object oriented protoconcept of $\mathbb{K}$ if and only if (A^c, B) is a protoconcept of $\mathbb{K}^c$.

Proof. $A_R^{\square\Diamond} = B_R^{\Diamond}$ if and only if $A_{-R}^{c''c} = B_{-R}'^c$ by 3 of Theorem 4, and $A_{-R}^{c''c} = B_{-R}'^c$ if and only if $A_{-R}^{c''} = B_{-R}'$. $\square$

Recall the operations $\sqcup, \sqcap, \neg, \lrcorner, \top, \bot$ defined in Sect. 3 that made the set $\mathfrak{S}(\mathbb{K})$ of object oriented semiconcepts a pure dBa (Theorem 5). The set $\mathfrak{R}(\mathbb{K})$ of object oriented protoconcepts turns out to be closed with respect to the same operations. In fact,

Theorem 8. (i) $\underline{\mathfrak{R}}(\mathbb{K}) := (\mathfrak{R}(\mathbb{K}), \sqcup, \sqcap, \neg, \lrcorner, \top, \bot)$ is a dBa and (ii) $\underline{\mathfrak{R}}(\mathbb{K})$ is isomorphic to $(\mathfrak{P}(\mathbb{K}^c), \sqcup, \sqcap, \neg, \lrcorner, \top, \bot)$.

Proof. (i) The proof is a routine check.
(ii) $h : \mathfrak{R}(\mathbb{K}) \to \mathfrak{P}(\mathbb{K}^c)$ defined as $h((A, B)) := (A^c, B)$ for any (A, B) in $\mathfrak{R}(\mathbb{K})$, is an isomorphism between the two dBas. h is well defined and onto by Proposition 5. Injectivity of h follows trivially. Verifying that h is a homomorphism is a routine check. $\square$

The dBa $\underline{\mathfrak{R}}(\mathbb{K})$ is called the *algebra of object oriented protoconcepts*.

Observation 4. Using the definition of the quasi-order in a dBa, one sees that for any $(A, B), (C, D) \in \mathfrak{R}(\mathbb{K})$, $(A, B) \sqsubseteq (C, D)$ if and only if $C \subseteq A$ and $D \subseteq B$.

Using Theorems 2 and 8 and Lemma 2, we get a representation result for dBas.

Theorem 9. For a dBa **D**, the map $h : D \rightarrow \mathfrak{R}(\mathbb{K}^c(\mathbf{D}))$ defined by $h(x) := (F_{\neg x}, I_x)$ for any $x \in D$, is a quasi-injective dBa homomorphism.

5   Logics for dBas and Pure dBas

We now give a sequent calculus **DBL** for the class of dBas, and extend it to **PDBL** to get a sequent calculus for pure dBas. The language of **DBL** consists of propositional constants $\bot, \top$, a set **VAR** of propositional variables, and logical connectives $\sqcup, \sqcap, \neg, \lrcorner$. The set $\mathfrak{F}$ of all formulae of **DBL** is given by the scheme:

$$\bot \mid \top \mid p \mid \alpha \sqcup \beta \mid \alpha \sqcap \beta \mid \neg \alpha \mid \lrcorner \alpha,$$

where $p \in \mathbf{VAR}$. $\vee$ and $\wedge$ are definable connectives: $\alpha \vee \beta := \neg(\neg \alpha \sqcap \neg \beta)$ and $\alpha \wedge \beta := \lrcorner(\lrcorner \alpha \sqcup \lrcorner \beta)$, for $\alpha, \beta \in \mathfrak{F}$.

A *sequent* in **DBL** with formulae $\alpha, \beta \in \mathfrak{F}$ is denoted in the usual manner as $\alpha \vdash \beta$. $\alpha \dashv\vdash \beta$ will be used as abbreviation for $(\alpha \vdash \beta$ and $\beta \vdash \alpha)$.

The axioms of **DBL** are given by the following schemes.

1a $\bot \vdash \alpha$	1b $\alpha \vdash \top$
2a $\alpha \sqcap \beta \vdash \alpha$	2b $\alpha \vdash \alpha \sqcup \beta$
3a $\alpha \sqcap \beta \vdash \beta$	3b $\beta \vdash \alpha \sqcup \beta$
4a $\alpha \sqcap \beta \vdash (\alpha \sqcap \beta) \sqcap (\alpha \sqcap \beta)$	4b $(\alpha \sqcup \beta) \sqcup (\alpha \sqcup \beta) \vdash \alpha \sqcup \beta$
5a $\alpha \sqcap \alpha \vdash \alpha \sqcap (\alpha \sqcup \beta)$	5b $\alpha \sqcup (\alpha \sqcap \beta) \vdash \alpha \sqcup \alpha$
6a $\neg(\alpha \sqcap \alpha) \vdash \neg \alpha$	6b $\lrcorner \alpha \vdash \lrcorner(\alpha \sqcup \alpha)$
7a $\alpha \sqcap \neg \alpha \vdash \bot$	7b $\top \vdash \alpha \sqcup \lrcorner \alpha$
8a $\neg \bot \dashv\vdash \top \sqcap \top$	8b $\lrcorner \top \dashv\vdash \bot \sqcup \bot$
9a $\alpha \sqcap \alpha \vdash \alpha \sqcap (\alpha \vee \beta)$	9b $\alpha \sqcup (\alpha \wedge \beta) \vdash \alpha \sqcup \alpha$
10a $\alpha \sqcap (\beta \vee \gamma) \dashv\vdash (\alpha \sqcap \beta) \vee (\alpha \sqcap \gamma)$	10b $\alpha \sqcup (\beta \wedge \gamma) \dashv\vdash (\alpha \sqcup \beta) \wedge (\alpha \sqcup \gamma)$
11a $\neg\neg(\alpha \sqcap \beta) \dashv\vdash (\alpha \sqcap \beta)$	11b $\lrcorner\lrcorner(\alpha \sqcup \beta) \dashv\vdash (\alpha \sqcup \beta)$
12a $\neg \top \vdash \bot$	12b $\top \vdash \lrcorner \bot$

13 $\alpha \vdash \alpha$

14 $(\alpha \sqcup \alpha) \sqcap (\alpha \sqcup \alpha) \vdash (\alpha \sqcap \alpha) \sqcup (\alpha \sqcap \alpha)$

Rules of inference:

$$\frac{\alpha \vdash \beta}{\alpha \sqcap \gamma \vdash \beta \sqcap \gamma} \ (R1)$$

$$\frac{\alpha \vdash \beta}{\alpha \sqcup \gamma \vdash \beta \sqcup \gamma} \ (R2)$$

$$\frac{\alpha \vdash \beta}{\neg \beta \vdash \neg \alpha} \ (R3)$$

$$\frac{\alpha \vdash \beta, \beta \vdash \gamma}{\alpha \vdash \gamma} \ (R4)$$

$$\frac{\alpha \vdash \beta}{\gamma \sqcap \alpha \vdash \gamma \sqcap \beta} \ (R1')$$

$$\frac{\alpha \vdash \beta}{\gamma \sqcup \alpha \vdash \gamma \sqcup \beta} \ (R2)'$$

$$\frac{\alpha \vdash \beta}{\lrcorner \beta \vdash \lrcorner \alpha} \ (R5)$$

$$\frac{\alpha \sqcap \beta \dashv\vdash \alpha \sqcap \alpha \quad \alpha \sqcup \beta \dashv\vdash \beta \sqcup \beta}{\alpha \vdash \beta} \ (R6)$$

Provability of sequents in **DBL** is defined in the standard way.

Proposition 6. The following rules are derivable in **DBL**.

$$\frac{\alpha \vdash \beta \quad \alpha \vdash \gamma}{\alpha \sqcap \alpha \vdash \beta \sqcap \gamma} \ (R7)$$

$$\frac{\beta \vdash \alpha \quad \gamma \vdash \alpha}{\beta \sqcup \gamma \vdash \alpha \sqcup \alpha} \ (R8)$$

Proof. $(R7)$ is derived using $(R1), (R1')$ and $(R4)$, while for $(R8)$ one uses $(R2), (R2')$ and $(R4)$. □

Theorem 10. The following sequents are provable in **DBL**.

(1a) $(\alpha \sqcap \beta) \dashv\vdash (\beta \sqcap \alpha)$.
(1b) $\alpha \sqcup \beta \dashv\vdash \beta \sqcup \alpha$.
(2a) $\alpha \sqcap (\beta \sqcap \gamma) \dashv\vdash (\alpha \sqcap \beta) \sqcap \gamma$.
(2b) $\alpha \sqcup (\beta \sqcup \gamma) \dashv\vdash (\alpha \sqcup \beta) \sqcup \gamma$.
(3a) $(\alpha \sqcap \alpha) \sqcap \beta \dashv\vdash (\alpha \sqcap \beta)$.
(3b) $(\alpha \sqcup \alpha) \sqcup \beta \dashv\vdash \alpha \sqcup \beta$.
(4a) $\neg \alpha \vdash \neg(\alpha \sqcap \alpha)$.
(4b) $\lrcorner(\alpha \sqcup \alpha) \vdash \lrcorner \alpha$.
(5a) $\alpha \sqcap (\alpha \sqcup \beta) \vdash (\alpha \sqcap \alpha)$.
(5b) $\alpha \sqcup \alpha \vdash \alpha \sqcup (\alpha \sqcap \beta)$.
(6a) $\alpha \sqcap (\alpha \vee \beta) \vdash \alpha \sqcap \alpha$.
(6b) $\alpha \sqcup \alpha \vdash \alpha \sqcup (\alpha \wedge \beta)$.
(7a) $\bot \vdash \alpha \sqcap \neg \alpha$.
(7b) $\alpha \sqcup \lrcorner \alpha \vdash \top$.
(8a) $\bot \vdash \neg \top$.
(8b) $\lrcorner \bot \vdash \top$.
(9) $(\alpha \sqcap \alpha) \sqcup (\alpha \sqcap \alpha) \vdash (\alpha \sqcup \alpha) \sqcap (\alpha \sqcup \alpha)$.

Proof. The proofs are straightforward. For instance, $(4a)$ follows from axiom $(2a)$ and $(R3)$. The proofs of (ib) are dual to those of (ia) for $i = 1, 2, 3, 4, 5, 6$. $(7a), (8a)$ follow from axiom (1a), and $7b, 8b$ follow from axiom (1b). Proposition 6 is also used in some of the proofs. □

PDBL is the logic obtained from **DBL** by adding the following axiom:
15.   for any $\alpha \in \mathfrak{F}$, either $\alpha \vdash \alpha \sqcap \alpha$ or $\alpha \sqcup \alpha \vdash \alpha$.
   Due to axioms (2a) and (2b), we get

Proposition 7. In **PDBL**, for any $\alpha \in \mathfrak{F}$, either $\alpha \dashv\vdash \alpha \sqcap \alpha$ or $\alpha \dashv\vdash \alpha \sqcup \alpha$.

Now we define the notion of validity for **DBL (PDBL)** with respect to the class of dBas (pure dBas).

Definition 12. Let **D** be a (pure) dBa. A *valuation* $v : \mathfrak{F} \to D$ in **D** is a map such that for all $\alpha, \beta \in \mathfrak{F}$ we have the following.

1. $v(\alpha \sqcup \beta) := v(\alpha) \sqcup v(\beta)$.
2. $v(\lrcorner\alpha) := \lrcorner v(\alpha)$.
3. $v(\top) := \top$.

4. $v(\alpha \sqcap \beta) := v(\alpha) \sqcap v(\beta)$.
5. $v(\neg\alpha) := \neg v(\alpha)$.
6. $v(\bot) := \bot$.

Definition 13. A sequent $\alpha \vdash \beta$ is said to be *satisfied* by a valuation v in a (pure) dBa **D** if and only if $v(\alpha) \sqsubseteq v(\beta)$. A sequent $\alpha \vdash \beta$ is *true* in **D** if and only if for all valuations v in **D**, v satisfies $\alpha \vdash \beta$. A sequent $\alpha \vdash \beta$ is *valid* in the class of all (pure) dBas if and only if it is true in every (pure) dBa.

Theorem 11 (Soundness). If $\alpha \vdash \beta$ is provable in **DBL** (**PDBL**) then it is valid in the class of all dBas (pure dBas).

Proof. The proof that all the axioms of **DBL** are valid in the class of all dBas and that the rules of inference preserve validity, is straightforward. Proposition 2 giving properties of dBas, is utilized. The result applies to **PDBL** and pure dBas, as axiom (15) reflects the defining axiom of pure dBas (Definition 3). ☐

The completeness theorem is established in the standard way, using the Lindenbaum-Tarski algebras of the logics **DBL** and **PDBL**. We sketch the route taken by the proof. For $\alpha, \beta \in \mathfrak{F}$, a relation $\equiv_\vdash$ is defined on $\mathfrak{F}$ by: $\alpha \equiv_\vdash \beta$ if and only if $\alpha \dashv\vdash \beta$. $\equiv_\vdash$ is shown to be a congruence relation on $\mathfrak{F}$ with respect to $\sqcup$, $\sqcap$, $\neg$, $\lrcorner$. The quotient set $\mathfrak{F}/\equiv_\vdash$ induced by the relation $\equiv_\vdash$ and operations induced by the logical connectives, give the Lindenbaum-Tarski algebra $\mathcal{L}(\mathfrak{F}) := (\mathfrak{F}/\equiv_\vdash, \sqcup, \sqcap, \neg, \lrcorner, [\top], [\bot])$. The axioms in **DBL** (**PDBL**) show that $\mathcal{L}(\mathfrak{F})$ of the respective logic is a dBa (pure dBa). One can then establish

Proposition 8. The following statements are equivalent.

1. $\alpha \vdash \beta$ is provable in **DBL**.
2. $[\alpha] \sqsubseteq [\beta]$ in $\mathcal{L}(\mathfrak{F})$ of **DBL**.

The result can be extended to the case of **PDBL**. Using these and the canonical map $v : \mathfrak{F} \to \mathfrak{F}/\equiv_\vdash$ defined as $v(\gamma) := [\gamma]$ for any $\gamma \in \mathfrak{F}$, one obtains

Theorem 12 (Completeness). If a sequent $\alpha \vdash \beta$ is valid in the class of all dBas (pure dBas) then it is provable in **DBL** (**PDBL**).

6   Object Oriented Protoconcept and Semiconcept Semantics for the Logics

In this section, we define object oriented protoconcept semantics for **DBL** and object oriented semiconcept semantics for **PDBL**, and show that the logics are sound and complete with respect to these semantics.

Definition 14. A *model* for **DBL** is a pair $\mathbb{M} := (\mathbb{K}, v)$, where $\mathbb{K}$ is a context and v is a map from the set $\mathfrak{F}$ of **DBL**-formulae to the set $\mathfrak{R}(\mathbb{K})$ of all object oriented protoconcepts of $\mathbb{K}$ satisfying the following conditions:

1. $v(\alpha \sqcup \beta) := v(\alpha) \sqcup v(\beta)$.
2. $v(\lrcorner\alpha) := \lrcorner v(\alpha)$.
3. $v(\top) := \top$.

4. $v(\alpha \sqcap \beta) := v(\alpha) \sqcap v(\beta)$.
5. $v(\neg\alpha) := \neg v(\alpha)$.
6. $v(\bot) := \bot$.

$\mathbb{M} := (\mathbb{K}, v)$ is a model for **PDBL** if v is a map from $\mathfrak{F}$ to the set $\mathfrak{S}(\mathbb{K})$ of all object oriented semiconcepts of $\mathbb{K}$. The properties satisfied by v remain the same as above.

Notation 8. For the class $\mathcal{K}$ of all contexts, we define $\mathfrak{R}(\mathcal{K}) := \{\mathfrak{R}(\mathbb{K}) \mid \mathbb{K} \in \mathcal{K}\}$ and $\mathfrak{S}(\mathcal{K}) := \{\mathfrak{S}(\mathbb{K}) \mid \mathbb{K} \in \mathcal{K}\}$.

Definition 15. A sequent $\alpha \vdash \beta$ is said to be *satisfied in a model* $\mathbb{M}$ for **DBL** if and only if $v(\alpha) \sqsubseteq v(\beta)$ in $\mathfrak{R}(\mathbb{K})$. A sequent $\alpha \vdash \beta$ is *true in* $\mathfrak{R}(\mathbb{K})$ of a context $\mathbb{K}$ if and only if every model $\mathbb{M}$ based on the context $\mathbb{K}$ satisfies the sequent $\alpha \vdash \beta$. A sequent $\alpha \vdash \beta$ is *valid in* $\mathfrak{R}(\mathcal{K})$ if and only if it is true in every $\mathfrak{R}(\mathbb{K})$ of $\mathfrak{R}(\mathcal{K})$.

Replacing $\mathfrak{R}(\mathbb{K})$ and $\mathfrak{R}(\mathcal{K})$ by $\mathfrak{S}(\mathbb{K})$ and $\mathfrak{S}(\mathcal{K})$ respectively in the above, we get the definitions for the case of **PDBL**.

Theorem 13 (Soundness). For any α and β in $\mathfrak{F}$,

(a) If $\alpha \vdash \beta$ is provable in **DBL** then it is valid in $\mathfrak{R}(\mathcal{K})$,
(b) If $\alpha \vdash \beta$ is provable in **PDBL** then it is valid in $\mathfrak{S}(\mathcal{K})$.

Proof. (a) As for any context $\mathbb{K} \in \mathcal{K}$ the set $\mathfrak{R}(\mathbb{K})$ of object oriented protoconcepts of $\mathbb{K}$ forms a dBa, and for any model $\mathbb{M} := (\mathbb{K}, v)$, v is a valuation according to Definition 12, Theorem 11 gives us the result.
(b) Replace $\mathfrak{R}(\mathbb{K})$ by $\mathfrak{S}(\mathbb{K})$ and dBa by pure dBa in the argument of (a).   □

Theorem 14 (Completeness). For any α and β in $\mathfrak{F}$,

(a) If a sequent $\alpha \vdash \beta$ is valid in $\mathfrak{R}(\mathcal{K})$ then it is provable in **DBL**,
(b) If a sequent $\alpha \vdash \beta$ is valid in $\mathfrak{S}(\mathcal{K})$ then it is provable in **PDBL**.

Proof. (a) If possible, suppose $\alpha \vdash \beta$ is not provable in **DBL**. Then by Proposition 8, $[\alpha] \not\sqsubseteq [\beta]$ in $\mathcal{L}(\mathfrak{F})$. Therefore by Proposition 1 we have either $[\alpha] \sqcap [\alpha] \not\sqsubseteq_\sqcap$ $[\beta] \sqcap [\beta]$ or $[\alpha] \sqcup [\alpha] \not\sqsubseteq_\sqcap [\beta] \sqcup [\beta]$. Now we consider the Lindenbaum-Tarski algebra $\mathcal{L}(\mathfrak{F})$ of **DBL** and the context $\mathbb{K}^c(\mathcal{L}(\mathfrak{F})) := (\mathcal{F}_p(\mathcal{L}(\mathfrak{F})), \mathcal{I}_p(\mathcal{L}(\mathfrak{F})), -\Delta)$. By the representation Theorem 9, there will then exist a quasi-injective homomorphism $h : \mathcal{L}(\mathfrak{F}) \to \mathfrak{R}(\mathbb{K}^c(\mathcal{L}(\mathfrak{F})))$ such that $h(x) := (F_{\neg x}, I_x)$ for all $x \in \mathcal{L}(\mathfrak{F})$. Define the valuation $i : \mathfrak{F} \to \mathcal{L}(\mathfrak{F})$ by $i(\gamma) := [\gamma]$, for any $\gamma \in \mathfrak{F}$. Therefore composition of the two maps $v := h \circ i$ gives a valuation from $\mathfrak{F}$ to $\mathfrak{R}(\mathbb{K}^c(\mathcal{L}(\mathfrak{F})))$. So we have a model $\mathbb{M} := (\mathbb{K}^c(\mathcal{L}(\mathfrak{F})), v)$.

Now if $[\alpha] \sqcap [\alpha] \not\sqsubseteq_\sqcap [\beta] \sqcap [\beta]$, there exists a prime filter F_0 in $\mathcal{L}(\mathfrak{F})_\sqcap$ (a Boolean algebra, by Proposition 1) such that $[\alpha] \sqcap [\alpha] \in F_0$ and $[\beta] \sqcap [\beta] \notin F_0$. Therefore by Lemma 1 there exists a filter F in $\mathcal{L}(\mathfrak{F})$ such that $F \cap \mathcal{L}(\mathfrak{F})_\sqcap = F_0$ and as F_0 is prime, $F \in \mathcal{F}_p(\mathcal{L}(\mathfrak{F}))$. As $[\alpha] \sqcap [\alpha] \in F_0$, $[\alpha] \sqcap [\alpha] \in F$ and $[\beta] \sqcap [\beta] \notin$ F as $[\beta] \sqcap [\beta] \notin F_0$. So $[\alpha] \in F$ as $[\alpha] \sqcap [\alpha] \sqsubseteq [\alpha]$, and $[\beta] \notin F$ otherwise $[\beta] \sqcap [\beta] \in F$. This gives $F \notin F_{\neg[\alpha]}$ and $F \in F_{\neg[\beta]}$, whence $F_{\neg[\beta]} \not\sqsubseteq F_{\neg[\alpha]}$. So $v(\alpha) = (F_{\neg[\alpha]}, I_{[\alpha]}) \not\sqsubseteq (F_{\neg[\beta]}, I_{[\beta]}) = v(\beta)$ by Proposition 4.

In case $[\alpha]\sqcup[\alpha] \not\sqsubseteq_\sqcap [\beta]\sqcup[\beta]$, we can dually show that there exists $I \in \mathcal{I}_p(\mathcal{L}(\mathfrak{F}))$ such that $[\alpha] \notin I$ and $[\beta] \in I$ giving $I_{[\beta]} \not\sqsubseteq I_{[\alpha]}$.

So $\alpha \vdash \beta$ is not true in $\mathfrak{R}(\mathbb{K}^c(\mathcal{L}(\mathfrak{F})))$, which is not possible as $\alpha \vdash \beta$ is valid in $\mathfrak{R}(\mathcal{K})$. Hence we get a contradiction.

A similar argument using the result for **PDBL** corresponding to Proposition 8 and the representation Theorem 6, gives (b). □

7   Conclusion

This work proposes the notion of object oriented protoconcepts, and characterizes them in terms of approximations by object oriented concepts. A representation result is obtained, showing that any double Boolean algebra is quasi-embeddable in an algebra of object oriented protoconcepts. A logic **DBL** for the class of object oriented protoconcepts is defined and extended to a logic **PDBL** for the class of pure double Boolean algebras. Using the representation results for double and pure double Boolean algebras, the logics are shown to be sound and complete with respect to the class of object oriented protoconcepts and semiconcepts over the class of all contexts respectively. As further work, one can investigate representation results for the algebras that yield isomorphisms. On the side of the logics, other semantics could be explored – for instance, a Kripke-style semantics that may be in the line of work done in [2].

References

1. Balbiani, P.: Deciding the word problem in pure double Boolean algebras. J. Appl. Logic **10**, 260–273 (2012)
2. Conradie, W., Frittella, S., Palmigiano, A., Piazzai, M., Tzimoulis, A., Wijnberg, N.M.: Toward an epistemic-logical theory of categorization. In: Electronic Proceedings in Theoretical Computer Science, EPTCS, TARK 2017, vol. 251, pp. 167–186 (2017)
3. Davey, B.A., Priestley, H.A.: Introduction to Lattices and Order. Cambridge University Press, Cambridge (2002)
4. Düntsch, I., Gediga, G.: Modal-style operators in qualitative data analysis. In: Proceedings of 2002 IEEE International Conference on Data Mining, pp. 155–162. IEEE (2002)
5. Ganter, B., Wille, R.: Formal Concept Analysis: Mathematical Foundations. Springer, Heidelberg (2012)
6. Howlader, P., Banerjee, M.: Algebras from semiconcepts in rough set theory. In: Nguyen, H.S., Ha, Q.-T., Li, T., Przybyła-Kasperek, M. (eds.) IJCRS 2018. LNCS (LNAI), vol. 11103, pp. 440–454. Springer, Cham (2018). https://doi.org/10.1007/978-3-319-99368-3_34
7. Kwuida, L.: Prime ideal theorem for double Boolean algebras. Discuss. Math.-General Algebra Appl. **27**(2), 263–275 (2007)
8. Vormbrock, B.: A solution of the word problem for free double boolean algebras. In: Kuznetsov, S.O., Schmidt, S. (eds.) ICFCA 2007. LNCS (LNAI), vol. 4390, pp. 240–270. Springer, Heidelberg (2007). https://doi.org/10.1007/978-3-540-70901-5_16

9. Vormbrock, B., Wille, R.: Semiconcept and protoconcept algebras: the basic theorems. In: Ganter, B., Stumme, G., Wille, R. (eds.) Formal Concept Analysis. LNCS (LNAI), vol. 3626, pp. 34–48. Springer, Heidelberg (2005). https://doi.org/10.1007/11528784_2

10. Wille, R.: Concept lattices and conceptual knowledge systems. Comput. Math. Appl. **23**(6–9), 493–515 (1992)

11. Wille, R.: Boolean concept logic. In: Ganter, B., Mineau, G.W. (eds.) ICCS-ConceptStruct 2000. LNCS (LNAI), vol. 1867, pp. 317–331. Springer, Heidelberg (2000). https://doi.org/10.1007/10722280_22

12. Yang, L., Luoshan, X.: On rough concept lattices. Electron. Notes Theor. Comput. Sci. **257**, 117–133 (2009)

13. Yao, Y.: A comparative study of formal concept analysis and rough set theory in data analysis. In: Tsumoto, S., Słowiński, R., Komorowski, J., Grzymała-Busse, J.W. (eds.) RSCTC 2004. LNCS (LNAI), vol. 3066, pp. 59–68. Springer, Heidelberg (2004). https://doi.org/10.1007/978-3-540-25929-9_6

14. Yao, Y.: Concept lattices in rough set theory. In: 2004 Annual Meeting of the North American Fuzzy Information Processing Society, vol. 2, pp. 796–801. IEEE (2004)

15. Yao, Y., Chen, Y.: Rough set approximations in formal concept analysis. In: Peters, J.F., Skowron, A. (eds.) Transactions on Rough Sets V. LNCS, vol. 4100, pp. 285–305. Springer, Heidelberg (2006). https://doi.org/10.1007/11847465_14

Fuzzy FCA Attribute Reduction Properties in Rough Set Theory

M. José Benítez-Caballero$^{(\boxtimes)}$ [iD], Jesús Medina [iD], and Eloísa Ramírez-Poussa [iD]

Department of Mathematics, University of Cádiz, Cádiz, Spain
{mariajose.benitez,jesus.medina,eloisa.ramirez}@uca.es

Abstract. Formal concept analysis and rough set theory are two of the most important mathematical tools for the treatment of information collected on relational systems. In particular, the idea of reducing the size of a database is widely studied in both theories separately. There are some papers that studied the reduction of a formal context by means of reducts from rough set. In this paper, we are focused in the reduction obtained in an information system considering the FCA reduction mechanism.

Keywords: Reduct · Formal concept analysis · Rough set · Size reduction

1 Introduction

Databases have been used in order to collect and store information in many fields of the everyday life as medicine, industry, criminology and more. The importance of databases has led to the development of mathematical tools for their study and management. Two powerful and useful mathematical tools are Formal Concept Analysis (FCA) [11] and Rough Set Theory (RST) [14]. Moreover, these two frameworks can be connected, among other ways, through the way in which databases are represented. Relational systems are composed by a set of attributes, a set of objects and a relationship between them.

One of the most important goal in both theories is the reduction of the size of databases. In order to do that, the notion of reduct arose. As a general definition, a reduct is a minimal subset of attributes keeping the original information. In a FCA framework, a reduct is a subset of attributes with which an isomorphic lattice to the original one is constructed [6,9,10]. In the case of RST, the minimal subset of attributes keeping the indiscernibility objects is called a reduct [8,16,17]. There are some papers connecting the reduction mechanisms in

Partially supported by the the 2014-2020 ERDF Operational Programme in collaboration with the State Research Agency (AEI) in projects TIN2016-76653-P and PID2019-108991GB-I00, and with the Department of Economy, Knowledge, Business and University of the Regional Government of Andalusia. in project FEDER-UCA18-108612, and by the European Cooperation in Science & Technology (COST) Action CA17124

both theories. A new reduction mechanism in FCA by means of reducts of RST is proposed in [2]. In the aforementioned paper, the results and properties are developed in a classical environment. Additionally, this study was extended in paper [3] to a fuzzy environment. Furthermore, a comparative study with other mechanism is presented. In these papers, the process starts in a context, an associated information system is constructed and the reductions in such information system are computed. Then, the lattice of reduced context with the calculated reducts is built. Therefore, these reduction mechanism provides a reduction in FCA considering the reducts of RST.

On the other hand, a reduction mechanism in an information system considering the FCA philosophy is studied in [1]. In this paper, we will define a fuzzy formal context and frame associated with an information system and the fuzzy environment for reduction in the associated context will be applied. Moreover, different properties will be studies in order to improve the FCA classification theorems in the RST framework.

Since this paper is focused on extend the study presented in [1], we recall the notions and results needed in Sect. 2. After that, the main contributions are presented in Sect. 3. Finally, in Sect. 4, we summarize the results in the conclusions and we present our future work.

2   Preliminaries

RST and FCA are the two mathematical tools considered in this paper in order to reduce the size of a relational database. Also, the relation between the reduction procedure in these frameworks is studied. Due to this fact, the notions and results needed from these two environments will be recalled in this section.

2.1   Rough Set Theory

Rough Set Theory was developed in order to treat and manage incomplete information. In this framework, information systems and decision systems are used to present the data [14,15]. These systems are composed by a set of objects, a set of attributes and a relation between them. In the particular case of the decision system, an specific attribute is considered to make an action over the objects. Since information systems are the only one considered in this paper, its definition is recalled:

Definition 1. *An* information system $(U, \mathcal{A})$ *is a tuple, where the set of objects* $U = \{x_1, x_2, \ldots, x_n\}$ *and the set of attributes* $\mathcal{A} = \{a_1, a_2, \ldots, a_m\}$ *are finite and non-empty sets. Each* $a \in \mathcal{A}$ *corresponds to a mapping* $\bar{a} : U \rightarrow V_a$, *where* V_a *is the value set of the attribute* a *over* U.

Moreover, for every subset D *of* $\mathcal{A}$, *the* D-*indiscernibility relation,* $Ind(D)$, *is defined by the following equivalence relation:*

$$Ind(D) = \{(x_i, x_j) \in U \times U \mid \text{ for all } a \in D, \bar{a}(x_i) = \bar{a}(x_j)\}$$

where each equivalence class is written as $[x]_D = \{x_i \in U \mid (x, x_i) \in Ind(D)\}$. $Ind(D)$ *produces a partition on* U *denoted as* $U/Ind(D) = \{[x]_D \mid x \in U\}$.

The following definition presents a useful tool for selecting the attributes which discern two objects [15], the discernibility matrix.

Definition 2. *Given an information system* $(U, \mathcal{A})$, *its discernibility matrix is a matrix with order* $|U| \times |U|$, *denoted by* $M_{\mathcal{A}}$, *in which the element* $M_{\mathcal{A}}(x, y)$ *for each pair of objects* (x, y) *is defined by:*

$$M_{\mathcal{A}}(x, y) = \{a \in \mathcal{A} \mid \bar{a}(x) \neq \bar{a}(y)\}$$

Moreover, in order to generalize this notions to a fuzzy environment, a fuzzy indiscernibility relationship can be considered to compare the objects instead of a classical one. This indiscernibility relation can be defined over any poset P, that is, $R \colon U \times U \to P$, as it was introduced in [8]. Due to the study if this paper is focused on the properties of the fuzzy FCA reduction to RST, the interval $[0, 1]$ will be considered as the poset used to define the indiscernibility relation.

From this point forward, we are going to consider the definition of the information system presented in Definition 1 together with the fuzzy indiscernibility relation $R \colon U \times U \to [0, 1]$ defined as:

$$R(x_i, x_j) = @(R_{a_1}(x_i, x_j), \ldots, R_{a_n}(x_i, x_j))$$

for every pair of objects $x_i, x_j \in U$, where $@ \colon [0, 1]^n \to [0, 1]$ is an aggregation operator and $R_{a_k} \colon U \times U \to [0, 1]$ is a tolerance relation with respect to the attribute a_k, for all $k \in \{1, \ldots, n\}$.

2.2 Multi-adjoint Formal Concept Analysis

The considered operators in order to define the concept-forming operators are the adjoint triples, which are generalizations of a triangular norm and its residuated implication [12].

Definition 3. *Let* $(P_1, \leq_1)$, $(P_2, \leq_2)$, $(P_3, \leq_3)$ *be posets and* $\& \colon P_1 \times P_2 \to P_3$, $\swarrow \colon P_3 \times P_2 \to P_1$, $\nwarrow \colon P_3 \times P_1 \to P_2$ *be mappings, then* $(\&, \swarrow, \nwarrow)$ *is an* adjoint triple *with respect to* P_1, P_2, P_3 *if the following double equivalence holds:*

$$x \leq_1 z \swarrow y \quad \text{iff} \quad x \& y \leq_3 z \quad \text{iff} \quad y \leq_2 z \nwarrow x \tag{1}$$

for all $x \in P_1$, $y \in P_2$ *and* $z \in P_3$. *This double equivalence is called* adjoint property.

Next, we will present the boundaries properties verified by operators of an adjoint triple.

Proposition 1. *Given an adjoint triple* $(\&, \swarrow, \nwarrow)$ *with respect to the posets* $(P_1, \leq_1, \perp_1, \top_1)$, $(P_2, \leq_2, \perp_2, \top_2)$ *and* $(P_3, \leq_3, \perp_3, \top_3)$, *the following boundary conditions are hold:*

1. $\perp_1 \,\&\, y = \perp_3 \; y \; x \,\&\, \perp_2 = \perp_3$, *for all* $x \in P_1$, $y \in P_2$.
2. $z \diagdown \perp_1 = \top_2 \; y \; z \diagup \perp_2 = \top_1$, *for all* $z \in P_3$.
3. $\top_3 \diagdown x = \top_2 \; y \; \top_3 \diagup y = \top_1$, *for all* $x \in P_1$, $y \in P_2$.

The following result presents a relation between boundary conditions which was proved in [4].

Proposition 2. *Let us consider an adjoint triple* $(\&, \diagup, \diagdown)$ *with respect to the posets* $(P_1, \leq_1, \perp_1, \top_1)$, $(P_2, \leq_2, \perp_2, \top_2)$ *and* $(P_3, \leq_3, \perp_3, \top_3)$*. If* $P_1 = P_3$*, we have that the following equivalence is provided:*

$$z \diagdown \top_2 = z, \text{ for all } z \in P_3 \text{ if and only if } x \,\&\, \top_2 = x, \text{ for all } x \in P_1$$

In the concept lattice settings, we need to consider that $(P_1, \leq_1)$ and $(P_2, \leq_2)$ are complete lattices [13]. In the following, we are going to recall the notion of multi-adjoint frame.

Definition 4. *A* multi-adjoint frame *$\mathcal{L}$ is a tuple:*

$$(L_1, L_2, P, \preceq_1, \preceq_2, \leq, \&_1, \diagup^1, \diagdown_1, \ldots, \&_n, \diagup^n, \diagdown_n)$$

where $(L_1, \preceq_1)$ *and* $(L_2, \preceq_2)$ *are complete lattices,* $(P, \leq)$ *is a poset and, for all* $i = 1, \ldots, n$*,* $(\&_i, \diagup^i, \diagdown_i)$ *is an adjoint triple with respect to* L_1, L_2, P*. Multi-adjoint frames are denoted as* $(L_1, L_2, P, \&_1, \ldots, \&_n)$*.*

Given a frame, a *multi-adjoint context* is a tuple consisting of sets of objects, attributes and a fuzzy relation among them; in addition, the multi-adjoint approach also includes a function which assigns an adjoint triple to each pair of objects and attributes.

Definition 5. *Let* $(L_1, L_2, P, \&_1, \ldots, \&_n)$ *be a multi-adjoint frame, a* context *is a tuple* (A, B, R, σ) *such that* A *and* B *are non-empty sets (usually interpreted as attributes and objects, respectively),* R *is a P-fuzzy relation* $R \colon A \times B \to P$ *and* $\sigma \colon A \times B \to \{1, \ldots, n\}$ *is a mapping which associates any element in* $A \times B$ *with some particular adjoint triple in the frame.*

Given a multi-adjoint frame and a context for that frame, the concept-forming operators are denoted as $\uparrow^\sigma \colon L_2^B \longrightarrow L_1^A$ and $\downarrow^\sigma \colon L_1^A \longrightarrow L_2^B$, L_2^B and L_1^A denote the set of fuzzy subsets $g \colon B \to L_2$ and $f \colon A \to L_1$, respectively, and are defined, for all $g \in L_2^B$, $f \in L_1^A$ and $a \in A$, $b \in B$, as:

$$g^{\uparrow_\sigma}(a) = \inf\{R(a, b) \diagup^{\sigma(a, b)} g(b) \mid b \in B\} \tag{2}$$

$$f^{\downarrow^\sigma}(b) = \inf\{R(a, b) \diagdown_{\sigma(a, b)} f(a) \mid a \in A\} \tag{3}$$

These two arrows form a Galois connection [13]. Hence, the notion of concept is defined as usual: a *multi-adjoint concept* is a pair $\langle g, f \rangle$ satisfying that $g \in L_2^B$, $f \in L_1^A$ and that $g^{\uparrow_\sigma} = f$ and $f^{\downarrow^\sigma} = g$; with $(\uparrow_\sigma, \downarrow^\sigma)$ being the Galois connection defined above.

Given $g \in L_2^B$ (resp. $f \in L_1^A$), we will call the concept $\langle g^{\uparrow_\sigma \downarrow^\sigma}, g^{\uparrow_\sigma} \rangle$ (resp. $\langle f^{\downarrow^\sigma}, f^{\downarrow^\sigma \uparrow_\sigma} \rangle$) the *concept generated by* g *(resp.* f*).*

Definition 6. *The* multi-adjoint concept lattice *associated with a multi-adjoint frame* $(L_1, L_2, P, \&_1, \ldots, \&_n)$ *and a context* (A, B, R, σ) *is the set*

$$\mathcal{M}(A, B, R, \sigma) = \{\langle g, f \rangle \mid g \in L_2^B, f \in L_1^A \text{ and } g^{\uparrow_\sigma} = f, f^{\downarrow^\sigma} = g\}$$

in which the ordering is defined by $\langle g_1, f_1 \rangle \preceq \langle g_2, f_2 \rangle$ *if and only if* $g_1 \preceq_2 g_2$ *(equivalently* $f_2 \preceq_1 f_1$*).*

The ordering just defined above provides $\mathcal{M}$ with the structure of a complete lattice [13]. From now on, in order to simplify the notation, we will write $\uparrow$ and $\downarrow$ instead of $\uparrow_\sigma$ and $\downarrow^\sigma$, respectively.

A theory for attribute reduction in multi-adjoint concept lattices will be introduced. From now on, a multi-adjoint frame $(L_1, L_2, P, \&_1, \ldots, \&_n)$ and a context (A, B, R, σ) will be fixed.

The following definition presents the most natural extension of a consistent set in the multi-adjoint framework, keeping the definitions considered in Rough Set Theory [14].

Definition 7. *A set of attributes* $Y \subseteq A$ *is called a* consistent set *of* (A, B, R, σ) *if* $\mathcal{M}(Y, B, R_Y, \sigma_{Y \times B}) \cong_E \mathcal{M}(A, B, R, \sigma)$. *This is equivalent to say that, for all* $\langle g, f \rangle \in \mathcal{M}(A, B, R, \sigma)$, *there exists a concept* $\langle g', f' \rangle \in \mathcal{M}(Y, B, R_Y, \sigma_{Y \times B})$ *such that* $g = g'$.

Moreover, if $\mathcal{M}(Y \setminus \{a\}, B, R_{Y \setminus \{a\}}, \sigma_{Y \setminus \{a\} \times B}) \not\cong_E \mathcal{M}(A, B, R, \sigma)$, *for all* $a \in Y$, *then* Y *is called a* reduct *of* (A, B, R, σ).

The core *of* (A, B, R, σ) *is the intersection of all the reducts of* (A, B, R, σ).

The main idea in attribute reduction in formal concept analysis is to classify the attributes from the irreducible elements in the concept lattice. Therefore, the definition of irreducible element of a lattice must be introduced.

Definition 8. *Given a lattice* $(L, \preceq)$, *such that* $\wedge, \vee$ *are the meet and the join operators. An element* $x \in L$ *verifying that*

1. *If* L *has a top element* $\top$, *then* $x \neq \top$.
2. *If* $x = y \wedge z$, *then* $x = y$ *or* $x = z$, *for all* $y, z \in L$.

is called meet-irreducible *($\wedge$-irreducible) element of* L. *Condition* (2) *is equivalent to*

2'. *If* $x < y$ *and* $x < z$, *then* $x < y \wedge z$, *for all* $y, z \in L$.

Hence, if x *is* $\wedge$-*irreducible, then it cannot be represented as the infimum of strictly greatest elements.*

A join-irreducible *($\vee$-irreducible) element of* L *is defined dually.*

A characterization of the meet-irreducible elements of a multi-adjoint concept lattice is introduced in this section. A similar result can be given to the join-irreducible elements.

Hence, we will consider a multi-adjoint concept lattice $(\mathcal{M}, \preceq)$ associated with a multi-adjoint frame $(L_1, L_2, P, \&_1, \ldots, \&_n)$, a context (A, B, R, σ), where L_1, L_2, P, A and B are finite and the following specific family of fuzzy subsets of L_1^A:

Definition 9. *For each $a \in A$, the fuzzy subsets of attributes $\phi_{a,x} \in L_1^A$ defined, for all $x \in L_1$, as*

$$\phi_{a,x}(a') = \begin{cases} x & if\ a' = a \\ 0 & if\ a' \neq a \end{cases}$$

will be called fuzzy-attributes. *The set of all fuzzy-attributes will be denoted as $\Phi = \{\phi_{a,x} \mid a \in A, x \in L_1\}$.*

Note that these mappings are generalizations of the crisp attributes and they were also assumed in the proof of representation theorem of several fuzzy concept lattices.

Once the technical results are introduced, the characterization of the $\wedge$-irreducible elements of multi-adjoint concept lattices can be proven. This theorem shows that the $\wedge$-irreducible elements are concepts generated by fuzzy-attributes and no more concepts can be $\wedge$-irreducible elements.

Theorem 1 [5]. *The set of $\wedge$-irreducible elements of $\mathcal{M}$, $M_F(A, B, R, \sigma)$, is:*

$$\left\{ \langle \phi_{a,x}^\downarrow, \phi_{a,x}^{\downarrow\uparrow} \rangle \mid \phi_{a,x}^\downarrow \neq \bigwedge \{\phi_{a_i,x_i}^\downarrow \mid \phi_{a_i,x_i} \in \Phi, \phi_{a,x}^\downarrow \prec_2 \phi_{a_i,x_i}^\downarrow\} \text{ and } \phi_{a,x}^\downarrow \neq g_\top \right\}$$

where $\top$ is the maximum element in L_2 and $g_\top \colon B \to L_2$ is the fuzzy subset defined as $g_\top(b) = \top$, for all $b \in B$.

The following definition presents a notion needed to classify the attributes of a context by means of the attributes used to generate a concept.

Definition 10. *Given a multi-adjoint frame $(L_1, L_2, P, \&_1, \dots, \&_n)$, a context (A, B, R, σ) associated with the concept lattice $(\mathcal{M}, \preceq)$ and a concept C of $(\mathcal{M}, \preceq)$, the set of attributes generating C is defined as the set:*

$$Atg(C) = \{a \in A \mid \text{there exists } x \in L_1 \text{ such that } \langle \phi_{a,x}^\downarrow, \phi_{a,x}^{\downarrow\uparrow} \rangle = C\}$$

As in this paper we are using the philosophy of fuzzy FCA in order to reduce the set of attributes, we will need some results that characterize the attributes from the meet-irreducible elements of the multi-adjoint concept lattice. The results presented are an adaptation of the attribute classification theorems introduced in [5], by using the Definition 10. This improvement in results simplifies the notation and facilitates its application. The following result characterizes the absolutely necessary attributes, by means of Definition 10.

Theorem 2 [7]. *Given an attribute $a \in A$, then $a \in C_f$ if and only if there exists a meet-irreducible concept C of $(\mathcal{M}, \preceq)$ satisfying that $a \in Atg(C)$ and $card(Atg(C)) = 1$.*

Finally, the next proposition shows the characterization of the absolutely unnecessary attributes considering the attributes generating a concept. To this group belong those attributes that are neither absolutely necessary nor relatively necessary.

Theorem 3 [7]. *Given an attribute $a \in A$, then $a \in I_f$ if and only if, for any $C \in M_F(A)$, $a \notin Atg(C)$, or if $a \in Atg(C)$ then $(A \setminus Atg(C)) \cup \{a\}$ is not a consistent set.*

3   Main Properties of Fuzzy Attribute Reduction in Information Systems with FCA Philosophy

This section will recall some notions and results presented in [1] and will study some properties of the attribute reduction of an information system provided by the FCA mechanism. First of all, we define a formal context from an information system, due to the fact that we are starting in a rough set framework.

Definition 11. *Given an information system* $(U, \mathcal{A})$ *and the indiscernibility relation* $R \colon U \times U \to L$ *defined on* $(U, \mathcal{A})$, *the associated fuzzy context is the context* (U, U, R).

Usually, L is the unit interval, although it can be, for example, a non-linear lattice or a granularity of the unit interval, such as $[0, 1] = \{0, 1/100, \dots, 99/100, 1\}$. Additionally, we need to fixe an associated framework $(L_1, L_2, P, \&_1, \dots, \&_n)$. In this case, the lattices L_1 and L_2 should be equal since they will represent the truth values associated with the elements of the same set. Moreover, we will need the adjoint conjunctors $\&_i$, for all $i \in \{1, \dots, n\}$, satisfy the boundary condition with the top element:

$$\top_1 \& x = x \tag{4}$$

Although this condition may seem very strict, there are really many adjoint triples that satisfy it, like t-norms (e.g., Gödel, Łukasiewicz and product) and other more general operators such as $x \& y = x^2 \cdot y$.

Furthermore, since Proposition 2 will be used later, we also need that $L_1 = P$. As a consequence, the sets L_1, L_2, P must be equal and depending on the truth values set considered in the indiscernibility relation R. In this paper, we will consider linear lattices.

The following example shows how to obtain an associated fuzzy context from a given information system. This example will be used in order to illustrate the properties presented in this paper.

Example 1. In this example, we will consider the information system $(U, \mathcal{A})$ presented in [1], composed of the set of objects $U = \{x_1, x_2, x_3, x_4, x_5\}$, the set of attributes $\mathcal{A} = \{a_1, a_2, a_3, a_4\}$ and the following table showing the relation between attributes and objects, by a truth value in $[0, 1]_{100}$:

	x_1	x_2	x_3	x_4	x_5
a_1	0.34	0.21	0.52	0.84	0.83
a_2	0.13	0.09	0.36	0.16	0.15
a_3	0.31	0.71	0.93	0.69	0.69
a_4	0.75	0.5	1	1	1

In order to build the discernibility matrix needed to define the relation in the context, we will consider the following fuzzy tolerance relation between objects for any attribute $a_i \in \mathcal{A}$:

$$R_{a_i}(x, y) = 1 - |\bar{a}_i(x) - \bar{a}_i(y)|$$

Then, we will take into account the aggregation operator $@(l_1, l_2.l_3, l_4) = \frac{1}{6}(l_1 + l_2 + 2(l_3 + l_4))$ to compute the values of the discernibility relation, which compares every couple of objects $x, y \in U$ as follows:

$$R_A(x, y) = @(R_{a_1}(x, y), R_{a_2}(x, y), R_{a_3}(x, y), R_{a_4}(x, y)) \tag{5}$$

Therefore, the following discernibility matrix is obtained:

$$\begin{pmatrix} 1 & 0.8 & 0.6 & 0.7 & 0.7 \\ 0.8 & 1 & 0.7 & 0.7 & 0.7 \\ 0.6 & 0.7 & 1 & 0.8 & 0.8 \\ 0.7 & 0.7 & 0.8 & 1 & 1 \\ 0.7 & 0.7 & 0.8 & 1 & 1 \end{pmatrix}$$

Now, we are able to built the associated fuzzy context (U, U, R), where R is the indiscernibility relation described in the discernibility matrix, and we consider the frame $([0, 1]_{10}, [0, 1]_{10}, [0, 1]_{10}, \&_G)$ where $\&_G$ is the Gödel conjunctor.

In order to stress the usefulness of the proposed attribute reduction mechanism in RST, based on FCA, diverse interesting properties will be studied next, relating elements of FCA and RST. The first result presents the value taken by the intent over the attribute generating the fuzzy-attribute over a general context.

Proposition 3. *Let (A, B, R, σ) be a multi-adjoint context, a multi-adjoint frame $(L_1, L_2, P, \&_1, \ldots, \&_n)$, an attribute $a \in A$ and a truth value $\beta \in L_1$. If the inequality $\beta \leq R(a, b)$ holds, for all $b \in B$, then*

$$\phi_{a,\beta}^{\downarrow\uparrow}(a) = \inf\{R(a, b) \mid b \in B\}$$

Proof. Taking into account the definitions of concept-forming operators presented in Expressions (2) and (3), we have for an attribute $a \in A$ and a truth value $\beta \in L_1$ that:

$$\phi_{a,\beta}^{\downarrow\uparrow}(a) = \inf\{R(a, b) \swarrow (R(a, b) \diagdown \beta) \mid b \in B\}$$

$$\overset{(*)}{=} \inf\{R(a, b) \swarrow 1 \mid b \in B\}$$

$$\overset{(**)}{=} \inf\{R(a, b) \mid b \in B\}$$

In order to justify Equality $(*)$ in the above chain of equalities, we know that $\beta \leq R(a, b)$, and considering the boundary condition presented in Eq. (4), we have that $\beta \& 1 \leq R(a, b)$. Then, taking into consideration the adjoint property described in Expression (1), we obtain that $1 \leq R(a, b) \diagdown \beta$, that is, $R(a, b) \diagdown \beta = 1$.

On the other hand, we have Equality $(**)$ thanks to Proposition 2 and boundary condition (4). Hence, we obtain that

$$\phi_{a,\beta}^{\downarrow\uparrow}(a) = \inf\{R(a, b) \mid b \in B\} \qquad \square$$

The following example illustrates this idea using the context presented in Example 1.

Example 2. Let us consider the formal context and the multi-adjoint frame presented in Example 1. Let us consider the attribute $x_2 \in U$ and the truth value $\beta = 0.5$. We have that $0.5 \leq R(x_2, x_i)$, for all $i \in \{1, 2, 3, 4, 5\}$, we will calculate $\phi_{x_2, 0.5}^{\downarrow}(x_2)$. By definition of concept-forming operator presented in Expression 3, we have that:

$$\phi_{x_2, 0.5}^{\downarrow}(x_2) = \inf\{R(x_i, x_2) \diagdown \phi_{x_2, 0.5}(x_i) \mid x_i \in U\}$$
$$= \inf\{0.8 \diagdown 0, 1 \diagdown 0.5, 0.7 \diagdown 0, 0.7 \diagdown 0, 0.7 \diagdown 0\}$$
$$= \inf\{1, 1, 1, 1, 1\} = 1$$

Applying the same calculations for all the objects, we obtain that $\phi_{x_2, 0.5}^{\downarrow} = \{1/x_1, 1/x_2, 1/x_3, 1/x_4, 1/x_5\}$. Considering now, the definition of the other concept-forming operator represented in Expression 2, we have that:

$$\phi_{x_2, 0.5}^{\downarrow\uparrow}(x_2) = \inf\{R(x_i, x_2) \diagup \phi_{x_2, 0.5}^{\downarrow}(x_i) \mid x_i \in U\}$$
$$= \inf\{0.8 \diagup 1, 1 \diagup 1, 0.7 \diagup 1, 0.7 \diagup 1, 0.7 \diagup 1\}$$
$$= \inf\{0.8, 1, 0.7, 0.7, 0.7\}$$
$$= \inf\{R(x_1, x_2), R(x_2, x_2), R(x_3, x_2), R(x_4, x_2), R(x_5, x_2)\} \quad \square$$

The following result highlights the important role the relationship R have for computing intents of formal concepts.

Proposition 4. *Given a multi-adjoint context (A, B, R, σ), a multi-adjoint framework $(L_1, L_2, P, \&_1, \ldots, \&_n)$, an attribute $a \in A$ and two truth values $\beta, \beta' \in L_1$, if $\beta, \beta' \leq R(a, b)$, for all $b \in B$, then*

$$\phi_{a, \beta}^{\downarrow\uparrow} = \phi_{a, \beta'}^{\downarrow\uparrow}$$

Proof. Given an attribute $a' \in A$, if we consider the truth values $\beta, \beta' \leq R(a, b)$, we have that

$$\phi_{a, \beta}^{\downarrow\uparrow}(a') = \inf\{R(a', b) \diagup (R(a, b) \diagdown \beta) \mid b \in B\}$$
$$= \inf\{R(a', b) \diagup 1 \mid b \in B\}$$
$$= \inf\{R(a', b) \mid b \in B\}$$

due to $\beta \leq R(a, b)$. Analogously, taking into account the truth value β', we have that

$$\phi_{a, \beta'}^{\downarrow\uparrow}(a') = \inf\{R(a', b) \mid b \in B\}$$

Therefore, for all $a' \in A$, we have that

$$\phi_{a, \beta}^{\downarrow\uparrow}(a') = \inf\{R(a', b) \mid b \in B\} = \phi_{a, \beta'}^{\downarrow\uparrow}(a')$$

Hence, we obtain that $\phi_{a, \beta}^{\downarrow\uparrow} = \phi_{a, \beta'}^{\downarrow\uparrow}$. $\quad\square$

From now on, we will consider the associated context from an information system $(U, \mathcal{A})$, obtained as Definition 11 describes. Notice that, as R is built with the indiscernibility relation, the relation R verifies the reflexive property.

Proposition 5. *Let us consider an information system* $(U, \mathcal{A})$, *its associated fuzzy context* (U, U, R), *its associated frame* $(L, L, L, \&)$ *and an object* $x_1 \in U$, *we have that*

$$\phi^{\downarrow\uparrow}_{x_1,\beta}(x) \leq R(x, x_1)$$

for all truth value $\beta \in L_1$ *and object* $x \in U$.

Proof. By definitions of the concept-forming operator presented in Expressions (2) and (3), given $x, x_1 \in U$, we have that

$$\phi^{\downarrow\uparrow}_{x_1,\beta}(x) = \inf\{R(x, x_2) \swarrow (R(x_1, x_2) \searrow \beta) \mid x_2 \in U\}$$

If $x_2 = x_1$, we obtain by Proposition 2 and boundary condition (4) that:

$$\phi^{\downarrow\uparrow}_{x_1,\beta}(x) \leq R(x, x_1) \swarrow (R(x_1, x_1) \searrow \beta)$$
$$= R(x, x_1) \swarrow (1 \searrow \beta)$$
$$= R(x, x_1) \swarrow 1$$
$$= R(x, x_1)$$

which proves the result. □

Next, the following proposition characterizes the value taken by the relation over two objects which generate the same intent of a concept.

Proposition 6. *Let* $(U, \mathcal{A})$ *be an information system,* (U, U, R) *its associated fuzzy context, its associated frame* $(L, L, L, \&)$, *two objects* $x_i, x_j \in U$ *and two truth values* $\beta, \beta' \in L$. *If the equality* $\phi^{\downarrow\uparrow}_{x_i,\beta} = \phi^{\downarrow\uparrow}_{x_j,\beta'}$ *holds, then*

$$\beta, \beta' \leq R(x_i, x_j)$$

Proof. By Proposition 5 and the hypothesis, evaluating over the object x_j, we have that

$$\beta' \leq \phi^{\downarrow\uparrow}_{x_j,\beta'}(x_j) = \phi^{\downarrow\uparrow}_{x_i,\beta}(x_j) \leq R(x_j, x_i)$$

Analogously, considering now the object x_i, we obtain that

$$\beta \leq \phi^{\downarrow\uparrow}_{x_i,\beta}(x_i) = \phi^{\downarrow\uparrow}_{x_j,\beta'}(x_i) \leq R(x_i, x_j)$$

Therefore, the result holds by the symmetry of R. □

The following consequence from the proposition above and Proposition 3 shows interesting lower and upper bounds associated with intents generated by different objects.

Corollary 1. *Given an information system* $(U, \mathcal{A})$, *its associated fuzzy context* (U, U, R), *its associated frame* $(L, L, L, \&)$, *two objects* $x_i, x_j \in U$ *and two truth values* $\beta, \beta' \in L$, *verifying that* $\phi_{x_i, \beta}^{\downarrow\uparrow} = \phi_{x_j, \beta'}^{\downarrow\uparrow}$, *then, the following chain of inequalities holds*

$$\inf\{R(x_i, x) \mid x \in U\} \leq \phi_{x_i, \beta}^{\downarrow\uparrow}(x_i) \leq R(x_i, x_j)$$

The upper truth value threshold can be determined to every object.

Proposition 7. *Let us consider an information system* $(U, \mathcal{A})$, *its associated fuzzy context* (U, U, R), *its associated frame* $(L, L, L, \&)$, *two objects* $x_i, x_j \in U$ *and two truth values* $\beta, \beta' \in L$. *If the equality* $\phi_{x_i, \beta}^{\downarrow\uparrow} = \phi_{x_j, \beta'}^{\downarrow\uparrow}$ *holds, then we have that*

$$\phi_{x_i, \beta}^{\downarrow\uparrow}(x) \leq R(x, x_i) \wedge R(x, x_j) \leq R(x_i, x_j)$$

for all object $x \in U$.

Proof. Follows from Proposition 5, the equality $\phi_{x_i, \beta_1}^{\downarrow\uparrow}(x) = \phi_{x_j, \beta_2}^{\downarrow\uparrow}(x)$, for all $x \in U$, and the transitivity property of R.

As a consequence, the following corollary arises.

Corollary 2. *Let* $(U, \mathcal{A})$ *be an information system,* (U, U, R) *its associated fuzzy context, its associated frame* $(L, L, L, \&)$, *two objects* $x_i, x_j \in U$ *and two truth values* $\beta, \beta' \in L$. *If the equality* $\phi_{x_i, \beta}^{\downarrow\uparrow} = \phi_{x_j, \beta'}^{\downarrow\uparrow}$ *holds, then*

$$\phi_{x_i, R(x_i, x_j)}^{\downarrow\uparrow}(x_i) = R(x_i, x_j)$$

By Proposition 6 we have that if $x_i, x_j \in \text{Atg}(C)$, then the values $\beta, \beta' \in L$, such that $\phi_{x_i, \beta}^{\downarrow\uparrow} = \phi_{x_j, \beta'}^{\downarrow\uparrow}$ must be less or equal to $R(x_i, x_j)$. Therefore, since

$$[x_i]_\alpha = \{x \in U \mid \alpha \leq R(x_i, x)\} \tag{6}$$

we cannot establish a relationship between α-block and the classification of attributes, as it was given in [1]. We can provide the following improvement of the FCA attribute classification on the RST framework.

Corollary 3. *Let* $(U, \mathcal{A})$ *be an information system,* (U, U, R) *its associated fuzzy context, an object* $x_i \in U$. *If* $\phi_{x_i, R(x_i, x)}^{\downarrow\uparrow}(x_i) \neq R(x_i, x)$, *for all* $x \in U$, *then* $x_i \in C_f$ *or* $x_i \in I_f$.

This consequence notably reduce the number of computations in order to know whether an object is absolutely necessary. Another reduction for checking whether an object is a core element is the following.

Corollary 4. *Let* $(U, \mathcal{A})$ *be an information system,* $x_i \in U$, *and* (U, U, R) *its associated fuzzy context. If there exists* $\beta \nleq \sup\{R(x_i, x) \mid x \in U\}$, *such as,* $\phi_{x_i, \beta}^{\downarrow\uparrow}$ *is the intent of a meet-irreducible, that is,* $\phi_{x_i, \beta}^{\downarrow\uparrow} = \text{Ext}(C)$, *with* $C \in M_F(U, U, R)$, *then* $x_i \in C_f$.

Therefore, in order to apply Theorem 2 we need to begin from the values $\beta \nleq \sup\{R(x_i, x) \mid x \in U\}$. This computation does not provide extra calculations but already required ones, and offers a sufficient condition to ensure whether $x_i \in C_f$. More improvements will be detailed in the future.

4   Conclusions and Future Work

In this paper, we have deepened in the study of the properties that the mechanism of reduction of an information system possesses considering the philosophy of FCA presented in [1]. We have proved different boundary properties, where the fuzzy relation (indiscernibility relation when RST is considered) plays an important role. For example, relevant boundary results have been proven in the particular case of two different objects generate the same formal concept, such as the upper bound is given by the relationship between both objects. As a consequence of these properties, we have provided several improvements over the attribute classification theorems [5].

As future work, more enhancements in the classification theorems over the objects will be studied, in order to provide a size reduction in an information system using the reduction mechanism of FCA. Also, we will study other ways to connect the reduction mechanism in FCA and RST, as the consideration of the bireducts of RST to reduce the size of a relational database of FCA.

References

1. Benítez-Caballero, M.J., Medina, J., Ramírez-Poussa, E.: FCA attribute reduction in information systems. In: Medina, J., et al. (eds.) IPMU 2018. Communications in Computer and Information Science, vol. 853, pp. 549–561. Springer, Cham (2018). https://doi.org/10.1007/978-3-319-91473-2_47
2. Benítez-Caballero, M.J., Medina, J., Ramírez-Poussa, E., Ślęzak, D.: A computational procedure for variable selection preserving different initial conditions. Int. J. Comput. Math. **97**, 387–404 (2019)
3. Benítez-Caballero, M.J., Medina, J., Ramírez-Poussa, E., Ślęzak, D.: Rough-set-driven approach for attribute reduction in fuzzy formal concept analysis. Fuzzy Sets Syst. **391**, 117–138 (2019)
4. Cornejo, M.E., Medina, J., Ramírez-Poussa, E.: A comparative study of adjoint triples. Fuzzy Sets Syst. **211**, 1–14 (2013)
5. Cornejo, M.E., Medina, J., Ramírez-Poussa, E.: Attribute reduction in multi-adjoint concept lattices. Inf. Sci. **294**, 41–56 (2015)
6. Cornejo, M.E., Medina, J., Ramírez-Poussa, E.: Attribute and size reduction mechanisms in multi-adjoint concept lattices. J. Comput. Appl. Math. **318**, 388–402 (2017). Computational and Mathematical Methods in Science and Engineering CMMSE-2015
7. Cornejo, M.E., Medina, J., Ramírez-Poussa, E.: Characterizing reducts in multi-adjoint concept lattices. Inf. Sci. **422**, 364–376 (2018)
8. Cornelis, C., Medina, J., Verbiest, N.: Multi-adjoint fuzzy rough sets: definition, properties and attribute selection. Int. J. Approx. Reason. **55**, 412–426 (2014)
9. Davey, B., Priestley, H.: Introduction to Lattices and Order, 2nd edn. Cambridge University Press, Cambridge (2002)
10. Dias, S.M., Vieira, N.J.: A methodology for analysis of concept lattice reduction. Inf. Sci. **396**, 202–217 (2017)
11. Ganter, B., Wille, R.: Formal Concept Analysis: Mathematical Foundation. Springer, Heidelberg (1999). https://doi.org/10.1007/978-3-642-59830-2

12. Hájek, P.: Metamathematics of Fuzzy Logic. Trends in Logic. Kluwer Academic, Dordrecht (1998)
13. Medina, J., Ojeda-Aciego, M., Ruiz-Calviño, J.: Formal concept analysis via multi-adjoint concept lattices. Fuzzy Sets Syst. **160**(2), 130–144 (2009)
14. Pawlak, Z.: Information systems theoretical foundations. Inf. Syst. **6**(3), 205–218 (1981)
15. Skowron, A., Rauszer, C.: The discernibility matrices and functions in information systems. In: Słowiński, R. (ed.) Intelligent Decision Support: Handbook of Applications and Advances of the Rough Sets Theory, pp. 331–362. Kluwer Academic Publishers, Dordrecht (1992)
16. Yao, Y., Zhao, Y.: Attribute reduction in decision-theoretic rough set models. Inf. Sci. **178**(17), 3356–3373 (2008)
17. Zhao, Y., Yao, Y., Luo, F.: Data analysis based on discernibility and indiscernibility. Inf. Sci. **177**(22), 4959–4976 (2007)

Algebraic Structure of Adjoint Triples Generating a Weak Negation on the Unit Interval

M. Eugenia Cornejo[✉], Jesús Medina, and Eloísa Ramírez-Poussa

Department of Mathematics, University of Cádiz, Cádiz, Spain
{mariaeugenia.cornejo,jesus.medina,eloisa.ramirez}@uca.es

Abstract. This paper presents a theoretical research about the relationship between weak negations and adjoint negations. Adjoint negations are a generalization of residuated negations built from the implications of an adjoint triple. Specifically, this work shows how to build adjoint triples on the unit interval such that their adjoint negations coincide with a given weak negation. Moreover, the algebraic structure formed by these adjoint triples is also investigated.

Keywords: Aggregation operator · Adjoint triple · Adjoint negation · Weak negation

1 Introduction

Non-monotonic operators play an important role in different applications [7,9,29,31, 38]. The need to use these operators in real applications has promoted the study and development of novel operators capable of addressing new challenges [3,4,8,14,18,38, 39]. Weak negations were introduced in [23,24,26,43] and they are one of the most versatile negation operators. For that reason, a generalization of weak negations was given in [14]. Specifically, it was proven that weak negations can be defined from the implications of adjoint triples.

Adjoint triples were introduced in [11–13] as a flexible tool to generalize the operators usually considered in residuated frameworks. One of the main advantages provided by these operators is the capability of being applied in non-associative or commutative settings. This fact has given rise to make more flexible frameworks such as logic programming [9,34–36], formal concept analysis [15,33], rough set theory [17], fuzzy relation equations [10,20,21] and fuzzy mathematical morphology [1,2,30].

In this paper, we will continue studying the relationship between weak negations and adjoint negations. Given a weak negation on the unit interval, we will show that different adjoint triples can be defined on the unit interval satisfying that its corresponding adjoint negations coincide with the weak negation. In addition, we will establish

Partially supported by the State Research Agency (AEI) and the European Regional Development Fund (ERDF) project TIN2016-76653-P, European Cooperation in Science & Technology (COST) Action CA17124.

© Springer Nature Switzerland AG 2020
R. Bello et al. (Eds.): IJCRS 2020, LNAI 12179, pp. 337–348, 2020.
https://doi.org/10.1007/978-3-030-52705-1_25

two different procedures to define these adjoint triples. We will also define an ordering relation on adjoint triples generating a given weak negation and we will prove that the set composed of these adjoint triples has the structure of a complete join-semilattice with maximum element.

The paper is organized as follows: Sect. 2 recalls the basic definitions and properties associated with adjoint triples, adjoint negations and weak negations. Given a weak negation, in Sect. 3, we propose two different mechanisms to define adjoint triples on $[0, 1]$ whose adjoint negations are equal to the considered weak negation. Furthermore, we analyze the algebraic structure formed by these adjoint triples. The contribution is accompanied by examples in order to illustrate some of the developed technical results. Some conclusions and prospects for future work are presented in Sect. 4.

2 Preliminaries

Adjoint triples provide an interesting generalization of the well-known adjoint property satisfied by a t-norm and its residuated implication, since they preserve the main properties usually assumed in residuated frameworks, dismissing for example the commutativity and the associativity [19].

Definition 1. *Let $(P_1, \leq_1)$, $(P_2, \leq_2)$, $(P_3, \leq_3)$ be posets and $\& : P_1 \times P_2 \to P_3$, $\swarrow : P_3 \times P_2 \to P_1$, $\nwarrow : P_3 \times P_1 \to P_2$ be mappings. We say that $(\&, \swarrow, \nwarrow)$ is an* adjoint triple *with respect to P_1, P_2, P_3 if the following double equivalence is satisfied:*

$$x \leq_1 z \swarrow y \quad \textit{iff} \quad x \& y \leq_3 z \quad \textit{iff} \quad y \leq_2 z \nwarrow x \tag{1}$$

for all $x \in P_1$, $y \in P_2$ and $z \in P_3$. The previous double equivalence is called adjoint property.

Interesting properties related to the monotonicity of the operators $\&$, $\swarrow$, $\nwarrow$, the boundary conditions and the preservation of the infimum and/or supremum, among others, can be deduced from the adjoint property. The following propositions show alternative ways to verify that the operators $\&$, $\swarrow$ and $\nwarrow$ form an adjoint triple, when they are defined on complete lattices.

Proposition 1 [13]. *Given the complete lattices $(L_1, \leq_1)$, $(L_2, \leq_2)$, $(L_3, \leq_3)$, an arbitrary operator $\& : L_1 \times L_2 \to L_3$ and the mappings $\swarrow : L_3 \times L_2 \to L_1$, $\nwarrow : L_3 \times L_1 \to L_2$, defined as $z \swarrow y = \sup\{x \in L_1 \mid x \& y \leq_3 z\}$ and $z \nwarrow x = \sup\{y \in L_2 \mid x \& y \leq_3 z\}$, respectively, for all $x \in L_1$, $y \in L_2$ and $z \in L_3$, the the following statements are equivalent:*

1. *$(\&, \swarrow, \nwarrow)$ is an adjoint triple with respect to L_1, L_2, L_3.*

2. *$\left(\bigvee_{x_i \in X} x_i \right) \& y = \bigvee_{x_i \in X} (x_i \& y)$, for any $X \subseteq L_1$ and $y \in L_2$.*

   $x \& \left(\bigvee_{y_i \in Y} y_i \right) = \bigvee_{y_i \in Y} (x \& y_i)$, for any $Y \subseteq L_2$ and $x \in L_1$.

3. $z \swarrow y = \max\{x \in L_1 \mid x \& y \leq_3 z\}$ and $z \nwarrow x = \max\{y \in L_2 \mid x \& y \leq_3 z\}$ for all $x \in L_1$, $y \in L_2$ and $z \in L_3$, where & is an order-preserving operator in both arguments.

Proposition 2 [13]. *Given three complete lattices* $(L_1, \leq_1)$, $(L_2, \leq_2)$, $(L_3, \leq_3)$, *the arbitrary operators* $\swarrow: L_3 \times L_2 \to L_1$, $\nwarrow: L_3 \times L_1 \to L_2$ *and the mapping* $\&: L_1 \times L_2 \to L_3$ *defined as* $x \& y = \inf\{z \in L_3 \mid x \leq_1 z \swarrow y\} = \inf\{z \in L_3 \mid y \leq_2 z \nwarrow x\}$, *for all* $x \in L_1$ *and* $y \in L_2$, *the following statements are equivalent:*

1. $(\&, \swarrow, \nwarrow)$ *is an adjoint triple with respect to* L_1, L_2, L_3.

2. $\left(\bigwedge_{z_i \in Z} z_i \right) \swarrow y = \bigwedge_{z_i \in Z} (z_i \swarrow y)$, *for all* $Z \subseteq L_3$ *and* $y \in L_2$.

   $\left(\bigwedge_{z_i \in Z} z_i \right) \nwarrow x = \bigwedge_{z_i \in Z} (z_i \nwarrow x)$, *for all* $Z \subseteq L_3$ *and* $x \in L_1$.

3. $x \& y = \min\{z \in L_3 \mid x \leq_1 z \swarrow y\} = \min\{z \in L_3 \mid y \leq_2 z \nwarrow x\}$, *for all* $x \in L_1$ *and* $y \in L_2$, *where* $\swarrow$ *and* $\nwarrow$ *are order-preserving operators in the first argument.*

A detailed study of adjoint triples can be found in [11, 13]. These operators were also considered to generalize residuated negations [6, 25, 40]. Specifically in [14], adjoint negations were defined from the implications of an adjoint triple. The formal definition of adjoint negations is given below.

Definition 2. *Let* $(P_1, \leq_1)$ *and* $(P_2, \leq_2)$ *be two posets,* $(P_3, \leq_3, \perp_3)$ *be a lower bounded poset and* $(\&, \swarrow, \nwarrow)$ *an adjoint triple with respect to* P_1, P_2 *and* P_3. *The mappings* $n_n: P_1 \to P_2$ *and* $n_s: P_2 \to P_1$ *defined, for all* $x \in P_1$, $y \in P_2$ *as:*

$$n_n(x) = \perp_3 \nwarrow x \qquad n_s(y) = \perp_3 \swarrow y$$

are called adjoint negations *with respect to* P_1 *and* P_2. *The operators* n_s *and* n_n *satisfying that* $x = n_s(n_n(x))$ *and* $y = n_n(n_s(y))$, *for all* $x \in P_1$ *and* $y \in P_2$, *are called* strong adjoint negations.

Now, we will show the notion of weak negation which is one of the most general negation operators given in the literature [23, 24, 26, 43].

Definition 3. *Given a mapping* $n: [0, 1] \to [0, 1]$ *is said to be a* weak negation *if the following conditions hold, for all* $x, y \in [0, 1]$:

1. $n(1) = 0$;
2. *if* $x \leq y$ *then* $n(y) \leq n(x)$;
3. $x \leq n(n(x))$.

We say that n *is a* strong negation *if the equality* $x = n(n(x))$ *holds, for all* $x \in [0, 1]$.

Once the notion of weak negation has been introduced, we can recall the relationship between adjoint negations and weak negations. The following result, which was proven in [14], shows that adjoint negations are more general than weak negations.

Theorem 1 [14]. *If the mapping* $n: [0, 1] \to [0, 1]$ *is a weak negation, then there exists an adjoint triple* $(\&, \swarrow, \nwarrow)$ *with respect to the poset* $([0, 1], \leq)$ *satisfying* $n = n_s = n_n$.

Notice that, the previous theorem shows that weak negations can be obtained from the implication operators of an adjoint triple. However, we cannot guarantee the unicity of the adjoint triple which allows us to ensure that each weak negation is actually an adjoint negation. This fact and the notions introduced in the current section will be illustrated in the following example.

Example 1. The most usual adjoint triples with respect to $([0, 1], \leq)$ are those defined from the Gödel, product and Łukasiewicz t-norms together with their residuated implications. Due to these t-norms are commutative, we have that $\swarrow^G = \nwarrow_G$, $\swarrow^P = \nwarrow_P$ and $\swarrow^L = \nwarrow_L$. As a consequence, the adjoint negations defined from these implications verify that $n_{s_G} = n_{n_G}$, $n_{s_P} = n_{n_P}$ and $n_{s_L} = n_{s_L}$. In order to simplify the notation, we will use n_G, n_P and n_L to refer to the adjoint negations obtained from the Gödel, product and Łukasiewicz implications, respectively. The mentioned adjoint triples are given below:

$$\&_G(x, y) = \min\{x, y\} \qquad z \swarrow^G y = \begin{cases} 1 & \text{if } y \leq z \\ z & \text{otherwise} \end{cases}$$

$$\&_P(x, y) = x \cdot y \qquad z \swarrow^P y = \min\{1, z/y\}$$

$$\&_L(x, y) = \max\{0, x + y - 1\} \qquad z \swarrow^L y = \min\{1, 1 - y + z\}$$

Taking into account the of definition these operators, we obtain that the adjoint negations associated with the Gödel and product residuated implications are defined as:

$$n_G(x) = n_P(x) = \begin{cases} 1 & \text{if } x = 0 \\ 0 & \text{otherwise} \end{cases}$$

for all $x \in [0, 1]$. From now on, this negation operator will be called product negation. In addition, the adjoint negation obtained from the Łukasiewicz residuated implication is defined as follows:

$$n_L(x) = 1 - x$$

for all $x \in [0, 1]$, and it is commonly known in the literature as the standard negation.

It is easy to see that the product negation is a weak negation whereas the standard negation is a strong negation. Obviously, we can ensure that the Gödel and product adjoint triples $(\&_G, \swarrow^G, \nwarrow_G)$ and $(\&_P, \swarrow^P, \nwarrow_P)$ verify Theorem 1, for the weak negation n_P. Hence, we can conclude that there exist at least two different adjoint triples whose adjoint negations coincide with the weak negation n_P.

Next section studies how to define adjoint triples such that their adjoint negations are equal to a given weak negation. Furthermore, the algebraic structure formed by these adjoint triples is analyzed.

3   Adjoint Triples Generating a Weak Negation

This section presents two different procedures to define adjoint triples whose adjoint negations coincide with a given weak negation. Besides, an ordering relation is defined on the whole set of adjoint triples generating a given weak negation. From this ordering relation, the algebraic structure composed of the aforementioned set of adjoint triples is obtained.

3.1   Adjoint Triples Associated with Weak Negations from Adjoint Triples

The first procedure to define adjoint triples generating a given weak negation n is presented in the following proposition. This procedure is based on the use of an adjoint triple $(\&, \swarrow, \nwarrow)$ with respect to $([0, 1], \leq)$ verifying the inequalities $0 \swarrow y \leq n(y)$ and $0 \nwarrow x \leq n(x)$, for all $x, y \in [0, 1]$. From now on, given a weak negation n, the set of all adjoint triples with respect to $([0, 1], \leq)$ such that their adjoint negations coincide with the weak negation n, we will denoted as $\mathcal{T}_n$.

Theorem 2. *Let n be a weak negation and $(\&, \swarrow, \nwarrow)$ be an adjoint triple with respect to $([0, 1], \leq)$ such that $0 \swarrow y \leq n(y)$ and $0 \nwarrow x \leq n(x)$, for all $x, y \in [0, 1]$. The mappings $\&_n, \swarrow^n, \nwarrow_n : [0, 1] \times [0, 1] \to [0, 1]$ defined, for all $x, y, z \in [0, 1]$, as:*

$$x \,\&_n\, y = \begin{cases} x \,\&\, y & \text{if } \quad x \nleq n(y) \\ 0 & \text{if } \quad x \leq n(y) \end{cases}$$

$$z \swarrow^n y = \max\{z \swarrow y, n(y)\} \qquad z \nwarrow_n x = \max\{z \nwarrow x, n(x)\}$$

form an adjoint triple with respect to $([0, 1], \leq)$ belonging to $\mathcal{T}_n$, that is, they satisfy that $n = n_{s_n} = n_{n_n}$, where n_{s_n} and n_{n_n} are the adjoint negations defined from the implications $\swarrow^n$ and $\nwarrow_n$, respectively.

Notice that, the previous result follows the idea presented in [5] for the construction of left continuous t-norms from a given weak negation. Specifically, Theorem 2 extends Lemma 1 introduced in [5] to the framework of adjoint triples.

Example 2. We will consider different adjoint triples with respect to $([0, 1], \leq)$ generating the weak negation n_P. We have considered this negation operator due to its simplicity and that it is not a strong negation. The first adjoint triple that we use in this example was already considered in previous works [11,32]. This adjoint triple $(\&, \swarrow, \nwarrow)$ with respect to $([0, 1], \leq)$ is defined as follows:

$$x \,\&\, y = x^2 \cdot y$$

$$z \swarrow y = \begin{cases} 1 & \text{if } y = 0 \\ \min\left\{\sqrt{\dfrac{z}{y}}, 1\right\} & \text{otherwise} \end{cases} \qquad z \nwarrow x = \begin{cases} 1 & \text{if } x = 0 \\ \min\left\{\dfrac{z}{x^2}, 1\right\} & \text{otherwise} \end{cases}$$

for all $x, y, z \in [0, 1]$. It is easy to check that the following inequalities $0 \swarrow y \leq n_P(y)$ and $0 \nwarrow x \leq n_P(x)$ are satisfied, for all $x, y \in [0, 1]$. Indeed, the inequalities are equalities in this case. Hence, by using Theorem 2, we can define the following operators:

$$x \,\&_{n_p}^1 y = \begin{cases} x \,\& \, y & \text{if } x \nleq n_P(y) \\ 0 & \text{if } x \leq n_P(y) \end{cases} = \begin{cases} x^2 \cdot y & \text{if } x \nleq n_P(y) \\ 0 & \text{if } x \leq n_P(y) \end{cases} = x^2 \cdot y$$

$$z \,\swarrow^{n_{P1}} y = \max\{z \swarrow y, n_P(y)\} = \begin{cases} 1 & \text{if } y = 0 \\ \min\left\{\sqrt{\dfrac{z}{y}}, 1\right\} & \text{otherwise} \end{cases}$$

$$z \,\nwarrow_{n_{P1}} x = \max\{z \nwarrow x, n_P(x)\} = \begin{cases} 1 & \text{if } x = 0 \\ \min\left\{\dfrac{z}{x^2}, 1\right\} & \text{otherwise} \end{cases}$$

Clearly, the adjoint triple $(\&_{n_p}^1, \swarrow^{n_{P1}}, \nwarrow_{n_{P1}})$ defined as in Theorem 2 coincides with $(\&, \swarrow, \nwarrow)$. Notice that, the adjoint triple $(\&_{n_p}^1, \swarrow^{n_{P1}}, \nwarrow_{n_{P1}})$ belongs to $\mathcal{T}_{n_p}$ since the following chains of equalities hold, for all $x, y \in [0, 1]$:

$$n_{S_{n_{P1}}}(y) = 0 \swarrow^{n_{P1}} y = \begin{cases} 1 & \text{if } y = 0 \\ 0 & \text{otherwise} \end{cases} = n_P(y)$$

$$n_{n_{n_{P1}}}(x) = 0 \nwarrow_{n_{P1}} x = \begin{cases} 1 & \text{if } x = 0 \\ 0 & \text{otherwise} \end{cases} = n_P(x)$$

The implications associated with the Gödel adjoint triple $(\&_G, \swarrow^G, \nwarrow_G)$ also verify the hypothesis required in Theorem 2, that is, the inequalities $0 \swarrow^G y = n_G(y) \leq n_P(y)$ and $0 \nwarrow_G x = n_G(x) \leq n_P(x)$ trivially hold, for all $x, y \in [0, 1]$. Consequently, applying Theorem 2, we can define the following operators:

$$x \,\&_{n_p}^2 y = \begin{cases} x \,\&_G \, y & \text{if } x \nleq n_P(y) \\ 0 & \text{if } x \leq n_P(y) \end{cases} = \begin{cases} \min\{x, y\} & \text{if } x \nleq n_P(y) \\ 0 & \text{if } x \leq n_P(y) \end{cases} = \min\{x, y\}$$

$$z \,\swarrow^{n_{P2}} y = \max\{z \swarrow^G y, n_P(y)\} = \begin{cases} 1 & \text{if } y \leq z \\ z & \text{otherwise} \end{cases}$$

Therefore, we also have that $(\&_{n_p}^2, \swarrow^{n_{P2}}, \nwarrow_{n_{P2}})$ defined as in Theorem 2 coincides with $(\&_G, \swarrow^G, \nwarrow_G)$. Indeed, this fact will arise to every adjoint triple satisfying the hypotheses in Theorem 2, due to the restrictive definition of the negation operator n_P. As previously, the following chain of equalities is satisfied, for all $y \in [0, 1]$:

$$n_{S_{n_{P2}}}(y) = 0 \swarrow^{n_{P2}} y = n_P(y) = 0 \nwarrow_{n_{P2}} y = n_{n_{n_{P2}}}(y)$$

Thus, $(\&_{n_p}^2, \swarrow^{n_{P2}}, \nwarrow_{n_{P2}}) \in \mathcal{T}_{n_p}$, that is, it is an adjoint triple generating the weak negation n_P. Obviously, the product adjoint triple $(\&_P, \swarrow^P, \nwarrow_P)$ also belongs to $\mathcal{T}_{n_p}$. Notice that, the Łukasiewicz adjoint triple $(\&_L, \swarrow^L, \nwarrow_L)$ cannot be considered to build an adjoint triple whose adjoint negations coincide with the weak negation n_P. This fact is due to that the hypothesis required in Theorem 2 are not satisfied. For example, if we consider $y = 0.2$ and $z = 0$, we have that:

$$0 \swarrow^L 0.2 = \min\{1, 1 - 0.2 + 0\} = 0.8 \nleq 0 = n_P(0.2)$$

3.2   Adjoint Triples Associated with Weak Negations from Sup-Homomorphisms

The following result establishes the second procedure to define adjoint triples generating a given weak negation. This second mechanism weakens the required conditions in the first procedure and considers more general operators than adjoint triples. In particular, the proposed mechanism is based on the use of mappings preserving the supremum of non-empty sets, which are called supremum-homomorphisms on lattice theory.

Theorem 3. *Let n be a weak negation, $f, g, h\colon [0,1] \times [0,1] \to [0,1]$ three mappings such that f preserves the supremum of non-empty sets in both arguments, g is defined as $g(z,y) = \sup\{x \in [0,1] \mid f(x,y) \le z\}$ satisfying that $g(0,y) \le n(y)$ and h is defined as $h(z,x) = \sup\{y \in [0,1] \mid f(x,y) \le z\}$ satisfying that $h(0,x) \le n(x)$, for all $x, y, z \in [0,1]$. The triple $(\&_n, \swarrow^n, \nwarrow_n)$ composed of the following operators:*

$$x \mathbin{\&_n} y = \begin{cases} f(x,y) & \text{if } x \nleq n(y) \\ 0 & \text{if } x \le n(y) \end{cases}$$

$$z \swarrow^n y = \max\{g(z,y), n(y)\} \qquad z \nwarrow_n x = \max\{h(z,x), n(x)\}$$

is an adjoint triple with respect to $([0,1], \le)$ of $\mathcal{T}_n$.

As a consequence of this result, general operators can be considered to define adjoint triples in $\mathcal{T}_n$, such as uninorms [13,22,27,41,42,44]. This fact notably increases the number of operators that can be considered for obtaining triples in $\mathcal{T}_n$, which has a direct consequence in the flexibility for using these operators in real cases.

Example 3. In this example we will consider the uninorm $f\colon [0,1] \times [0,1] \to [0,1]$, defined for all $x, y \in [0,1]$ as follows.

$$f(x,y) = \begin{cases} \min\{x,y\} & \text{if } x \le \dfrac{1}{2} \text{ and } y \le \dfrac{1}{2} \\ \max\{x,y\} & \text{otherwise} \end{cases}$$

It is easy to check that f preserves the supremum of non-empty sets in both arguments. Moreover, from f, we can define two mappings $g, h\colon [0,1] \times [0,1] \to [0,1]$ as $g(z,y) = \sup\{x \in [0,1] \mid f(x,y) \le z\}$ and $h(z,x) = \sup\{y \in [0,1] \mid f(x,y) \le z\}$, for all $x, y, z \in [0,1]$. Notice that, f is a commutative mapping and therefore $g = h$. The analytic expression of the mapping g is displayed below:

$$g(z,y) = \sup\{x \in [0,1] \mid f(x,y) \le z\} = \begin{cases} \dfrac{1}{2} & \text{if } y \le z \le \dfrac{1}{2} \\ z & \text{if } y \le z \text{ and } z > \dfrac{1}{2} \\ 0 & \text{otherwise} \end{cases}$$

Clearly, the inequality $g(0,y) \le n_P(y)$ holds for all $y \in [0,1]$, and consequently $h(0,x) \le n_P(x)$, for all $x \in [0,1]$. Under the hypothesis of Theorem 3, we can define an adjoint triple $(\&^3_{n_P}, \swarrow^{n_{P3}}, \nwarrow_{n_{P3}})$ from the mappings f, g and h such that it belongs to $\mathcal{T}_{n_P}$. Specifically, the analytical expression of the conjunctor $\&^3_{n_P}$ is:

$$
x \&^3_{n_P} y = \begin{cases} f(x,y) & \text{if } x \nleq n_P(y) \\ 0 & \text{if } x \leq n_P(y) \end{cases} = \begin{cases} 0 & \text{if } x = 0 \text{ or } y = 0 \\ \min\{x,y\} & \text{if } x, y \in \left(0, \frac{1}{2}\right] \\ \max\{x,y\} & \text{otherwise} \end{cases}
$$

In this case, $\&^3_{n_P}$ does not coincided with f. As the conjunctor $\&^3_{n_P}$ is commutative, we have that $\swarrow^{n_{P3}} = \nwarrow_{n_{P3}}$. For all $y, z \in [0, 1]$, the implication $\swarrow^{n_{P3}}$ is defined as:

$$
z \swarrow^{n_{P3}} y = \max\{g(z,y), n_P(y)\} = \begin{cases} 1 & \text{if } y = 0 \\ z & \text{if } 0 < y \leq z \text{ and } z > \frac{1}{2} \\ \frac{1}{2} & \text{if } 0 < y \leq z \leq \frac{1}{2} \\ 0 & \text{otherwise} \end{cases}
$$

As we mentioned above, $(\&^3_{n_P}, \swarrow^{n_{P3}}, \nwarrow_{n_{P3}}) \in \mathcal{T}_{n_P}$ since the following chain of equalities is verified, for all $y \in [0, 1]$:

$$
n_{S_{n_{P3}}}(y) = 0 \swarrow^{n_{P3}} y = n_P(y) = 0 \nwarrow_{n_{P3}} y = n_{n_{n_{P3}}}(y)
$$

It is important to emphasize that f does not preserve the supremum of non-empty sets in both arguments. For instance, when $X = \varnothing$ and $y = 1$, we have that:

$$
f\left(\bigvee_{x_i \in X} x_i, y\right) = f(0, 1) = 1 \neq 0 = \bigvee_{x_i \in X} f(x_i, 1) = \bigvee_{x_i \in X} f(x_i, y)
$$

As a consequence, f cannot be the conjunctor of an adjoint triple. This fact allows us to ensure that the mechanism given in Theorem 3 provides adjoint triples built from more general operators.

3.3 Algebraic Structure of $\mathcal{T}_n$

The following theorem includes a point-wise ordering relation defined on the conjunctors of adjoint triples generating a given weak negation. This ordering relation provides the set of adjoint triples, whose adjoint negations coincide with such a weak negation, with the structure of a complete join-semilattice.

Theorem 4. *Given a weak negation n, we have that the pair $(\mathcal{T}_n, \sqsubseteq)$ forms a complete join-semilattice, where $\sqsubseteq$ is the ordering relation defined as:*

$$
(\&^j_n, \swarrow^{n_j}, \nwarrow_{n_j}) \sqsubseteq (\&^k_n, \swarrow^{n_k}, \nwarrow_{n_k}) \qquad \text{iff} \qquad x \&^j_n y \leq x \&^k_n y
$$

for all $x, y \in [0, 1]$ *and* $(\&_n^j, \swarrow^{n_j}, \nwarrow_{n_j}), (\&_n^k, \swarrow^{n_k}, \nwarrow_{n_k}) \in \mathcal{T}_n$. *The greatest element in* $\mathcal{T}_n$ *is the adjoint triple* $(\&_n^g, \swarrow^{n_g}, \nwarrow_{n_g})$ *such that* $\swarrow^{n_g} = \nwarrow_{n_g}$, *which is defined as follows:*

$$
x \,\&_n^g\, y = \begin{cases} 1 & \text{if} \quad x \nleq n(y) \\ 0 & \text{if} \quad x \leq n(y) \end{cases}
\qquad
z \,\swarrow^{n_g}\, y = \begin{cases} n(y) & \text{if} \quad z \neq 1 \\ 1 & \text{if} \quad z = 1 \end{cases}
$$

for all $x, y, z \in [0, 1]$.

If there exist two different adjoint triples $(\&_n^j, \swarrow^{n_j}, \nwarrow_{n_j}), (\&_n^k, \swarrow^{n_k}, \nwarrow_{n_k}) \in \mathcal{T}_n$ such that $(\&_n^j, \swarrow^{n_j}, \nwarrow_{n_j}) \not\sqsubseteq (\&_n^k, \swarrow^{n_k}, \nwarrow_{n_k})$ and $(\&_n^k, \swarrow^{n_k}, \nwarrow_{n_k}) \not\sqsubseteq (\&_n^j, \swarrow^{n_j}, \nwarrow_{n_j})$, then we will say that these adjoint triples are incomparable. In this case, we will write that $(\&_n^j, \swarrow^{n_j}, \nwarrow_{n_j}) \| (\&_n^k, \swarrow^{n_k}, \nwarrow_{n_k})$.

Finally, we clarify the previous result by means of the following example. This example will be used to illustrate that the set $(\mathcal{T}_n, \sqsubseteq)$ has not the structure of a complete lattice, since the infimum of the elements in $\mathcal{T}_n$ could not necessarily belong to $\mathcal{T}_n$.

Example 4. Given the weak negation n_{P}, we will establish a hierarchy among the proposed adjoint triples in $\mathcal{T}_{n_{\mathrm{P}}}$. According to the ordering relation introduced in Theorem 4, we obtain that:

$$
(\&_{n_{\mathrm{P}}}^1, \swarrow^{n_{P1}}, \nwarrow_{n_{P1}}) \sqsubseteq (\&_{\mathrm{P}}, \swarrow^{\mathrm{P}}, \nwarrow_{\mathrm{P}}) \sqsubseteq (\&_{n_{\mathrm{P}}}^2, \swarrow^{n_{P2}}, \nwarrow_{n_{P2}}) \sqsubseteq (\&_{n_{\mathrm{P}}}^3, \swarrow^{n_{P3}}, \nwarrow_{n_{P3}})
$$

Although we can find other adjoint triples belonging to $\mathcal{T}_{n_{\mathrm{P}}}$ greater than the previous ones, by Theorem 4, we can ensure that the greatest adjoint triple in $\mathcal{T}_{n_{\mathrm{P}}}$ is the triple $(\&_g, \swarrow^g, \nwarrow_g)$ such that $\swarrow^g = \nwarrow_g$, which is defined, for all $x, y, z \in [0, 1]$, as follows:

$$
x \,\&_g\, y = \begin{cases} 1 & \text{if} \quad x \nleq n_{\mathrm{P}}(y) \\ 0 & \text{if} \quad x \leq n_{\mathrm{P}}(y) \end{cases}
\qquad
z \swarrow^g y = \begin{cases} n_{\mathrm{P}}(y) & \text{if} \quad z \neq 1 \\ 1 & \text{if} \quad z = 1 \end{cases}
$$

Finally, we will show that there exist incomparable adjoint triples in $\mathcal{T}_{n_{\mathrm{P}}}$ and, therefore, the complete join-semilattice $(\mathcal{T}_{n_{\mathrm{P}}}, \sqsubseteq)$ is not linear. Given $a \in (0, 1)$, we can define the operators $\&_a, \swarrow^a, \nwarrow_a$ on the unit interval such that $\swarrow^a = \nwarrow_a$ as follows:

$$
x \,\&_a\, y = \begin{cases} a & \text{if} \quad x \nleq n_{\mathrm{P}}(y) \\ 0 & \text{if} \quad x \leq n_{\mathrm{P}}(y) \end{cases}
\qquad
z \swarrow^a y = \begin{cases} n_{\mathrm{P}}(y) & \text{if} \quad a \nleq z \\ 1 & \text{if} \quad a \leq z \end{cases}
$$

In particular, these triples are incomparable with the adjoint triples previously defined $(\&_{\mathrm{P}}, \swarrow^{\mathrm{P}}, \nwarrow_{\mathrm{P}}), (\&_{n_{\mathrm{P}}}^2, \swarrow^{n_{P2}}, \nwarrow_{n_{P2}})$ and $(\&_{n_{\mathrm{P}}}^3, \swarrow^{n_{P3}}, \nwarrow_{n_{P3}})$. Considering $a = 0.45$, then the adjoint triple $(\&_{0.45}, \swarrow^{0.45}, \nwarrow_{0.45})$ verifies that:

$$
\left.\begin{array}{l} 0.8 \,\&_{0.45}\, 0.6 = 0.45 \leq 0.48 = 0.8 \,\&_{\mathrm{P}}\, 0.6 \\ 0.5 \,\&_{\mathrm{P}}\, 0.6 = 0.3 \leq 0.45 = 0.5 \,\&_{0.45}\, 0.6 \end{array}\right\} \text{ then } (\&_{\mathrm{P}}, \swarrow^{\mathrm{P}}, \nwarrow_{\mathrm{P}}) \| (\&_{0.45}, \swarrow^{0.45}, \nwarrow_{0.45})
$$

$$
\left.\begin{array}{l} 0.5 \,\&_{0.45}\, 0.6 = 0.45 \leq 0.5 = 0.5 \,\&_{n_{\mathrm{P}}}^2\, 0.6 \\ 0.7 \,\&_{n_{\mathrm{P}}}^2\, 0.4 = 0.4 \leq 0.45 = 0.7 \,\&_{0.45}\, 0.4 \end{array}\right\} \text{ then } (\&_{n_{\mathrm{P}}}^2, \swarrow^{n_{P2}}, \nwarrow_{n_{P2}}) \| (\&_{0.45}, \swarrow^{0.45}, \nwarrow_{0.45})
$$

$$
\left.\begin{array}{l} 0.7 \,\&_{0.45}\, 0.4 = 0.45 \leq 0.7 = 0.7 \,\&_{n_{\mathrm{P}}}^3\, 0.4 \\ 0.5 \,\&_{n_{\mathrm{P}}}^3\, 0.3 = 0.3 \leq 0.45 = 0.5 \,\&_{0.45}\, 0.3 \end{array}\right\} \text{ then } (\&_{n_{\mathrm{P}}}^3, \swarrow^{n_{P3}}, \nwarrow_{n_{P3}}) \| (\&_{0.45}, \swarrow^{0.45}, \nwarrow_{0.45})
$$

Thus, when the weak negation n_P is considered, a proper non linear complete join-semilattice arises. Moreover, $(\mathcal{T}_{n_P}, \sqsubseteq)$ is not a complete lattice since the infimum of the subset $\{(\&_a, \swarrow^a, \nwarrow_a) \mid a \in (0, 1]\} \subseteq \mathcal{T}_{n_P}$ is the adjoint triple whose conjunctor is constantly zero, which is not an adjoint conjunctor of an adjoint triple in $\mathcal{T}_{n_P}$.

4 Conclusions and Future Work

We have extended the studied carried out in [14], providing different procedures to determine adjoint triples on the unit interval, whose adjoint negations are actually a previously fixed weak negation. We have also defined an ordering on which the set of these adjoint triples forms a complete join-semilattice. In addition, we have characterized the maximum element of the mentioned complete join-semilattice. In order to clarify the developed theory in this paper, we have included some illustrative examples.

As a future work, we will apply the developed theoretical results to different frameworks such as formal concept analysis, fuzzy relation equations and rough set theory. For example, these results will be fundamental for studying families of adjoint triples for defining preferences on objects or/and attributes in relational datasets [16, 28, 37]. This fact will allow us to address real problems related to image processing and digital forensic analysis.

Acknowledgement. Partially supported by the 2014-2020 ERDF Operational Programme in collaboration with the State Research Agency (AEI) in projects TIN2016-76653-P and PID2019-108991GB-I00, and with the Department of Economy, Knowledge, Business and University of the Regional Government of Andalusia in project FEDER-UCA18-108612, and by the European Cooperation in Science & Technology (COST) Action CA17124.

References

1. Alcalde, C., Burusco, A., Díaz-Moreno, J., Fuentes-González, R., Medina, J.: Fuzzy property-oriented concept lattices in morphological image and signal processing. In: Rojas, I., Joya, G., Cabestany, J. (eds.) IWANN 2013. LNCS, pp. 246–253. Springer, Heidelberg (2013). https://doi.org/10.1007/978-3-642-38682-4_28
2. Alcalde, C., Burusco, A., Díaz-Moreno, J.C., Medina, J.: Fuzzy concept lattices and fuzzy relation equations in the retrieval processing of images and signals. Int. J. Uncertain. Fuzziness Knowl.-Based Syst. **25**(Supplement–1), 99–120 (2017)
3. Asiain, M.J., Bustince, H., Mesiar, R., Kolesárová, A., Takáč, Z.: Negations with respect to admissible orders in the interval-valued fuzzy set theory. IEEE Trans. Fuzzy Syst. **26**(2), 556–568 (2018)
4. Chajda, I.: A representation of residuated lattices satisfying the double negation law. Soft. Comput. **22**(6), 1773–1776 (2017)
5. Cignoli, R., Esteva, F., Godo, L., Montagna, F.: On a class of left-continuous t-norms. Fuzzy Sets Syst. **131**(3), 283–296 (2002)
6. Cintula, P., Klement, E.P., Mesiar, R., Navara, M.: Residuated logics based on strict triangular norms with an involutive negation. Math. Log. Q. **52**(3), 269–282 (2006)
7. Cintula, P., Klement, E.P., Mesiar, R., Navara, M.: Fuzzy logics with an additional involutive negation. Fuzzy Sets Syst. **161**(3), 390–411 (2010). Fuzzy Logics and Related Structures

8. Cornejo, M.E., Esteva, F., Medina, J., Ramírez-Poussa, E.: Relating adjoint negations with strong adjoint negations. In: Kóczy, J.M.L. (ed.) Proceedings 7th European Symposium on Computational Intelligence and Mathematics, pp. 66–71 (2015)
9. Cornejo, M.E., Lobo, D., Medina, J.: Syntax and semantics of multi-adjoint normal logic programming. Fuzzy Sets Syst. **345**, 41–62 (2018)
10. Cornejo, M.E., Lobo, D., Medina, J.: On the solvability of bipolar max-product fuzzy relation equations with the product negation. J. Comput. Appl. Math. **354**, 520–532 (2019)
11. Cornejo, M.E., Medina, J., Ramírez-Poussa, E.: A comparative study of adjoint triples. Fuzzy Sets Syst. **211**, 1–14 (2013)
12. Cornejo, M.E., Medina, J., Ramírez-Poussa, E.: Multi-adjoint algebras versus extended-order algebras. Appl. Math. Inf. Sci. **9**(2L), 365–372 (2015)
13. Cornejo, M.E., Medina, J., Ramírez-Poussa, E.: Multi-adjoint algebras versus non-commutative residuated structures. Int. J. Approx. Reason. **66**, 119–138 (2015)
14. Cornejo, M.E., Medina, J., Ramírez-Poussa, E.: Adjoint negations, more than residuated negations. Inf. Sci. **345**, 355–371 (2016)
15. Cornejo, M.E., Medina, J., Ramírez-Poussa, E.: Characterizing reducts in multi-adjoint concept lattices. Inf. Sci. **422**, 364–376 (2018)
16. Cornejo, M.E., Medina, J., Ramírez-Poussa, E., Rubio-Manzano, C.: Multi-adjoint concept lattices, preferences and Bousi Prolog. In: Flores, V., et al. (eds.) IJCRS 2016. LNCS, vol. 9920, pp. 331–341. Springer, Cham (2016). https://doi.org/10.1007/978-3-319-47160-0_30
17. Cornelis, C., Medina, J., Verbiest, N.: Multi-adjoint fuzzy rough sets: Definition, properties and attribute selection. Int. J. Approx. Reason. **55**, 412–426 (2014)
18. Della Stella, M.E., Guido, C.: Associativity, commutativity and symmetry in residuated structures. Order **30**(2), 363–401 (2013)
19. Demirli, K., De Baets, B.: Basic properties of implicators in a residual framework. Tatra Mount. Math. Publ. **16**, 31–46 (1999)
20. Díaz-Moreno, J.C., Medina, J.: Multi-adjoint relation equations: definition, properties and solutions using concept lattices. Inf. Sci. **253**, 100–109 (2013)
21. Díaz-Moreno, J.C., Medina, J.: Using concept lattice theory to obtain the set of solutions of multi-adjoint relation equations. Inf. Sci. **266**, 218–225 (2014)
22. Duan, Q., Zhao, B.: Maximal chains on the interval [0, 1] with respect to t-norm-partial orders and uninorm-partial orders. Inf. Sci. **516**, 419–428 (2020)
23. Esteva, F.: Negaciones en retículos completos. Stochastica **I**, 49–66 (1975)
24. Esteva, F., Domingo, X.: Sobre funciones de negación en [0,1]. Stochastica **IV**, 141–166 (1980)
25. Esteva, F., Godo, L., Hájek, P., Navara, M.: Residuated fuzzy logics with an involutive negation. Arch. Math. Log. **39**(2), 103–124 (2000)
26. Esteva, F., Trillas, E., Domingo, X.: Weak and strong negation functions in fuzzy set theory. In: Proceedings of the XI International Symposium on Multivalued Logic, pp. 23–26 (1981)
27. Jenei, S.: Introducing group-like uninorms-construction and characterization. In: Kóczy, L., Medina-Moreno, J., Ramírez-Poussa, E., Šostak, A. (eds.) Computational Intelligence and Mathematics for Tackling Complex Problems. Studies in Computational Intelligence, vol. 819, pp. 51–57. Springer, Cham (2020). https://doi.org/10.1007/978-3-030-16024-1_7
28. Khan, M.A.: Formal reasoning in preference-based multiple-source rough set model. Inf. Sci. **334–335**, 122–143 (2016)
29. Madrid, N., Ojeda-Aciego, M.: Measuring inconsistency in fuzzy answer set semantics. IEEE Trans. Fuzzy Syst. **19**(4), 605–622 (2011)
30. Madrid, N., Ojeda-Aciego, M., Medina, J., Perfilieva, I.: L-fuzzy relational mathematical morphology based on adjoint triples. Inf. Sci. **474**, 75–89 (2019)
31. Massanet, S., Recasens, J., Torrens, J.: Fuzzy implication functions based on powers of continuous t-norms. Int. J. Approx. Reason. **83**, 265–279 (2017)

32. Medina, J., Ojeda-Aciego, M.: Multi-adjoint t-concept lattices. Inf. Sci. **180**(5), 712–725 (2010)
33. Medina, J., Ojeda-Aciego, M., Ruiz-Calviño, J.: Formal concept analysis via multi-adjoint concept lattices. Fuzzy Sets Syst. **160**(2), 130–144 (2009)
34. Medina, J., Ojeda-Aciego, M., Vojtáš, P.: Multi-adjoint logic programming with continuous semantics. In: Eiter, T., Faber, W., Truszczyński, M. (eds.) LPNMR 2001. LNCS, vol. 2173, pp. 351–364. Springer, Berlin, Heidelberg (2001). https://doi.org/10.1007/3-540-45402-0_26
35. Medina, J., Ojeda-Aciego, M., Vojtáš, P.: Similarity-based unification: a multi-adjoint approach. Fuzzy Sets Syst. **146**, 43–62 (2004)
36. Moreno, G., Penabad, J., Vázquez, C.: Beyond multi-adjoint logic programming. Int. J. Comput. Math. **92**(9), 1956–1975 (2015)
37. Pan, W., She, K., Wei, P.: Multi-granulation fuzzy preference relation rough set for ordinal decision system. Fuzzy Sets Syst. **312**, 87–108 (2017). Theme: Fuzzy Rough Sets
38. Pradera, A., Beliakov, G., Bustince, H., Baets, B.D.: A review of the relationships between implication, negation and aggregation functions from the point of view of material implication. Inf. Sci. **329**, 357–380 (2016)
39. Pradera, A., Massanet, S., Ruiz-Aguilera, D., Torrens, J.: The non-contradiction principle related to natural negations of fuzzy implication functions. Fuzzy Sets Syst. **359**, 3–21 (2019)
40. San-Min, W.: Logics for residuated pseudo-uninorms and their residua. Fuzzy Sets Syst. **218**, 24–31 (2013)
41. Su, Y., Liu, H.-W., Pedrycz, W.: The distributivity equations of semi-uninorms. Int. J. Uncertain. Fuzziness Knowl.-Based Syst. **27**(02), 329–349 (2019)
42. Su, Y., Riera, J., Ruiz-Aguilera, D., Torrens, J.: The modularity condition for uninorms revisited. Fuzzy Sets Syst. **357**, 27–46 (2019). Theme: Aggregation Functions
43. Trillas, E.: Sobre negaciones en la teoría de conjuntos difusos. Stochastica **III**, 47–60 (1979)
44. Zong, W., Su, Y., Liu, H.-W., Baets, B.D.: On the construction of uninorms by paving. Int. J. Approx. Reason. **118**, 96–111 (2020)

Representative Set of Objects in Rough Sets Based on Galois Connections

Nicolás Madrid[1] and Eloísa Ramírez-Poussa[2(✉)]

[1] Department of Applied Mathematics, Universidad de Málaga, Málaga, Spain
nicolas.madrid@uma.es
[2] Department of Mathematics, Universidad de Cádiz, Cádiz, Spain
eloisa.ramirez@uca.es

Abstract. This paper introduces a novel definition, called representative set of objects of a decision class, in the framework of decision systems based on rough sets. The idea behind such a notion is to consider subsets of objects that characterize the different classes given by a decision system. Besides the formal definition of representative set of objects of a decision class, we present different mathematical properties of such sets and a relationship with classification tasks based on rough sets.

Keywords: Rough Set Theory · Attribute reduction · Object reduction reduct · Decision systems

1 Introduction

Rough Set Theory (RST) is a mathematical theory that have shown its suitability for practical tasks [11,24]. The search to increase its range of application has given rise to different generalizations of this theory [6,7,10,28] as well as the relationships with other theories [4,18,25].

There exist different procedures to define the basic operators of rough sets, i.e., the lower and upper approximation, as those based on element operators, granular classes or subsystems [26]. In this work, we consider the approximation operators given by interior and closure operators obtained from the composition of operators in an isotone Galois connection [3,8], which has been built from a slight modification of the operators in [27]. This idea has been already considered in other works, as [14,17,20,21], and has two important advantages:

N. Madrid—Partially supported by the Spanish Ministry of Sciences project PGC2018-095869-B-I00, by the Junta de Andalucíca project UMA2018-FEDERJA-001 (European Regional Development Funds) and by the European Cooperation in Science & Technology (COST) Action CA17124.

E. Ramírez-Poussa—Partially supported by the 2014-2020 ERDF Operational Programme in collaboration with the State Research Agency (AEI) in projects TIN2016-76653-P and PID2019-108991GB-I00, and with the Department of Economy, Knowledge, Business and University of the Regional Government of Andalusia in project FEDER-UCA18-108612, and by the European Cooperation in Science & Technology (COST) Action CA17124.

© Springer Nature Switzerland AG 2020
R. Bello et al. (Eds.): IJCRS 2020, LNAI 12179, pp. 349–361, 2020.
https://doi.org/10.1007/978-3-030-52705-1_26

- the use of the operators introduced in [27] could lead us to the following situation: the lower approximation of a set may not be contained in the set and its upper approximation may not contain the set. The consideration of interior and closure operators avoids such a situation.
- the approximation operators obtained from the interior and closure operator are more accurate that those in [27] (see [17]).

In this paper we focus on the reduction of objects for a classification task. This kind of reduction has been seldom considered by the research community, which has mainly focused on attribute reduction [2,5,12,18]. Some examples of the study of object reductions are [15], which analyses the reduction of objects oriented to keep the original attribute reducts and [1,13,16,22,23] that reducts objects and attributes in parallel. The present paper is oriented in a different way than the existing approaches dealing with object reduction. We show that when the indiscernibility relation is not an equivalence relation, the objects in the different classes of a classification task can be characterized by only few objects in the class; we call that objects representative of the decision class. In such a way, the representative objects of the decision classes can be used as clusters in classification tasks. In this paper, we provide the formal definition of the set of representative objects of a decision class and analyze its mathematical properties.

The paper is organized as follows: Sect. 2 introduces the definitions of the approximation operators based on isotone Galois connections considered in this work, together with some results needed to understand this work. In Sect. 3, we present the formal definition of the set of representative objects of a decision class and analyze its mathematical properties. Section 4 provides the conclusions and presents some prospect for future work.

2   Preliminaries

In this section we recall some basic notions in order to make the contribution as self-contained as possible.

The first notion we have to recall is the notion of approximation space.

Definition 1. *An* approximation space *is a pair (U, R), where U is a set (called universe) and R is a binary relation over U.*

In this work, we consider approximation spaces whose relation R can be an arbitrary relation. This fact leads us to distinguish between left and right relationships, and to generalize the standard definition of R-foreset.

Definition 2. *Let (U, R) be an approximation space, the sets defined as:*

$$xR = \{y \in U \mid (x, y) \in R\} \ and \ Ry = \{x \in U \mid (x, y) \in R\}$$

are the R-right-foreset of $x \in U$ and the R-left-foreset of $y \in U$, respectively.

From the previous generalization, four different approximation operators arises.

Definition 3. *Let (U, R) be an approximation space and $A \subseteq U$. We define the following operators:*

- $R{\downarrow}^r A = \{x \in U \mid xR \subseteq A\}$
- $R{\uparrow}_r A = \{x \in U \mid xR \cap A \neq \varnothing\}$
- $R{\downarrow}^\ell A = \{y \in U \mid Ry \subseteq A\}$
- $R{\uparrow}_\ell A = \{y \in U \mid Ry \cap A \neq \varnothing\}$.

It is important to highlight that the approximation operators $R{\downarrow}^\ell$ and $R{\uparrow}_\ell$ coincide with those presented in [27]. Additionally, the equalities $R{\downarrow}^r = R{\downarrow}^\ell = R{\downarrow}$ and $R{\uparrow}_r = R{\uparrow}_\ell = R{\uparrow}$ are satisfied, when the relation is symmetric. For such a reason, hereafter, if R is symmetric, we will write $R{\downarrow}$ and $R{\uparrow}$ instead of $R{\downarrow}^r$, $R{\downarrow}^\ell$ and $R{\uparrow}_r$, $R{\uparrow}_\ell$, respectively.

On the other hand, the pairs $(R{\uparrow}_r, R{\downarrow}^\ell)$ and $(R{\uparrow}_\ell, R{\downarrow}^r)$ are isotone Galois connections [8,9], whose definition is recalled below.

Definition 4. *Let $(P, \leq_P)$ and $(Q, \leq_Q)$ be posets. A pair (φ, ψ) of mappings $\varphi \colon P \to Q$, $\psi \colon Q \to P$ is called* isotone Galois connection *between P and Q if the following equivalence is satisfied, for all $p \in P$ and $q \in Q$:*

$$\varphi(p) \leq_Q q \quad \text{if and only if} \quad p \leq_P \psi(q).$$

This notion is also called adjunction. *The mapping φ is called* lower (or left) adjoint *of ψ and the mapping ψ* upper (or right) adjoint *of φ.*

At this point, it is important to point out that in the case of considering arbitrary relations, the operators $R{\downarrow}_r$ and $R{\downarrow}_\ell$ may be unsuitable to represent lower approximations, since the inequalities $R{\downarrow}_r(A) \subseteq A$ or $R{\downarrow}_\ell(A) \subseteq A$ may not hold for some set $A \subseteq U$. Similarly, $R{\uparrow}_r$ and $R{\uparrow}_\ell$ may be unsuitable to represent upper approximations, since $A \subseteq R{\uparrow}_r(A)$ or $A \subseteq R{\uparrow}_\ell(A)$ could not be satisfied for some set $A \subseteq U$. However, the compositions $R{\uparrow}_r(R{\downarrow}^\ell(A))$ and $R{\uparrow}_\ell(R{\downarrow}^r(A))$ are always contained in A, whereas A is always contained in the composition $R{\downarrow}^\ell(R{\uparrow}_r(A))$ and $R{\downarrow}^r(R{\uparrow}_\ell(A))$.

Certainly, the inequalities $R{\downarrow}_r(A) \subseteq A$, $R{\downarrow}_\ell(A) \subseteq A$, $A \subseteq R{\uparrow}_r(A)$ and $A \subseteq R{\uparrow}_\ell(A)$ are satisfied for reflexive relations. But even in that case, the composition of these operators provide better approximations than considering simply $R{\downarrow}_r$, $R{\downarrow}_\ell$, $R{\uparrow}_r$ and $R{\uparrow}_\ell$. That is, if R is reflexive, we have

$$R{\downarrow}^\ell(A) \subseteq R{\uparrow}_r(R{\downarrow}^\ell(A)) \subseteq A \subseteq R{\downarrow}^\ell(R{\uparrow}_r(A)) \subseteq R{\uparrow}_r(A).$$

and

$$R{\downarrow}^r(A) \subseteq R{\uparrow}_\ell(R{\downarrow}^r(A)) \subseteq A \subseteq R{\downarrow}^r(R{\uparrow}_\ell(A)) \subseteq R{\uparrow}_\ell(A).$$

for all $A \subseteq U$.

In such a way, the notion of rough set is defined for arbitrary relations by using the following definition.

Definition 5. *Let* (U, R) *be an approximation space and* $A \subseteq U$. *The* lower approximations *of* A *are defined as:*

$$R\uparrow_r (R\downarrow^\ell (A)) \quad and \quad R\uparrow_\ell (R\downarrow^r (A))$$

and the upper approximations *of* A *are defined as:*

$$R\downarrow^\ell (R\uparrow_r (A)) \quad and \quad R\downarrow^r (R\uparrow_\ell (A)).$$

A set $A \subseteq U$ *is called* a generalized rough set *if it is different from the two lower approximations and from the two upper approximations.*

The following theorem summarizes some basic properties of such compositions.

Theorem 1. *Let* (U, R) *be an approximation space and* $A, B \subseteq U$, *then:*

- *If* $A \subseteq B$ *then* $R\uparrow_r (R\downarrow^\ell (A)) \subseteq R\uparrow_r (R\downarrow^\ell (B))$
- *If* $A \subseteq B$ *then* $R\uparrow_\ell (R\downarrow^r (A)) \subseteq R\uparrow_\ell (R\downarrow^r (B))$
- *If* $A \subseteq B$ *then* $R\downarrow^\ell (R\uparrow_r (A)) \subseteq R\downarrow^\ell (R\uparrow_r (B))$
- *If* $A \subseteq B$ *then* $R\downarrow^r (R\uparrow_\ell (A)) \subseteq R\downarrow^r (R\uparrow_\ell (B))$
- $R\uparrow_r (R\downarrow^\ell (A)) \subseteq A \subseteq R\downarrow^r (R\uparrow_\ell (A))$
- $R\uparrow_r (R\downarrow^\ell (A)) \subseteq A \subseteq R\downarrow^\ell (R\uparrow_r (A))$
- $R\uparrow_\ell (R\downarrow^r (A)) \subseteq A \subseteq R\downarrow^\ell (R\uparrow_r (A))$
- $R\uparrow_\ell (R\downarrow^r (A)) \subseteq A \subseteq R\downarrow^r (R\uparrow_\ell (A))$
- $R\uparrow_r (R\downarrow^\ell (R\uparrow_r (R\downarrow^\ell (A)))) = R\uparrow_r (R\downarrow^\ell (A))$
- $R\uparrow_\ell (R\downarrow^r (R\uparrow_\ell (R\downarrow^r (A)))) = R\uparrow_\ell (R\downarrow^r (A))$
- $R\downarrow^\ell (R\uparrow_r (R\downarrow^\ell (R\uparrow_r (A)))) = R\downarrow^\ell (R\uparrow_r (A))$
- $R\downarrow^r (R\uparrow_\ell (R\downarrow^r (R\uparrow_\ell (A)))) = R\downarrow^r (R\uparrow_\ell (A))$.

In [19] is stated that the approximation operators in Definition 5 coincide with those of [27] when the relation is a preorder.

Theorem 2 [19, Theorem 1]. *Let* (U, R) *be an approximation space. The following items are equivalent:*

- $R\uparrow_\ell (R\downarrow^r (A)) = R\downarrow^r (A)$, *for all* $A \subseteq U$.
- $R\uparrow_r (R\downarrow^\ell (A)) = R\downarrow^\ell (A)$, *for all* $A \subseteq U$.
- R *is a preorder (i.e.* R *is reflexive and transitive).*

In our approach, we intend to use more general relations than equivalence relations and different from preorder relations. Below, we recall a non-transitive indiscernibility relation that will be considered in this work. But first, we need to recall the notion of information system.

Definition 6. *An* information system $(U, \mathcal{A})$ *is a tuple, such that* $U = \{x_1, x_2, \ldots, x_n\}$ *and* $\mathcal{A} = \{a_1, a_2, \ldots, a_m\}$ *are finite, non-empty sets of objects and attributes, respectively. Each* $a \in \mathcal{A}$ *is associated with a mapping* $\bar{a} \colon U \to V_a$, *where* V_a *is the value set of* a *over* U.

If $V_a = \{0, 1\}$ *for each* $a \in \mathcal{A}$, *we say that* $(U, \mathcal{A})$ *is a* Boolean information system.

Now, we introduce the notion of s-indiscernibility relation.

Definition 7. *Given an information system* $(U, \mathcal{A})$, $s \in \mathbb{N}$ *and* $B \subseteq \mathcal{A}$, *the s-indiscernibility relation with respect to* B, R_B^s, *is defined as follows.*

Two objects $x, y \in U$ *belongs to* R_B^s *if and only if there are at most s attributes* $\{a_1, \ldots, a_s\} \subseteq B$ *such that* $\bar{a}_k(x) \neq \bar{a}_k(y)$ *for all* $k \in \{1, \ldots, s\}$.

If $(x, y) \in R_B^s$, *we say that* x *and* y *are s-indiscernible in* B. *When* $B = \mathcal{A}$, *we simply say that* x *and* y *are s-indiscernible and the relation is denoted as* R^s.

In this paper, we focus on the study of a special kind of information system called decision system.

Definition 8. *A decision system* $(U, \mathcal{A} \cup \{d\})$ *is a kind of information system in which* $d \notin \mathcal{A}$ *is called the decision attribute.*

In this framework, the notions of positive region and the degree of dependency are generalized as follows.

Definition 9. *Let* $(U, \mathcal{A} \cup \{d\})$ *be a decision system,* $B \subseteq \mathcal{A}$ *and* (U, R_B) *a derived approximation space. The* R_B-*left positive and* R_B-*right positive regions with respect to* R_B, *denoted as* $POS_{R_B}^\ell$ *and* $POS_{R_B}^r$ *respectively, are defined as:*

$$POS_{R_B}^\ell = \bigcup_{x \in U} R_B \uparrow_r (R_B \downarrow^\ell [x]_d)$$

$$POS_{R_B}^r = \bigcup_{x \in U} R_B \uparrow_\ell (R_B \downarrow^r [x]_d)$$

and the degree of dependency of d *over* R_B, $\gamma_{R_B}^*$, *as:*

$$\gamma_{R_B}^* = \frac{\max\left\{\mathrm{Card}(POS_{R_B}^\ell), \mathrm{Card}(POS_{R_B}^r)\right\}}{\mathrm{Card}(U)}$$

where $[x]_d$ *represents the equivalence class of the object* $x \in U$ *with respect to the indiscernibility relation* Ind_d *given by*

$$\mathrm{Ind}_d = \{(x, y) \in U \times U \mid \bar{d}(x) = \bar{d}(y)\}$$

Remark 1. The degree of dependency $\gamma_{R_B}^* = 1$ plays a remarkable role in decision systems since in such a case, a perfect classification can be performed taking into account the information provided by the approximation space (U, R_B). Additionally, note that if $\gamma_{R_B}^* = 1$, we have

$$\max\left\{\mathrm{Card}(POS_{R_B}^\ell), \mathrm{Card}(POS_{R_B}^r)\right\} = \mathrm{Card}(U).$$

In other words, $POS_{R_B}^\ell = U$ or $POS_{R_B}^r = U$. Moreover, by Theorem 1 we have that $R_B \uparrow_r (R_B \downarrow^\ell [x]_d) \subseteq [x]_d$ and $R_B \uparrow_\ell (R_B \downarrow^r [x]_d) \subseteq [x]_d$, as a result, if $POS_{R_B}^\ell = U$ we have that $R_B \uparrow_r (R_B \downarrow^\ell [x]_d) = [x]_d$, for all $x \in U$, and when $POS_{R_B}^r = U$ then $R_B \uparrow_\ell (R_B \downarrow^r [x]_d) = [x]_d$, for all $x \in U$.

3 Representative Set of Objects of a Decision Class

In this section, we introduce a novel class of objects, called representative. The underlying idea in such a definition is to determine a subset of objects that characterizes a certain decision class.

Definition 10. *Given a decision system $(U, \mathcal{A} \cup \{d\})$ and $x \in U$, we say that a subset of objects $X \subseteq U$ is:*

- *a left-representative set of the decision class $[x]_d$ if $R \uparrow_\ell (X) = [x]_d$.*
- *a right-representative set of the decision class $[x]_d$ if $R \uparrow_r (X) = [x]_d$.*

We will denote as $ROS^\ell([x]_d)$ and $ROS^r([x]_d)$ to the set of left-representative sets and right-representative sets of the decision class $[x]_d$, respectively.

Notice that when the relation R is symmetric, the left-representative sets coincide with the right-representative sets. In such a case, we call that sets *representative sets of a decision class* $[x]_d$, and denote the set formed by them as $ROS([x]_d)$.

Note also that if X is a representative set of a decision class $[x]_d$ for certain $x \in U$, then every element in the class of $[x]_d$ is related at least with one element in X and moreover, all the elements in X are related only to elements of $[x]_d$. In other words, we can characterize the elements in the class $[x]_d$ by checking which objects in U are related (or not) to elements in X.

The following example illustrates the previous definition.

Example 1. Consider a decision system $(U, \mathcal{A} \cup \{d\})$ composed of the set of objects $U = \{x_1, x_2, x_3, x_4, x_5, x_6\}$, the set of attributes $\mathcal{A} = \{a_1, a_2, a_3, a_4, a_5\}$ related between them as the following table shows:

	a_1	a_2	a_3	a_4	a_5	d
x_1	x		x	x		x
x_2		x	x		x	
x_3	x	x	x	x		x
x_4	x	x	x		x	
x_5		x	x	x	x	
x_6	x	x	x			x
x_7		x	x		x	

Note that, in this case, the obtained decision classes are:

$$[x_1]_d = \{x_1, x_3, x_6\}$$
$$[x_2]_d = \{x_2, x_4, x_5\}$$

We consider $B = \mathcal{A}$ and the s-indiscernibility relation with $s = 1$, that is, $R_{\mathcal{A}}^1$. The results obtained from the considered s-indiscernibility relation is shown in the table below.

R_A^1	x_1	x_2	x_3	x_4	x_5	x_6	x_7
x_1	x		x				
x_2		x		x	x		x
x_3	x		x			x	
x_4		x		x		x	x
x_5		x			x		x
x_6			x	x		x	
x_7		x		x	x		x

According to the previous table, we obtain that:

$$R \uparrow (\{x_3\}) = \{x_1, x_3, x_6\} = [x_1]_d$$
$$R \uparrow (\{x_2\}) = \{x_2, x_4, x_5\} = [x_2]_d$$

Therefore, we can assert that the set $\{x_3\}$ is a representative set of the decision class $[x_1]_d$ and the set $\{x_2\}$ is a representative set of the decision class $[x_2]_d$. But $\{x_3\}$ is not the only representative set of the decision class $[x_1]_d$, we can also find a set, composed of more than one object, that represents the same decision class as, for example, the set $\{x_1, x_3\}$; it is easy to check that $R \uparrow (\{x_1, x_3\}) = [x_1]_d$. However, it is important to note that the set $\{x_1\}$ is not a representative set of it decision class $[x_1]_d$ because $x_6 \notin R \uparrow (\{x_1\})$. In addition, the set $\{x_3, x_6\}$ is not a representative set of the decision class $[x_1]_d$ either, since $x_4 \in R \uparrow (\{x_3, x_6\})$ and $x_4 \notin [x_1]_d$.   □

In the previous example, we have shown that adding or removing objects from a given representative set of a decision class may change such a feature. Therefore, it looks interesting to study the structure of the set composed of all representative sets of a certain decision class.

In order to present the first result related to the structure of the representative sets, we need to introduce the following definition.

Definition 11. *Given an approximation space (U, R), we say that*

- *$x \in U$ is a left-isolated object of the relation R if there is no element $y \in U$ satisfying that $(x, y) \in R$. The set composed of all the left-isolated objects is denoted as $\mathrm{Is}^\ell(R)$*
- *$y \in U$ is a right-isolated object of the relation R if there is no element $x \in U$ satisfying that $(x, y) \in R$. The set composed of all the right-isolated objects is denoted as $\mathrm{Is}^r(R)$.*
- *$x \in U$ is a isolated object of the relation R if it is left-isolated and right-isolated. The set composed of all the isolated objects is denoted as $\mathrm{Is}(R)$.*

The first result shows that the set of left-representative (right-representative) sets of two different decision classes are disjoint except for isolated objects.

Proposition 1. *Let $(U, \mathcal{A} \cup \{d\})$ be a decision system, R a discernibility relation and $X, Y \subseteq U$ two left-representative (right-representative) sets of two different decision classes, then the set $X \bigcap Y$ is contained in the set of left-isolated (right-isolated) objects of the relation R.*

Proof. Let $x, y \in U$ such that $[x]_d \neq [y]_d$. Then, necessarily $[x]_d \bigcap [y]_d = \varnothing$. Let X and Y be left-representative sets of the classes $[x]_d$ and $[y]_d$, respectively, and let us prove that $X \bigcap Y \subseteq \mathrm{Is}^\ell(R)$. We consider $x \in X \bigcap Y$, then we have that $R\uparrow_\ell (x) \subseteq R\uparrow_\ell (X \bigcap Y)$. In addition, since $(R\uparrow_\ell, R\downarrow^r)$ is a Galois connection, we have that

$$R\uparrow_\ell (x) \subseteq R\uparrow_\ell \left(X \bigcap Y \right) \subseteq R\uparrow_\ell (X) \bigcap R\uparrow_\ell (Y) = [x]_d \bigcap [y]_d = \varnothing.$$

Therefore, we have that $R\uparrow_\ell (x) = \varnothing$. Then, according to Definition 3, there is not $y \in U$ such that $(x, y) \in R$, that is, x is a left-isolated object of the relation R. Hence, we have that $X \bigcap Y \subseteq \mathrm{Is}^\ell(R)$.

The proof with right-representative sets is developed in an analogous way. □

The following result shows that the set of representative sets of a certain decision class has the structure of a join-semilattice with respect to the standard ordering between subsets.

Proposition 2. *Let $(U, \mathcal{A} \cup \{d\})$ be a decision system and let $X, Y \subseteq U$ such that X is a left-representative (right-representative) set of a decision class $[x]_d$, with $x \in U$, and $R\uparrow_\ell (Y) \subseteq [x]_d$ (respectively $R\uparrow_r (Y) \subseteq [x]_d$). Then $X \bigcup Y$ is also a left-representative (right-representative) set of $[x]_d$.*

Proof. Let $X, Y \subseteq U$ such that X is a left-representative set of a decision class $[x]_d$, with $x \in U$, and $R\uparrow_\ell (Y) \subseteq [x]_d$. Since $(R\uparrow_\ell, R\downarrow^r)$ is a Galois connection, we have that

$$R\uparrow_\ell \left(X \bigcup Y \right) = R\uparrow_\ell (X) \bigcup R\uparrow_\ell (Y) = [x]_d \bigcup R\uparrow_\ell (Y) = [x]_d.$$

In other words, $X \bigcup Y$ is a left-representative set of $[x]_d$.

The proof follows similarly for the right-representative sets. □

The following consequence of the previous proposition shows that the construction of left-representative sets and right-representative sets can be done by singletons.

Corollary 1. *Let $(U, \mathcal{A} \cup \{d\})$ be a decision system and let $X, Y \subseteq U$ such that X and $X \bigcup Y$ are left-representative (right-representative) sets of a decision class $[x]_d$, with $x \in U$. Then $X \bigcup \{y\}$ is also a left-representative (right-representative) set of $[x]_d$, for all $y \in Y$.*

The following result shows that the set of representative sets of a certain decision class has the structure of join-semilattice with respect to the standard ordering between subsets.

Corollary 2. *Let $(U, \mathcal{A} \cup \{d\})$ be a decision system and let $X, Y \subseteq U$ be two left-representative (right-representative) sets of the same decision class $[x]_d$, with $x \in U$. Then $X \bigcup Y$ is also a left-representative (right-representative) set of $[x]_d$.*

Example 2. In Example 1, we have two representative sets for the class $[x_1]_d$:

$$ROS([x_1]_d) = \big\{\{x_3\}, \{x_1, x_3\}\big\}.$$

On the other hand, we have six representative sets for the class $[x_2]_d$, namely:

$$ROS([x_2]_d) = \big\{\{x_2\}, \{x_7\}, \{x_2, x_7\}, \{x_2, x_5\}, \{x_7, x_5\}, \{x_2, x_5, x_7\}\big\}.$$

Note that $ROS([x_1]_d)$ and $ROS([x_2]_d)$ are disjoint, because there is not isolated elements in the considered discernibility relation, as Proposition 1 asserts.

On the other hand, according to Proposition 2, it is easy to check the join of arbitrary representative sets is also a representative for the respective class. Specifically, we can observe that the representative sets for the class $[x_1]_d$ has a lattice structure, but the set of representative sets for the class $[x_2]_d$ only has the structure of a join-semilattice. That fact can be seen, for example, in the intersection of the sets $\{x_2, x_5\}$ and $\{x_7, x_5\}$ which is the singleton $\{x_5\}$ that is not a representative set of the class $[x_2]_d$. $\Box$

Let us analyze now the minimal and maximal representative sets of decision classes.

Definition 12. *Let $(U, \mathcal{A} \cup \{d\})$ be a decision system and $X \subseteq U$ a representative set of a decision class $[x]_d$, with $x \in U$. We say that:*

- *X is a minimal left-representative (right-representative) set of the decision class $[x]_d$ if $X \setminus \{x'\}$ is not a left-representative (right-representative) set of $[x]_d$, for all $x' \in X$.*
- *X is a maximal left-representative (right-representative) set of the decision class $[x]_d$ if there is no object $x' \in U \setminus X$ such that $X \bigcup \{x'\}$ is a left-representative (right-representative) set of $[x]_d$.*

Thanks to the join-semilattice structure of the set of representative sets of a decision class (Proposition 2), we can directly infer that the maximal representative set of a decision class is unique; i.e., it is a maximum, as the following corollary states.

Corollary 3. *Let $(U, \mathcal{A} \cup \{d\})$ be a decision system and $x \in U$. If there exists a left-representative (right-representative) set of a decision class $[x]_d$ then, there is a unique maximal left-representative (right-representative) set of that decision class.*

The unicity stated by the previous result does not hold for minimal representative sets; i.e., it may exists several minimal representative sets of a decision class. The following example shows that fact.

Example 3. Coming back to Example 2, according to Definition 12, we have that the sets $\{x_3\}$ is the only minimal representative set of the decision class $[x_1]_d$. However, there are two minimal representative sets of the decision class $[x_2]_d$, namely $\{x_2\}$ and $\{x_7\}$.

On the other hand, it can be proved easily that $\{x_1, x_3\}$ and $\{x_2, x_5, x_7\}$ are the two maximal representative sets of $[x_1]_d$ and $[x_2]_d$, respectively.   □

The following result determines the maximal left-representative set of a decision class, if it exists.

Theorem 3. *Let $(U, \mathcal{A} \cup \{d\})$ be a decision system, $B \subseteq \mathcal{A}$, (U, R_B) an approximation space and $x \in U$. Then:*

- *If there exists a right-representative set of the decision class $[x]_d$, then the set $R \downarrow^\ell ([x]_d)$ is the maximum right-representative set of the decision class $[x]_d$.*
- *If there exists a left-representative set of the decision class $[x]_d$, then the set $R \downarrow^r ([x]_d)$ is the maximum left-representative set of the decision class $[x]_d$.*

Proof. Let $x \in U$ such that there exists a right-representative set of its decision class. By Corollary 3, we have that there exists the maximum right-representative set of the decision class $[x]_d$, denoted by $X \subseteq U$. By Theorem 1, we have that $R_B \uparrow_r (R_B \downarrow^\ell [x]_d) \subseteq [x]_d$. Then, since X is a right-representative set of the decision class $[x]_d$, by Proposition 2, we have that $X \cup R_B \downarrow^\ell [x]_d$ is a representative set of the decision class $[x]_d$ as well. As a result, by the maximality of the set X, we have that $R_B \downarrow^\ell [x]_d \subseteq X$.

Let us prove now that $X \subseteq R_B \downarrow^\ell [x]_d$. Consider $y \in X$, since X is a right-representative set of $[x]_d$ and by the monotonicity of $R \uparrow_r$, we have that $R \uparrow_r (\{y\}) \subseteq [x]_d$. Therefore, by definition of $R \uparrow_r$, if we consider $z \in U$ satisfying that $zR \cap \{y\} \neq \varnothing$, then $z \in [x]_d$. As a consequence, for all $z \in U$ such that $(z, y) \in R$ we have that $z \in [x]_d$, which is equivalent to say that $Ry \subseteq [x]_d$ and then, $y \in R_B \downarrow^\ell [x]_d$. In other words $X \subseteq R_B \downarrow^\ell [x]_d$.

Finally, we can assert that $R_B \downarrow^\ell [x]_d$ is the maximal right-representative set of $[x]_d$.

The proof for the maximal left-representative set of $[x]_d$ follows analogously. □

In the last result we relate the representative sets of a decision class to the degree of dependency $\gamma^*_{R_B}$. Specifically, we show that $\gamma^*_{R_B} = 1$ is equivalent to assert the existence of left-representative sets for each decision class or right-representative sets for each decision class.

Theorem 4. *Let $(U, \mathcal{A} \cup \{d\})$ be a decision system, $B \subseteq \mathcal{A}$ and (U, R_B) an approximation space. R_B satisfies that $\gamma^*_{R_B} = 1$ if and only if*

- *there exists at least one left-representative set for each decision class,*
- *or there exists at least one right-representative set for each decision class.*

Proof. Let us assume that R_B satisfies that $\gamma^*_{R_B} = 1$. Then, $R_B \uparrow_r \left(R_B \downarrow^\ell [x]_d\right) = [x]_d$, for all $x \in U$ or $R_B \uparrow_\ell (R_B \downarrow^r [x]_d) = [x]_d$ for all $x \in U$ (see Remark 1). Therefore, we have that $R_B \downarrow^\ell [x]_d$ is a right-representative set of $[x]_d$, for all $x \in U$, or $R_B \downarrow^r [x]_d$ is a left-representative set of $[x]_d$, for all $x \in U$.

Now, let us prove the converse. Without loss of generality, let us assume that there exists at least one left-representative set for each decision class. Then, by Theorem 3, we have that $R \downarrow^r ([x]_d)$ is the maximal left-representative set of the class $[x]_d$, that is, $R_B \uparrow_\ell (R_B \downarrow^r [x]_d) = [x]_d$ for all $x \in U$. As a consequence:

$$POS^r_{R_B} = \bigcup_{x \in U} R_B \uparrow_\ell (R_B \downarrow^r [x]_d) = \bigcup_{x \in U} [x]_d = U$$

and therefore, $\gamma^*_{R_B} = 1$. □

4   Conclusions and Future Work

In this paper we have provided the formal definition of the notion of representative set of objects of a decision class. Moreover, we have presented some mathematical properties of such kind of sets and shown its connection with a classification task based on rough sets.

There are different future lines based on the notion of representative set of objects. Firstly, the obtention of more mathematical properties about the objects forming representative sets is interesting for several purposes; for example, for its construction or for determining minimal representative sets. Secondly, analyzing the relationship between reducts of attributes and the set of representative sets of objects has our attention as well. Last but not least, the construction of a classification procedure based on representative sets of objects seems to be appropriated when the dataset is involved with uncertainty; for example when we need to classify an object that is discernible with all the objects in the training dataset.

References

1. Benítez-Caballero, M.J., Medina, J., Ramírez-Poussa, E., Ślęzak, D.: Bireducts with tolerance relations. Inf. Sci. **435**, 26–39 (2018)
2. Benítez-Caballero, M.J., Medina, J., Ramírez-Poussa, E., Ślęzak,D.: Rough-set-driven approach for attribute reduction in fuzzy formal concept analysis. Fuzzy Sets Syst. (2019)
3. Birkhoff, G.: Lattice Theory, 3rd edn. American Mathematical Society, Providence (1967)
4. Bloch, I.: On links between mathematical morphology and rough sets. Pattern Recogn. **33**, 1487–1496 (2000)
5. Cornelis, C., Jensen, R., Hurtado, G., Ślęzak, D.: Attribute selection with fuzzy decision reducts. Inf. Sci. **180**, 209–224 (2010)
6. Cornelis, C., Medina, J., Verbiest, N.: Multi-adjoint fuzzy rough sets: definition, properties and attribute selection. Int. J. Approx. Reason. **55**, 412–426 (2014)

7. Couso, I., Dubois, D.: Rough sets, coverings and incomplete information. Fundam. Inf. **108**(3–4), 223–247 (2011)
8. Davey, B., Priestley, H.: Introduction to Lattices and Order, 2nd edn. Cambridge University Press, Cambridge (2002)
9. Denecke, K., Erné, M., Wismath, S.L. (eds.): Galois Connections and Applications. Kluwer Academic Publishers, Dordrecht (2004)
10. Han, S.-E.: Roughness measures of locally finite covering rough sets. Int. J. Approx. Reason. **105**, 368–385 (2019)
11. Hassanien, A.E., Schaefer, G., Darwish, A.: Computational intelligence in speech and audio processing: recent advances. In: Gao, X.-Z., Gaspar-Cunha, A., Köppen, M., Schaefer, G., Wang, J. (eds.) Soft Computing in Industrial Applications. Advances in Intelligent and Soft Computing, vol. 75, pp. 303–311. Springer, Heidelberg (2010). https://doi.org/10.1007/978-3-642-11282-9_32
12. Janusz, A., Ślęzak, D.: Rough set methods for attribute clustering and selection. Appl. Artif. Intell. **28**(3), 220–242 (2014)
13. Janusz, A., Ślęzak, D., Nguyen, H.S.: Unsupervised similarity learning from textual data. Fundam. Inf. **119**, 319–336 (2012)
14. Järvinen, J., Radeleczki, S., Veres, L.: Rough sets determined by quasiorders. Order **26**(4), 337–355 (2009)
15. Kudo,Y., Murai, T.: An attempt of object reduction in rough set theory. In: 2018 Joint 10th International Conference on Soft Computing and Intelligent Systems (SCIS) and 19th International Symposium on Advanced Intelligent Systems (ISIS), pp. 33–36, December 2018
16. Mac Parthalain, N., Jensen, R.: Simultaneous feature and instance selection using fuzzy-rough bireducts. In: 2013 IEEE International Conference on Fuzzy Systems (FUZZ-IEEE 2013), pp. 1–8, July 2013
17. Madrid, N., Medina, J., Ramírez-Poussa, E.: Rough sets based on galois connections. Int. J. Appl. Math. (2020, accepted)
18. Medina, J.: Relating attribute reduction in formal, object-oriented and property-oriented concept lattices. Comput. Math. Appl. **64**(6), 1992–2002 (2012)
19. Pagliani, P.: The relational construction of conceptual patterns - tools, implementation and theory. In: Kryszkiewicz, M., Cornelis, C., Ciucci, D., Medina-Moreno, J., Motoda, H., Raś, Z.W. (eds.) Rough Sets and Intelligent Systems Paradigms. LNCS, vol. 8537, pp. 14–27. Springer, Cham (2014). https://doi.org/10.1007/978-3-319-08729-0_2
20. Pagliani, P.: Covering rough sets and formal topology - a uniform approach through intensional and extensional constructors. In: Peters, J., Skowron, A. (eds.) Transactions on Rough Sets XX. LNCS, vol. 10020, pp. 109–145. Springer, Heidelberg (2016). https://doi.org/10.1007/978-3-662-53611-7_4
21. Shao, M.-W., Liu, M., Zhang, W.-X.: Set approximations in fuzzy formal concept analysis. Fuzzy Sets Syst. **158**(23), 2627–2640 (2007). Theme: Logic
22. Stawicki, S., Ślęzak, D.: Recent advances in decision bireducts: complexity, heuristics and streams. In: Lingras, P., Wolski, M., Cornelis, C., Mitra, S., Wasilewski, P. (eds.) RSKT 2013. LNCS, vol. 8171, pp. 200–212. Springer, Heidelberg (2013). https://doi.org/10.1007/978-3-642-41299-8_19
23. Stawicki, S., Ślęzak, D., Janusz, A., Widz, S.: Decision bireducts and decision reducts - a comparison. Int. J. Approx. Reason. **84**, 75–109 (2017)
24. Varma, P.R.K., Kumari, V.V., Kumar, S.S.: A novel rough set attribute reduction based on ant colony optimisation. Int. J. Intell. Syst. Technol. Appl. **14**(3–4), 330–353 (2015)

25. Yao, Y., Lingras, P.: Interpretations of belief functions in the theory of rough sets. Inf. Sci. **104**(1), 81–106 (1998)
26. Yao, Y., Yao, B.: Covering based rough set approximations. Inf. Sci. **200**, 91–107 (2012)
27. Yao, Y.Y.: Two views of the theory of rough sets in finite universes. Int. J. Approx. Reason. **15**(4), 291–317 (1996). Rough Sets
28. Ziarko, W.: Probabilistic approach to rough sets. Int. J. Approx. Reason. **49**(2), 272–284 (2008). Special Section on Probabilistic Rough Sets and Special Section on PGM'06

Data Summarization

Discovering Fails in Software Projects Planning Based on Linguistic Summaries

Iliana Pérez Pupo[1](✉) ⓘ, Pedro Y. Piñero Pérez[1] ⓘ,
Roberto García Vacacela[2] ⓘ, Rafael Bello[3] ⓘ,
and Luis Alvarado Acuña[4] ⓘ

[1] Departamento de Investigaciones en Gestión de Proyectos, Universidad de las
Ciencias Informáticas, 54830 La Habana, CP, Cuba
{iperez,ppp}@uci.cu
[2] Facultad de Especialidades Empresariales, Universidad Católica De Santiago
de Guayaquil, Guayaquil, Ecuador
roberto.garcia@cu.ucsg.edu.ec
[3] Centro de Investigación en Informática, Universidad Central Marta Abreu de
las Villas, Santa Clara, Cuba
rbellop@uclv.edu.cu
[4] Departamento de Ingeniería de la Construcción, Universidad Católica del
Norte, Antofagasta, Chile
lualvar@ucn.cl

Abstract. Linguistic data summarization techniques help to discover complex relationships between variables and to present the information in natural language. There are some investigations associated to algorithms to build linguistic summaries. But the literature does no report investigations concerned with combination linguistic data summarization techniques and outliers' mining applied to planning of software project. In particular, outliers' mining is a datamining technique, useful in errors and fraud detection. In this work authors present new algorithms to build linguistic data summaries from outliers in software project planning context. Besides, authors compare different outliers' detection algorithms in software project planning context. The main motivation of this work is to detect planning errors in projects, to avoid high cost and time delays. Authors consider that the combination of outliers' mining and linguistic data summarization support project managers to decision-making process in the software project planning. Finally, authors present the interpretation of obtained summaries and comment about its impact.

Keywords: Linguistic data summarization · Outliers mining · Project management · Software project planning

1 Introduction

During the planning of software project, managers continuously have to take decisions to avoid delays and the elevation of project's cost. There are standards and authors that reflect best practices in project management. Some of them stand out: The Capability Maturity Model Integration (CMMI) [1], the guide of Project Management Body of

© Springer Nature Switzerland AG 2020
R. Bello et al. (Eds.): IJCRS 2020, LNAI 12179, pp. 365–375, 2020.
https://doi.org/10.1007/978-3-030-52705-1_27

Knowledge (PMBOK) [2], the ISO 21500 [3], Pressman [4] and Wilson Padua [5]. Despite the existence of these guides, there are still numerous difficulties that are reflected in successful, failed and renegotiated projects. The indexes of successful, failed and renegotiated projects have moved slightly around 29%, 19% and 52% respectively.

The main causes in project failings include planning errors, errors in human resources management and low control and monitoring level [6, 7]. In organizations that develop software projects, planning errors often appear, such as:

- Errors in the cost estimate.
- Errors in the estimation of resources.
- Errors in the estimation of the duration of activities.

Errors manual detection in software project planning constitutes a high time consuming work [8], which affects the projects correct operation. Automatic or semiautomatic detection of errors helps to reduce the cost during projects execution, projects planning and the total cost at the project end.

Planning errors can be identified as derived data from the projects plans. In this sense, it is identified in this investigation early detection of software project planning errors and linguistic data summarization techniques with using outlier mining, will help project managers to correct difficulties. In general, different authors have given their outliers definition [9, 10] among which Hawkins' definition stands out. Hawkins defines in page 2 of [11] that "Outlier is an observation that deviates greatly from the rest of the observations, appearing as a suspicious observation that could have been generated by mechanisms different from the rest of the data" [12].

Nevertheless, it should be perceived that there are not enough publications about outliers' mining in software project planning. In addition, errors presentation and negative impact factors in projects in natural language leads project managers to a better situations understanding and making quick decisions [13].

The objective of this work is to present different algorithms for detect errors in software project planning and construction of linguistic summaries that represent the errors' behavior in this discipline. The work is organized in sections as follows: Second section presents a brief analysis of outliers mining and linguistic data summarization art state. In third section, authors present linguistic data summarization algorithms based on outliers' mining in software project planning processes. The four section aims at the results obtained by the application of proposed algorithms in software project planning environments. Last section presents the conclusions.

2 Algorithm for Discovering Fails in Software Projects Planning Based on Linguistic Summaries

2.1 Brief Analysis of Linguistic Summaries and Outlier Mining

Most of the authors classify the outliers in three categories: punctual outlier's values, collective outliers or contextual outliers [9, 14]. On the other hand, authors classify the outlier detection algorithms following different criteria. In this work, the authors consider the approach proposed by Aggarwal [9], who establishes the following categories for outliers' detection: supervised, unsupervised and semi-supervised methods.

Unsupervised methods include: statistical techniques, techniques based on proximity and spatial data analysis [15]. Methods based on statistical techniques are based on: descriptive statistics [16], linear regression [17] and in the principal components' analysis [18]. These methods are not efficient when increasing the data set or dimensionality. Proximity-based methods include: distance-based on methods [19], clusters [20] and density-based methods [19].

Distance-based methods usually establish a ranking where the first elements in ranking represent data with high probability of being outliers [21]. In distance-based methods the distance function has a high relevance; for example, different authors refer that Mahalanobis-distance reports better results than Euclidean distance. But data sceneries are different in each case. Authors should test with different methods to discover the best technique. Density-based methods focus on identifying regions of space as a function of their data density, and they are very useful for their interpretation. Among the best known methods of this approach are: local anomalous data factor (LOF) method [22] and local integral correlation method (LOCI). Finally, clustering methods are further subdivided into hierarchical methods, partition-based methods, grid-based methods and constraint-based methods [17]. In this context, the question "what is the best method: cluster algorithm or proximity-based method?" does not have a unique answer. Researchers should analyze data nature in most of the situations and apply empirical tests in every one of the sceneries in order to recognize the algorithms with best results.

On the other hand, supervised methods in outliers mining represent traditional approach based on objects classification by having objects previously classified. In this sense different approaches are presented such as: decision trees, vector support techniques [23], rule-based systems [24], neural networks [25] and the use of meta-heuristics [26]. However, these methods usually do not report the best results in outlier's detection because the outliers' mining usually represent a problem with unbalanced classes or with completely unknown classes. For this reason, supervised methods are frequently combined with unsupervised techniques.

In this paper, summaries are generated from outliers. The authors of this work discuss different linguistic data summarization techniques. In [13] defined summary as "using few words to give the most important information about something".

Kacprzyk and Zadrożny are recognized authors in Linguistic data summarization techniques. They define a set of six protoforms that describe linguistic summaries structure and the queries for their search [27]. In this paper, the authors group six protoforms into two basic structures [27, 28] in order to build the linguistic summaries. The elements contained in summary are described in Table 1. Examples:

Table 1. Elements contained in summary.

Elements	Meaning
Q	Represents quantifiers such as: most, some, a few, etc.
R	Represents filters for example: "high planned material resources"
y	Represent the object of study for example "outlier projects"
S	Represents summarizer such as: "very high"
T	Represents measures to evaluate the linguistic summaries quality

First: summaries without filters $Qy's\ are\ S$, representing relationships such as:

$$T\,(\text{Most employees have low pay}) = 0.7$$

Second: summaries with filters $QRy's\ are\ S$, describes relationships such as:

$$T\,(\text{Most young employees have low pay}) = 0.7$$

There are different approaches to generate linguistic summaries; the simplest protoforms can be obtained by combining fuzzy logic with descriptive statistics or by combining fuzzy logic with sql database query language [29]. But in this work, authors concentrate on summaries generation that represent more complex protoforms and associated to outlier's detection. In this context, basic techniques are not appropriated. More complex protoforms can be built by using mining of fuzzy association rules, Kacprzyk [30] or by using genetic algorithms [31]. These strategies focus on linguistic summaries that represent most of the objects in database. Nevertheless, in this paper authors are in focus of outliers, rare elements and hard difficult detecting elements by using association rules or meta heuristics. For this reason, authors propose a new algorithm in next section.

2.2   A New Algorithm for Generating Summaries from Outliers in Software Project Planning

In this section, an algorithm is proposed for the construction of linguistic summaries from outliers. The following is a hybrid algorithm that combines clustering techniques with distance-based methods to detect outliers and to build linguistic summaries from the outliers detected.

Algorithm's name: *Outlier_Hybrid_LDS.*
  Notation
    O: outliers set.
    B_0: threshold based on the b_0 compact assembly concept.
    Ranking_outlier(*S*): returns elements from *S* set, sorted
      in descending order according to distance.
    R: set of linguistic summaries obtained.
    SetFuzzyVar: set of linguistic variables, one for each
      attribute that describes data behavior.
  Inputs:
    D: data set associated to software project planning.
    C: seeded center sets;
    Distance (*d,S*): distance function from *d* to the set of
      points *S*.
    P: percentile used for the determination of the outliers
      (the 0.92 percentile was taken).
    Q: linguistic variable that describes the quantifiers
      of the summaries.
    Threshold: threshold (ε) is used for the calculation of
      the *T* and for quantifying the default value as 0.3
    ParT-S_norm: Aggregation operators, T-norm pair and S-
      norm.

begin
  1. $O = \{\}$;
  2. *clusters* = Cluster(*D*, *centers=C*)
  3. *centers* = clusters.centers
  4. For each *cluster$_i$* in clusters, make
     4.1 B_0 = Calculate_threshold(*clusters$_i$*)
     4.2 O = clusters.out_centers_B_o
    End of the cycle
  5. O = Ranking_outlier(*O, P*)
  6. O_f = Transforms elements in *O*, into linguistic values
     by using the *SetFuzzyVar* variables
  7. $R = \{\}$ //initializing rule base
  8. For each O_{fi} in O_f
     8.1 If does not exist rule in *R* that cover O_{fi}
     (see Definition 1) then
     8.2 R_k = Build rule from O_{fi}
     8.3 $R = R \cup \{R_k\}$
    End of the cycle
  9. S = Build a summary from each rule in *R*
  10. S_f = Complete summaries *S* with quantifiers *Q*
  11. Calculate truth grade *T* for each summary in S_f
  12. Refine summaries S_f using active learning techniques
  13. Return S_f sorted, considering *T* values calculated
End

Definition 1: An object X is cover by a rule $G = (P, C)$ with P antecedents and C consequent if and only if for each attribute of $x_i \in X$, $\exists$ $(P_k \in P$ or $C_t \in C)$: $x_i \equiv P_k$ or $x_i \equiv C_t$ (operator $\equiv$ means equivalents).

This algorithm could be applied with different clustering methods. Selection of appropriate clustering algorithm depends on data nature. For example, for numerical data could be used *kmeans* cluster algorithm; although, the use of *kmeans* themselves create clusters forming hyper spheres. In each cluster the objects furthest from the center can represent potential outliers.

These objects are detected by using distance methods. In this sense, algorithm can be implemented by using different distance methods, with different threshold values too. In step 10, outlier's data are transformed into linguistic values by using the *SetFuzzyVar* variables defined for each variable and the maximum membership principle. The algorithm continues creating fuzzy rules from detected outliers, and for each fuzzy rule, it creates a candidate linguistic summary. After that, each candidate linguistic summary is completed with quantifiers calculated.

3 Application, Results and Discussion

This algorithm was applied to help projects' managers in software project planning, and to understand projects evolution and projects' human resources behavior. Authors was compared different combinations of algorithms in multiple project management databases. The algorithms are compared by analyzing their performance with the following databases: "mul_plan", "mul_rate", "mul_mix", "alone_rate" and "col_mix" from "170905_gp_eval_proy_fuzzy" Research Database Repository of Project Management Research Group [32]. Each database contains 8430 records with 19 attributes. Different attributes are modified to convert them in outliers. The modification is applied following a supervised way. Later, during test, authors calculate the quality of each algorithm setting in outlier, see Table 2.

Table 2. Description of the databases used in the experimentation.

Database	Meaning	Percent of outliers
alone_rate	rate_rrhh	5% of the modified
mul_plan	serv_plan_quantity, rrhh_plan_quantity, eqp_plan_quantity, inf_plan_quantity, mat_plan_quantity	5% of the modified
mul_rate	rate_equipment, rate_rrhh, rate_service, rate_material	5% of the modified
mul_mix	rate_rrhh, rrhh_plan_quantity, rate_material, mat_plan_quantity, rrhh_plan_quantity, rrhh_real_quantity	5% of the modified
col_mix	rate_rrhh, rrhh_plan_quantity, rate_material	95% of the records in each project transformed to be collective outlier

Table 3. Comparison of multiple algorithms respect to efficacy in outlier detection (numbers joined to each algorithm name, represent parameters).

Group	col_mix	Alone	mult_mix	mul_rate	mult_plan
a	Distance_Mahalanobis_3_0.92 Outlier_Hibrid_LDS_0.92	Outlier_Hibrid_LDS_0.92 Distance_Mahalanobis_3_0.92	Outlier_Hibrid_LDS_0.92	Outlier_Hibrid_LDS_0.92 Kmodr_3	Outlier_Hibrid_LDS_0.92
b	Angle_5_0.95 Kmodr_3_0	Kmodr_3_0 Angle_5_0.95	Distance_Mahalanobis_3_0.92 Angle_5_0.95	Angle_5_0.95 Distance_Mahalanobis_3_0.92	Kmodr_3_0
c	Crossclustering_5_3 Distance_Euclidean_9_0.92	Crossclustering_5_3 Distance_Euclidean_9_0.92	Kmodr_3 Distance_Euclidean_9_0.92 Crossclustering_5_3	Crossclustering_5_3_0	Angle_5_0.95 Distance_Mahalanobis_3_0.92
d				Distance_Euclidean_9_0.92	Crossclustering_5_3 Distance_Euclidean_9_0.92

For each of these databases, 20 partitions are built using cross validation techniques. The algorithms are then compared using non-parametric test of Wilcoxon for two samples related to 95% confidence interval. The following algorithms were used in comparisons: Angle algorithm [33] based on the spatial data analysis approach, cross-clustering algorithm [34] based on partial clustering with automatic estimation of clusters number and outliers' identification, Kmodr algorithm [35] and *Outlier_Hybrid_LDS* based on kmeans (with k = 5), *Distance_Mahalanobis* [36] and *Distance_Euclidean* [9]. Table 3 resumes the comparisons result among the algorithms.

In the comparison, the algorithms groups are organized according to results quality, such as "group a" > "group b" > "group c" > "group d". The algorithms in the same group have no significant differences between them. In most of these databases, *Outlier_Hybrid_LDS* algorithm obtained good results except in the collective anomalous database (col_mix), where *Distance_Mahalanobis* algorithm is slightly superior. The worst result was *Distance_Euclidean_9_0.92*. Regarding efficiency, the best results are found with distance-based methods.

Outlier_Hybrid_LDS detected 450 outliers, representing 95.27% of real outlier's total number. This algorithm generates 44 rules that were unified by considering logical relations and finally 11 linguistic summaries were generated. All summaries were evaluated by using active learning techniques, by project management specialists. The following 5 summaries were identified as the most relevant for project management decisions:

1. Around 50% "outlier projects" have a "very high human resources' plan". T (0.76, 0.44, 0.69, 0.22, 1, 0.62).
2. Around 30% "outlier projects" have "Very high rate of human resources". T (0.5, 0.86, 0.55, 0.06, 1, 0.59).
3. Around 30% "outlier projects" have "Very high material resources' plan". T (0.53, 0.26, 0.27, 0.15, 1, 0.44).
4. Some "outlier projects" with "High material resources plan" have "High rate equipment resource". T (0.78, 0.16, 0.79, 0.33, 1, 0.61).
5. Around 30% "outlier projects" with "High human resources' plan" have "Very high human resources real plan". T (0.95, 0.49, 0.44, 0.23, 1, 0.62).

T vector means the evaluation of summaries by considering the traditional T values defined by Zadeh [37]. In order to get more legible linguistic summaries, algorithm introduces English language words such as "with" and "have" to connect filters and summarizers.

First linguistic summary means, around 50% of "outlier projects" have over-planned the human resources required. The second summary represents that human resources cost of around 30% "outlier projects" are over-planned. The third and fourth summaries represent that some "outlier projects" have over-planned the material resources, and some of them, have over-planned equipment cost rate. From the fifth summary it is interpreted that, in some cases, the number of human resources was planned below the actual number of human resources used. All these summaries help projects managers, correct errors in project management and scheduling.

4 Conclusions

From the results of this investigation, we can reach the following conclusions:

- In used databases, the most detected outliers deal with overestimation of human resources in project tasks.
- Around 30% of outlier projects incur higher costs for using more resources than planned.
- Around 30% "outlier projects" over-planned material resources and some of them contains over-planned equipment's cost-rate.
- Summaries detected from outliers help to project managers to fix errors on project scheduling and to detect project's over-cost.
- In used database the best outlier detection algorithm was the combination of "Kmeans" method with Mahalanobis distance.
- Mahalanobis distance method reports better results than the Euclidean distance in the context of this investigation.
- The experimentation demonstrated that is possible the errors' detection in software project planning from combination of techniques, such as outliers mining and linguistic data summarization.

References

1. Chrissis, M.B., Konrad, M., Shrum, S.: CMMI Guidlines for Process Integration and Product Improvement. Addison-Wesley Longman Publishing Co., Inc., Boston (2003)
2. PMI: A guide to the project management body of knowledge (PMBOK guide) Sixth Edition/Project Management Institute. Project Management Institute, Inc. Newtown Square, Pennsylvania 19073-3299, USA (2017)
3. Grau, N., Bodea, C.-N.: ISO 21500 Project Management Standard: Characteristics, Comparison and Implementation. VShaker Verlag GmbH, Germany (2014)
4. Pressman, R.: Ingeniería del Software Un Enfoque Práctico, 7ma edn. University of Connecticut, Storrs (2010)
5. Pádua, W.: Measuring complexity, effectiveness and efficiency in software course projects. In: Proceedings of the 32nd ACM/IEEE International Conference on Software Engineering, vol. 1, pp. 545–554. IEEE (2010)
6. Gimeno Alonso, J.Á.: Fallos en proyectos: investigación sobre causas generales. In: XVI Congreso Internacional de Ingeniería de Proyectos (2012)
7. Hussain, A., Mkpojiogu, E.O., Kamal, F.M.: The role of requirements in the success or failure of software projects. Int. Rev. Manage. Mark. 6, 306–311 (2016)
8. Pérez Pupo, I., García Vacacela, R., Piñero Pérez, P., Sadeq, G., Peña Abreu, M.: Experiencias en el uso de técnicas de soft-computing en la evaluación de proyectos de software. Rev. Invest. Oper. 41(1), 106–117 (2015)
9. Aggarwal, C.H.C.: Datos anómalos analysis. IBM T.J.: Watson Research Center Yorktown Heights, New York, USA (2013)
10. Gupta, M., Gao, J., Aggarwal, C.C., Han, J.: Outlier detection for temporal data: a survey. IEEE Trans. Knowl. Data Eng. 26, 2250–2267 (2014)

11. Williams, G., Baxter, R., He, H., Hawkins, S., Gu, L.: A comparative study of RNN for outlier detection in data mining. In: Proceedings of 2002 IEEE International Conference on Data Mining, ICDM 2003, pp. 709–712. IEEE (2002)
12. Hawkins, D.M.: Identification of Outliers. Springer, Heidelberg (1980). https://doi.org/10.1007/978-94-015-3994-4
13. Degtiarev, K.Y., Remnev, N.V.: Linguistic resumes in software engineering: the case of trend summarization in mobile crash reporting systems. Proc. Comput. Sci. **102**, 121–128 (2016)
14. Castro Aguilar, G.F., Pérez Pupo, I., Piñero Pérez, P.Y., Martínez, N., Crúz Castillo, Y.: Aplicación de la minería de datos anómalos en organizaciones orientadas a proyectos. Rev. Cubana Ciencias Inf. **10**, 195–209 (2016)
15. Hubert, M., Rousseeuw, P.J., Segaert, P.: Multivariate functional outlier detection. Stat. Methods Appl. **24**, 177–202 (2015)
16. Templ, M., Gussenbauer, J., Filzmoser, P.: Evaluation of robust outlier detection methods for zero-inflated complex data. J. Appl. Stat. 1–24 (2019)
17. Patel, S.P., Shah, V., Vala, J.: Outlier detection in dataset using hybrid approach. Int. J. Comput. Appl. (2015)
18. Bro, R., Smilde, A.K.: Principal component analysis. In: Analytical Methods, pp. 2812–2831 (2014)
19. Kamble, B., Doke, K.: Outlier detection approaches in data mining. J. Eng. Technol. (IRJET) **4**, 634–638 (2017). International Research
20. Ranga Suri, N.N.R., Murty, M.N., Athithan, G.: Research issues in outlier detection. In: Ranga Suri, N.N.R., Murty, M.N., Athithan, G. (eds.) Outlier Detection: Techniques and Applications: A Data Mining Perspective, vol. 155, pp. 29–51. Springer, Cham (2019). https://doi.org/10.1007/978-3-030-05127-3_3
21. Radovanović, M., Nanopoulos, A., Ivanović, M.: Reverse nearest neighbors in unsupervised distance-based outlier detection. IEEE Trans. Knowl. Data Eng. **27**, 1369–1382 (2015)
22. Mishra, S., Chawla, M.: A comparative study of local outlier factor algorithms for outliers detection in data streams. In: Abraham, A., Dutta, P., Mandal, J.K., Bhattacharya, A., Dutta, S. (eds.) Emerging Technologies in Data Mining and Information Security. Advances in Intelligent Systems and Computing, vol. 183, pp. 347–356. Springer, Singapore (2019). https://doi.org/10.1007/978-981-13-1498-8_31
23. Abdulalla, F.Q., Abduljabar, A.S., Shaker, S.H.: A survey of human face detection methods. J. Al-Qadisiyah Comput. Sci. Math. 108–117 (2018)
24. Rajeswari, A., Sridevi, M., Deisy, C.: Outliers detection on educational data using fuzzy association rule mining. In: Proceedings of International Conference on Advanced in Computer Communication and Information Science (ACCIS-14), pp. 1–9 (2014)
25. Jain, L.C., Martin, N.: Fusion of Neural Networks, Fuzzy Systems and Genetic Algorithms: Industrial Applications. CRC Press, Boca Raton (2020)
26. Stützle, T., López-Ibáñez, M.: Automated design of metaheuristic algorithms. In: Gendreau, M., Potvin, J.-Y. (eds.) Handbook of Metaheuristics, vol. 272, pp. 541–579. Springer, Cham (2019)
27. Kacprzyk, J., Zadrożny, S.: Linguistic summarization of the contents of web server logs via the Ordered Weighted Averaging (OWA) operators. Fuzzy Sets Syst. **285**, 182–198 (2016)
28. Donis-Diaz, C., Muro, A., Bello-Pérez, R., Morales, E.V.: A hybrid model of genetic algorithm with local search to discover linguistic data summaries from creep data. Expert Syst. Appl. **41**, 2035–2042 (2014)
29. Kacprzyk, J., Zadrożny, S.: Fquery for access: fuzzy querying for a Windows based DBMS. In: Bosc, P., Kacprzyk, J. (eds.) Fuzziness in Database Management Systems. Studies in Fuzziness, vol. 5, pp. 415–433. Springer, Heidelberg (1995). https://doi.org/10.1007/978-3-7908-1897-0_18

30. Kacprzyk, J., Zadrozny, S.: Linguistic summarization of data sets using association rules. In: The 12th IEEE International Conference on Fuzzy Systems, FUZZ 2003, pp. 702–707. IEEE (2003)
31. Donis-Diaz, C.A., Bello, R., Kacprzyk, J., et al.: Linguistic data summarization using an enhanced genetic algorithm. Czasopismo Tech. 3–12 (2014)
32. Pérez, P.P., Pupo, I.P., Rivero Hechavarría, C.C., Lusardo, C.R., Sosa, R.G., López, S.T.: Repositorio de datos para investigaciones en gestión de proyectos. Rev. Cubana Ciencias Inf. 176–191 (2019). https://gespro.uci.cu/projects/
33. Jimenez, J.: Angle-based outlier detection (2016). https://cran.r-project.org/
34. Hennig, C., Meila, M., Murtagh, F., Rocci, R.: Handbook of Cluster Analysis. CRC Press, Boca Raton (2015)
35. Howe, D.C.: K-Means with simultaneous outlier detection (2016)
36. Rakhe, S.S., Vaidya, A.S.: Enhanced outlier detection for high dimensional data using different neighbor metrics. Int. J. Eng. Sci. (2016)
37. Zadeh, L.A.: A computational approach to fuzzy quantifiers in natural languages. Comput. Math. Appl. 149–184 (1983)

Approximate Decision Tree Induction over Approximately Engineered Data Features

Dominik Ślęzak[1(✉)] and Agnieszka Chądzyńska-Krasowska[2]

[1] Institute of Informatics, University of Warsaw, Warsaw, Poland
slezak@mimuw.edu.pl
[2] Polish-Japanese Academy of Information Technology, Warsaw, Poland

Abstract. We propose a simple SQL-based decision tree induction algorithm which makes its heuristic choices how to split the data basing on the results of automatically generated analytical queries. We run this algorithm using standard SQL and the approximate SQL engine which works on granulated data summaries. We compare the accuracy of trees obtained in these two modes on the real-world dataset provided to participants of the Suspicious Network Event Recognition competition organized at IEEE BigData 2019. We investigate whether trees induced using approximate SQL queries – although execution of such queries is incomparably faster – may yield poorer accuracy than in the standard scenario. Next, we investigate features – inputs to the decision tree induction algorithm – derived using SQL from a bigger associated data table which was provided in the aforementioned competition too. As before, we run standard and approximate SQL, although again, that latter mode needs to be checked with respect to the accuracy of trees learnt over the data with approximately extracted features.

Keywords: SQL-based decision tree induction · SQL-based feature engineering · Approximate SQL engines · Granulated data summarization · Big data analytics · Cybersecurity analytics

1 Introduction

Every typical KDD process consists of several stages, such as data preparation, attribute construction and selection, decision model induction and more [8, 21]. Given the growing sizes of data required to be mined, there are a number of approaches attempting to utilize higher-level interfaces to data storage and data processing systems instead of operating directly on raw data sources. With this respect, employment of relational database systems and SQL is one of intensively examined opportunities [13, 22].

In our research, we often refer to KDD methods based on standard SQL queries supported by most of database vendors. We rewrite some of algorithms which are well-known in the KDD domain to illustrate how basic SQL procedures can replace lower-level computations. This way KDD solutions can gain important data management and computational scalability features of modern database systems. Moreover, users who are familiar with SQL can easily introduce changes into previous implementations, at the level which is specific to declarative rather than imperative languages.

© Springer Nature Switzerland AG 2020
R. Bello et al. (Eds.): IJCRS 2020, LNAI 12179, pp. 376–384, 2020.
https://doi.org/10.1007/978-3-030-52705-1_28

In this paper, for the purpose of illustrating the role that database systems can play in KDD, we introduce a very simple new version of SQL-based decision tree induction algorithm. This particular algorithm is dedicated to datasets with numeric attributes and binary decisions. It makes its heuristic "attribute greater/lower than value" split choices by basing on the results of automatically generated aggregate queries. Such approach is surely not novel [7, 10]. Still, our contribution is twofold. First, we run our algorithm using one of approximate SQL engines available in the market [17], in order to verify whether decision trees constructed using approximate queries may yield poorer accuracy than while basing on classical exact SQL. Second, given the multi-table characteristics of the considered real-world dataset [5], we investigate whether newly engineered attributes – added to the main training table by executing analytical SQL statements over another available data table – could be derived using approximate queries instead of exact ones, with no harm to the efficiency of further learning mechanisms.

Both above aspects reflect the same challenge, although they refer to different KDD stages – attribute engineering (exemplified by SQL-based usage of one-to-many relation between data tables) and decision model construction (exemplified by our decision tree induction algorithm). The question is whether approximate SQL – which can be orders of magnitude faster than exact SQL over big datasets – is able to drive KDD processes accurately enough, so acceleration is achieved without losing too much quality. Indeed, one could suspect that the quality of the aforementioned splits made during decision tree construction is potentially worse if their heuristic evaluation relies on not-fully-precise calculations over the training data. Analogously, one may be afraid of using impre-cisely derived values of newly created attributes as the input to any machine learning algorithm, no matter whether that algorithm itself is based on exact or approximate computations. Our goal is to illustrate to what extent such worries are justified.

The rest of the paper is structured as follows. Section 2 refers to some related works. Section 3 describes the dataset used in our studies. Section 4 outlines the proposed deci-sion tree induction algorithm. Section 5 reports our experimental results in four modes: running our algorithm using classical or approximate SQL, over the data derived using classical or approximate SQL. Section 6 concludes our work.

2   Related Work

Let us begin with the literature on SQL-based machine learning/data mining. We have already cited papers [13, 22] (related to SVM and k-NN methods) and [7, 10] (related to decision tree induction). For further research in this field we refer to [3, 11, 14, 15] (fea-ture selection, data clustering, association rules and more details about decision trees). An interesting additional aspect of applying SQL in KDD corresponds to relational – single-table or multi-table – feature/attribute engineering [6, 20].

We refer also to approximate query engines which become popular because of big data analytics challenges [9, 12]. We work with the first-ever engine based entirely on the concept of data summarization [16, 17], which was successfully deployed in indus-try[1]. This engine, whereby query execution operations are designed as transformations of granulated data summaries, can be used as if it was standard PostgreSQL. It delivers

---

[1] securityondemand.com/solutions/superscale-analytics-threat-detection/.

Table 1. SQL-based features. 'x' identifies a record for which a new feature value is calculated.

devicetype_cd(x)	select count(distinct devicetype) from localizedalerts where threatwatchalertid = x;
devicevendor_cd(x)	select count(distinct devicevendor) from localizedalerts where threatwatchalertid = x;
direction_cd(x)	select count(distinct direction) from localizedalerts where threatwatchalertid = x;
domain_cd(x)	select count(distinct domain) from localizedalerts where threatwatchalertid = x;
dstip_cd(x)	select count(distinct dstip) from localizedalerts where threatwatchalertid = x;
dstipcategory_cd(x)	select count(distinct dstipcategory) from localizedalerts where threatwatchalertid = x;
dstport_cd(x)	select count(distinct dstport) from localizedalerts where threatwatchalertid = x;
eventname_cd(x)	select count(distinct eventname) from localizedalerts where threatwatchalertid = x;
p6(x)	select count(distinct alerttype) from localizedalerts where threatwatchalertid = x;
p9(x)	select count(*) from localizedalerts where
	alerttype like 'Suspicious Outbound Anomaly%' and threatwatchalertid = x;
protocol_cd(x)	select count(distinct protocol) from localizedalerts where threatwatchalertid = x;
reportingdevice_cd(x)	select count(distinct reportingdevice) from localizedalerts where threatwatchalertid = x;
severity_cd(x)	select count(distinct severity) from localizedalerts where threatwatchalertid = x;
srcip_cd(x)	select count(distinct srcip) from localizedalerts where threatwatchalertid = x;
srcipcategory_cd(x)	select count(distinct srcipcategory) from localizedalerts where threatwatchalertid = x;
srcport_cd(x)	select count(distinct srcport) from localizedalerts where threatwatchalertid = x;
username_cd(x)	select count(distinct username) from localizedalerts where threatwatchalertid = x;

accurate results even for highly selective queries involving combinations of numeric and alphanumeric columns, such as those in Table 1. We will apply it for both SQL-based decision tree induction and the above-mentioned attribute engineering.

As we run our experiments on the data disclosed in an online machine learning competition, let us emphasize the importance of such events for development of both academic and commercial research. The most widely recognized platform in this area is Kaggle[2], although there are also others, such as KnowledgePit[3]. The reader can find more details about machine learning competitions held on KnowledgePit in [4,5].

3   The Data from the IEEE BigData 2019 Competition

We conduct experiments on the dataset made available at one of machine learning competitions held at IEEE BigData 2019. This competition was organized jointly by Security On-Demand (SOD)[4] and QED Software[5], at aforementioned KnowledgePit[6].

The data was provided by SOD in three tables. The first one contains nearly 60,000 records corresponding to so-called threatwatch alerts investigated by the security team at SOD in Q4 of 2018 and Q1 of 2019. Alerts are described by 61 columns and represent information that is available to security analysts during their decision-making processes. For each record, it is indicated whether the given alert was considered as *serious* by an analyst and therefore, whether the given SOD's client was notified about it.

The second table includes so-called localized alerts registered by SOD. For each record in the first table, there is a series of associated localized alerts. This table contains about 8,700,000 records described by a mixture of 20 numeric and symbolic features. It provides more detailed information about the network traffic and devices related

---

[2] www.kaggle.com.

[3] www.knowledgepit.ml.

[4] www.securityondemand.com.

[5] www.qed.pl.

[6] www.knowledgepit.ml/suspicious-network-event-recognition/.

to threatwatch alerts evaluated by security analysts. In particular, the severity of each localized alert is automatically assessed by expert-made heuristics designed by SOD.

The third table is an extract from raw network event logs that are continually captured by SOD using so-called collectors. This table is considerably larger than the previous ones. Its fragment disclosed to competition participants consisted of nearly 9,000,000,000 anonymized records described by 26 features. More information about this data source can be found in [16]. For more information about the discussed machine learning competition and its results we refer to [5].

In this paper, we concentrate on the two first tables. During the competition, participants did their best to utilize localized alerts to extract new attributes describing threatwatch alerts. The task was to learn – basing on the historical data labeled by SOD's analysts – how to distinguish between threatwatch alerts requiring and not requiring client notifications. Thus, aggregations derived for threatwatch alerts from their associated collections of localized alerts could be helpful.

For our experiments, we selected 8 numeric features from the first table and 17 new features generated from the second one. Our selection was based on SOD's expertise and on some of successful competition solutions. Given one-to-many relation between tables, all new features were derived using *SELECT COUNT* or *COUNT DISTINCT* queries, so they can be treated as numeric too. As a result, we obtain a dataset with nearly 60,000 records, 25 numeric columns, and the binary decision attribute (which can be referred also as the target variable) corresponding to client notifications.

The 8 original features are: *parentcategory, overallseverity, correlatedcount, isiptrusted, untrustscore, trustscore, flowscore, enforcementscore*. The 17 derived features are listed in Table 1 together with SQL statements executed to compute them. The meanings of columns in considered data tables are quite typical for the area of cybersecurity [1, 19], although SOD's way of calculating their values is unique.

4 Naïve SQL-Based Decision Tree Induction

The aim of the algorithm introduced below is to establish a framework for investigating the quality-related differences between decision models derived using classical and approximate SQL statements. The algorithm itself is extremely simplified and we refer the reader to other aforementioned publications for more sophisticated ideas how to take advantage of relational database systems in decision tree induction [7, 10, 15, 21]. We actually wanted to keep it so simple to concentrate mainly on the classical versus approximate SQL comparison. In future, analogous comparisons can be studied for other SQL-based machine learning implementations as well.

The algorithm works with standard tabular data input, i.e., each attribute corresponds to a separate column. The decision attribute (target variable) is declared as the binary column *DECISION*. Conditional attributes (dependent variables) are assumed to be numeric, although there is also an interesting interpretation of our algorithm for binary columns. We denote attributes-columns as $a1,...,an$, where n is the number of conditional attributes in the training set. The algorithm can be triggered with parameter $K > 0$ which stands for the maximum depth of decision tree induced from the data.

Let us recall that the considered approximate engine can be queried as if it was a typical instance of PostgreSQL, where the only difference is that results of analytical *SELECT* statements are not guaranteed to be fully precise (and on the other hand, one obtains those results incomparably faster than in the case of any standard engine because of the ability to work entirely on granulated data summaries) [16,17]. Therefore, we scripted our algorithm in standard PL/pgSQL. Moreover, it is straightforward to run it in the same way on the considered approximate engine and on classical PostgreSQL.

The algorithm is constructing a tree in a typical greedy way, whereby the heuristic binary split evaluation function is triggered recursively for every current leaf unless it satisfies one of three stop conditions illustrated in Fig. 1. At the beginning the following statement is executed:

```
SELECT DECISION, COUNT(*), AVG(a1)...AVG(an) FROM DATA GROUP BY DECISION;
```

Then, ai with the highest difference between its average values on records dropping into decision classes 0 and 1 is selected. Precisely, using notation in Fig. 1, we choose ai with the maximum ratio $| AVG0(ai) - AVG1(ai) | / (|AVG0(ai)| + |AVG1(ai)|)$, where denominator is used to compare more fairly between attributes with varying scales. New nodes are created with the cut $(AVG0(ai) + AVG1(ai)) / 2$, i.e., records satisfying conditions $ai < (AVG0(ai) + AVG1(ai)) / 2$ and $ai \geq (AVG0(ai) + AVG1(ai)) / 2$ are assigned to left and right nodes, respectively. The procedure is repeated with each of these nodes independently, whereby the only difference is that the previously chosen ai

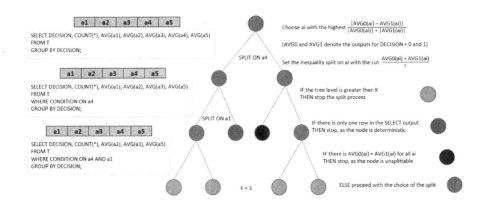

Fig. 1. High-level illustration of our naïve decision tree induction algorithm, with the maximum tree depth fixed as $K = 3$. The node statuses: Green – ready for further splits; Red – the maximum depth reached; Blue – deterministic leaf pointing at a single decision, no further splits needed; Black – available attributes do not provide sufficient discrimination between decisions, no further splits make sense. Left-side green/red statuses reflect whether particular numeric attributes have been already used in the given tree path (our naïve implementation does not allow to reuse an attribute in a path). Quantities $AVG0(ai)$ and $AVG1(ai)$ refer to *SELECT ... GROUP BY ...* results – they denote the average value of the i-th attribute in the given node, for decisions 0 and 1, respectively. The *COUNT(*)* component is used to derive decision class distributions. (Color figure online)

does not occur on the *SELECT* list any longer while the *WHERE* part is extended by the above-specified inequality conditions over column *ai*.

Surely, removing once chosen *ai* from *SELECT* lists in subsequent phases of splitting nodes is a simplification. Another simplification corresponds to one of the stop criteria – the *black* one in Fig. 1. The fact that quantities *AVG0(ai)* and *AVG1(ai)* are the same does not always mean that it is impossible to set up a useful condition on *ai*, although indeed attributes yielding larger differences between *AVG0* and *AVG1* can be regarded as more informative. Nevertheless, the whole algorithm is quite efficient, as it executes only a single SQL query per node. The *blue* stop criterion in Fig. 1 is particularly elegant, as a single-tuple output of the considered *SELECT* statement means that the corresponding node drops fully into one of decision classes.

Let us also note that the attribute choice criterion driven by the above queries neglects probabilities of decision classes in particular tree nodes. Indeed, quantities *AVG0(ai)* and *AVG1(ai)* are simply compared to each other, no matter how many records were taken into account while deriving them. It may happen that *AVG0(ai)* is the average value of attribute *ai* calculated on, e.g., 100 records with *DECISION* = 0 while there is only one record satisfying condition *DECISION* = 1. Then, the "greater/lower than value" split on *ai* is fixed as a completely non-weighted mean of *AVG0(ai)* and *AVG1(ai)*. This kind of *Bayesian* approach to decision tree induction was first proposed in [2] (whereby yet another SQL-based data mining methodology was employed) and further formalized with respect to arbitrary feature-based data partitions in [18] (whereby a decision tree induces a special case of data partition).

5   Experimental Results

As mentioned in Sect. 1, we report experiments conducted in four (two times two) following modes: running our naïve decision tree induction algorithm using two variants of SQL, i.e., classical PostgreSQL versus approximate engine [16,17], and over two versions of the considered dataset, whereby features displayed in Table 1 were computed using – again – classical or approximate *SELECT* statements.

More precisely, in its both versions, the dataset discussed in the end of Sect. 3 has nearly 60,000 records (corresponding to the same set of threatwatch alerts investigated by SOD's analysts), 25 numeric attributes and the same binary decision. The only difference between these two versions is the way of creating 17 new features from the associated data table that stores localized alerts. That table was actually loaded both into PostgreSQL and the considered approximate query engine, which internally replaced its 8,700,000 records with far lower number of multidimensional data summaries.

Experimental results are summarized in Table 2. The difference between two above data versions is indicated by the column *new columns*. Both those datasets were loaded into both PostgreSQL and our approximate engine, in order to run two considered variants of decision tree induction – indicated by column *tree induction*. For example, combination *approximate-exact* means that a tree was learnt using classical SQL but on the dataset with 17 features calculated using approximate SQL.

Outcomes for three different maximum tree depth levels K are presented. The number of leaves grows when longer root-to-leaf paths are allowed, although it can be noticed that the algorithm introduced in Sect. 4 tends to produce more compact trees when working with approximate SQL (for both *approximate* and *exact* versions of new features). The last two columns of Table 2 report additionally average intensities of occurrence of original and derived attributes in tree paths.

Table 2. Characteristics of decision trees induced using different settings of our procedure.

New columns	Tree induction	K level	R score	# of nodes	Derived	Original
Exact	Exact	5	0.251	45	0.195	0.102
Exact	Approximate	5	0.324	63	0.063	0.492
Approximate	Exact	5	0.181	59	0.162	0.266
Approximate	Approximate	5	0.288	63	0.080	0.469
Exact	Exact	10	0.408	473	0.259	0.153
Exact	Approximate	10	0.394	875	0.190	0.586
Approximate	Exact	10	0.369	771	0.276	0.324
Approximate	Approximate	10	0.332	805	0.195	0.553
Exact	Exact	15	0.559	2573	0.271	0.165
Exact	Approximate	15	0.454	3339	0.210	0.595
Approximate	Exact	15	0.421	2793	0.295	0.334
Approximate	Approximate	15	0.349	2599	0.210	0.561

It is also important to evaluate the quality of induced trees. Herein, we follow the aforementioned approach which was developed in [2, 18] to assess data partitions (induced by subsets of attributes or collections of root-to-leaf tree paths) with respect to the level of information that they provide about decisions. The considered methodology is based on the following *relative information gain* measure $R(tree) =$

$$\sum_{leaves} \max_j \frac{\text{\# of records in leaf with } DECISION = j}{\text{\# of records in dataset with } DECISION = j} - 1 \qquad (1)$$

Measure R has values ranging from 0 to 1, whereby equality $R(tree) = 1$ holds, if and only if all tree leaves are deterministic (i.e. they support single decision classes). Moreover, R is monotonic – splitting any leaf onto two new leaves cannot decrease its value – and it is generally perceived as a good indicator in the case of analyzing highly imbalanced datasets, such as the one disclosed in the considered machine learning competition at the IEEE BigData 2019 conference.

6   Conclusions

We introduced a naïve SQL-based decision tree induction algorithm, with the aim to compare classical PostgreSQL and the approximate query engine working on granulated data summaries with respect to the quality of trees derived from the data.

Our experiments focused on real-world dataset made publicly available in frame of the Suspicious Network Event Recognition competition held at the IEEE BigData 2019 conference. In particular, we studied two out of three data tables disclosed in the competition and additionally, we utilized the considered approximate engine to investigate opportunities of approximate-SQL-driven feature engineering.

In future, besides improvements of the above-mentioned algorithm, we intend to extend our decision-tree-related research onto the third data source associated with the discussed machine learning competition. Given its huge volume, this data source needs approximate analytical methods to the highest extent. Let us also point out that the experimental framework developed in this paper can serve as a useful environment for testing enhancements of our approximate engine and other analogous solutions.

References

1. Garcia-Teodoro, P., Díaz-Verdejo, J.E., Maciá-Fernández, G., Vázquez, E.: Anomaly-based network intrusion detection: techniques, system challenges. Comput. Secur. **28**(1–2), 18–28 (2009)
2. Grant, A., et al.: Examination of routine practice patterns in the hospital information data warehouse: use of OLAP and rough set analysis with clinician feedback. In: AMIA 2001, p. 916 (2001)
3. Hu, X., Lin, T.Y., Han, J.: A new rough sets model based on database systems. Fundam. Inform. **59**(2–3), 135–152 (2004)
4. Janusz, A., Grad, Ł., Grzegorowski, M.: Clash royale challenge: how to select training decks for win-rate prediction. In: FedCSIS 2019, pp. 3–6 (2019)
5. Janusz, A., Kałuża, D., Chądzyńska-Krasowska, A., Konarski, B., Holland, J., Ślęzak, D.: IEEE BigData 2019 cup: suspicious network event recognition. In: IEEE BigData (2019)
6. Kobdani, H., Schütze, H., Burkovski, A., Kessler, W., Heidemann, G.: Relational feature engineering of natural language processing. In: CIKM 2010, pp. 1705–1708 (2010)
7. Kowalski, M., Stawicki, S.: SQL-based heuristics for selected KDD tasks over large data sets. In: FedCSIS 2012, pp. 303–310 (2012)
8. Kurgan, L.A., Musílek, P.: A survey of knowledge discovery and data mining process models. Knowl. Eng. Rev. **21**(1), 1–24 (2006)
9. Mozafari, B., Niu, N.: A handbook for building an approximate query engine. IEEE Data Eng. Bull. **38**(3), 3–29 (2015)
10. Nguyen, H.S., Nguyen, S.H.: Fast split selection method and its application in decision tree construction from large databases. Int. J. Hybrid Intell. Syst. **2**(2), 149–160 (2005)
11. Ordonez, C., Cereghini, P.: SQLEM: fast clustering in SQL using the EM algorithm. In: SIGMOD 2000, pp. 559–570 (2000)
12. Orr, L., Suciu, D., Balazinńska, M.: Probabilistic database summarization for interactive data exploration. PVLDB **10**(10), 1154–1165 (2017)
13. Rüping, S.: Support vector machines in relational databases. In: Lee, S.W., Verri, A. (eds.) SVM 2002. LNCS, vol. 2388, pp. 310–320. Springer, Heidelberg (2002). https://doi.org/10.1007/3-540-45665-1_24
14. Sarawagi, S., Thomas, S., Agrawal, R.: Integrating association rule mining with relational database systems: alternatives and implications. Data Min. Knowl. Discov. **4**(2–3), 89–125 (2000)
15. Sattler, K., Dunemann, O.: SQL database primitives for decision tree classifiers. In: CIKM 2001, pp. 379–386 (2001)

16. Ślęzak, D., Chądzyńska-Krasowska, A., Holland, J., Synak, P., Glick, R., Perkowski, M.: Scalable cyber-security analytics with a new summary-based approximate query engine. In: IEEE BigData 2017, pp. 1840–1849 (2017)
17. Ślęzak, D., Glick, R., Betliński, P., Synak, P.: A new approximate query engine based on intelligent capture and fast transformations of granulated data summaries. J. Intell. Inf. Syst. **50**(2), 385–414 (2018)
18. Ślęzak, D., Ziarko, W.: The investigation of the bayesian rough set model. Int. J. Approx. Reason. **40**(1–2), 81–91 (2005)
19. Sommer, R., Paxson, V.: Outside the closed world: on using machine learning for network intrusion detection. In: IEEE S&P 2010, pp. 305–316 (2010)
20. Wróblewski, J.: Analyzing relational databases using rough set based methods. In: IPMU 2000, vol. 1, pp. 256–262 (2000)
21. Wróblewski, J., Stawicki, S.: SQL-based KDD with infobright's RDBMS: attributes, reducts, trees. In: Kryszkiewicz, M., Cornelis, C., Ciucci, D., Medina-Moreno, J., Motoda, H., Raś, Z.W. (eds.) Rough Sets and Intelligent Systems Paradigms. LNCS, vol. 8537, pp. 28–41. Springer, Cham (2014). https://doi.org/10.1007/978-3-319-08729-0_3
22. Yao, B., Li, F., Kumar, P.: K nearest neighbor queries and kNN-joins in large relational databases (almost) for free. In: ICDE 2010, pp. 4–15 (2010)

Linguistic Summaries Generation with Hybridization Method Based on Rough and Fuzzy Sets

Iliana Pérez Pupo[1]([⊠]) [iD], Pedro Y. Piñero Pérez[1] [iD],
Rafael Bello[2] [iD], Luis Alvarado Acuña[3] [iD],
and Roberto García Vacacela[4] [iD]

[1] Departamento de Investigaciones en Gestión de Proyectos,
Universidad de las Ciencias Informáticas, 54830 La Habana, CP, Cuba
{iperez,ppp}@uci.cu
[2] Centro de Investigación en Informática, Universidad Central Marta
Abreu de las Villas, Santa Clara, Cuba
rbellop@uclv.edu.cu
[3] Departamento de Ingeniería de la Construcción, Universidad Católica
del Norte, Antofagasta, Chile
lualvar@ucn.cl
[4] Facultad de Especialidades Empresariales, Universidad Católica de Santiago
de Guayaquil, Guayaquil, Ecuador
roberto.garcia@cu.ucsg.edu.ec

Abstract. In this paper authors propose a new algorithm for linguistic data summarization based on hybridization of rough sets and fuzzy sets techniques. The new algorithm applies rough sets theory for feature selection in early stages of linguistic summaries' generation. The rough sets theory was used to reduce on significant way, the amount on summaries obtained by others algorithms. The algorithm combines lower approximation, k grade dependency and fuzzy sets to get linguistic summaries. The results of proposed algorithm are compared with association rules approach. In order to validate the algorithm proposed, authors apply both qualitative and quantitative methods. Authors used two databases in order to validate the algorithm; theses databases belong to "Repository of Project Management Research". The first database is associated to personality traits and human performance in software projects. The second database is associated to analysis of revenue assurance in different organization. Considering quantitative approach, the algorithm proposed, obtains better results than the algorithm based on association rules; while regards execution time, the best algorithm was the algorithm based on association rules, because rough sets theory was high time-consuming technique.

Keywords: Linguistic data summarization · Rough sets · Project management · Human resources

© Springer Nature Switzerland AG 2020
R. Bello et al. (Eds.): IJCRS 2020, LNAI 12179, pp. 385–397, 2020.
https://doi.org/10.1007/978-3-030-52705-1_29

1 Introduction

Linguistic data summarization was first introduced by Yager in 1982's [1] and it has been applied in real scenarios such as: autonomy of things, medicine and others real scenarios. Different investigations associated to this technique have been developed in the last two decades following three main work lines:

- Conceptualization of linguistic summaries and its structure.
- Indicators to evaluate the quality of linguistic summaries.
- Algorithms to generate linguistic summaries from data.

About the structure, the summaries are classified considering different "protoforms" [2, 3]:

- Classic protoforms, to summarize attributes [1, 4].
- Time series protoforms [5].
- Events representation protoforms [6].

But the most used are protoforms with the following syntax:

- Overviews whose structure is *Qy's are S*, which describe relationships such as the following:
  T(Most employees have low pay) = 0.7.
- Summaries structured as *QRy's are S*, describing relationships such as:
  T(Most young employees have low pay) = 0.7.

Kacprzyk and Zadrożny classified in [7] six protoforms that described the structure of summaries and the queries for their search, see Table 1.

Table 1. Classification of protoforms of LDS [7].

Type	Protoform	Given	Sought
0	QRy's are S	All	Validity T
1	Qy's are S	S	Q
2	QRy's are S	S and R	Q
3	Qy's are S	Q and structure of S	Linguistic values in S
4	QRy's are S	Q, R and structure of S	Linguistic values in S
5	QRy's are S	Nothing	S, R and Q

About the indicators to measure the quality of the summary, several authors have been proposed different *T* indicators. For example, in [1] Yager proposed six indicators, called as *T* values *T1, T2, T3, T4, T5, T6*, as follows:

- Degree of truth (*T1*): called the measure of validity of the summary, provides an indication of how compatible the linguistic summary is with the database.
- Degree of imprecision (*T2*): is important validity criterion, measure of both uncertainty and vagueness concepts. This indicator depends on the form of the summary, not on the database.
- The degree of coverage (*T3*): measures how many objects in database are supported for linguistic summary.
- Degree of appropriateness (*T4*): This degree describes how characteristic is the summary for the particular database. It degree permits to distinguish between trivial summaries, having full validity (truth), and really important summaries. The summary found reflects an interesting, not fully excepted relation in our data [8].
- The length of an overview (*T5*) measure of the length or summaries, how many elements conform the summary.
- An indicator for resume the quality evaluation of a particular linguistic overview (*T6*), is defined as the weighted average of the previous 5° of validity.

There are other measures, for example in [5] Kacprzyk and Wilbik proposed a set of indicators specifically for time series scenarios. In [9], the authors proposed a set of indicators to extend the Yager's indicators based on degree of indeterminacy information in summaries. About the algorithms to generate summaries there are different trends too, such as:

- Linguistic summaries generated form sql queries [10].
- Generation of summaries by using association rules [11, 12].
- Generation of summaries through meta-heuristics [13].
- Generation of summaries by using clustering techniques [6].
- Other approaches that combine previous works [14–17].

But most of algorithms, to generate summaries reported in bibliography, not use appropriately, the information associated to the attributes relationships. In order to improve the summaries' generation methods, authors of this work proposed a new algorithm for linguistic data summarization based on hybridization of rough sets and fuzzy sets.

This work is organized in the following sections. The second section presents a brief analysis of rough sets concepts and its adoption in the algorithm proposed. The third section presents the results of algorithm in a human resource problem. Finally, the conclusions of the work are presented.

2   A New Algorithm for Linguistic Data Summarization

In linguistic summaries generation is very important to discover the attributes relationships. The linguistic summaries consist on filters and summarizers, in general these components can be represented from an information system $S = (U, A \cup D)$, where filters belong to set A while summarizers belong to decision attributes D. In this sense, the authors of this paper propose the application of rough sets theory to discover the attributes relationships. The authors adopted some concepts of rough sets theory in the new algorithm, in the next paragraphs we explain main concepts of this theory.

The rough sets theory was proposed in 1986 by Pawlak for application in data inconsistency. Usually, rough sets are used in two alternatives: to discreet data [18] based on equivalence relationships or to extended indiscernibility relationships [19, 20]. Different extensions of rough sets applications were reported in [18, 21, 22].

Given an information system $S = (U, A \cup D)$, let $X \subseteq U$ a set of objects and $B \subseteq A$, a selected set of attributes, from the information contain in B, X can be approximate like following:

- The lower approximation of X with respect to B is:

$$B_*(X) = \{x \in U : B(x) \subseteq X\} \tag{1}$$

- The upper approximation of X respect to B is:

$$B^*(X) = \{x \in U : B(x) \cap X = \Phi\} \tag{2}$$

- The boundary region we can define as:

$$BNB(X) = B*(X) - B*(X) \tag{3}$$

- The negative region of decision d with respect to B is:

$$NEGB(X) = U - B*(X) \tag{4}$$

- Indiscernibility relation: defines an equivalence relation $INDB$ [23, 24], and this relation is denoted by:

$$IND(B)B = \{(x, y) \in U \times U : a(x) = a(y) \text{ } for \text{ } every \text{ } a \in B\} \tag{5}$$

- The positive region of decision d with respect to B is:

$$POS_B(d) = \cup \{B_*(X) : X \in U/IND(d), d \in D\} \tag{6}$$

Other useful concept useful is k grade dependency, that we explain in next paragraph.

Definition 1: Intuitively, a set of decision attributes D, depends totally on a set of B attributes, denoted by $B \Rightarrow D$, if all the values of the D attributes are univocally determined by the values of the attribute in B. In other words, D depends totally on B, if there is a functional dependency between the values of D and B [23]. D depends on B in a k grade where $k \in [0,1]$, and denoted by $B \Rightarrow_k D$, see Eq. (7). If $k = 1$ then D depends totally on B, while if $k < 1$ then D depends partially on B.

$$k = \frac{|POS_B(D)|}{|U|} \tag{7}$$

Where:

$$POS_B(D) = \bigcup_{X \in \frac{U}{D}} B_*(X) \tag{8}$$

2.1   LDS_RoughSet Algorithm

In this section we propose an algorithm for the construction of linguistic summaries of data, generating them from hybridization of rough sets and association rules. The authors established a linguistic variable associated to quantifiers for the construction of the summaries, see Fig. 1:

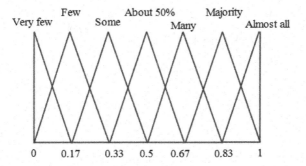

Fig. 1. Linguistic variable associated to quantifiers.

The proposed algorithm and its parameters are presented below.

```
Algorithm's name: LDS_RoughSet.
  Inputs:
    Information system S = (U, A ∪ D)
    U: dataset for analysis.
    A, D: attributes in the information system.
    Fuzzifiers: Set of fuzzy variables and its fuzzy sets
      for each attribute aᵢ ∈ A ∪ D.
    αₖ: alpha cut k to limit the low rough set
      dependency.
    Q: linguistic variable that represent the quantifiers
      of the summaries.
  Begin
    1. Transform S = (U, A ∪ D) in linguistic dataset by
       considering fuzzifiers set and maximal membership
       principles.
    2. A_ItemSet = attributes_to_sets(A∪D)
    3. D_ItemSet = A_ItemSet
    4. Stackset.push(A_ItemSet)
    5. CS = {}
    6. while not Stackset.empty
    6.1.   A_ItemSet = Stackset.pop
    6.2.   for each B ⊆ A_ItemSet
    6.3.      for each X ⊆ D_ItemSet: B ∩ X = ∅
```

$$6.4. \qquad POS_B(D) = \bigcup_{X \in U/D} B_*(X), \quad k = \frac{|POS_B(D)|}{|U|}, \quad B \Rightarrow_k D, \; k \geq \alpha_k$$

```
    6.5.         for each Oₜ(B,X) ∈ POS_B(D)
    6.6.            if (there is total dependency
                            B⇒ₖ₌₁X in Oₜ(B,X))
    6.7.               CS = CS ∪ Oₜ(B,X)
    6.8.            else
    6.9.               Stackset.push({B ∪ X})
    6.10.           end
                 end
              end
           end
       end while
    7. LSUM = BuildSummaries(CS, Q)
    8. Calculate T values for each summary s ∈ LSUM
    9. Reorder LSUM considering T values
    10. Return LSUM
```

In step 2, each element $a \in A \cup D$ it is transformed in a set $\{a\}$ and insert into $A\_ItemSet$ set, example: $\{a_1, a_2, a_3\} => \{\{a_1\}, \{a_2\}, \{a_3\}\}$.

In step 4, a stack called *Stackset* is used for working with item sets to build the summaries.

In step 5, *CS* variable is a set to storage the candidate summaries.

In step 6.4 the positive region $POS_B(D)$ is calculated according to Definition 1 and Eqs. (7), (8). The $k \geq \alpha_k$ condition, help to find summaries between partial dependency attributes, but considering the sets of attributes that have minimum relationship level. Later, in 6.6 step, $O_t \in U$, $O_t(B,X)$ represents the values of attributes $a \in B \cup X$ in O_t object; if there is total dependency in O_t context, then $O_t(B, X)$ attributes values will be used as candidates summaries.

Afterward, in 6.9 step the algorithm searches other attributes combination to generate summaries with more filters. But no superset attributes of any low dependency attributes set, should be considered for summary generation. In order to prune the attributes combination, just must be consider the O_t objects that $O_t \in POS_B(D)$.

Finally, the summaries are sorted according to T values and then, submitted for evaluation of an experts group in this thematic area. The active learning method is used to identify and validate the best summaries from semantic point of view. This step is important in the final selection of summaries for decision-making.

3 Results

In order to validate the algorithm, authors applied qualitative and quantitative methods as follow in this section. Authors consider project management scenarios because projects are practically organized in all areas of society with a high social and economic impact [26, 27]. The demonstration of linguistic summaries applicability in project management help to show the high applicability of these algorithm in wide areas of human activity.

The qualitative evaluation of the algorithm was based on discovering of relationships between personality traits and human performance in software projects, see Sect. 3.1. Particularly in software projects, human resources are the main resource because these projects depend in large extent on professional skills, creativity and motivation of their resources.

The quantitative evaluation of the algorithm was based on the comparison of *LDS_RoughSet* algorithm with an algorithm to generate linguistic summaries based on association rules (*LDS_AssociationRules*), see Sect. 3.2.

3.1 Qualitative Evaluation Based on Real Case Study Application

In this section authors present the application of algorithm to discovering of relationships between personality traits and human performance in software projects. The principal motivations were:

- In order to achieve an adequate selection and conformation of teams, it is important to elaborate a personality profile according to the daily situations of the personnel. Personality is composed of several cognitive characteristics and behavioral trends that determine the similarities and differences in thoughts, feelings and behaviors of individuals [11].
- There are several tests that are used extensively and for multiple purposes. Most of these tests were standardized in correspondence with different populations and, there is some consensus on the application and interpretation of results that they provide.
- In order to acquire human resources for software project, authors consider the combination of: sociological, technical and quality of life test, is very important.
- In general, it is considered that, the characterization of these resources with respect to learning styles and personality traits, is essential for the formation of balanced teams, for the increase of efficiency and effectiveness in the development of projects. But most research only focuses on explaining the importance of psychological characteristics' analysis, but does not establish mechanisms to identify relationships between personality traits and job performance in projects.

The experiment was applied to a population with 62 professionals for whom information on job performance is available in different roles and projects over a period of 3 years. Each person in the experiment population complete four questionnaires to known about its personality traits [28]:

- Instrument: Questionnaire on Leadership Styles.
- Instrument: Questionnaire on leadership styles using word computation.
- Instrument: Personality Inventory 16 PF Form C [29].
- Instrument: BFQ, Big Five Questionnaire [30].

The information obtained for each test was extended with the performance evaluations of the respondents and authors conform four datasets [31]. Finally, the algorithm was applied to each dataset and the following linguistic summaries were obtained:

Results in analysis of dataset "An Instrument Questionnaire on Leadership Styles":

1. The specialists with high performance in programmer role are characterized by being passive and task oriented. In addition, they can perform tasks in architect role.
2. The project members with high performance in a third role as implementer are characterized as passive people in normal conditions and can perform tasks in programmer role.
3. The specialists with an average performance in a second role as analyst are people-oriented under normal conditions and can perform tasks in programmer role.

Results in analysis of dataset "B Instrument: Questionnaire on leadership styles using word computation":

1. Programmers with high performance are characterized by being passive and people-oriented under both normal and stressful conditions. Under normal conditions, they have a technical mix, so they consider themselves to be exact, precise, calm and logical people, they complete important tasks following proven methods and do not like to take risks.
2. The specialists who work in second role as architects, with high performance, are characterized by being passive in stressful conditions, and can perform tasks in programmer role.
3. The specialists who work in second role as analysts, with average performance, under normal conditions are passive and people-oriented. Under stress conditions they are also passive. They can perform tasks as programmers.
4. The specialists who work in a third role as implanter, with a medium performance, are people-oriented under normal conditions and tasks-oriented under stress conditions.
5. The specialists who work in quality with high performance are people-oriented under normal conditions and tasks-oriented under stressed conditions.

Results in analysis of dataset "C Instrument: Personality Inventory 16 PF Form C".

1. The specialists with high performance in programmer role work in group and strengthen their ego. In a group, they consider themselves suspicious, complicate themselves and they act with premeditation. In addition, they have a lot of strength in their ego, so they are characterized by being emotionally stable, calm, mature, realistic, balanced and able to maintain solid group morale.
2. The specialists who work in quality with high performance, agreed as group on normal values in animation, sensitivity, abstraction and socialization.
3. High implanter: They point to cunning, are considered cunning, calculating, insightful, subtle and lucid people. His approach is intellectual and unsentimental.
4. The specialists with an average performance in a second role as analysts, agreed as group on normal values in animation, apprehension or security and socialization.
5. The specialists with high performance in a second role as architect, agreed as group on normal values in perfectionism and attention to standards.

Results in analysis of dataset "D Instrument: BFQ, Big Five Questionnaire":

1. The specialists with high performance in programmer role, are characterized by being moderately meticulous, precise, responsible, orderly and able to master their emotions. They are also unsympathetic and tolerant.
2. The specialists with high performance in quality tasks are characterized by being moderately creative, informed and open to cultural interests. Also they are quite responsible, orderly, cooperative and affectionate.
3. Professionals with a high degree of implanting competence are considered responsible and orderly people and can take on programmer role.
4. The specialists with high performance in a second role as architects are characterized by being moderately responsible, orderly and diligent. With little peace and quiet and patience. In addition, they are very inactive and can perform tasks in programmer role.

5. The specialists with high performance in a second role as analysts are characterized by being moderately creative, knowledgeable, understanding and tolerant. In addition, they have some positive bias in their responses, so they tend to deny their personal shortcomings or they are particularly naive.
6. Specialists with average performance in a second role as analysts are characterized by being moderately creative, informed, meticulous, precise, open to new ideas and values different from their own. Also, they are unsympathetic, tolerant and affectionate.

From the analysis of the summaries obtained, it is identified that the runs carried out on the different databases yielded the following common results:

1. The specialists with high performance in programmer role, agreed as group on normal values in animation, security, self-sufficiency and extroversion. In addition, they are passive people under normal conditions.
2. The specialists with average performance in a second role as analysts, agreed as group on normal values in animation, safety, extroversion and attention to standards.

These results were presented to the respondents, and they were asked to evaluate them without specifying the test that generated them. It was concluded that the majority of respondents considered the Big Five test to be the most appropriate for their personal characteristics. On the other hand, the Management Styles Test, that use computer with words techniques for the evaluation, gave better results than the variant using discrete variables. This research results were used in processes of acquisition and formation of software development teams in the organization where this research was applied.

In this investigation, it was possible to identify the characteristics associated to personality traits, as well their relationship with high performance in a given role; that is why we can predict, with some certainty, from the results of a personality analysis test, in which position a new employee will have better results. However, we emphasize that these results must be combined with professional skills for the correct assignment of roles.

3.2 Quantitative Evaluation from Comparison with Other Algorithm

In this section authors compare *LDS_RoughSet* algorithm and *LDS_AssociationRules* algorithm based on association rules techniques. Authors apply the two algorithms to database "200226_gp_eval_proyfinal" from "Repository of Project Management Research" [31]. This database contains 202 records and 8 attributes: 4 nominal attributes and 4 numeric attributes. Each record represents an organization with information about province location, economical affectation types and amount of economical affectation in several moneys.

For comparison, authors propose the following metrics: number of summaries obtained, execution time and a set of statistical metrics for each T value. For each T value ($T1$, $T2$, $T3$, $T4$ $T5$) of linguistic summaries authors calculate: mean, standard deviation, minimum and maximum values.

Table 2. Linguistic summaries evaluation using T indicators.

Algorithm		$T1$	$T2$	$T3$	$T4$	$T5$
LDS_RoughSets	Min	0.968	0.027	0.833	0.005	0.25
	Max	0.99	0.067	1	0.97	0.5
Execution time 2.6987 ms	StdDev	0.0058	0.011	0.029	0.31	0.062
Summaries amount 46	Mean	0.987	0.031	0.99	0.72	0.48
LDS_AssociationRules	Min	0.775	0.027	0.338	0	0.063
	Max	0.99	0.257	1	0.88	0.5
Execution time 2.2607 ms	StdDev	0.053	0.068	0.14	0.31	0.147
Summaries amount 76	Mean	0.96	0.057	0.94	0.56	0.409

Table 2 shows that the algorithm proposed obtains better results than the algorithm based on association rules regards the following metrics: summaries' amount, $T1$, $T3$, $T4$ and $T5$. But the algorithm based on association rule is better in execution time; because rough sets theory is high time-consuming. There are not significant differences between algorithms regards $T2$ indicator.

The analysis of T indicators revel that in future works traditional T indicators could be extended by considering other elements like indeterminacy and falsity.

4 Conclusions

The proposed procedure allows the identification of relationships between personality traits and the performance evaluation index in the roles assigned in software projects. This method has been tested in software projects, but its conception allows its application in several scenarios.

In the application of personality instruments, participants reported Big Five questionnaire as suitable instrument for their characteristics.

In the experiment it was found that specialists with high performance in programmer role, is characterized by being moderately meticulous, accurate, responsible, orderly and able to master their emotions. However, they are not very tolerant and project managers need to be aware of these characteristics in order to facilitate communication within the project and avoid interpersonal conflicts.

The research results allow the identification of personal characteristics of professionals that facilitate the communication with them of managers and avoid conflicts in work teams.

It is also identified that personality traits suitable for analyst role are creative, informed, meticulous, precise, open to new things, ideas and values different from their own and these characteristics help their work performance and exchange with clients.

Considering quantitative analysis, the proposed algorithm obtains better results than the algorithm based on association rules in the most of the indicators. In particular, LDS_RoughSet algorithm was superior regards indicators: summaries' amount, $T1$, $T3$, $T4$ and $T5$. While, the algorithm based on association rules was the best considering execution time, because of rough sets theory is high time-consuming. There are not significant differences between algorithms regards $T2$ indicator.

References

1. Yager, R.R.: A new approach to the summarization of data. Inf. Sci. **28**(1), 69–86 (1982). https://doi.org/10.1016/0020-0255(82)90033-0
2. Ramos-Soto, A., Martin-Rodillab, P.: Enriching linguistic descriptions of data: a framework for composite protoforms. Fuzzy Sets Syst. **26** (2019). https://doi.org/10.1016/j.fss.2019.11.013
3. Hudec, M., Bednárová, E., Holzinger, A.: Augmenting statistical data dissemination by short quantified sentences of natural language. J. Off. Stat. **34**(4), 981–1010 (2018). https://doi.org/10.2478/jos-2018-0048
4. Kacprzyk, J., Zadrożny, S.: Linguistic database summaries and their protoforms: towards natural language based knowledge discovery tools. Inf. Sci. **173**(4), 281–304 (2005). https://doi.org/10.1016/j.ins.2005.03.002
5. Kacprzyk, J., Wilbik, A.: Linguistic summaries of time series: on some extended aggregation techniques. Stud. Mater. Polskiego Stowarzyszenia Zarzdzania Wiedza **2010**(31), 326–337 (2010)
6. Wilbik, A., Dijkman, R.M.: On the generation of useful linguistic summaries of sequences. In 2016 IEEE International Conference on Fuzzy Systems (FUZZ-IEEE), pp. 555–562. IEEE (2016). https://doi.org/10.1109/fuzz-ieee.2016.7737736
7. Kacprzyk, J., Zadrożny, S.: Linguistic database summaries using fuzzy logic, towards a human-consistent data mining tool (20), 10 (2009)
8. Kacprzyk, J., Zadrozny, S.: Linguistic data summarization: a high scalability through the use of natural language? In: Scalable Fuzzy Algorithms for Data Management and Analysis: Methods and Design, pp. 214–237. IGI Global (2010)
9. Pérez Pupo, I., Piñero Pérez, P.Y., García Vacacela, R., Bello, R., Santos Acosta, O., Leyva Vázquez, M.Y.: Extensions to linguistic summaries indicators based on neutrosophic theory, applications in project management decisions. Neutrosophic Sets Syst. **22**, 87–100 (2018)
10. Kacprzyk, J., Zadrożny, S., Dziedzic, M.: A novel view of bipolarity in linguistic data summaries. In: Kóczy, L.T., Pozna, C.R., Kacprzyk, J. (eds.) Issues and Challenges of Intelligent Systems and Computational Intelligence. SCI, vol. 530, pp. 215–229. Springer, Cham (2014). https://doi.org/10.1007/978-3-319-03206-1_16
11. Pérez Pupo, I., García Vacacela, R., Piñero Pérez, P., Sadeq, G., Peña Abreu, M.: Experiencias en el uso de técnicas de soft-computing en la evaluación de proyectos de software. Rev. Invest. Oper. **41**(1), 106–117 (2020)
12. Wilbik, A., Kaymak, U., Dijkman, R. M.: A method for improving the generation of linguistic summaries. In: 2017 IEEE International Conference on Fuzzy Systems (FUZZ-IEEE): pp. 1–6. IEEE (2017). https://doi.org/10.1109/fuzz-ieee.2017.8015752
13. Donis-Díaz, C.A., Bello, R., Kacprzyk, J.: Using ant colony optimization and genetic algorithms for the linguistic summarization of creep data. In: Angelov, P., et al. (eds.) Intelligent Systems'2014. AISC, vol. 322, pp. 81–92. Springer, Cham (2015). https://doi.org/10.1007/978-3-319-11313-5_8
14. Kaczmarek-Majer, K., Hryniewicz, O., Dominiak, M., Święcicki, Ł.: Personalized linguistic summaries in smartphone-based monitoring of bipolar disorder patients. Presented at the 2019 Conference of the International Fuzzy Systems Association and the European Society for Fuzzy Logic and Technology (EUSFLAT 2019). Atlantis Press (2019). https://doi.org/10.2991/eusflat-19.2019.56
15. Jain, A., Popescu, M., Keller, J., Rantz, M., Markway, B.: Linguistic summarization of in-home sensor data. J. Biomed. Inf. **96** (2019). https://doi.org/10.1016/j.jbi.2019.103240

16. Smits, G., Nerzic, P., Lesot, M.-J., Pivert, O.: FRELS: fast and reliable estimated linguistic summaries. In: 2019 IEEE International Conference on Fuzzy Systems (FUZZ-IEEE), pp. 1–6. IEEE (2019). https://doi.org/10.1109/fuzz-ieee.2019.8858836

17. Smits, G., Nerzic, P., Pivert, O., Lesot, M.-J.: Efficient generation of reliable estimated linguistic summaries. In: 2018 IEEE International Conference on Fuzzy Systems (FUZZ-IEEE), pp. 1–8 (2018). https://doi.org/10.1109/fuzz-ieee.2018.8491604

18. Pawlak, Z., Skowron, A.: Rough sets: some extensions. Inf. Sci. **177**(1), 28–40 (2007). https://doi.org/10.1016/j.ins.2006.06.006

19. Vanderpooten, D.: Similarity relation as a basis for rough approximations. Adv. Mach. Intell. Soft Comput. **4**, 17–33 (1997)

20. Bello, R.: Uncertainty Management with Fuzzy and Rough Sets: Recent Advances and Applications, vol. 377. Springer, Heidelberg (2019). http://www.springer.com/series/2941

21. Wang, G.: Extension of rough sets under incomplete information systems. In: 2002 IEEE World Congress on Computational Intelligence. In: 2002 IEEE International Conference on Fuzzy Systems, FUZZ-IEEE 2002. Proceedings (Cat. No.02CH37291), vol. 2, pp. 1098–1103 (2002). https://doi.org/10.1109/fuzz.2002.1006657

22. Zhai, J., Zhang, S., Zhang, Y.: An extension of rough fuzzy set. J. Intell. Fuzzy Syst. **30**(6), 3311–3320 (2016). https://doi.org/10.3233/IFS-152079

23. Pawlak, Z.: Rough sets. Int. J. Comput. Inform. Sci. **11**(5), 341–356 (1982). https://doi.org/10.1007/BF01001956

24. Bello, R., Verdegay, J.L.: Rough sets in the soft computing environment. Inf. Sci. **212**, 1–14 (2012). https://doi.org/10.1016/j.ins.2012.04.041

25. Wasilewska, A.: Apriori algorithm. Lecture Notes (2007). https://www3.cs.stonybrook.edu/cse634/lecture_notes/07apriori.pdf

26. PMI: A guide to the project management body of knowledge (PMBOK guide) Sixth Edition/Project Management Institute. Project Management Institute, Inc., Newtown Square, Pennsylvania 19073-3299, USA (2017)

27. Varajão, J., Colomo-Palacios, R., Silva, H.: ISO 21500: 2012 and PMBoK 5 processes in information systems project management. Comput. Stand. Interfaces **50**, 216–222 (2017)

28. López, P.: Procedimiento para la aplicacion de test de personalidad como apoyo a la gestion de recursos humanos en proyectos informáticos (Maestría en Gestión de Proyectos). Universidad de las Ciencias Informáticas, La Habana, Cuba (2017)

29. Furnham, A.: Personality and occupational success: 16PF correlates of cabin crew performance. Pers. Individ. Differ. **12**(1), 87–90 (1991)

30. Caprara, G.V., Barbaranelli, C., Borgogni, L., Perugini, M.: The "big five questionnaire": a new questionnaire to assess the five factor model. Pers. Individ. Differ. **15**(3), 281–288 (1993)

31. Pérez, P.P., Pupo, I.P., Rivero Hechavarría, C.C., Lusardo, C.R., Sosa, R.G., López, S.T.: Repositorio de datos para investigaciones en gestión de proyectos. Rev. Cubana Ciencias Inf. **13**(1), 176–191 (2019). https://gespro.uci.cu/projects/

Community Detection

Rough Net Approach for Community Detection Analysis in Complex Networks

Ivett Fuentes[1,2(✉)], Arian Pina[3], Gonzalo Nápoles[2], Leticia Arco[4], and Koen Vanhoof[2]

[1] Computer Science Department, Central University of Las Villas, Santa Clara, Cuba
`ivett@uclv.cu`
[2] Faculty of Business Economics, Hasselt University, Hasselt, Belgium
[3] Faculty of Technical Sciences, University of Sancti Spíritus, Sancti Spíritus, Cuba
[4] AI Lab, Computer Science Department, Vrije Universiteit Brussel, Brussels, Belgium

Abstract. Rough set theory has many interesting applications in circumstances characterized by vagueness. In this paper, the applications of rough set theory in community detection analysis are discussed based on the Rough Net definition. We will focus the application of Rough Net on community detection validity in both *monoplex* and *multiplex* networks. Also, the topological evolution estimation between adjacent layers in *dynamic* networks is discussed and a new community interaction visualization approach combining both complex network representation and Rough Net definition is adopted to interpret the community structure. We provide some examples that illustrate how the Rough Net definition can be used to analyze the properties of the community structure in real-world networks, including *dynamic* networks.

Keywords: Extended rough set theory · Community detection analysis · Monoplex complex networks · Multiplex complex networks

1 Introduction

Complex networks have proved to be a useful tool to model a variety of complex systems in different domains including sociology, biology, ethology and computer science. Most studies until recently have focused on analyzing simple static networks, named *monoplex* networks [7,17,18]. However, most of real-world complex networks are dynamics. For that reason, *multiplex* networks have been recently proposed as a mean to capture this high level complexity in real-world complex systems over time [19]. In both *monoplex* and *multiplex* networks the key feature of the analysis is the community structure detection [11,19].

Community detection (CD) analysis consists of identifying dense subgraphs whose nodes are densely connected within itself, but sparsely connected with the rest of the network [9]. CD in *monoplex* networks is a very similar task to classical clustering, with one main difference though. When considering complex

© Springer Nature Switzerland AG 2020
R. Bello et al. (Eds.): IJCRS 2020, LNAI 12179, pp. 401–415, 2020.
https://doi.org/10.1007/978-3-030-52705-1_30

networks, the objects of interest are nodes, and the information used to perform the partition is the network topology. In other words, instead of considering some individual information (attributes) like for clustering analysis, CD algorithms take advantage of the relational one (links). However, the result is the same in both: a partition of objects (nodes), which is called community structure [9].

Several CD methods have been proposed for *monoplex* networks [7,8,12,16–18]. Also, different approaches have been recently emerged to cope with this problem in the context of *multiplex* networks [10,11] with the purpose of obtaining a unique community structure involving all interactions throughout the layers. We can classify latter existing approaches into two broad classes: (I) by transforming into a problem of CD in simple networks [6,9] or (II) by extending existing algorithms to deal directly with *multiplex* networks [3,10]. However, the high-level complexity in real-world networks in terms of the number of nodes, links and layers, and the unknown reference of classification in real domain convert the evaluation of CD in a very difficult task. To solve this problem, several quality measures (internal and external) have emerged [2,13]. Due to the performance may be judged differently depending on which measure is used, several measures should be used to be more confident in results. Although, the modularity is the most widely used, it suffers the resolution limit problem [9]. Another goal of the CD analysis is the understanding of the structure evolution in *dynamic* networks, which is a special type of *multiplex* that requires not only discovering the structure but also offering interpretability about the structure changes.

Rough Set Theory (RST), introduced by Pawlak [15], has often proved to be an excellent tool for analyzing the quality of information, which means inconsistency or ambiguity that follows from information granulation in a knowledge system [14]. To apply the advantages of RST in some fields of CD analysis, the goal of our research is to define the new Rough Net concept. Rough Net is defined starting from a community structure discovered by CD algorithms applied to *monoplex* or *multiplex* networks. This concept allows us obtaining the upper and lower approximations of each community, as well as, their accuracy and quality. In this paper, we will focus the application of the Rough Net concept on CD validity and topological evolution estimation in *dynamic* networks. Also, this concept supports visualizing the interactions of the detected communities.

This paper is organized as follows. Section 2 presents the general concepts about the extended RST and its measures for evaluating decision systems. We propose the definition of Rough Net in Sect. 3. Section 4 explains the applications of Rough Net in the community detection analysis in complex networks. Besides, a new approach for visualizing the interactions between communities based on Rough Net is provided in Sect. 4. In Sect. 5, we illustrate how the Rough Net definition can be used to analyze the properties of the community structure in real-world networks, including *dynamic* networks. Finally, Sect. 6 concludes the paper and discusses future research.

2   Extended Rough Set Theory

The rough sets philosophy is based on the assumption that with every object of the universe U there is associated a certain amount of knowledge expressed through some attributes A used for object description. Objects having the same description are indiscernible with respect to the available information. The indiscernibility relation R induces a partition of the universe into blocks of indiscernible objects resulting in information granulation, that can be used to build knowledge. The extended RST considers that objects which are not indiscernible but similar can be grouped in the same class [14]. The aim is to construct a similarity relation R' from the relation R by relaxing the original indiscernibility conditions. This relaxation can be performed in many ways, thus giving many possible definitions for similarity. Due to that R' is not imposed to be symmetric and transitive, an object may belong to different similarity classes simultaneously. It means that R' induces a covering on U instead of a partition. However, any similarity relation is reflexive. The rough approximation of a set $X \subseteq U$, using the similarity relation R', has been introduced as a pair of sets called R'-lower (R'_*) and R'-upper (R'^*) approximations of X. A general definition of these approximations which can handle any reflexive R' are defined respectively by Eqs. (1) and (2).

$$R'_*(X) = \{x \in X : R'(x) \subseteq X\} \tag{1}$$

$$R'^*(X) = \bigcup_{x \in X} R'(x) \tag{2}$$

$$\alpha(X) = \frac{|R'_*(X)|}{|R'^*(X)|} \tag{3}$$

The extended RST offers some measures to analyze decision systems, such as the accuracy and quality of approximation and quality of classification measures. The accuracy of approximation of a rough set X, where $|X|$ denotes the cardinality of $X \neq \varnothing$, offers a numerical characterization of X. Equation (3) formalizes this measure such that $0 \leq \alpha(X) \leq 1$. If $\alpha(X) = 1$, X is crisp (exact) with respect to the set of attributes, if $\alpha(X) < 1$, X is rough (vague) with respect to the set of attributes. The quality of approximation formalized in Eq. (4) expresses the percentage of objects which can be correctly classified into the class X. Note that $0 \leq \alpha(X) \leq \gamma(X) \leq 1$, and $\gamma(X) \leq 0$ if $\alpha(X) \leq 0$, while $\gamma(X) \leq 1$ if $\alpha(X) \leq 1$ [14]. Quality of classification expresses the proportion of objects which can be correctly classified in the system; Equation (5) formalizes this coefficient where $C_1, \cdots, C_m$ correspond to the decision classes of the decision system DS. Notice that if the quality of classification value is equal to 1, then DS is consistent, otherwise is inconsistent [14]. Equation (6) shows the accuracy of classification, which measures the average the accuracy per classes with different importance levels. Its weighted version is formalized in Eq (7) [4].

$$\gamma(X) = \frac{|R'_*(X)|}{|X|} \tag{4}$$

$$\gamma(DS) = \frac{\sum_{i=1}^{m} |R'_*(C_i)|}{|U|} \tag{5}$$

$$\alpha(DS) = \frac{\sum_{i=1}^{m} \alpha(C_i)}{m} \tag{6}$$

$$\alpha_w(DS) = \frac{\sum_{i=1}^{m} (\alpha(C_i) \cdot |C_i|)}{\sum_{i=1}^{m} |C_i|} \tag{7}$$

3   Rough Net Definition

Monoplex (simple) networks can be represented as graphs $G = (V, E)$ where V represents the vertices (nodes) and E represents the edges (interactions) between these nodes in the network. *Multiplex* networks have multiple layers, where each one is a *monoplex* network. Formally, a *multiplex* network can be defined as a triplet $< V, E, L >$ where $E = \bigcup E_i$ such that E_i corresponds to the interactions on layer i-th and L is the number of layers. This extension of graph model is powerful enough though to allow modeling different types of networks including *dynamic* and *attributed* networks [9]. CD algorithms exploit the topological structure for discovering a collection of dense subgraphs (communities). Several *multiplex* CD approaches emphasize on how to obtain a unique community structure throughout all layers, by considering as similar nodes that ones with the same behavior in most of the layers [3,10]. In the context of *dynamic* networks, the goal is to detect the conformation by layers for characterizing the evolutionary or stationary properties of the CD structures. Due to the quality of the community structure may be judged differently depending on which measure is used, to be more confident in results several measures should be used [9]. In this section, we recall some basic notions related to the definition of the extension of RST in complex networks. Also, we will focus on the introduction of the Rough Net concept by extrapolating these notions to the analysis of the consistency of the detected communities in complex networks. This concept supports to validate, visualize, interpret and understand the communities and also their evolution. Besides, it has a potential application in labeling and refining the detected communities. As was mentioned, it is necessary to start from the definition of the decision system, the similarity relation, and the basic concepts of lower and upper approximations.

Definition 1. *(Rough Net): Given a complex network G, where V represents the nodes. Let $s : V \times V \to R'$ be a function that measures the similarity between nodes of V. The Rough Net comprises the combination of the topological structure and the CD results as a decision system $DS = (V, A \cup d)$, where A is a finite set of topological (i.e., the adjacency tensor of G) or non-topological features which may additionally be available if the network is attributed and $d \notin A$ is the decision attribute resulting from the detected communities.*

We use a similarity relation R' in our definition of Rough Net, because two nodes of V can be similar but not equal. The similarity class of the node x is denoted by $R'(x)$, as shown in Eq. (8). The R'-lower and R'-upper approximations for each similarity class are computed by Eqs. (1) and (2) respectively. There is a variety of distances and similarities for comparing nodes [1], such as Salton, Hub Depressed Index (HDI), Hub Promoted Index (HPI), similarities based on the topological structure, and Dice and Cosine coefficients which capture the attribute relations. In this paper, we use the Jaccard similarity for computing the similarities based on the topological structure because it has the attraction of simplicity and normalization. The Jaccard similarity, which also allows us to emphasize the network topology necessary to apply RST in complex networks, is defined in Eq. (9), where $\Gamma(X)$ denotes the neighborhood of the node x including it.

$$R'(x) = \{y \in V : yR'x, \textit{iff } s(x,y) \geq \xi\} \qquad (8)$$

$$s(x,y) = \frac{|\Gamma(x) \cap \Gamma(y)|}{|\Gamma(x) \cup \Gamma(y)|} \qquad (9)$$

3.1   Decision System for Applying RST on *monoplex* Networks

An adjacency tensor for a *monoplex* (i.e., single layer) network can be reduced to an adjacency matrix. The topological relation between nodes comprises an $|V| \times |V|$ adjacency matrix M, in which each entry $M_{i,j}$ indicates the relationships between nodes i and j weighted or not. The weight can be obtained as a result of the application of both a flattening process in a multi-relational network or a network construction schema when we want to apply network-based learning methods to vector-based datasets. If we apply some CD algorithm to this adjacency matrix, then we can consider the combination of the topological structure and the CD results as a decision system $DS_{monoplex} = (V, A \cup d)$, where A is a finite set of topological or non-topological features and $d \notin A$ is the decision attribute resulting from the detected communities over the network.

3.2   Decision System for Applying RST on *multiplex* Networks

Multiplex are powerful enough though to allow modeling different types of networks including *multi-relational*, *attributed* and *dynamic* networks [11]. Note that *multiplex* networks explicitly incorporate multiple channels of connectivity in which entities can have a different set of neighbors in each layer. In a *dynamic* network each layer corresponds to the network state at a given time-stamp (or each layer represents a snapshot). Like a time-series analysis, if attributes are captured in each time, a complex network can be represented as a *dynamic* network [19].

An adjacency tensor for a *dynamic* network with dimension L, which corresponds to the number of layers, represents a collection of adjacency matrices. The topological interaction between nodes within each layer k-th of a *multiplex* network comprises an $|V| \times |V|$ adjacency matrix M_k, in which each entry M_{ij}^k indicates the relationships between nodes i and j in the k-th layer. If we apply a CD algorithm to the whole *multiplex* network topology by considering *multiplex* CD approaches [10,19] in order to compute the unique final community structure, then we can consider the application of RST concepts over the *multiplex* network as the aggregation of the application of the RST concepts over each layer k-th. Consequently, the decision system for the k-th layer is the combination of the topological structure M_k and the CD results, formalized as $DS_{layer_k} = (V, A_k \cup d)$, where A_k is a finite set of topological or non-topological features in the k-th layer and $d \notin A$ is the decision attribute resulting from the detected communities in the *multiplex* topology (i.e., each node and their counterpart in each layer represent a unique node that belongs to a specific community). Besides, it is possible to transform a *multiplex* into a *monoplex* network by a flattening process. The main flatten approaches are the binary flatten, the weighted flatten and another based on deep learning [10]. Taking into account these variants, we can consider the combination of the topological structure of the transformed network and the CD results as a decision system $DS_{monoplex} = (V, A \cup d)$, where $A = \bigcup_{k \in L} A_k$ is a finite set of topological or non-topological features that characterize the networks and $d \notin A$ is the decision attribute resulting from the detected communities. The multiple instance or ensemble similarity measures are powerful for computing the similarity between nodes taking into account the similarity per layers (contexts).

4 The Application of Rough Net in the Community Detection Analysis

In this section, we describe the application of Rough Net in important tasks of the CD analysis: the validation and visualization of detected communities and their interactions, and the evolutionary estimation in *dynamic* networks.

4.1 Community Detection Validity

A community can be defined as a subgraph whose nodes are densely connected within itself, but sparsely connected with the rest of the network, though other patterns are possible. The existence of communities implies that nodes interact more strongly with the other members of their community than they do with nodes of the other communities. Consequently, there is a preferential linking pattern between nodes of the same community (being modularity [13] one of the most used internal measures [9]). This is the reason why link densities end up being higher within communities than between them. Although the modularity is the most widely quality measure used in complex networks, it suffers the resolution limit problem [9] and, therefore, is unable to judge in a correct way community structure of the networks with small communities or where communities may be very heterogeneous in size, especially if the network is large. Several methods and measures have been proposed to detect and evaluate communities in both *monoplex* and *multiplex* networks [2,3,13]. As well as modularity, Normalized Mutual Information (NMI), Adjusted Rand (AR), Rand, Variation of Information (VI) measures [2] are widely used, but the latter ones need an

Algorithm 1. Community Detection Validity

Input: A *Monoplex* or *multiplex* network G (attributed or not), detected communities, a threshold ξ and a similarity s (topological or non-topological features)
Output: Values of quality, accuracy and weighted accuracy of classification measures

1: **if** G is a *monoplex* network **then**
2:     $DS[1] \leftarrow DS_{monoplex}$ (see Section 3.1)
3:     $C[1] \leftarrow communities(G, d)$
4: **else if** G is a *multiplex* network **then**
5:     **for** k in L **do**
6:         $DS[k] \leftarrow DS_{layer_k}$ (see Section 3.2)
7:         $C[k] \leftarrow communities(layer(G, k), d)$
8:     **end for**
9: **end if**
10: **for** k in $(1 : size(DS))$ **do**
11:     Obtain the similarity class $R'_k(x)$ based on Equation (8)
12:     **for** X in $C[k]$ **do**
13:         Calculate $R'_{k*}(X)$ and $R'^{*}_k(X)$ approximations (see equations (1)–(2))
14:         Calculate $\alpha(X)$ and $\gamma(X)$ approximation measures (see equations (3)–(4))
15:     **end for**
16:     Calculate $\gamma(DS_k)$, $\alpha(DS_k)$ and $\alpha_w(DS_k)$ in DS_k (see equations (4)–(7))
17:     $\gamma_G(DS)+ = \gamma(DS_k)$, $\alpha_G(DS)+ = \alpha(DS_k)$ and $\alpha_{w_G}(DS)+ = \alpha_w(DS_k)$
18: **end for**
19: $\gamma_G(DS) = \gamma_G(DS)/L$, $\alpha_G(DS) = \alpha_G(DS)/L$ and $\alpha_{w_G}(DS) = \alpha_{w_G}(DS)/L$

external reference classification to produce a result. However, it is very difficult to evaluate a community result because the major of complex networks occur in real world situations since reference classifications are usually not available. We propose to use quality, accuracy and weighted accuracy of classification measures described in Sect. 2 to validate community results, taking into account the application of accuracy and quality of approximation measures to validate each community structure. Aiming at providing more insights about the validation, we provide a general procedure based on Rough Net. Notice that $R'_k(x)$ is computed by considering the attributes or topological features of networks in the k-th layer, by using Eq. (8). Algorithm 1 allows us to measure the quality of the community structure using Rough Net, by considering the quality and precision of each community. Rough Net allows judging the quality of the CD by measuring the vagueness of each community. For that reason, if boundary regions are smaller, then we will obtain better results of quality, accuracy and weighted accuracy of classification measures.

4.2 The Evolutionary Estimation in Dynamic Networks

A huge of real-world complex networks are dynamic in nature and change over time. The change can be usually observed in the birth or death of interactions within the network over time. In a *dynamic* network is expected that nodes of the same community have a higher probability to form links with their partners than

Algorithm 2. Evolution Analysis

Input: Two-consecutive layers CL of G, a threshold ξ and a similarity s
Output: The evolutionary estimations
1: $d_{k-1} \leftarrow CD(layer(G, k-1))$
2: $d_k \leftarrow CD(layer(G, k))$, $d_{k-0} \leftarrow d_k$
3: $DS_{k-1} \leftarrow (V, A_{k-1} \cup d_k)$ (see Section 3.2)
4: $DS_k \leftarrow (V, A_k \cup d_{k-1})$, $DS_{k-0} \leftarrow DS_k$ (see Section 3.2)
5: **for** i in $(0:1)$ **do**
6:    **for** X in d_{k-i} **do**
7:        Obtain the similarity class $R'_{k-i}(x)$ based on Equation (8)
8:        Calculate $R'_{k-i} * (X)$ and $R'^*_{k-i}(X)$ approximations (see equations (1)–(2))
9:        Calculate $\alpha_{k-i}(X)$ and $\gamma_{k-i}(X)$ measures (see equations (3)–(4))
10:   **end for**
11:   $\gamma_i = \gamma(DS_{k-i})$, $\alpha_i = \alpha(DS_{k-i})$ and $\alpha_{wi} = \alpha_w(DS_{k-i})$ in DS_{k-i}
12: **end for**
13: $\gamma_{CL} = (\gamma_0 + \gamma_1)/2$, $\alpha_{TC} = (\alpha_0 + \alpha_1)/2$ and $\alpha_{w_{CL}} = (\alpha_{w0} + \alpha_{w1})/2$

---

with other nodes [19]. For that reason, the key feature of the community detection analysis in *dynamic* networks is the evolution of communities over time. Several methods have been proposed to detect these communities over time for specific time-stamp windows [3,10]. Often more than one community structure is required to judge if the network topology has suffered transformation over time for specific window size. To the best of our knowledge, there is no measure able which captures this aspect. For that reason, in this paper, we propose measures based on the average of quality, accuracy and weighted accuracy of classification for estimating in a real number the change level during a specific window time-stamp. We need to consider two-consecutive layers for computing the quality, accuracy and weighted accuracy of classification measures in the evolutionary estimation (see Algorithm 2). For that reason, we need to apply twice the Rough Net concept for each pair of layers. The former Rough Net application is based on the decision system $DS = (V, A_k \cup d_{k-1})$, where A_k is a set of topological attributes in the layer k and $d_{k-1} \notin A_k$ is the result of the community detection algorithm in the layer $k-1$ (decision attribute). The latter Rough Net application is based on the decision system $DS = (V, A_{k-1} \cup d_k)$, where A_{k-1} is a set of topological attributes in the layer $k-1$ and $d_k \notin A_{k-1}$ is the result of the community detection algorithm in the layer k (decision attribute). The measures can be applied over a window size K by considering the aggregation of the quality classification between all pairs of consecutive (adjacency) layers. Values nearer to 0 express the topology is evolving over time.

4.3 Discovering Interactions Between Communities

In many applications more than a unique real value that expresses the quality of the community conformation is required for the understanding of the interactions throughout the networks. Besides, real-world complex networks usually are

Algorithm 3. Visualization for Community Structure Analysis

Input: A complex network G, detected communities, a threshold ξ and a similarity s
Output: Community network representation
1: Create an empty network $G'(V', E')$
2: **for** x in V **do**
3:     Obtain the similarity class $R'(x)$ based on Equation (8)
4: **end for**
5: **for** X in $communities(G, d)$ **do**
6:     Calculate $R'_*(X)$ and $R'^*(X)$ approximations (see equations (1)–(2))
7:     Calculate $\alpha(X)$ and $\gamma(X)$ approximation measures (see equations (3)–(4))
8:     Add a new node X where the size corresponds to quality or accuracy
9: **end for**
10: **for** X, Y in $communities(G)$, $X \neq Y$ **do**
11:     Calculate the similarity s_{BN} between communities X-th and Y-th
12:     Add a new edge (I, Y, w_{XY}) where the weighted $w_{ij} = s_{BN}(X, Y)$
13: **end for**

---

composed by many nodes, edges, and communities, making difficult to interpret the obtained results. Thus, we propose a new approach for visualizing the interactions between communities taking into account the quality of the community structure by using the combination of the Rough Net definition and the complex network representation. Our proposal, formalized in Algorithm 3, allows us to represent the quality of the community structure in an interpretable way.

The similarity measure used for weighted the interactions between communities in the network representation is formalized in Eq. (10). The $s_{BN}(X, Y)$ captures the proportion of nodes members of the community X, which cannot be unambiguously classified into this community but belong to the community Y and vice-versa. The above idea is computed based on the boundary region BN of both communities X and Y. The Rough Net approach allows us to evaluate the interaction between the communities and its visualization facilitates interpretability. In turn, it helps experts redistribute communities and change granularity based on the application domain requirements.

$$s_{BN}(X, Y) = \frac{\frac{|BN(X) \cap Y|}{|BN(X)|} + \frac{|BN(Y) \cap X|}{|BN(Y)|}}{2} \qquad (10)$$

5   Illustrative Examples

For illustrating the performance of the Rough Net definition in the community detection analysis, we apply it to three networks, two known to have *monoplex* topology and the third *multiplex* one. To be more confident in results, we should use several measures for judging the performance of a CD algorithm [2,5].Thus, we compare our approach to validate detected communities (i.e., accuracy and quality of classification) with the most popular internal and external measures used for community detection validity: modularity, AR, NMI, Rand, VI [2]. Modularity [13] quantifies when the division is a good one, in the sense of having many within-community edges. It takes its largest value (1) in the trivial case where all nodes belong to a single community. A value near to 1 indicates strong community structure in the network. All other mentioned measures need external references for operating. All measures except VI, express the best result though values near to 1. For that reason, we use the notation VIC for denoting the complement of VI measure (i.e., $VIC = 1 - VI$).

5.1   Zachary Network

Zachary is the much-discussed network[1] of friendships between 34 members of a karate-club at a US university. Figure 1 shows the community structures reported by the application of the standard CD algorithms Label Propagation (LP), Multilevel Louvain (LV), Fast Greedy Optimization (FGO), Leading Eigenvector (EV), Infomap (IM) and Walktrap (WT) to the Zachary network. Each community has been identified with a different colour. These algorithms detect communities, which mostly not correspond perfectly to the reference communities, except the LP algorithm which identically matches. For that reason, we can affirm that the LP algorithm reported the best division. However, in Fig. 2 we can observe that the modularity values not distinguish the LP as the best conformation of nodes into communities, while the proposed accuracy and quality

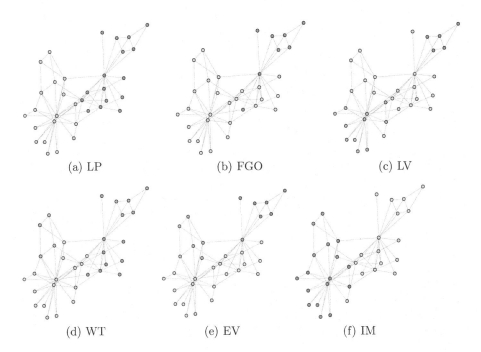

Fig. 1. Communities detected by different algorithms in the Zachary network.

[1] http://networkrepository.com/ucidata-zachary.php.

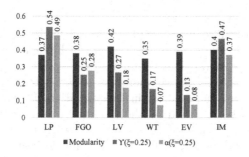

Fig. 2. Performance of the internal measures on the Zachary CD evaluation.

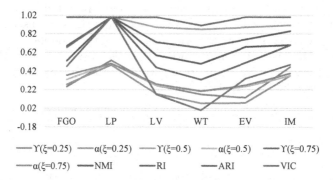

Fig. 3. Performance of the proposed measures on the Zachary CD evaluation.

of classification measures based on the Rough Net definition, assign the higher value to the LP conformation regardless of the used threshold. On the other hand, our measures grant the lowest quality results for the community structure obtained by the EV algorithm as expected. Notice that FGO and EV assign the orange node with high centrality in the orange community structure in a wrong manner. We can notice that most neighbors of this node are in another community. Indeed, the FGO and WT are the following lowest results reported by our measures. Figure 3 shows the performance reported by the application of the standard community detection algorithms before mentioned by using the proposed quality measures and the external ones. All measures exhibit the same monotony behaviors with independence of the selected similarity threshold ξ. Our measures have the advantage that are internal and behave similarly to external measures.

5.2 Jazz Network

The Jazz network[2] represents the collaboration between jazz musicians, where each node represents a jazz musician and interactions denote that two musicians

---

[2] http://konect.cc/networks/arenas-jazz/.

are playing together in a band. Six CD algorithms were applied to this network with the objective of subsequently exploring the behavior of validity measures. Figure 4 displays that LP obtains a partition in which the number of interactions shared between nodes of different communities is smaller than the number of interactions shared between the communities obtained by the FGO algorithm. However, this behavior is not reflected in the estimation of the modularity values, while it manages to be captured by the proposed quality measures, as shown in Fig. 5. Besides, the number of interactions shared between the communities detected by the algorithms LV, FGO, and EV is much greater than the number of interactions shared between the communities detected by the algorithms LP, WT, and IM. Therefore, this behavior was expected to be captured through the Rough Net definition. Figure 5 shows that the results reported by our measures coincide with the expected results. On the one hand, we can observe that our quality measures exhibit a better performance than the modularity measure in this example. Our measures also capture the presence of outliers, this is the reason why the community structure reported by the WT algorithm is higher than the obtained by the LP algorithm.

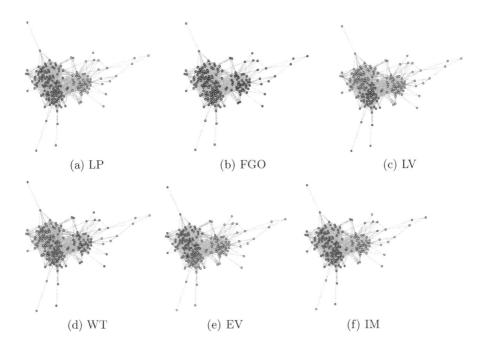

(a) LP   (b) FGO   (c) LV

(d) WT   (e) EV   (f) IM

Fig. 4. Communities detected by different algorithms in the Jazz network.

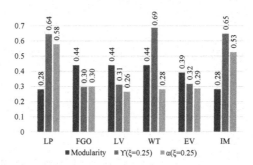

Fig. 5. Performance of the internal measures on the Jazz CD evaluation.

5.3   CElegans Network

Caenorhabditis elegans connectome (CElegans) is a *multiplex* network[3] that consists of layers corresponding to different synaptic junctions: electric (ElectrJ), chemical monadic (MonoSyn), and polyadic (PolySyn). Figure 6 shows the mapping of the community structure in each network layer, which has been obtained by the application of the MuxLod CD algorithm [10]. Notice that a strong community structure result must correspond to a structure of densely connected subgraphs in each network layer. This reflexion property is not evident for these communities in the CElegans network. For that reason, both the modularity and the proposed quality community detection measures obtain low results ($Modularity$ = 0.07, $\alpha(\xi = 0.25)$ = 0.24 and $\gamma(\xi = 0.25)$ = 0.14). Figure 7 shows the interactions between the communities in each layer by considering the MuxLod community structure and the algorithm described in Sect. 4.3. The community networks show high interconnections and as expected, the results of the quality measures are low. Figure 7 shows that the topologies of the PolySyn and ElectrJ layers do not match exactly. In this sense, let us suppose without loss of generalization, that we want to estimate if there has been a change in

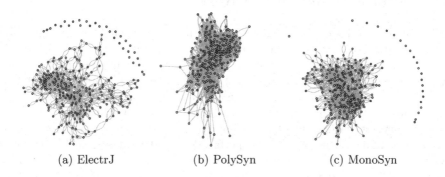

(a) ElectrJ   (b) PolySyn   (c) MonoSyn

Fig. 6. Application of the MuxLod CD to the CElegans network.

---

[3] http://deim.urv.cat/~alexandre.arenas/data/welcome.htm.

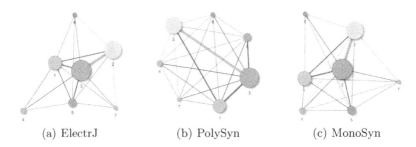

(a) ElectrJ   (b) PolySyn   (c) MonoSyn

Fig. 7. Visualization of community quality based on the MuxLod CD to the CElegans network.

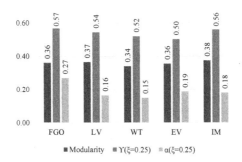

Fig. 8. Performance of the internal measures on the CElegans CD evaluation.

the topology considering these layers as consecutive. To estimate these results, we apply the algorithm described in Sect. 4.2. Figure 8 shows the modularity, accuracy and quality of classification obtained values, which reflect that the community structure between layers does not completely match, so it can be concluded that the topology has evolved (changed).

6   Conclusions and Future Work

In this paper, we have described new quality measures for exploratory analysis of community structure in both *monoplex* and *multiplex* networks based on the Rough Net definition. The applications of Rough Net in community detection analysis demonstrate the potential of the proposed measures for judging the community detection quality. Rough Net allows us to asses the detected communities without requiring the referenced structure. Besides, the proposed evolutionary estimation and the new approach for discovering the interactions between communities allows to the experts a deep understanding of complex real systems mainly based on the visualization of interactions. For the future work, we propose to extend the applications of Rough Net to the estimation of the community structure in the next time-stamp based on the refinement between adjacent layers in *dynamic* networks.

References

1. Ahajjam, S., El Haddad, M., Badir, H.: A new scalable leader-community detection approach for community detection in social networks. Soc. Netw. **54**, 41–49 (2018)
2. Aldecoa, R., Marín, I.: Surprise maximization reveals the community structure of complex networks. Sci. Rep. **3**(1), 1–9 (2013)
3. Amelio, A., Pizzuti, C.: Community detection in multidimensional networks. In: 26th International Conference on Tools with Artificial Intelligence, pp. 352–359. IEEE (2014)
4. Arco, L., Bello, R., Caballero, Y., Falcón, R.: Rough text assiting text mining: focus on document clustering validity. In: Bello, R., Falcón, R., Pedrycz, W., Kacprzyk, J. (eds.) Granular Computing: at the Junction of Rough Sets and Fuzzy Sets. Studies in Fuzziness and Soft Computing, vol. 224, pp. 229–248. Springer, Heidelberg (2008). https://doi.org/10.1007/978-3-540-76973-6_15
5. Bai, L., Liang, J., Du, H., Guo, Y.: A novel community detection algorithm based on simplification of complex networks. Knowl.-Based Syst. **143**, 58–64 (2018)
6. Berlingerio, M., Pinelli, F., Calabrese, F.: ABACUS: frequent pattern mining-based community discovery in multidimensional networks. Data Min. Knowl. Disc. **27**(3), 294–320 (2013)
7. Blondel, V.D., Guillaume, J.L., Lambiotte, R., Lefebvre, E.: Fast unfolding of communities in large networks. J. Stat. Mech: Theory Exp. **2008**(10), P10008 (2008)
8. Clauset, A., Newman, M.E.J., Moore, C.: Finding community structure in very large networks. Phys. Rev. E **70**, 066111 (2004)
9. De Domenico, M., et al.: Mathematical formulation of multilayer networks. Phys. Rev. X **3**(4), 041022 (2013)
10. Falih, I., Kanawati, R.: MUMA: a multiplex network analysis library. In: IEEE/ACM International Conference on Advances in Social Networks Analysis and Mining (ASONAM), pp. 757–760. IEEE (2015)
11. Kanawati, R.: Multiplex network mining: a brief survey. IEEE Intell. Inf. Bull. **16**, 24–28 (2015)
12. Newman, M.E.: Finding community structure in networks using the eigenvectors of matrices. Phys. Rev. E **74**(3), 036104 (2006)
13. Newman, M.E., Leicht, E.A.: Mixture models and exploratory analysis in networks. Proc. Natl. Acad. Sci. **104**(23), 9564–9569 (2007)
14. Orlowska, E.: Incomplete Information: Rough Set Analysis, vol. 13. Physica (2013)
15. Pawlak, Z.: Rough set theory and its applications to data analysis. Cybern. Syst. **29**(7), 661–688 (1998)
16. Pons, P., Latapy, M.: Computing communities in large networks using random walks. In: Yolum, P., Güngör, T., Gürgen, F., Özturan, C. (eds.) ISCIS 2005. LNCS, vol. 3733, pp. 284–293. Springer, Heidelberg (2005). https://doi.org/10.1007/11569596_31
17. Raghavan, U.N., Albert, R., Kumara, S.: Near linear time algorithm to detect community structures in large-scale networks. Phys. Rev. E **76**, 036106 (2007)
18. Rosvall, M., Bergstrom, C.T.: Maps of information flow reveal community structure in complex networks. arXiv preprint physics.soc-ph/0707.0609 (2007)
19. Zou, Y., Donner, R.V., Marwan, N., Donges, J.F., Kurths, J.: Complex network approaches to nonlinear time series analysis. Phys., Rep. **787**, 1–97 (2019)

Overlapping Community Detection Using Multi-objective Approach and Rough Clustering

Darian Horacio Grass-Boada[1]([⊠]), Airel Pérez-Suárez[1], Leticia Arco[2],
Rafael Bello[3], and Alejandro Rosete[4]

[1] Advanced Technologies Application Center (CENATAV), Havana, Cuba
{dgrass,asuarez}@cenatav.co.cu
[2] AI Lab, Computer Science Department, Vrije Universiteit Brussel,
Brussels, Belgium
larcogar@vub.be
[3] Department of Computer Science, Universidad Central "Marta Abreu" de Las
Villas, Santa Clara, Cuba
rbellop@uclv.edu.cu
[4] Facultad de Ingeniería Informática, Universidad Tecnológica de la Habana "José
Antonio Echeverría" (Cujae), Havana, Cuba
rosete@ceis.cujae.edu.cu

Abstract. The detection of overlapping communities in Social Networks has been successfully applied in several contexts. Taking into account the high computational complexity of this problem as well as the drawbacks of single-objective approaches, community detection has been recently addressed as Multi-objective Optimization Evolutionary Algorithms (MOEAs). One of the challenges is to attain a final solution from the set of non-dominated solutions obtained by the MOEAs. In this paper, an algorithm to build a covering of the network based on the principles of the Rough Clustering is proposed. The experiments in a synthetic networks showed that our proposal is promising and effective for overlapping community detection in social networks.

Keywords: Social network analysis · Community detection · Multi-objective Optimization · Rough clustering

1 Introduction

The Analysis of Social Networks has received a lot of attention due to its wide range of applications in several contexts [1]. Specifically, in Social Network Analysis, the Community Detection Problem (CDP) plays an important role [5]. Community detection in social networks aims to organize the nodes of the network in groups or communities such that nodes belonging to the same community are densely interconnected but sparsely connected with the remaining nodes in the network [2]. Even though most of the community detection algorithms assume

© Springer Nature Switzerland AG 2020
R. Bello et al. (Eds.): IJCRS 2020, LNAI 12179, pp. 416–431, 2020.
https://doi.org/10.1007/978-3-030-52705-1_31

that communities are disjoint, according to Palla *et al.* in [6], most real-world networks have overlapping community structure, that is, a node can belong to more than one community.

On the other hand, since the community detection problem has an NP-hard nature, most reported approaches use heuristics to search for a set of nodes that optimises an objective function which captures the intuition of community, these single-objective optimization approaches face two main difficulties: a) the optimization of only one function confines the solution to a particular community structure, and b) returning one single partition may not be suitable when the network has many potential structures. To overcome the aforementioned problems, many community detection algorithms model the problem as a Multi-objective Optimization Problem, and specifically, they use Multi-objective Optimization Evolutionary Algorithms (MOEAs) to solve them.

Once the set of non-dominated solutions is obtained by the MOEAs, one of the main challenges is to accomplish a final solution. Most of the proposed algorithms [5, 7–9] use the internal criteria (e.g., Modularity Index [10]) or the external criteria (e.g., Normalized Mutual Information (NMI) [3]) to select the final solution. The drawbacks of these approaches are that the internal criteria does not often correspond to the objective function used by MOEAs and the external criteria uses the ground truth of the network, which it is not always known. Also, the selected final solutions obtained by both approaches do not use the knowledge of the overlapping communities (Pareto set) obtained by MOEAs.

Rough Set Theory (RST) may be used to evaluate significance of attributes, to deal with inconsistent data, and to describe dependencies among attributes, to mention just some uses in machine learning and data mining [22].

The main advantage of Rough Set Theory in data analysis is that it does not need any preliminary or additional information about data [17]. RST allows to approximate a rough concept by a pair of exact concepts, called the lower and upper approximations. The lower approximation is the set of objects definitely belonging to a vague concept, whereas the upper approximation is the set of objects possibly belonging to the mentioned vague concept [17]. The upper and lower approximations can be used in a broader context such as clustering, denoted as Rough Clustering [13].

In our proposal, we focus on describing the relationship between the elements of the network (vertices) only taking into consideration their belonging to the communities of the Pareto Set. Then, we use Rough Clustering to obtain a final covering of the network, that describes the communities with their lower and upper approximations. The lower approximation is the set of vertices belonging to the community without uncertainty, whereas the upper approximation is the set of vertices possibly belonging to this community, therefore located at the boundary of it. Hence, the selected final solution uses the knowledge of the overlapping communities (Pareto set) obtained by MOEAs.

In this paper, we propose an Overlapping Community Detection Algorithm using Multi-objective approach and Rough Clustering, denoted as MOOCD-RC. Our algorithm allows selecting the final solution based on the subjective information as the number of vertices located in the cores or boundaries of the communities. As a consequence, it helps decision-makers (DM) incorporate their domain knowledge into the community detection process. Our main contributions are as follows:

1. We define an indiscernibility relationship between vertices of the network by taking the number of communities in the Pareto Set where they match.
2. We use the Rough Clustering foundation to build and describe the final covering of the network through the lower and upper approximations of the communities.

This paper is arranged as follows. Section 2 briefly introduces the necessary notions of multi-objective community detection problem and Rough Clustering. In Sect. 3, we introduce our proposal. Section 4 presents the experimental evaluation of our proposal and compared against other related state-of-the-art algorithms over synthetic networks. Finally, Sect. 5 gives the conclusions and some ideas about future work.

2   Background

This section introduces the necessary background knowledge for understanding the proposed method. First, the definition of multi-objective community detection problem and multi-objective algorithms of the related work are presented. Next, we will give the basics about Rough Set Theory and Rough Clustering.

2.1   Multi-objective Community Detection Problem

Let $G = (V, E)$ be a given network, where V is the set of vertices and E is the set of edges among the vertices. A multi-objective community detection problem aims to search for a partition P^* of G such that:

$$F(P^*) = min_{P \in \Omega} (f_1(P), f_2(P), \ldots, f_r(P)), \tag{1}$$

where P is a partition of G, Ω is the set of feasible partitions, r is the number of objective functions, f_i is the ith objective function and $min(\cdot)$ is the minimum value obtained by a partition P taking into account all the objective functions. With the introduction of the multiple objective functions, there is usually no absolute optimal solution, thus, the goal is to find a set of *Pareto* optimal solutions [2]. A commonly used way to solve a multi-objective community detection problem is by using MOEAs [9].

The first algorithm using MOEAs for detecting overlapping communities is named Multiobjective Evolutionary Algorithm to solve CDP (MEA_CDP) [5]. MEA_CDP uses an undirected representation of the solution and the classical Nondominated Sorting Genetic Algorithm II (NSGA-II) with the reverse operator to search for the solutions optimising the average community fitness, the average community separation and the overlapping degree among communities.

On the other hand, the Improved Multiobjective Evolutionary Algorithm to solve CDP (iMEA_CDP) [7] uses the same representation and optimization framework of MEA_CDP but it proposes to employ the PMX crossover operator and the simple mutation operator as evolutionary operators. iMEA_CDPs employs the Modularity function [10] and a combination of the average community separation and overlapping degree as its objective functions.

The Overlapping Community Detection Algorithm based on MOEA (MOEA-OCD) [9] uses the classical NSGA-II optimization framework and a representation based on adjacents among edges of the network. On the other hand, MOEA-OCD uses the negative fitness sum and the unfitness sum as objective functions. Unlike previously mentioned algorithms, in MOEA-OCD algorithm, a local expansion strategy is introduced into the initialization process to improve the quality of initial solutions.

Another algorithm is the Maximal Clique based on MOEA (MCMOEA) [8] which first detects the set of maximal cliques of the network and then it builds the maximal-clique graph. Starting from this transformation, MCMOEA uses a representation based on labels and the Multiobjective Evolutionary Algorithm based on Decomposition (MOEA/D) in order to detect the communities optimising the Radio Cut (RC) and Kernel K-Means (KKM) objective functions [11].

In [16] the authors combine Granular Computing and a multi-objective optimization approach for discovering overlapping communities in social networks. This algorithm, denoted as MOGR-OV, starts by building a set of seeds that is afterwards processed for building overlapping communities, using three introduced steps, named *expansion, improving* and *merging*.

Most of the exiting works focus on developing MOEAs to detect overlapping communities but not addresses the problem of selecting a final solution from the set of the obtained non-dominated solutions.

2.2   Foundations of Rough Clustering

The main components in the Rough Set Theory are an information system and an indiscernibility relation [17]. The classical RST was originally proposed using on a particular type of indiscernibility relations called equivalence relations (i.e., those that are symmetric, reflexive and transitive). Yao et al. [19] described various generalizations of rough sets by relaxing the assumptions of an underlying equivalence relation.

RST takes a pair of precise concepts to study the vagueness of a concept, named the lower and upper approximations. The lower approximation composes of all objects which surely belong to the concept, whereas the upper approximation contains all objects which perhaps belong to the concept. The boundary region of the vague concept is the difference between the upper and the lower approximations [18].

Lingras et al. [15] define another generalization of the approximate sets, seeing them as interval sets. The authors propose the rough k-means algorithm, where the concept of k-means is extended by viewing each cluster as an interval or rough set. The core idea is to separate discernible from indiscernible objects and to assign objects to lower $\underline{A}(X)$ and upper $\overline{A}(X)$ approximations of a set X. This proposal allows overlaps between clusters [20]. The upper and lower approximation concepts require to follow some of the basic rough set properties such as [14]:

1. An object v can be part of at most one lower approximation. This implies that any two lower approximations do not overlap.
2. An object v that is member of a lower approximation of a set is also part of its upper approximation. This implies that a lower approximation of a set is a subset of its corresponding upper approximation.
3. If an object v is not part of any lower approximation it belongs to two or more upper approximations. This implies that an object cannot only belong to a single boundary region.

The way to incorporate rough sets into k-means clustering requires adapting the calculation of the centroids and deciding whether an object is assigned to a lower or upper approximation of a cluster. In the first moment, the centroids of clusters are calculated including the effects of lower as well as upper approximations. Next, an object is assigned to the lower approximation of a cluster when the distance (similarity) between the object and the particular cluster center is smaller than the distances to the remaining other cluster centers [14].

3   Proposal

The proposed algorithm obtains a final covering through two steps. It starts building sets of indiscernible (similar) objects that form basic granules of knowledge on the network $G = (V, E)$, where V represents the set of nodes and E represents the set of edges which connect nodes. Thus, a partition of the set V is obtained allowing us to define an equivalence relation in V. From our point of view, two vertices should be related if they share many communities at the Pareto Set. Next, through the Rough Clustering foundations, specifically the rough k-means algorithm ideas [15], we build the final covering of the network by viewing each community as a rough set, which allows us to obtain overlapping communities.

3.1   First Step: Build the Granules of Indiscernible Objects

In this step, we build a set of granules which represents a partition of V. First of all, we describe a series of useful concepts that we are applying in our proposal.

Definition 1 (Thresholded similarity graph). *Let $V = \{v_1, v_2, \ldots, v_n\}$ be the set of vertices of the network $G = (V, E)$, β a user-defined parameter and $S(v_i, v_j)$ a symmetric similarity function between vertices v_i and v_j, a thresholded similarity graph is an undirected graph $G_\beta = (V, E_\beta)$ where $(v_i, v_j) \in E_\beta$ if and only if $S(v_i, v_j) \geq \beta$.*

Definition 2 (Subgraph). *Let $G_1 = (V_1, E_1)$ and $G_2 = (V_2, E_2)$ be two graphs. $G_1 = (V_1, E_1)$ is a subgraph of $G_2 = (V_2, E_2)$, denoted as $G_1 \subseteq G_2$, if and only if $V_1 \subseteq V_2$ and $E_1 \subseteq E_2$.*

Definition 3 (Induced subgraph). *Let $V' \subseteq V$ be a set of vertices, the subgraph of G induced by V' is $G' = (V', E')$, such that $E' = \{(v_i, v_j) \in E \mid v_i, v_j \in V'\}$.*

Definition 4 (β-Connected component). *Let $G_\beta = (V, E_\beta)$ be a thresholded similarity graph and $G' = (V', E')$ a subgraph of G_β. The subgraph G' is a β-connected component in G_β if and only if satisfies the following conditions:*

1. *$\forall u, v \in V', u \neq v$, exists $v_1, v_2, \ldots, v_q \in V'$, such that $\forall i = 1 \ldots q$, $(v_i, v_{i+1}) \in E'$ and also $v_1 = u$ and $v_q = v$ or $v_1 = v$ and $v_q = u$.*
2. *do not exist another subgraph of G_β, $G_1 = (V_1, E_1)$ with $G_1 \neq G'$, that pleases the condition 1 and also $G' \subseteq G_1$.*

Let $S_{ps}(v_i, v_j)$ be the similarity function between v_i and v_j. $S_{ps}(v_i, v_j)$ employs the solutions in the Pareto Set, denoted as PS. Let CV_i be a solution of PS and G_{v_i} the set of communities where v_i belongs. Let $mc(v_i, v_j)$ be the number of matching clusters between v_i and v_j in CV_i. The function $S_{ps}(v_i, v_j)$ is defined as follows:

$$S_{ps}(v_i, v_j) = \frac{\sum_{CV_i \in PS} match(v_i, v_j)}{ps} \tag{2}$$

where ps is the number of solutions in PS and $match(v_i, v_j) = \frac{mc(v_i,v_j)}{|G_{v_i}| \cdot |G_{v_j}|}$.

We build the thresholded similarity graph $G_\beta = (V, E_\beta)$ based on Eq. 2 and the user-defined parameter β ($\beta \in [0, 1]$). Let $G'_r = \{G'_{r_1}, G'_{r_2}, \ldots, G'_{r_q}\}$ be the β-connected component set. By definition, the connected component set in a graph constitutes a partition of the set of vertices.

We will say that a vertex $v_i \in V$ is related with a vertex $v_j \in V$, denoted as $v_i R_{ps} v_j$, if and only if $\exists G'_{r_i} \in G'_r$ such that $v_i, v_j \in G'_{r_i}$, being R_{ps} a equivalence relation. The set built from all the vertices related to a vertex v_i forms the so called *equivalence class* of v_i, denoted as $[v_i]_{R_{ps}}$. Therefore, $[v_i]_{R_{ps}}$ is the set of $v_j \in V$ such that share the same connected component G'_{r_i}. This means that the vertices belonging to the same connected component have a strong relationship in terms of sharing the equal communities of PS. This strong relationship is measured by $S_{ps}(v_i, v_j)$.

Let $EC = \{[v_1]_{R_{ps}}, [v_2]_{R_{ps}}, \ldots, [v_q]_{R_{ps}}\}$ be a set of equivalence classes under the indiscernibility relation R_{ps}. The elements of EC are disjoint sets. Let $G_r = \{G_{r_1}, G_{r_2}, \ldots, G_{r_q}\}$ be the set of subgraphs induced by EC on $G = (V, E)$. Hence, G_{r_i} is a subgraph on $G = (V, E)$ induced from $[v_i]_{R_{ps}}$. Therefore, G_r is viewed as granules of indistinguishable elements which do not share vertices. These granules constitutes our initial granularity criterion [21], and also we will use them to build the final covering of the network.

3.2   Second Step: Build the Final Covering of $G = (V, E)$

We take the k biggest granules, $G_{r_i} \in G_r$, according to the number of vertices, as prototypes of clusters and the remaining of them are assigned to those selected ones. Therefore, the foundation is to initially covering the network with those granules of indistinguishable vertices that give greater coverage of the network. The variable $k, 1 \leq k \leq q$ receives the median value of the number of clusters that form the solutions at the Pareto Set. For this purpose, we define a similarity function between any two granules $G_{r_i}, G_{r_j} \in G_r$. This function is defined as follows:

$$S_{G_r}(G_{r_i}, G_{r_j}) = \frac{\sum_{v_i \in G_{r_i}} \sum_{v_j \in G_{r_j}} S_{ps}(v_i, v_j)}{|G_{r_i}| \cdot |G_{r_j}|} \tag{3}$$

As described in Sect. 2, the use of k-means clustering in Rough Clustering requires adapting the calculation of the centroids (cluster prototype) and decides whether an object is assigned to a lower or upper approximation of a cluster. In our case, we selected as prototypes of communities the k biggest granules, according to their number of vertices. Next, the remaining granules are assigned to those selected ones. A granule G_{r_i} is assigned to the lower approximation of a community when the similarity between G_{r_i} and the particular prototype of the community $G_{r_j}, 1 \leq j \leq k$, is much greater than the similarity to the remaining other prototypes. In this case, the similarity function defined in the Eq. 3 is used for deciding whether the remained granules are assigned to a lower or upper approximation of the selected k granules.

Worth noting that in this step, the assignation process uses the granules obtained in the previous step, $G_r = \{G_{r_1}, G_{r_2}, \ldots, G_{r_q}\}$. The selected k biggest granules represent the initial communities of network and also the lower approximations of them. The remaining granules $G_{r_i}, k < i \leq q$ will be part of the lower or upper approximations of the communities according to the similarity S_{G_r} and the γ user-defined parameter ($\gamma \in [0, 1]$).

The pseudocode of MOOCD-RC is shown in Algorithm 1. It is important to notice that the used Pareto Set is the result of using the MOGR-OV algorithm [16]. In MOOCD-RC, initially the cover CV is formed by the k greatest granules in G_r, which ones represent the lower approximations of the communities. These k selected granules represent the prototypes of communities to be built. Afterly, the remaining granules are included in the lower or upper approximations of the communities in CV according to S_{G_r}. Worth noting that the lower approximation of those communities are formed by the vertices that definitely belong to them, whereas the upper approximations are formed by the vertices that are located at the boundary of the communities. These vertices represent the overlapping in themselves.

In the first step, the building of the equivalence classes is tightly bound to the thresholded similarity graph $G_\beta = (V, E_\beta)$, which in turn depends on the β user-defined parameter. The higher the value of β the smaller granules will be obtained and vice versa. On the other hand, in the second step the dimensions of the lower and upper approximations of the communities depend on γ user-defined

---

Algorithm 1: MOOCD-RC algorithm

Input: $G = (V, E)$, Pareto Set with overlapping communities (*PSetOC*)
Output: Covering of the network $CV = \{CV_1, CV_2, \ldots, CV_k\}$
First Step: build the granules of indiscernible objects
for $v_i, v_j \in V$ **do**
 └ Take *PSetOC* and compute $S_{p_s}(v_i, v_j)$;

Build a thresholded similarity graph $G_\beta = (V, E_\beta)$;
Identify a β-connected component in G_β;
Compute $[v_i]_{R_{p_s}}$ for each $v_i \in V$;
Build the set $G_r = \{G_{r_1}, G_{r_2}, \ldots, G_{r_q}\}$, subgraphs induced by each
$[v_i]_{R_{p_s}}$, $v_i \in V$;
Second Step: build the final covering of $G = (V, E)$
Sort descending Gr by number of vertices;
Select the first k granules $G_{r_i} \in Gr$ as prototypes of communities $CV_i \in CV$;
for $i = 1$ *to* k **do**
 └ $CV_i \leftarrow G_{r_i}$;

for $j = k + 1$ *to* q **do**
 │ Determine the most similarity between G_{r_j} and the k granules $G_{r_i} \in Gr$:
 │ $G_{r_{max}} \leftarrow \max\limits_{1 \leq i \leq k} S_{G_r}(G_{r_j}, G_{r_i})$;
 │ $T \leftarrow \{\}$;
 │
 │ **for** $i = 1$ *to* k **do**
 │ │ **if** $S_{G_r}(G_{r_j}, G_{r_i}) / S_{G_r}(G_{r_j}, G_{r_{max}}) \leq \gamma$ **then**
 │ │ └ Add G_{r_i} to T;
 │
 │ **if** $|T| > 1$ **then**
 │ │ $\forall G_{r_i} \in T$ take the community CV_i associated;
 │ └ Add G_{r_j} to $\overline{CV_i}$;
 │ **else**
 │ │ Take take the community CV_i associate to $G_{r_{max}}$;
 │ └ Add G_{r_j} to $\underline{CV_i}$ and $\overline{CV_i}$;

return CV

---

parameter. In the way of this parameter changes we will obtain boundaries of communities more or less tight.

The parameters β and γ allow decision-makers to obtain a final covering of the network by adjusting the cores or boundaries of the communities. In our experiments, we set $\beta = 0.75$ and $\gamma = 0.1$. We chose these values according to the related works [13,14,20].

4   Experimental Results

In this section, we conduct several experiments for evaluating the effectiveness of our proposal. Since the built-in communities in benchmark networks are already known, we use the Normalized Mutual Information external evaluation measure to test the performances of different community detection algorithms.

Hence, the experiments were focused on evaluating the accuracy attained by our proposal in terms of the NMI value. Our algorithm was applied to synthetic networks generated from the Lancichinetti–Fortunato–Radicchi (LFR) benchmark dataset [4]. Its performances were compared against the one attained by MEA_CDP [5], iMEA_CDP [7], MCMOEA [8] and MOEA-OCD [9] algorithms, described in Sect. 2.

The algorithms of the related works do not build a final covering from the communities of the Pareto Set. Thus, we choose the best solution in the Pareto Set, according to the NMI, and compare this solution with respect to the ones obtained by our algorithm.

The NMI takes values in [0, 1] and it evaluates a set of communities based on how much these communities resemble a set of communities manually labeled by experts, where 1 means identical results and 0 completely different results.

In LFR benchmark networks, both node degrees and community sizes follow the power-law distribution and they are regulated using the parameters τ_1 and τ_2. Besides, the significance of the community structure is controlled by a mixing parameter μ, which denotes the average fraction of edges each vertex has with others from other communities in the network. The smaller the value of μ, the more significant community structure the LFR benchmark network has. The parameter O_n is specially defined for controlling the overlapping rate of communities in the network. O_n is the number of overlapping nodes, evaluating overlapping density among communities. Similar to μ, the higher the value of O_n, the more ambiguous the community structure is.

In the first part of the experiment, we set the network size to $N = 1000$, $\tau_1 = 2$, $\tau_2 = 1$, the node degree is in [0, 50] with an average value of 20, whilst the community sizes vary from 10 to 50 elements. Using previous parameter values we vary μ from 0.1 to 0.6 with an increment of 0.05. After, we set $\mu = 0.1$ and $\mu = 0.5$, and we vary the percent of overlapping nodes existing in the network (parameter O_n of LFR Benchmark) from $0.1N$ to $0.5N$ with an increment of 0.1; the other parameters remain the same as the first experiment.

The average NMI value attained for each algorithm over the LFR benchmark when μ varies from 0.1 to 0.6 with an increment of 0.05, as show in Fig. 1. As the value of μ increases the performance of each algorithm deteriorates, being both MOEA-OCD and MOOCD-RC those that performing the best. As the mixing parameter μ exceeds 0.5, the MOEA-OCD algorithm begins to decline in its performance and it is outperformed by MOOCD-RC. Figure 1 shows the good performance of our method.

For summarizing the above results, we evaluated the statistical significance of the NMI values using the Friedman test as Non-Parametric Statistic Procedure included in the KEEL Software Tool. Also, we used the Holms and Finner as post hoc methods. Table 1 shows the average ranks obtained by each method in the Friedman test. Our method ranks second, however, Table 2 shows the overall performance of MOEA-OCD with respect to the remaining algorithms, where

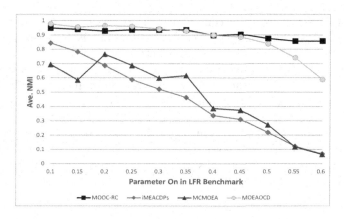

Fig. 1. Average NMI value attained by each algorithm on LFR benchmark networks when μ varies from 0.1 to 0.6 with an increment of 0.05.

Table 1. LFR benchmark networks when μ varies from 0.1 to 0.6. Average Rankings of the algorithms (Friedman).

Algorithm	Ranking
MOOCD-RC	1.5455
iMEACDPs	3.6364
MCMOEA	3.3636
MOEAOCD	1.4545

Table 2. LFR benchmark networks when μ varies from 0.1 to 0.6. Post Hoc comparison where $\alpha = 0.05$ (Friedman).

i	Algorithm	$z = (R_0 - R_i)/SE$	p	Holm	Finner
3	iMEACDPs	3.96347	0.000074	0.000222	0.000222
2	MCMOEA	3.468036	0.000524	0.001049	0.000786
1	MOOCD-RC	0.165145	0.86883	0.86883	0.86883

there is not statistically significance between our proposal and MOEA-OCD. The Friedman statistic value distributed according to chi-square with three degrees of freedom is 26.6727. Besides, the p-value computed by the Friedman test is 0.000007.

The structures of the networks are well defined in the second part of the experiment, as shown in Fig. 2. Our proposal and MOEA-OCD have a performance almost stable, independently of the number of overlapping nodes in the network, being MOEA-OCD the one that performs the best. On the other hand, when the structure of the communities is uncertain, the performance of the MOEA-OCD algorithm drops off when the overlapping in the network increases, being our proposal the one that performs better, as shown in Fig. 3.

Similar to the previous experiment, we evaluated the statistical significance of the NMI values. Table 3 shows the average ranks obtained by each algorithm in the Friedman test. The Friedman statistic value distributed according to chi-square with three degrees of freedom is 25.92. Besides, the p-value computed by the Friedman test is 0.00001. Our algorithm ranks second, however, like the previous experiment, Table 4 shows the overall performance of MOEA-OCD with respect to the remaining algorithms, where there is not statistically significance between our proposal and MOEA-OCD.

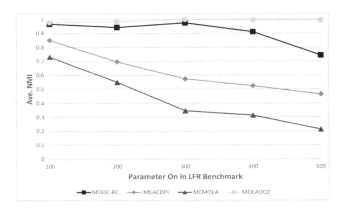

Fig. 2. Average NMI value attained by each algorithm on LFR benchmark networks when $\mu = 0.1$ and O_n varies from 100 to 500 with an increment of 100.

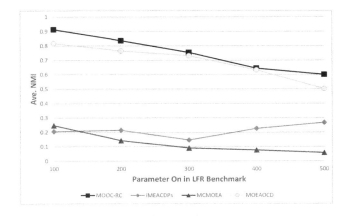

Fig. 3. Average NMI value attained by each algorithm on LFR benchmark networks when $\mu = 0.5$ and O_n varies from 100 to 500 with an increment of 100.

Table 3. LFR benchmark networks when $\mu = 0.1$, $\mu = 0.5$, and O_n varies from 100 to 500. Average Rankings of the algorithms (Friedman).

Algorithm	Ranking
MOOCD-RC	1.5
iMEA-CDPs	3.1
MCMOEA	3.9
MOEA-OCD	1.5

Table 4. LFR benchmark networks when $\mu = 0.1$, $\mu = 0.5$, and O_n varies from 100 to 500. Post Hoc comparison where $\alpha = 0.05$ (Friedman).

i	Algorithm	$z = (R_0 - R_i)/SE$	p	Holm	Finner
3	MCMOEA	4.156922	0.000032	0.000097	0.000097
2	iMEA-CDPs	2.771281	0.005584	0.011167	0.008364
1	MOOCD-RC	0	1	1	1

From the above experimental results, we can conclude that MOEA-OCD and our proposal have outstanding performances on LFR benchmark networks in most cases. However, our algorithm employs the information contained in the communities of Pareto Set to build a final covering of the network. Although the solutions of Pareto Set do not have overlapping communities, our proposal does not depend on this for building the final communities. Thus, our algorithm can be used by multi-objective evolutionary algorithms which build disjoint or overlapping community structures.

It should be noted that our proposal depends on the obtained non-dominated solutions. In these experiments we used the algorithm MOGR-OV [16] to generate the Pareto Set. On the other hand, the settings of β and γ have a narrow relationship over the obtained final covering. Following, we will give a brief description about this.

4.1 Community Structure Under Different Lower and Upper Approximation Scales

In the above experiments, the parameters β and γ are fixed to 0.75 and 0.1, respectively. We will have as results boundaries of communities more or less tight, depending on the way we change those parameters. Hence, both of them allow decision-makers to analyze the network according to the domain problem.

Using the synthetic network generated above with the parameters values $\mu = 0.1$ and $O_n = 0.1N$, we will show the overlapping communities with different lower and upper approximation scales. For that, we change the γ parameter and keep the same β value used in the experiments. The parameter γ allows to tune the boundaries of communities. Thus, the higher the value of γ is, the wider the boundaries are and vice versa, which means that there is going to be more or less overlapping vertices, respectively.

Furthermore, we build two coverings of the obtained synthetic network by considering $\gamma = 0.1$ and $\gamma = 0.25$. For a better comprehension of the studied network we used the graph analysis tool Gephi. It employs both the network properties (e.g., vertex degree) and also the identified communities in the network in the visualization process. Figures 4 and 5 showed next were obtained using the Force Atlas 2 [23] method belonging to Gephi.[1]

As shown in Figs. 4 and 5, the covering obtained using $\gamma = 0.25$ shows boundaries of communities wider than the covering obtained with $\gamma = 0.1$. Thus, the communities showed in Fig. 5 have more overlapping vertices than communities showed in Fig. 4. The overlapped vertices are bigger visualized than others and they are placed in the boundaries of communities. As described before, the parameter γ allows the DM from its own knowledge to tight or wide the boundaries of communities. In this way, the decision maker has a mechanism to weigh the importance of lower and upper approximations in the obtained communities. However, the adjustment of β and γ has a direct control over the final covering. Worth noting that our algorithm builds the final covering only using the information about the communities of the Pareto Set.

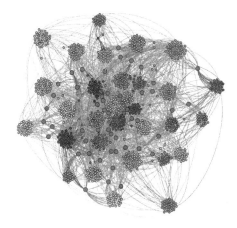

Fig. 4. Covering obtained over the obtained synthetic network based on the parameter values $\mu = 0.1$, $O_n = 0.1N$ and $\gamma = 0.1$.

---

[1] http://gephi.github.io/.

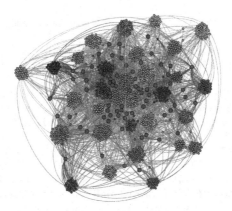

Fig. 5. Covering obtained over the obtained synthetic network based on the parameter values $\mu = 0.1$, $O_n = 0.1N$ and $\gamma = 0.25$.

5 Conclusions

In this paper, we proposed a new algorithm, named MOOCD-RC, for discovering overlapped communities through a combination of a multi-objective approach and Rough Clustering. It is composed of two steps: (a) build the granules of the indiscernible objects, and (b) build the final covering of network.

In the fist step, MOOCD-RC defined an equivalence relation between each pair of vertices of the network through the thresholded similarity graph. The obtained equivalence classes under the indiscernibility relation induce a granule set which constitutes our initial granularity criterion. We will also use them to build the final covering of the network. Afterward, in the second steps, the algorithm built the resulting communities through the Rough Clustering, taking the k greatest granules as prototypes of the communities; they also represent the lower approximations inside their own communities.

The MOOCD-RC algorithm was evaluated over synthetic networks in terms of its accuracy and it was compared against four algorithms of the related work. From the above experimental results, we can draw the conclusion that MOEA-OCD and our algorithm have outstanding performances on LFR benchmark networks in most cases. Moreover, this evaluation showed that MOOCD-RC is promising and effective for overlapping community detection in complex networks. As future work, we would like to make a more automatic adjustment to the β and γ parameters.

References

1. Maivizhi, R., Sendhilkumar, S., Mahalakshmi, G.S.: A survey of tools for community detection and mining in social networks. In: Proceedings of the International Conference on Informatics and Analytics. ACM (2016)

2. Shi, C., Yan, Z., Cai, Y., Wu, B.: Multi-objective community detection in complex networks. Appl. Soft Comput. **12**(2), 850–859 (2012)
3. Lancichinetti, A., Fortunato, S., Kertesz, J.: Detecting the overlapping and hierarchical community structure of complex networks. New J. Phys. **11**(3), 033015 (2009)
4. Lancichinetti, A., Fortunato, S., Radicchi, F.: Benchmark graphs for testing community detection algorithms. Phys. Rev. E **78**(4), 046110 (2008)
5. Liu, J., Zhong, W., Abbass, H., Green, D.G.: Separated and overlapping community detection in complex networks using multiobjective evolutionary algorithms. In: IEEE Congress on Evolutionary Computation (CEC), (2010)
6. Palla, G., Derényi, I., Farkas, I., Vicsek, T.: Uncovering the overlapping community structure of complex networks in nature and society. Nature **435**, 814–818 (2005)
7. Liu, C., Liu, J., Jiang, Z.: An improved multi-objective evolutionary algorithm for simultaneously detecting separated and overlapping communities. Int. Nat. Comput. **15**(4), 635–651 (2015). https://doi.org/10.1007/s11047-015-9529-y
8. Wen, X., et al.: A maximal clique based multiobjective evolutionary algorithm for overlapping community detection. IEEE Trans. Evol. Comput. **21**, 363–377 (2016)
9. Yuxin, Z., Shenghong, L., Feng, J.: Overlapping community detection in complex networks using multi-objective evolutionary algorithm. Comput. Appl. Math. **36**(1), 749–768 (2015). https://doi.org/10.1007/s40314-015-0260-1
10. Shen, H., Cheng, X., Cai, K., Hu, M.B.: Detect overlapping and hierarchical community structure in networks. Phys. A Stat. Mech. Appl. **388**(8), 1706–1712 (2009)
11. Gong, M., Cai, Q., Chen, X., Ma, L.: Complex network clustering by multiobjective discrete particle swarm optimisation based on decomposition. IEEE Trans. Evol. Comput. **18**(1), 82–97 (2014)
12. Zhou, A., Qu, B.Y., Li, H., Zhao, S.Z., Suganthan, P.N., Zhang, Q.: Rough-fuzzy collaborative clustering. IEEE Trans. Syst. Man Cybern. Part B (Cybern.) **36**(4), 795–805 (2006)
13. Lingras, P., Chen, M., Miao, D.: Qualitative and quantitative combinations of crisp and rough clustering schemes using dominance relations. Int. J. Approximate Reasoning **55**(1), 238–258 (2014)
14. Lingras, P., Chen, M., Miao, D.: Applying rough set concepts to clustering. In: Peters, G., Lingras, P., Ślezak, D., Yao, Y. (eds.) Rough Sets: Selected Methods and Applications in Management and Engineering. Advanced Information and Knowledge Processing, pp. 23–37. Springer, London (2012). https://doi.org/10.1007/978-1-4471-2760-4_2
15. Lingras, P., West, C.: Interval set clustering of Web users with rough k-means. Technical Report No. 2002–002, Department of Mathematics and Computer Science, St. Mary's University, Halifax, Canada (2002)
16. Grass-Boada, D.H., Pérez-Suárez, A., Bello, R., Rosete, A.: Multiobjective overlapping community detection algorithms using granular computing. In: Bello, R., Falcon, R., Verdegay, J.L. (eds.) Uncertainty Management with Fuzzy and Rough Sets. SFSC, vol. 377, pp. 233–256. Springer, Cham (2019). https://doi.org/10.1007/978-3-030-10463-4_13
17. Pawlak, Z.: Rough Sets: Theoretical Aspects of Reasoning About Data. Kluwer Academic, Dordrecht (1991)
18. Pawlak, Z., Andrzej, S.: Rough sets: some extensions. Inf. Sci. **177**, 28–40 (2006)
19. Yao, Y.Y., Lin, T.Y.: Generalization of rough sets using modal logic. Intell. Autom. Soft Comput. **2**(2), 103–120 (1996)
20. Mitra, S.: An evolutionary rough partitive clustering. Pattern Recogn. Lett. **25**(12), 1439–1449 (2004)

21. Mitra, S.: Granular computing: basic issues and possible solutions. In: Proceedings of the 5th Joint Conference on Information Sciences, no. 1, pp. 186–189 (2000)
22. Slowinski, R., Vanderpooten, D.: A generalized definition of rough approximations based on similarity. IEEE Trans. Knowl. Data Eng. $12(2)$, 331–336 (2000)
23. Mathieu, J., Tommaso, V., Sebastien, H., Mathieu, B.: ForceAtlas2, a continuous graph layout algorithm for handy network visualization designed for the Gephi software. PLoS ONE $9(6)$, e98679 (2014)

Detecting Overlapping Communities Using Distributed Neighbourhood Threshold in Social Networks

Rajesh Jaiswal and Sheela Ramanna[✉]

Department of Applied Computer Science, University of Winnipeg,
Winnipeg, MB R3B 2E9, Canada
jaiswal-r@webmail.uwinnipeg.ca, s.ramanna@uwinnipeg.ca

Abstract. In this work, we have proposed a simple overlapping community detection algorithm based on a distributed neighbourhood threshold method (DNTM). DNTM uses pre-partitioned disjoint communities and then analyzes the neighbourhood distribution of boundary nodes in disjoint communities to detect overlapping communities. It is a form of seed-based global method since boundary nodes are considered as seeds and become the starting point for detecting overlapping communities. Threshold value for each boundary node is used as minimum influence by the neighbours of a node in order to determine its belongingness to any community. The effectiveness of the DNTM algorithm has been demonstrated by testing on fifteen real-world datasets and compared with seven overlapping community detection algorithms. DNTM outperforms comparable algorithms with 10 out of 15 datasets and gives comparable results for the remaining 5 datasets in terms of the extended modularity Q_{ov} measure. Experiments with various disjoint algorithms on 15 datasets reveal that DNTM with tolerance community detection (TCD) as a preprocessing algorithm gives the best result.

Keywords: Community detection · Social networks analysis · Overlapping communities · Graph clustering

1 Introduction

There are a plethora of methods for detecting overlapping communities in social networks for both synthetic and real-world datasets starting from [19]. Classical strategies include: local expansion of seed nodes [20,22], label propagation [7,13,33], clique-based [26] and ensemble-based methods [3,4] to name a few. In this paper, we propose a new method based on detecting overlapping

This research has been supported by the Natural Sciences and Engineering Research Council of Canada (NSERC) Discovery grant 194376. Rajesh Jaiswal's research is supported by the UW Graduate Studies Scholarship and Linda and Vana Kirby Scholarship.

© Springer Nature Switzerland AG 2020
R. Bello et al. (Eds.): IJCRS 2020, LNAI 12179, pp. 432–445, 2020.
https://doi.org/10.1007/978-3-030-52705-1_32

communities by i) utilizing disjoint communities, and ii) analyzing the neighbourhood distribution of boundary nodes in disjoint communities to detect overlapping clusters. Our method is akin to the more recent class of ensemble methods [3] that uses disjoint methods as a starting point for development of overlapping method. In this paper, we propose a distributed neighbourhood threshold method (DNTM) which depends on the neighbourhood distribution of boundary nodes in disjoint communities. The threshold for each boundary node is used as minimum neighbour influence for a node to belong in any community. DNTM can be considered as global method since we are not performing any local expansion on a set of initial seed nodes for generating overlapping clusters. Instead, we are using boundary nodes and exploring the clusters external to the home clusters of boundary nodes to generate overlapping clusters. It is also a form of *seed-based* method since boundary nodes are considered as seeds and become the starting point for detecting overlapping clusters. There is only a user-defined maximum threshold (*tolerance*) criteria to form a neighbourbood. Four disjoint methods have been considered in this work with the primary method based on a tolerance community detection (TCD) [15]. The other partitioning methods include: Louvain [1], Girvan-Newman [10] and Greedy Modularity [5]. Typical metrics such as Overlapping Normalized Mutual Information (ONMI), Precision, Recall, or F-measure require *ground-truth* communities. However, ground-truth communities are readily available for large real networks. In their absence, computer generated benchmark networks with built-in ground-truth communities, called synthetic networks such as LFR [19] must be used, to first generate the ground-truth communities. In this paper, DNTM uses an extended modularity Q_{ov} measure introduced by Nicosia et al. [24] as a performance metric. The effectiveness of the DNTM algorithm has been demonstrated by testing on fifteen real-world datasets and compared with seven overlapping community detection algorithms.

The contribution of this paper is a simple algorithm which outperforms comparable algorithms with 10 out of 15 datasets and gives comparable results for the remaining 5 datasets in terms of extended modularity Q_{ov} measure. Another noteworthy feature of DNTM is that no optimization strategy such as satisfying some fitness function criteria has been used. Experiments with various partitioning methods on 15 datasets reveal that: TCD gives the best result with 7 datasets, Greedy Modularity method gives the best result with 4 datasets and both Louvain and Girvan-Newman methods with 4 datasets.

Our paper is organized as follows: In Sect. 2, we briefly review some representative overlapping community detection algorithms. In Sect. 3, we give a brief overview of definitions and cluster quality measure used in this paper. In Sect. 4, we give details of the proposed DNTM algorithm and its complexity. In Sect. 5, we present experimental results and analysis. Lastly, we give concluding remarks in Sect. 6.

2   Related Works

In this section, we briefly review some representative algorithms in terms of general strategies used by these algorithms.

2.1   Local Expansion

The general strategy is to start with a set of initial nodes as seeds and then expand to communities based on a fitness function criteria.

OSLOM [20]: Introduced in 2011 by Lancichinetti et al., this method was the first that detected communities based on their statistical significance that takes into account different types of graphs, edge direction, edge weights, overlapping communities, network hierarchy and to recognize the absence of community structure and/or the presence of randomness in graphs. It is based on a *local expansion and optimization strategy* where community expansion is performed by comparing the statistical significance of clusters defined with respect to a global null model (which is the configuration model).

LEMON [22]: This algorithm proposed in 2018 by Li et al., is based on the concepts of seed sets, local spectral diffusion, and local spectra. Here, a subspace around the initial seed sets called local spectra is explored using a short random walk also known as local spectral diffusion. Local spectra avoids computation burden by replacing a large number of singular vectors with short random walks. The running time of LEMON scales with the size of the community rather than that of the entire graph and has been tested on large networks.

2.2   Label Propagation

The general strategy is to label every node with a unique value and replace the node's label value with that of its most commonly detected *neighbour*. Once this process terminates, the nodes having the same label form a community.

COPRA [13]: Introduced in 2010, this method extends the label propagation algorithm(LPA) method by Raghavan et al. [27] to detect overlapping communities with a novel termination condition. This method is dependent on parameters such as node belonging coefficient and maximum number of communities a node can belong to, and can handle weighted and bipartite graphs. COPRA usually produces results that are better (in terms of modularity) for large networks.

SLPA [33]: This algorithm is based on speaker-listener mechanism to transfer the information known as labels between the nodes. Each node in this method maintains a list of labels and a randomly selected label from this list is propagated further to the node under consideration presently for detecting communities.

DEMON [7,8]: Label propagation algorithm is applied at the core of DEMON method to merge the locally generated clusters using merging function to obtain overlapping communities.

2.3   Ensemble Based

The general strategy here is to leverage disjoint clusters produced by various disjoint community detection algorithms to discover the overlapping communities.

MEDOC [4]: Introduced in 2016 by Chakraborty et al., this is the first ensemble based method for discovering overlapping communities by using meta-communities created from combining various similar clusters produced by disjoint communities detection methods. Further an association matrix which records the probability of a vertex belonging to a meta-community is utilized to generate both non-overlapping and overlapping communities.

EnCoD [3]: This method uses various disjoint community detection algorithms to generate disjoint clusters and further utilize the good qualities of these clusters to create an ensemble solution. This algorithm uses node membership as a feature and similarity of node pairs to form a network.

2.4   Others

CPM [26]: Introduce by Gergely Palla et al. in 2005, this classical algorithm is the first method to detect overlapping communities based on clique-percolation technique.

NECTAR [6]: It is a node-centric overlapping community detection algorithm in which the best communities for a given node are found using objective function and further this node is added to these communities to obtain the overlapping communities. In this method, Louvain's local search heuristic approach is generalized to discover overlapping communities. This algorithm tries to maximize the dynamically chosen objective function (i.e. WOCC and Q^E) by testing every possible existence of each node in it's neighbouring cluster in order to generate overlapping communities. All the clusters with a maximum value of objective function are considered to obtain the overlapping communities.

IEDC [14]: This algorithm provides an integrated framework for discovering both overlapping and non-overlapping communities. It uses a node-based criteria with a probabilistic model. It includes computation of internal associations (non-overlapping communities), computation of external associations (overlapping communities) using interaction matrix and a community propagation probability of its neighbours.

3   Preliminaries

Here, we give a brief overview of definitions and cluster quality measure used in this paper.

Undirected Graph: A graph G is defined as a pair of (V, E) where V is a set consisting all the nodes and E is set consisting all the edges $E \subseteq V \times V$.

Undirected graphs are such graphs in which if an edge $(x, y) \in E$ then edge (y, x) must also be in E. The *degree* of a node v is defined as the number of edges containing v. Two nodes are adjacent if they share a common edge.

Path: A path is composed of a series of nodes $P = (v_1, v_2, \ldots, v_n) \in V^n$ where $\forall i, 1 \leq i < n$, v_i is adjacent to v_{i+1}. The path length of P is measured as $n - 1$ where n is the total number of nodes in path P. It is also measured as the number edge(s) in that path. The path with minimum length (or number of edge(s)) from a source node s to a destination node d is called the shortest path sp from s to d.

Neighbourhood of a Node: The neighbourhood of a node x for a graph $G = (V, E)$ is defined as:

$$N_r(x) =_{def} \{y \in V : dist(x, y) < \varepsilon\} \tag{1}$$

where

$$dist(x, y) = \begin{cases} \infty \text{ if no } sp \text{ exists} \\ |sp| \text{ else} \end{cases} \tag{2}$$

ε is a user-defined positive real threshold value, sp is the shortest path from x to y and $|sp|$ is the number of edge(s) in sp. A *breadth first search* is used for traversing the graph in order to find the neighbourhood of any given node.

Neighbourhood Cluster of a Node: Let $C = \{C_1, C_2, \ldots C_n\}$ be a set of disjoint clusters that *cover* the graph G where $C_i = \{v_1, v_2, \ldots v_n\}$ is a cluster or community such that $v_i \in V$. Let $x \in C_j$ where C_j is the *home cluster*, then

$$NC(x) =_{def} \{C_i \in C \setminus C_j : \exists\, y \in C_i \wedge y \in N_r(x)\} \tag{3}$$

In Fig. 1, the neighbourhood cluster(s) for the *green node* belonging to cluster C_1 are: clusters C_2 and C_3. Note, for the *green node*, cluster C_1 is considered as the home cluster.

Distributed Neighbourhood Threshold: Equaion 4 defines this threshold as the ratio of total number of the neighbours of a given node v over the total number of neighbourhood clusters of v plus the home cluster of v.

$$D_t(v) =_{def} \left\lfloor \frac{|N_r(v)|}{|NC(v)| + 1} \right\rfloor \tag{4}$$

Overlapping Candidate Node: Let $v \in C_j$, then v is a candidate overlapping node if it satisfies the following equation:

$$O_{cn}(v) =_{def} NC(v) \neq \emptyset \tag{5}$$

Overlapping Node: Node v is a overlapping node if for any $C_i \in NC(v)$ it satisfies the following equation:

$$ON(v) =_{def} O_{cn}(v) \wedge (D_t(v) \leq |\{y : y \in N_r(v) \wedge y \in C_i\}|) \tag{6}$$

Example 1. In Fig. 1, the *green node* in cluster C_1 is an overlapping candidate node since it has neighbours in clusters C_2 and C_3. All nodes that have neighbours outside their home clusters are considered as *overlapping candidate nodes*. Using Eq. 4, $|N_r(green\ node)| = 8$ and $|NC(green\ node)| = 2$, hence $D_t(v) = 2$. In other words, $D_t(v)$ is considered as the minimum threshold value for a node v to be classified as *overlapping node*. As shown in Fig. 1 *green node* shares 3 edges with C_3 which also means $|N_r(green\ node)|$ in C_3 is 3. Since cluster C_3 includes neighbours of *green node* and $D_t(green\ node)$ meets the threshold requirement, the *green node* will be shared with C_3 as shown in Fig. 2.

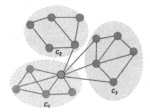

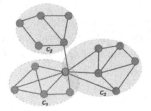

Fig. 1. Overlapping candidate node (Color figure online)

Fig. 2. Sample overlapping clusters (Color figure online)

Cluster Quality Measure: Extended Modularity: In this work we have used the extended modularity Q_{ov} measure introduced by Nicosia in [24,25] given in Eq. 7 where V is the set of nodes, $|V|$ represents the number of nodes, C represents the set of overlapping cluster, m is the total number of edges and $A_{i,j}$ is the *adjacency matrix* for the graph. We have chosen to use this measure since it does not require the ground-truth to measure the quality of the generated clusters. Generally, good quality overlapping clusters have higher Q_{ov} value. The value of Q_{ov} will be 0 when only one cluster is obtained with all the nodes in it. Details about various coefficients in Eq. 7 can also be found in [25].

$$Q_{ov} = \frac{1}{m} \sum_{c \in C} \sum_{i,j \in V} \left[\beta_{l(i,j),c} A_{i,j} - \frac{\beta_{l(i,j),c}^{out} k_i^{out} \beta_{l(i,j),c}^{in} k_j^{in}}{m} \right] \quad (7)$$

$$\beta_{l(i,j),c}^{in} = \frac{\sum_{i \in V} F(\alpha_{i,c}, \alpha_{j,c})}{|V|} \quad (8)$$

$$\beta_{l(i,j),c}^{out} = \frac{\sum_{j \in V} F(\alpha_{i,c}, \alpha_{j,c})}{|V|} \quad (9)$$

In overlapping communities, each node can belong to multiple communities but with different strengths of belonging. An array of such *belonging factor* $[\alpha_{i,1}, \alpha_{i,1}, \alpha_{i,1}, ...\alpha_{i,|C|}]$ is calculated and allotted to each node i in the graph G. The strength of node i belonging to community c is depicted by coefficient $\alpha_{i,c}$.

Since the belonging coefficient for each node is already defined, it is also possible to define the belonging coefficient to each community for edges incoming to or outgoing from a node. Belonging coefficient of edge $l = (i, j)$ with source node i and target node j to community c is represented by function $\beta_{l,c}$. Further, the belonging coefficient for link $l(i, j)$ pointing to a node going into the community c is represented by $\beta_{l(i,j),c}^{in}$ and given by Eq. 8 similarly the belonging coefficient for link $l(i, j)$ pointing to a node going out of the community c is obtained by using Eq. 9 and is represented by $\beta_{l(i,j),c}^{out}$. Extended Modularity measures for overlapping cluster depends on $F(\alpha_{i,c}, \alpha_{j,c})$ which is defined in the Eq. 10

$$F(\alpha_{i,c}, \alpha_{j,c}) = \frac{1}{(1 + e^{-f(\alpha_{i,c})})(1 + e^{-f(\alpha_{j,c})})} \tag{10}$$

where $f(\alpha_{i,c})$ is a simple linear scaling function given in Eq. 11 . The value of p is set to 30 in [25]. Generally, good quality overlapping clusters have higher Q_{ov} value. The value of Q_{ov} will be 0 when only one cluster is obtained with all the nodes in it.

$$f(x) = 2px - p, p \in \mathcal{R} \tag{11}$$

Datasets: Various sized real-world datasets were used in this study: Karate [34], Dolphin [23], Lesmis [16], Football [10], Polbooks [17], Jazz [11], Power grid [31], Durgnet [32], Highschool [18], Netscience [29], C.elegans [9], Bible-names [18], Protein [18], Internet-Route [21] and PGP [2].

4   Overlapping Community Detection Algorithm: DNTM

In Fig. 3, the flow of the DNTM algorithm is given where DNTM takes crisp partitioned clusters as input irrespective of the algorithm used. We first generate non-overlapping clusters and use these clusters to examine all such nodes which have neighbours in other clusters to find overlapping nodes. Once an overlapping node is found, we update the respective clusters by including this overlapping node to obtain the resultant overlapping clusters.

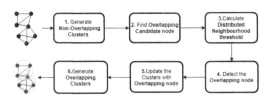

Fig. 3. Flow diagram of DNTM Algorithm

The main steps of DNTM algorithm are as follows: i) generate non-overlapping clusters, ii) find candidate overlapping nodes using Eq. 5, iii) calculate distributed neighbourhood threshold using Eq. 4, iv) filter overlapping

---

Algorithm 1: Find Overlapping Clusters

Input: G // Input graph.

L // List of non-overlapping clusters.

ϵ // Distance Threshold.

Output: $\{OC_1, OC_2, OC_3, \ldots \ldots OC_n\}$ // List of Overlapping clusters

1 **procedure** findOverlapCluster (G, L, ϵ)

2 $NC_{dic} \leftarrow \emptyset$

3 $L_{dic} \leftarrow \emptyset$

4 $L_o \leftarrow \emptyset$

5 $c_{id} \leftarrow 0$

6 **for each cluster** $C \in L$ **do**

7    $c_{id} \leftarrow c_{id} + 1$

8   **for each node** $v \in C$ **do**

9      $NC_{dic}[v] \leftarrow c_{id}$

10   $L_{dic}[c_{id}] \leftarrow C$

11 $C_oN_{dic} \leftarrow \emptyset$

12 $N_rN_{dic} \leftarrow \emptyset$

13 **for each cluster** $C \in L$ **do**

14   **for each node** $v \in C$ **do**

15     $N_r(v) \leftarrow BFS(G, v, \epsilon)$

16     $N_rN_{dic}[v] \leftarrow N_r(v)$

17     $N_r(v) \leftarrow N_r(v) - C$

18   **if** $N_r(v) \neq \emptyset$ **then**

19       $C_oN_{dic}[v] \leftarrow N_r(v)$

20 **for each** $v \in C_oN_{dic}.keys()$ **do**

21   $N_rC_{dic} \leftarrow \emptyset$

22   $N_r(v) \leftarrow C_oN_{dic}[v]$

23   **for each vertex** $v_n \in N_r(v)$ **do**

24     $c_{id} \leftarrow NC_{dic}[v_n]$

25     $N_rC_{dic}[c_{id}] \leftarrow \{v_n\}$

26   $NC(v) \leftarrow N_rC_{dic}.keys()$

27   $D_t(v) \leftarrow \frac{|N_rN_{dic}[v]|-1}{|NC(v)|+1}$

28   **for each clusterId** $c_{id} \in NC(v)$ **do**

29   **if** $Size(N_rC_{dic}[c_{id}]) \geq D_t(v)$ **then**

30       $L_{dic}[c_{id}] \leftarrow L_{dic}[c_{id}] \cup v$

31 **for each clusterId** $c_{id} \in L_{dic}.keys()$ **do**

32   $L_o.append(L_{dic}[c_{id}])$

33 **return** L_o

---

nodes using Eq. 6, and v) update the clusters with overlapping nodes to obtain the resultant overlapping clusters. Note, DNTM takes crisp partitioned clusters as input, irrespective of the algorithm used (see Fig. 6 and 7).

Algorithm 1 includes the following data structures: list of overlapping clusters L_o is used to store generated overlapping clusters, Node-Cluster Dictionary NC_{dic} to store cluster id of each node, Cluster-Node Dictionary L_{dic} to store nodes in each cluster, Neighbour Node-Cluster Dictionary N_rC_{dic} to store cluster id of neighbourhood nodes, Overlapping-Candidate-Node Dictionary C_oN_{dic} to store overlapping candidate nodes and its neighbours N_r from neighbourhood cluster NC, Node-Neighbour Dictionary N_rN_{dic} to store node and its neighbours.

4.1 Time Complexity

In DNTM algorithm for a graph $G(V, E)$, the time taken for pre-processing the disjoint clusters is $O(|L|.|C|)$ which is less than or equal to $O(|V|)$ where $|L|$ is the number of disjoint clusters, $|C|$ represents the number of nodes in a cluster C and $|V|$ represent total number of nodes in graph G. Running time of BFS is $O(b^d)$ where b is branching factor and d is maximum depth. In DNTM, we consider neighbours at depth 1, so time taken is $O(b)$. To find overlapping candidate nodes, the time consumed is $O(|L|.|C|).O(b) = O(|V|.b)$. To filter overlapping nodes, computation time is $O(|OCN|).O(|N_r| + |N_rC_{dic}|)$ where $|OCN|$ is the number of overlapping candidate nodes, $|N_r|$ is the number of neighbourhoods in other clusters and $|N_rC_{dic}|$ is the number of neighbourhood clusters. Since $|N_r| \geq |N_rC_{dic}|$, so the computation time will be $O(|OCN|.|N_r|)$. Finally it takes $O(|L|)$ time to generate overlapping clusters. So the obtained final time complexity is $O(|V|.b + |OCN|.|N_r|)$

5 Experiments and Results

To examine the performance of DNTM, 15 real world data-sets were used and compared with the following overlapping communities detection algorithms: CPM [26], OSLOM [20], COPRA [13], SLPA [33], Node Perception [30], DEMON [7,8] and CONGO [12] with h = 2 and h = 3. Except for OSLOM and COPRA, all other algorithms were taken from CDlib [28] Python package. Table 1 gives the results of our experiments where DNTM (TCD) is the proposed algorithm which uses TCD method to generate non-overlapping clusters with $\varepsilon = 2$ with source code made available by the authors. TCD method relies on a tolerance relation where a tolerance class represents members of the same community and uses an objective function based on two well-known quality functions, modularity and coverage.

Since most of the algorithms have a non-unique output for Q_{ov} for each execution, hence these algorithms were executed 10 times and the average of the 5 best scores for Q_{ov} was used in our reporting shown in Table 1 and bold values represent the best score for each dataset. In additon, the number of clusters generated by *majority* of the algorithms is used as *input* for those algorithms that require *number of clusters* as input.

Based on the results in Table 1 and Fig. 4 and Fig. 5, we can observe that the proposed DNTM algorithm outperforms comparable algorithms with 10 out of 15 datasets and gives comparable results for the remaining 5 datasets. The quality of generated overlapping clusters from DNTM is greatly affected by the number of disjoint clusters passed as input, generated by the initial disjoint algorithm. From Eq. 4 it can be observed that D_t has an inverse relation with

Table 1. Extended Modularity (Q_{ov}) values

Datasets	CPM	OSLOM	COPRA	SLPA	NodePer.	DEMON	CONGO h = 2	CONGO h = 3	DNTM (TCD)
Karate	0.51	0.7099	0.7228	0.5405	0.1944	0.38	0.3423	0.488	**0.7282**
Dolphins	0.66	0.7426	**0.7434**	0.7231	0.1947	0.457	0.4085	0.134	0.734
Lesmis	0.586	0.6908	0.7156	**0.7772**	0.3259	0.385	0.315	0.6586	0.755
Football	0.44	0.6674	0.6962	0.7052	0.072	0.353	0.4332	0.4955	**0.75**
Polbooks	0.786	0.8263	0.8226	**0.8286**	0.142	0.279	0.3468	0.4945	0.81
Jazz	0.096	0.5142	0.6626	**0.7401**	0.0438	0.382	0.24	0.22	0.6904
Power	0.15	0.3887	0.4842	0.6363	0.0970	0.077	0.8312	0.7878	**0.90**
Durgnet	0.207	0.1697	0.7664	0.6255	0.1355	0.155	0.235	0.235	**0.7853**
Highschool	0.056	0.6762	0.7064	0.6581	0.144	0.056	0.4612	0.7015	**0.755**
Netscience	0.0	0.7862	0.8444	0.8353	0.512	0.436	0.7547	0.7314	**0.953**
C.elegans	0.217	0.4551	0.212	0.4346	0.080	0.0279	0.07426	0.10357	**0.61**
Bible names	0.425	0.2965	0.4025	0.3657	0.0938	0.013	0.19	0.160	**0.6424**
Protein	0.16	0.1784	0.363	0.7402	0.1015	0.140	0.57221	0.5858	**0.7958**
Internet route	0.245	0.3475	0.102	**0.63**	0.0213	0.0045	0.1467	0.25482	0.5273
PGP	0.568	0.5364	0.775	0.737	0.2523	0.2024	0.5607	0.5563	**0.7963**

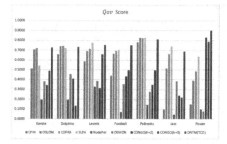

Fig. 4. Part 1: Q_{ov} results with 7 datasets

Fig. 5. Part 2: Q_{ov} results with 8 datasets

number of communities. D_t is highly sensitive and dependent on the number of communities. As a result, increasing number of communities, will decrease the value of D_t, which will in turn affect the overlap between the communities. In our experiments, the number of communities, range from 2 to 109. We also observed that in general, for the datasets, where the number of communities is greater than 4, DNTM achieves the best result. Also, DNTM depends on the boundary nodes in the disjoint clusters as well their internal and external links (edges). If the number of external links of a node is extremely less as compared to its internal links, this node is less likely to qualify the condition in Eq. 6 to be classified as an overlapping node. Most algorithms use an internal objective function to obtain *good* quality clusters which entails parameter selection. DNTM does not have this limitation as it does not use an internal objective function and the major computation is done for overlapping candidate nodes which is comparatively less than $|V|$. Hence DNTM is computationally efficient. Table 2 gives comparative results for Q_{ov} with the proposed DNTM algorithm where the input (disjoint clusters) was obtained using Louvain [1] DNTM (LN), Girvan-Newman [10] DNTM (GN) and Greedy Modularity [5] DNTM (GD) methods on all the datasets. It can be observed that DNTM (TCD) is giving best results in 7 out of 15 datasets and comparable with the other data sets (either second best or third best).

Table 2. DNTM results with different partitioning methods

Datasets	DNTM (TCD)	DNTM (LN)	DNTM (GN)	DNTM (GD)	Best in DNTM
Karate	**0.7282**	0.615	0.7185	0.5861	TCD
Dolphins	0.734	0.6193	0.7232	**0.7359**	GD
Lesmis	**0.755**	0.6644	0.2689	0.7034	TCD
Football	0.75	0.6563	**0.7777**	0.6493	GN
Polbooks	0.81	0.8138	0.8090	**0.825**	GD
Jazz	0.6904	**0.7064**	0.0379	0.7016	LN
Power	0.90	**0.9513**	0.8709	0.9511	LN
Durgnet	0.7853	0.7299	**0.8654**	0.7907	GN
Highschool	**0.755**	0.5909	0.5964	0.7329	TCD
Netscience	**0.953**	0.9154	0.8674	0.9256	TCD
C.elegans	**0.61**	0.3473	0.0756	0.5035	TCD
Bible names	**0.6424**	0.4156	0.1	0.5815	TCD
Protein	0.7958	0.8076	0.6095	**0.8171**	GD
Internet route	**0.5273**	0.4305	0.01519	0.4375	TCD
PGP	0.7963	0.8975	0.2042	**0.9082**	GD

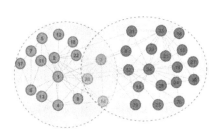

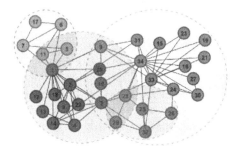

Fig. 6. DNTM clustering using disjoint clusters generated from Girvan-Newman method on the Karate dataset

Fig. 7. DNTM clustering using disjoint clusters generated from Louvain method on the Karate dataset

Figure 6 and 7 show overlapping clusters generated with the proposed DNTM algorithm where the input (disjoint clusters) was obtained using Louvain [1] and

Girvan-Newman [10] methods on the Karate dataset. In Fig. 6, three overlapping nodes [3, 14, 20] were detected, whereas using TCD as input method, five overlapping nodes [9, 10, 20, 29, 31] were detected. In Fig. 7, 12 overlapping nodes were detected including a hierarchical cluster where nodes [28, 29] are present in 3 clusters.

6   Conclusion and Future Work

In this paper, we have proposed a new overlapping community detection algorithm (DNTM) based on: i) utilizing disjoint communities produced by community detection algorithm(s), and ii) analyzing the neighbourhood distribution of boundary nodes of discovered disjoint communities to detect overlapping clusters. The effectiveness of the DNTM algorithm has been demonstrated by testing on fifteen real-world datasets and compared with seven overlapping community detection algorithms in terms of an extended modularity Q_{ov} measure. Three other well-known disjoint methods have been considered in this work with the primary method based on a tolerance community detection. DNTM outperforms comparable algorithms with 10 out of 15 datasets and gives comparable results for the remaining 5 datasets. Experiments with various disjoint algorithms on 15 datasets reveal that DNTM with TCD as a preprocessing algorithm gives the best result. Another noteworthy feature of DNTM is that no any optimization strategy has been used during or after the clustering process. Future work with DNTM will include: i) considering an ensemble mechanism to use various disjoint methods to select the best disjoint clusters in terms of quality and number of clusters as a preprocessing step to the DNTM algorithm, ii) defining an internal objective function to obtain good quality clusters, iii) testing and analyzing the behavior of DNTM on synthetic networks and iv) implementing a parallel DNTM to be able to handle datasets with larger nodes and communities.

References

1. Blondel, V.D., Guillaume, J.L., Lambiotte, R., Lefebvre, E.: Fast unfolding of communities in large networks. J. Stat. Mech: Theory Exp. **2008**(10), P10008 (2008)
2. Boguná, M., Pastor-Satorras, R., Díaz-Guilera, A., Arenas, A.: Models of social networks based on social distance attachment. Phys. Rev. E **70**(5), 056122 (2004)
3. Chakraborty, T., Ghosh, S., Park, N.: Ensemble-based overlapping community detection using disjoint community structures (2018)
4. Chakraborty, T., Park, N., Subrahmanian, V.: Ensemble-based algorithms to detect disjoint and overlapping communities in networks. In: 2016 IEEE/ACM International Conference on Advances in Social Networks Analysis and Mining (ASONAM), pp. 73–80. IEEE (2016)
5. Clauset, A., Newman, M.E., Moore, C.: Finding community structure in very large networks. Phys. Rev. E **70**(6), 066111 (2004)

6. Cohen, Y., Hendler, D., Rubin, A.: Node-centric detection of overlapping communities in social networks. In: Shmueli, E., Barzel, B., Puzis, R. (eds.) NetSci-X 2017. Springer Proceedings in Complexity, pp. 1–10. Springer, Cham (2017). https://doi.org/10.1007/978-3-319-55471-6_1

7. Coscia, M., Rossetti, G., Giannotti, F., Pedreschi, D.: DEMON: a local-first discovery method for overlapping communities. In: Proceedings of the 18th ACM SIGKDD International Conference on Knowledge Discovery and Data Mining, pp. 615–623. ACM (2012)

8. Coscia, M., Rossetti, G., Giannotti, F., Pedreschi, D.: Uncovering hierarchical and overlapping communities with a local-first approach. ACM Trans. Knowl. Discov. Data (TKDD) 9(1), 6 (2014)

9. Duch, J., Arenas, A.: Community detection in complex networks using extremal optimization. Phys. Rev. E 72(2), 027104 (2005)

10. Girvan, M., Newman, M.E.: Community structure in social and biological networks. Proc. Natl. Acad. Sci. 99(12), 7821–7826 (2002)

11. Gleiser, P.M., Danon, L.: Community structure in jazz. Adv. Complex Syst. 6(04), 565–573 (2003)

12. Gregory, S.: A fast algorithm to find overlapping communities in networks. In: Daelemans, W., Goethals, B., Morik, K. (eds.) ECML PKDD 2008. LNCS, vol. 5211, pp. 408–423. Springer, Heidelberg (2008). https://doi.org/10.1007/978-3-540-87479-9_45

13. Gregory, S.: Finding overlapping communities in networks by label propagation. New J. Phys. 12(10), 103018 (2010)

14. Hajiabadi, M., Zare, H., Bobarshad, H.: IEDC: an integrated approach for overlapping and non-overlapping community detection. Knowl.-Based Syst. 123, 188–199 (2017)

15. Kardan, V., et al.: Tolerance methods in graph clustering: application to community detection in social networks. In: Nguyen, H., Ha, Q.T., Li, T., Przybyła-Kasperek, M. (eds.) IJCRS 2018. LNCS, vol. 11103, pp. 73–87. Springer, Cham (2018). https://doi.org/10.1007/978-3-319-99368-3_6

16. Knuth, D.E.: The Stanford GraphBase: A Platform for Combinatorial Computing, vol. 37. Addison-Wesley Reading, Boston (1993)

17. Krebs, V.: Books about us politics. http://networkdata.ics.uci.edu/data.php?d=polbooks

18. Kunegis, J.: KONECT - the Koblenz network collection. In: Proceedings of International Conference on World Wide Web Companion, pp. 1343–1350 (2013). http://userpages.uni-koblenz.de/~kunegis/paper/kunegis-koblenz-network-collection.pdf

19. Lancichinetti, A., Fortunato, S.: Benchmarks for testing community detection algorithms on directed and weighted graphs with overlapping communities. Phys. Rev. E 80(1), 016118 (2009)

20. Lancichinetti, A., Radicchi, F., Ramasco, J.J., Fortunato, S.: Finding statistically significant communities in networks. PLoS One 6(4), e18961 (2011)

21. Leskovec, J., Kleinberg, J., Faloutsos, C.: Graph evolution: densification and shrinking diameters. ACM Trans. Knowl. Discov. Data 1(1), 1–40 (2007)

22. Li, Y., He, K., Kloster, K., Bindel, D., Hopcroft, J.: Local spectral clustering for overlapping community detection. ACM Trans. Knowl. Discov. Data (TKDD) 12(2), 17 (2018)

23. Lusseau, D., Newman, M.E.: Identifying the role that animals play in their social networks. Proc. R. Soc. London Ser. B: Biol. Sci. 271(suppl_6), S477–S481 (2004)

24. Nicosia, V., Mangioni, G., Carchiolo, V., Malgeri, M.: Extending the definition of modularity to directed graphs with overlapping communities. J. Stat. Mech: Theory Exp. **2009**(03), P03024 (2009)
25. Nicosia, V., Mangioni, G., Malgeri, M., Carchiolo, V.: Extending modularity definition for directed graphs with overlapping communities. Technical report (2008)
26. Palla, G., Derényi, I., Farkas, I., Vicsek, T.: Uncovering the overlapping community structure of complex networks in nature and society. Nature **435**(7043), 814 (2005)
27. Raghavan, U.N., Albert, R., Kumara, S.: Near linear time algorithm to detect community structures in large-scale networks. Phys. Rev. E **76**(3), 1–12 (2007)
28. Rossetti, G., Milli, L., Cazabet, R.: CDLIB: a Python library to extract, compare and evaluate communities from complex networks. Appl. Netw. Sci. **4**(1), 52 (2019)
29. Rossi, R.A., Ahmed, N.K.: The network data repository with interactive graph analytics and visualization. In: Proceedings of the Twenty-Ninth AAAI Conference on Artificial Intelligence (2015). http://networkrepository.com
30. Soundarajan, S., Hopcroft, J.E.: Use of local group information to identify communities in networks. ACM Trans. Knowl. Discov. Data (TKDD) **9**(3), 21 (2015)
31. Watts, D.J., Strogatz, S.H.: Collective dynamics of 'small-world' networks. Nature **393**(6684), 440 (1998)
32. Weeks, M.R., Clair, S., Borgatti, S.P., Radda, K., Schensul, J.J.: Social networks of drug users in high-risk sites: finding the connections. AIDS Behav. **6**(2), 193–206 (2002)
33. Xie, J., Szymanski, B.K., Liu, X.: SLPA: uncovering overlapping communities in social networks via a speaker-listener interaction dynamic process. In: 2011 IEEE 11th International Conference on Data Mining Workshops, pp. 344–349. IEEE (2011)
34. Zachary, W.W.: An information flow model for conflict and fission in small groups. J. Anthropol. Res. **33**(4), 452–473 (1977)

Fuzzy Cognitive Maps

On the Convergence of Input-Output Fuzzy Cognitive Maps

István Á. Harmati[1(✉)] and László T. Kóczy[2,3]

[1] Department of Mathematics and Computational Sciences, Széchenyi István University, Egyetem tér 1, Győr 9026, Hungary
harmati@sze.hu
[2] Department of Information Technology, Széchenyi István University, Egyetem tér 1, Győr 9026, Hungary
koczy@sze.hu
[3] Department of Telecommunication and Media Informatics, Budapest University of Technology and Economics, Magyar tudósok körútja 2, Budapest 1117, Hungary

Abstract. Fuzzy cognitive maps are recurrent neural networks, where the neurons have a well-defined meaning. In certain models, some neurons receive outer input, while other neurons produce the output of the system. According to this observation, some neurons are categorized as input neurons and the others are the state neurons and output neurons. The output of the system is provided as a limit of an iteration process, which may converge to an equilibrium point, but limit cycles or chaotic behaviour may also show up. In this paper, we examine the existence and uniqueness of fixed points for two types of input-output fuzzy cognitive maps. Moreover, we use network-based measures like in-degree, out-degree and connectivity, to express conditions for the convergence of the iteration process.

Keywords: Fuzzy cognitive map · Input-output fuzzy cognitive map · Stability · Convergence · Equilibrium point

1 Introduction

Fuzzy cognitive maps (FCMs) are decision support tools, based on the recurrent neural network modelling method. The essence is that the neurons have well-defined meaning, they represent specific factors or characteristics of the modelled system [14]. The structure of a fuzzy cognitive map is a weighted, directed graph. The weights are assigned to the edges from the interval $[-1, 1]$ to express the strength and direction of causal connections. The current states of the neurons (which are called concepts in FCM literature) are also described by values from the $[0, 1]$ interval (or from the interval $[-1, 1]$, see for example [15]). These are the activation values of the concepts [12].

The system can be described by the set of concepts $(C_1, C_2, \ldots, C_n)$; the current activation values of the concepts $(A_1, A_2, \ldots, A_n)$; the weight matrix

© Springer Nature Switzerland AG 2020
R. Bello et al. (Eds.): IJCRS 2020, LNAI 12179, pp. 449–461, 2020.
https://doi.org/10.1007/978-3-030-52705-1_33

W which assigns weight w_{ij} to each edge connecting the nodes C_i and C_j, expressing how strongly influenced is concept C_i by concept C_j. The sign of w_{ij} indicates whether the relationship between C_j and C_i is direct or inverse. So matrix W represents the weighted causal connections between the concepts. A transformation (or transfer, or threshold) function $f : \mathbb{R} \rightarrow [0,1]$ calculates the activation value of concepts at every time step of the iteration and keeps the activation values in the allowed range (sometimes a function $f : \mathbb{R} \rightarrow [-1,1]$ is applied).

The iteration rule which calculates the values of the concept at every step may or may not include self-feedback. In general form it can be written as

$$A_i(k) = f\left(\sum_{j=1, j\neq i}^{n} w_{ij} A_j(k-1) + d_i A_i(k-1) \right) \qquad (1)$$

where $A_i(k)$ is the value of concept C_i at discrete time k, w_{ij} is the weight of the connection from concept C_j to concept C_i and $0 \leq d_i \leq 1$ expresses the possible self-feedback. If $d_i = 0$, then there is no self-feedback. If we include the d_is into the diagonal of weight matrix W, the iteration equation can be rewritten in more compact style:

$$A_i(k+1) = f\left(\sum_{j=1}^{n} w_{ij} A_j(k) \right) = f(w_i A(k)), \qquad (2)$$

where $w_i = [w_{i1}, \ldots, w_{in}]$ is the ith row of W and $A(k) = [A_1(k), \ldots, A_n(k)]^T$ is the concept vector after k iterations. We apply dot product between them, so $w_i A^{(k)}$ is a real number.

Moreover, if we couple the coordinates of the concept vector together and denote by G the mapping $\mathbb{R}^n \rightarrow \mathbb{R}^n$ that generates the concept vector $A(k+1)$ from $A(k)$, then we have that:

$$A(k+1) = \begin{bmatrix} A_1(k+1) \\ \vdots \\ A_n(k+1) \end{bmatrix} = \begin{bmatrix} f(w_1 A(k)) \\ \vdots \\ f(w_n A(k)) \end{bmatrix} = G(A(k)). \qquad (3)$$

The iteration rule is repeated until either the FCM converges to an equilibrium state (fixed point) or the maximal number of iterations is reached. Mathematically, the FCM may converge to a fixed point, may arrive to a limit cycle or shows chaotic pattern [4, 10, 11].

Sufficient mathematical condition for the existence and uniqueness of fixed points of a special class of FCMs has been introduced in [2], expressed by the sum of the squared elements of W. This result was later generalized in [4]. In [7], the authors examined the problem of unique fixed points taking into consideration only the topology of the FCM, but not the weights. They pointed out that if the parameter of the sigmoid transfer function is small enough, then the FCM has exactly one fixed point. In [8] the global asymptotic stability of FCMs has been discussed via Lyapunov method.

Recently, various generalizations of FCMs have been introduced [1,3,9], where some concepts (neurons, nodes of the graph) are considered as inputs to the system, while some other (or all of the remaining) concepts form the output of the system. This article aims to provide sufficient converge conditions for these models, based on the weight structure and the parameter(s) of the transfer (threshold) function(s).

The rest of the paper is organized as follows. In Sect. 2 we recall the most important mathematical tools and notions applied in the proofs of our findings. In Sect. 3, we examine the behaviour of the generalized FCM model introduced in [1] and [3]: sufficient conditions for the existence and uniqueness of fixed points are provided. Moreover, we show that under certain mathematical conditions different input values may produce different steady-state concept vectors and different output values. In Sect. 4, sufficient condition for the convergence of FCM model introduced in [9] is given, and finally in Sect. 5 we summarize the main contributions of the paper.

2   Mathematical Background

In this section, we recall the most important definitions and results applied in the subsequent sections. First we recall the definition of contraction mapping [13]:

Definition 1. *Let (X, d) be a metric space. A mapping $G \colon X \to X$ is a contraction mapping or* contraction *if there exists a constant c (independent from x and y), with $0 \leq c < 1$, such that*

$$d\left(G(x), G(y)\right) \leq cd(x, y). \tag{4}$$

The notion of contraction is related to the distance metric d applied. It may happen that a function is a contraction w.r.t. one distance metric, but not a contraction w.r.t. another distance metric. The iterative process of an FCM may end at an equilibrium point, which is a so-called fixed point.

Let $G \colon X \to X$, then a point $x^* \in X$ such that $G(x^*) = x^*$ is a fixed point of G. The following theorem provides sufficient condition for the existence and uniqueness of a fixed point [13]. Moreover, if mapping that generates the iteration is a contraction, it ensures the stability of the iteration.

Theorem 1 *(Banach's fixed point theorem). If $G \colon X \to X$ is a contraction mapping on a nonempty complete metric space (X, d), then G has only one fixed point x^*. Moreover, x^* can be found as follows: start with an arbitrary $x_0 \in X$ and define the sequence $x_{n+1} = G(x_n)$, then $\lim_{n \to \infty} x_n = x^*$.*

Definition 2. *Let x^* be a fixed point of the iteration $x_{n+1} = G(x_n)$. x^* is locally asymptotically stable if there exist a neighborhood U of x^*, such that for each starting value $x_0 \in U$ we get that*

$$\lim_{n \to \infty} x_n = x^*. \tag{5}$$

If this neighborhood U is the entire domain of G, then x^* is a globally asymptotically stable fixed point.

Corollary 1. *If $G\colon X \to X$ is a contraction mapping on a nonempty complete metric space (X, d), then its unique fixed point x^* is globally asymptotically stable.*

The following property of the sigmoid function will be applied: The derivative of the sigmoid function $f\colon \mathbb{R} \to \mathbb{R}$, $f(x) = 1/(1 + e^{-\lambda x})$, $(\lambda > 0)$ is bounded by $\lambda/4$. Moreover, for every $x, y \in \mathbb{R}$ the following inequality holds

$$|f(x) - f(y)| \le \lambda/4 \cdot |x - y|. \tag{6}$$

Basic properties of the spectral radius of a matrix M [6]:

- The spectral radius of matrix $M \in \mathbb{R}^{n \times n}$ is given by

$$\rho(M) = \max\{|\lambda_i|\colon \lambda_i \text{ eigenvalue of } M\} \tag{7}$$

We should note that the spectral radius itself is not a norm.
- $\rho(M) = \inf\{\|M\|\colon \| * \| \text{ is a matrix norm on } \mathbb{R}^{n \times n}\}$
- Let matrix M have spectral radius $\rho(M)$. If $\varepsilon > 0$ is any positive number, then there exists a matrix norm $\|\cdot\|$, such that $\rho(M) \le \|M\| \le \rho(M) + \varepsilon$.
- If for matrices M_1, M_2 the entry-wise inequality $0 \le M_1 \le M_2$ holds, then $\rho(M_1) \le \rho(M_2)$.

3   Input-Output Fuzzy Cognitive Maps

In classical fuzzy cognitive map modelling, the concepts have their initial activation values and the final activation values are computed as the limit of the iteration (if the limit exists). In some cases, few features of the modelled system should not change during the simulation. From the FCM point of view, it means that values of some concepts should not change, but must remain the same for all steps of the iteration. This fact requires the re-thinking of the FCM-based modelling.

Based on the well-known discrete time linear time-invariant model:

$$\begin{aligned} x(k+1) &= Ax(k) + Bu(k) \\ y(k+1) &= Cx(k) + Du(k) \end{aligned} \tag{8}$$

Groumpos, Anninou et al. introduced the following FCM model [1,3]:

$$\begin{aligned} x(k+1) &= f\left(W_A x(k) + W_B u(k)\right) \\ y(k+1) &= f\left(W_C x(k) + W_D u(k)\right) \end{aligned} \tag{9}$$

where $u \in \mathbb{R}^r$, $x \in \mathbb{R}^p$, $y \in \mathbb{R}^m$. The matrices are extracted from the weight structure of the FCM:

- W_A describes the dynamics between the states (x);
- W_B describes the role of the inputs (u);
- W_C describes the role of x in the output (y);
- W_D describes the contribution of u to the output.

The following block scheme defines the weight matrix. The order of concepts: input, state, output (vertically and horizontally). Of course it is a bit redundant modell, since a state neuron can have input and state can be an output, too, but for the analogy with discrete linear systems we preserve these categories.

$$W = \begin{bmatrix} 0 & 0 & 0 \\ W_B & W_A & 0 \\ W_D & W_C & 0 \end{bmatrix} \tag{10}$$

The activation vector $A \in \mathbb{R}^{r+p+m}$ then $A = [u, x, y]^T$. The updating process is determined by the mapping $A(k+1) = G(A(k))$, where $G \colon \mathbb{R}^{r+p+m} \to \mathbb{R}^{r+p+m}$ is defined elementwise:

- Input variables: if $1 \leq i \leq r$, then $A_i = u_i$ (constant input signal for every input channel);
- State variables: if $r + 1 \leq i \leq r + p$, then

$$A_i(k + 1) = f(w_{A_i}x(k) + w_{B_i}u) = f(w_i A(k))$$

- Output variables: if $r + p + 1 \leq i \leq r + p + n$, then

$$A_i(k + 1) = f(w_{C_i}x(k) + w_{D_i}u) = f(w_i A(k))$$

Consequently,

$$A(k+1) = [\underbrace{A_1, \ldots, A_r}_{input}, \underbrace{A_{r+1}, \ldots, A_{r+p}}_{state}, \underbrace{A_{r+p+1}, \ldots, A_{r+p+m}}_{output}]^T$$

$$= [\underbrace{A_1, \ldots, A_r}_{u}, \underbrace{f(w_{A_i}x(k) + w_{B_i}u)}_{r+1 \leq i \leq r+p}, \underbrace{f(w_{C_i}x(k) + w_{D_i}u)}_{r+p+1 \leq i \leq r+p+m}]^T \tag{11}$$

3.1   Convergence Condition for the Input-Output FCM

In this subsection, a sufficient condition for the existence and uniqueness of fixed points of input-output FCMs will be stated. The fixed point is unique in the sense that for a given input, the FCM reaches the same fixed point (activation vector) regardless of the initial values of the other (state and output) concepts.

The condition is based on the Jacobian matrix (matrix of the partial derivatives) of the mapping G generating the iteration and the on the tight upper bound of the derivative of the sigmoid function.

The Jacobian of mapping G is the matrix $J_G(i,j) = \frac{\partial G_i}{\partial A_j}$, namely

– for input variables (u):

$$J_G(i,j) = \begin{cases} 1 \text{ if } i = j \\ 0 \text{ otherwise} \end{cases}$$

– for state variables (x):

$$J_G(i,j) = \begin{cases} \lambda \cdot w_{Bij} f(w_i A)(1 - f(w_i A)) \text{ , if } (i,j) \in (state, input) \\ \lambda \cdot w_{Aij} f(w_i A)(1 - f(w_i A)) \text{ , if } (i,j) \in (state, state) \end{cases}$$

– for output variables (y):

$$J_G(i,j) = \begin{cases} \lambda \cdot w_{Dij} f(w_i A)(1 - f(w_i A)) \text{ , if } (i,j) \in (output, input) \\ \lambda \cdot w_{Cij} f(w_i A)(1 - f(w_i A)) \text{ , if } (i,j) \in (output, state) \end{cases}$$

Since the input terms are constant values (do not change during the iteration), the iteration is convergent if and only if the dynamical terms generate a convergent sequence. The dynamics of this part is governed by the submatrix

$$W' = \left[\begin{array}{c|c} W_B & W_A \\ \hline W_D & W_C \end{array} \right] \tag{12}$$

The Jacobian of G belonging to this submatrix is

$$J'_G = \lambda \cdot \text{diag}[f(w_i A)(1 - f(w_i A))] \cdot W' \tag{13}$$

Since for any A and w_i, $f(w_i A)(1 - f(w_i A)) \leq 1/4$, the spectral radius of the Jacobian at any point:

$$\rho(J'_G) \leq \rho \left(\frac{\lambda}{4} W' \right) = \frac{\lambda}{4} \rho(W') \tag{14}$$

If the spectral radius over the whole space is less than one, then the iteration converges to a unique fixed point. So we can conclude the following theorem:

Theorem 2. *Consider the input-output FCM model described by Eq. 9 with a constant input vector u. Let W' be the matrix constructed by matrices W_A, W_B, W_C and W_D, according to Eq. 12. If*

$$\rho(W') < \frac{4}{\lambda} \tag{15}$$

then the iteration converges to a unique fixed point, regardless of the initial activation values of the state and output concepts.

The spectral radius of a matrix is less than any norm of this matrix, i.e. $\rho(W') \leq \|W'\|$. It means that if we express a condition for convergence to a unique stable equlibrium point using any norm of W', then we get weaker theorem. Nevertheless, in some cases a weaker a condition gives more comprehensible explanation.

Remark 1. We have concluded that if $\rho(W') < \frac{4}{\lambda}$, then the FCM has exactly one fixed point, i.e. the limit of the iteration process is the same, regardless to the initial values of non-input variables. It also means that this fixed point is globally asymptotically stable.

3.2 Further Convergence Conditions

In this subsection, we prove other conditions for the convergence of input-output FCMs. Although these conditions are weaker, they might be useful, since they are directly based on the weight structure of the FCM. First, we recall some definitions about the structure of the network.

Definition 3. *The weighted in-degree of concept C_j equals the sum of the absolute values of the weights of in-coming edges:*

$$deg_j^{in} = \sum_{i=1}^{n} |w_{ij}| \tag{16}$$

which is the sum of the absolute values of the entries of the jth column of W.

Definition 4. *The weighted out-degree of concept C_i equals the sum of the absolute values of the weights of out-going edges:*

$$deg_i^{out} = \sum_{j=1}^{n} |w_{ij}| \tag{17}$$

which is the sum of the absolute values of the entries of the ith row of W.

Although usually not considered graphically as a real edge, but self-feedback means self-loop in the graph. So if self-feedbacks are applied in the iteration, then the weights of the feedback are counted in the in-degree and the out-degree, too. It is the reason that we did not exclude $i = j$ from the summations above.

Definition 5. *The connectivity of an FCM is the ratio of the number of connections between concepts to the maximum number of such possible connections.*

In some sense, connectivity measures the 'density' of the network. If self-feedback is allowed, then the maximum number of connections is n^2, if not, then the maximum number of connections is $n(n-1)$.

Definition 6. *The weighted connectivity of an FCM is the ratio of the sum of absolute values of weights of connections between concepts to the maximum number of such possible connections.*

If self-feedback is allowed, then the weighted connectivity is

$$Con_w = \frac{\sum_{i=1}^{n} \sum_{j=1}^{n} |w_{ij}|}{n^2} \tag{18}$$

If self-feedback is not allowed, then the weighted connectivity is

$$Con_w = \frac{\sum_{i=1}^{n} \sum_{j=1}^{n} |w_{ij}|}{n(n-1)} \tag{19}$$

Theorem 3. *Let λ be the parameter of the sigmoid threshold function applied for every concept. If the maximal in-degree of the FCM (including possible feedback) is less than $4/\lambda$, then the FCM has one and only one fixed point.*

Proof. Using the definition of in-degree:

$$\max_{1 \leq j \leq n} deg_j^{in} = \max_{1 \leq j \leq n} \sum_{i=1}^{n} |w_{ij}| = \|W\|_1 \tag{20}$$

Since $\|W\|_1 \geq \rho(W)$, if $\|W\|_1 < 4/\lambda$, then $\rho(W) < 4/\lambda$, which ensures the convergence to a unique fixed point.

Theorem 4. *Let λ be the parameter of the sigmoid threshold function applied for every concept. If the maximal out-degree of the FCM (including possible feedback) is less than $4/\lambda$, then the FCM has one and only one fixed point.*

Proof. The proof goes similarly to the previous one, but instead of 1-norm we use the infinity norm.

$$\max_{1 \leq i \leq n} deg_i^{out} = \max_{1 \leq i \leq n} \sum_{j=1}^{n} |w_{ij}| = \|W\|_\infty \tag{21}$$

As in the previous case, if $\|W\|_\infty < 4/\lambda$, then $\rho(W) < 4/\lambda$, which ensures the convergence to a unique fixed point.

Theorem 5. *Let λ be the parameter of the sigmoid threshold function applied for every concept. If the weighted connectivity (Con_w) of the FCM small enough, namely*

1. if self-feedback is allowed:

$$Con_w < \frac{4}{\lambda n^2}, \tag{22}$$

2. if self-feedback is not allowed:

$$Con_w < \frac{4}{\lambda n(n-1)}, \tag{23}$$

then the FCM has one and only one fixed point.

Proof. Consider the following entry-wise matrix norm:

$$\sum_{i=1}^{n} \sum_{j=1}^{n} |w_{ij}| \tag{24}$$

(Unfortunately, the usual notation of this norm is $\| * \|_1$, which is confusing, since the 1-norm has the same notation.) We know that

$$\rho(W) \leq \sum_{i=1}^{n} \sum_{j=1}^{n} |w_{ij}| \tag{25}$$

So, if $\sum_{i=1}^{n} \sum_{j=1}^{n} |w_{ij}| < 4/\lambda$, then $\rho(W) < 4/\lambda$.

Consequently, if $\dfrac{\lambda}{4} \sum_{i=1}^{n} \sum_{j=1}^{n} w_{ij} < 1$, then the mapping is a contraction. It means that the iteration converges to a unique fixed point, regardless to the initial value. Rearranging this inequality and division both sides by n^2 (or $n(n-1)$) completes the proof.

The direct practical usability of this result is very limited, since it gives very weak condition. Nevertheless, it has an important mathematical statement: extremely weakly connected fuzzy cognitive maps always produce simple behaviour. Of course, the notion 'weakly' depends on n and λ.

3.3   Different Input - Different Output?

Under certain circumstances, classical FCMs may converge to the same equilibrium state (fixed point) from completely different initial values. This property is advantageous in some applications, for example, it ensures the system's robustness against noise, while it is not useful for example in pattern recognition problems. In this subsection, we examine input-output FCMs from this point of view.

Let us assume that the inputs are u_1 and u_2, and the iteration converges to a fixed point in both cases. Let's denote these fixed points by A_1^* and A_2^*, respectively. According to our assumption, both scenario lead to a steady state, i.e.:

$$A_1^* = [u_1, x_1^*, y_1^*]^T \in \mathbb{R}^{r+p+m}$$
$$A_2^* = [u_2, x_2^*, y_2^*]^T \in \mathbb{R}^{r+p+m} \tag{26}$$

Consequently, the steady state equations hold for $[u_1, x_1^*, y_1^*]^T$ and $[u_2, x_2^*, y_2^*]^T$:

$$\begin{aligned}
x_1^* &= f\left(W_A x_1^* + W_B u_1\right) & x_2^* &= f\left(W_A x_2^* + W_B u_2\right) \\
y_1^* &= f\left(W_C x_1^* + W_D u_1\right) & y_2^* &= f\left(W_C x_2^* + W_D u_2\right)
\end{aligned} \tag{27}$$

Let's assume, that $u_1 \neq u_2$, but $x_1^* = x_2^*$. From the equations and from the monotonicity of f we have

$$W_A x_1^* + W_B u_1 = W_A x_2^* + W_B u_2 \tag{28}$$

Rearranging the equation yields:

$$W_A(x_1^* - x_2^*) = W_B(u_2 - u_1) \tag{29}$$

According to our assumption, the left hand side is zero:

$$0 = W_B(u_2 - u_1) \tag{30}$$

Since $u_1 \neq u_2$, this equality holds if and only if $u_2 - u_1$ lies in the null-space of W_B. If W_B is of full rank, then $\dim Ker W_B = 0$, so every different input value generates different steady-state values ($Ker W_B$ denotes the null-space, a.k.a. kernel of W_B). When $\dim Ker W_B \neq 0$, and $u_1 - u_2 \in Ker W_B$, then u_1 and u_2 generate the same equilibrium state. Else, when $u_1 - u_2 \notin Ker W_B$, they produce different x_1^* and x_2^*. Similar arguments hold for y_1^* and y_2^*. $y_1^* = y_2^*$ implies that $u_2 - u_1$ lies in the null-space of W_D, but there are infinite number of cases when $u_1 - u_2 \notin Ker W_D$, and in these cases $u_1 \neq u_2$ yields $y_1^* \neq y_2^*$.

4   Hybrid Fuzzy Cognitive Maps

An other input-output model has been introduced by Napoles et al. [9] under the name hybrid FCM, with the following more general and highly flexible sigmoid threshold function defined for the ith concept :

$$f_i(x) = l_i + \frac{u_i - l_i}{1 + e^{-\lambda_i(x - h_i)}} \tag{31}$$

The topology of the proposed neural system is comprised of r input neurons and m output neurons, so there are no distinct inner state neurons. The weight matrix W is composed of two submatrices W_I and W_O. The first one contains the connections between the input concepts, while the second one contains the weights connecting the input neurons with the output ones. There are no connections from output neurons to input neurons.

Comparing this model to the previous one, we can observe that

- here is no difference between input and state concepts;
- inputs do not act directly on the output;
- the transfer functions f_i are highly customized to each neuron, ensuring more flexibility in modelling.

$$W = \begin{bmatrix} W_I & 0 \\ \hline W_O & 0 \end{bmatrix} \tag{32}$$

The dynamics of the system is determined by the input part W_I (but don't forget that in this model there is no difference between input and state concepts).

The general term of the Jacobian of the mapping that generates the updating process is the following:

$$J_G(i,j) = \frac{\partial G_i}{\partial A_j} = w_{ij}\lambda_i(u_i - l_i)\frac{1}{1 + e^{-\lambda_i(w_i A - h_i)}}\left(1 - \frac{1}{1 + e^{-\lambda_i(w_i A - h_i)}}\right) \quad (33)$$

With the shorthand $g_i = \frac{1}{1+e^{-\lambda_i(w_i A - h_i)}}$, the Jacobian is

$$J_G = \text{diag}[\lambda_i(u_i - l_i)]\text{diag}[g_i(1 - g_i)] \cdot W \quad (34)$$

Since $g_i(1 - g_i) \leq 1/4$, the following inequality holds for the spectral radius of the Jacobian at any point:

$$\rho(J_G) \leq \frac{1}{4}\rho\left(\text{diag}[\lambda_i(u_i - l_i)] \cdot W\right) \quad (35)$$

Moreover, because of the block structure of W, the spectral radius (largest absolute value of the eigenvalues) of W equals the spectral radius of W_I (it also proves that the dynamics of the system is determined by the input neurons and their weight structure). Consequently,

$$\rho(J_G) \leq \frac{1}{4}\rho\left(\text{diag}[\lambda_i(u_i - l_i)] \cdot W_I\right) \quad (36)$$

Similarly to the previous section, we get the following theorem:

Theorem 6. *Consider an FCM with weight structure described by Eq. 32 and transfer functions defined by Eq. 31. If*

$$\frac{1}{4}\rho\left(\text{diag}[\lambda_i(u_i - l_i)] \cdot W_I\right) < 1, \quad (37)$$

then the FCM has exactly one fixed point. This fixed point is the limit of the iteration from any starting point.

Remark 2. In a special case, when $l_i = -u_i$ and $h_i \equiv 0$, the concept vector $A = [0, \ldots, 0]^T$ is always a fixed point, but not always a fixed point attractor. If the inequality in Theorem 6 holds, then this point is a globally asymptotically stable equilibrium point. On the other hand, when the inequality does not hold, the iteration may lead to this fixed point from certain starting point(s) (these are the elements of $KerW$), but this fixed point is not stable. This problem was discussed for the case of hyperbolic tangent threshold function in [5].

5   Summary

In this paper, the input-output fuzzy cognitive map model has been examined from the viewpoint of unique fixed points. Based on the spectral radius of the weight matrix and with various matrix norms, several convergence conditions have been proved. Although the conditions expressed by matrix norms are weaker, they are might more understandable for the users of FCMs.

Classical FCMs may produce the same output for totally different initial activation values. Although this property is useful in some models, since it means a kind of robustness concerning noise, there are many applications (for example pattern recognition or classification problems), where this is a disadvantageous feature. As we have seen, the input-output model does not have this drawback, it can produce different fixed points for different outputs. On the other hand, there are cases when different input values yield the same output values.

Finally, convergence condition for another type of input-output FCM was introduced, expressed by the spectral radius of the submatrix containing the weight structure between the input neurons.

Acknowledgment. The research presented in this paper was carried out as part of the EFOP-3.6.2-16-2017-00016 project in the framework of the New Széchenyi Plan. The completion of this project is funded by the European Union and co-financed by the European Social Fund.

This research was supported in part by National Research, Development and Innovation Office (NKFIH) K124055.

References

1. Anninou, A.P., Groumpos, P.P., Poulios, P., Gkliatis, I.: A new approach of dynamic fuzzy cognitive knowledge networks in modelling diagnosing process of meniscus injury. IFAC PapersOnLine **50**(1), 5861–5866 (2017). https://doi.org/10.1016/j.ifacol.2017.08.1289

2. Boutalis, Y., Kottas, T.L., Christodoulou, M.: Adaptive estimation of fuzzy cognitive maps with proven stability and parameter convergence. IEEE Trans. Fuzzy Syst. **17**(4), 874–889 (2009). https://doi.org/10.1109/TFUZZ.2009.2017519

3. Groumpos, P.P., Anninou, A.P.: A critical overview of modelling methods and decision support systems for complex dynamic systems. Ann. Fac. Eng. Hunedoara Int. J. Eng. **15**(3), 17–26 (2017)

4. Harmati, I.Á., Hatwágner, M.F., Kóczy, L.T.: On the existence and uniqueness of fixed points of fuzzy cognitive maps. In: Medina, J., et al. (eds.) IPMU 2018. CCIS, vol. 853, pp. 490–500. Springer, Cham (2018). https://doi.org/10.1007/978-3-319-91473-2_42

5. Harmati, I.Á., Kóczy, L.T.: Notes on the dynamics of hyperbolic tangent fuzzy cognitive maps. In: 2019 IEEE International Conference on Fuzzy Systems (FUZZ-IEEE), pp. 1–6. IEEE (2019). https://doi.org/10.1109/FUZZ-IEEE.2019.8858950

6. Horn, R.A., Johnson, C.R.: Matrix Analysis, 2nd edn. Cambridge University Press, New York (2013). https://doi.org/10.1017/9781139020411

7. Knight, C.J., Lloyd, D.J., Penn, A.S.: Linear and sigmoidal fuzzy cognitive maps: an analysis of fixed points. Appl. Soft Comput. **15**, 193–202 (2014). https://doi.org/10.1016/j.asoc.2013.10.030

8. Lee, I.K., Kwon, S.H.: Design of sigmoid activation functions for fuzzy cognitive maps via Lyapunov stability analysis. IEICE Trans. Inf. Syst. **93**(10), 2883–2886 (2010). https://doi.org/10.1587/transinf.E93.D.2883

9. Nápoles, G., Jastrzebska, A., Mosquera, C., Vanhoof, K., Homenda, W.: Deterministic learning of hybrid fuzzy cognitive maps and network reduction approaches. Neural Netw. **124**, 258–268 (2020). https://doi.org/10.1016/j.neunet.2020.01.019

10. Nápoles, G., Papageorgiou, E., Bello, R., Vanhoof, K.: On the convergence of sigmoid fuzzy cognitive maps. Inf. Sci. **349–350**, 154–171 (2016). https://doi.org/10.1016/j.ins.2016.02.040

11. Nápoles, G., Papageorgiou, E., Bello, R., Vanhoof, K.: Learning and convergence of fuzzy cognitive maps used in pattern recognition. Neural Process. Lett. **45**(2), 431–444 (2017). https://doi.org/10.1007/s11063-016-9534-x

12. Papageorgiou, E.I., Salmeron, J.L.: Methods and algorithms for fuzzy cognitive map-based decision support. In: Papageorgiou, E.I. (ed.) Fuzzy Cognitive Maps for Applied Sciences and Engineering (2013). https://doi.org/10.1007/978-3-642-39739-4_1

13. Rudin, W., et al.: Principles of Mathematical Analysis, vol. 3. McGraw-Hill, New York (1964)

14. Stylios, C.D., Groumpos, P.P.: Modeling complex systems using fuzzy cognitive maps. IEEE Trans. Syst. Man Cybern. Part A Syst. Hum. **34**(1), 155–162 (2004). https://doi.org/10.1109/TSMCA.2003.818878

15. Tsadiras, A.K.: Comparing the inference capabilities of binary, trivalent and sigmoid fuzzy cognitive maps. Inf. Sci. **178**(20), 3880–3894 (2008). https://doi.org/10.1016/j.ins.2008.05.015

On the Behavior of Fuzzy Grey Cognitive Maps

Leonardo Concepción[1,2]([✉]), Gonzalo Nápoles[2,3], Rafael Bello[1],
and Koen Vanhoof[2]

[1] Department of Computer Science, Universidad Central de Las Villas,
Santa Clara, Cuba
lcperez@uclv.cu, rbellop@uclv.edu.cu
[2] Faculty of Business Economics, Hasselt Universiteit, Hasselt, Belgium
{gonzalo.napoles,koen.vanhoof}@uhasselt.be
[3] Department of Cognitive Science and Artificial Intelligence, Tilburg University,
Tilburg, The Netherlands

Abstract. Fuzzy Cognitive Maps (FCMs) are recurrent neural networks
made up of well-defined neurons and causal relations. Fuzzy Grey Cognitive Maps (FGCMs) are an extension of FCMs, intended to surpass the
intrinsic uncertainties modeling real-world problems by means of Grey
theory. Despite the rising number of studies about FGCM-based models, little has been investigated with regard to the convergence of such
networks. In this paper, we build a mathematical basis to uncover the
behavior FGCM-based models equipped with transfer F-functions. To
do so, we propose sufficient conditions for the existence and unicity of
fixed-point attractors. Also, the results reported in the literature on the
convergence of FGCMs, are compared with ours. Furthermore, we eluci-
date the reach and depth of our findings, especially and not exclusive to
the prediction of FCMs' behavior.

Keywords: Fuzzy Cognitive Maps · Fuzzy Grey Cognitive Maps ·
Convergence · Grey theory · Shrinking Grey State Vector · Limit grey
state

1 Introduction

Fuzzy Cognitive Maps (FCMs) are knowledge-based recurrent neural networks for
modeling complex systems [5] and an increasing number of FCM scientific articles have been published in the last few years [4,11–14]. Whether using FCMs or
not, the construction of models to face real-world problems always carry in intrinsic uncertainties. Sometimes the vast information contained in a complex system
cannot be represented only by means of crisp values. Grey numbers [20] emerged
as a way to shape these uncertainties and, given that FCMs are not exempt from
these issues, Fuzzy Grey Cognitive Maps (FGCMs) were proposed by [16]. These
networks are convenient for modeling human knowledge in decision-making process. FGCMs are considered a generalization of FCMs, since the latter is an FGCM

© Springer Nature Switzerland AG 2020
R. Bello et al. (Eds.): IJCRS 2020, LNAI 12179, pp. 462–476, 2020.
https://doi.org/10.1007/978-3-030-52705-1_34

with all the causal relations represented by white numbers. Some FGCMs' theoretical breakthroughs emerged in the last decade, as well as applications on time series forecasting [6], surveillance assets coordination [17], reliability engineering [18] and radiotherapy treatment planning [19].

Convergence analysis is the most discussed topic in theoretical studies on the FCM field [4,10–12]. The possible states of the FCMs' inference process are the same as in FGCMs, since both may reach a fixed point or a limit cycle, or exhibit chaotic behavior. Like in FCMs, the convergence in FGCMs is crucial because cycles and chaos make the network responses to be unstable. Unstable cases occur given that activation values always vary through iterations, while remaining stable when a fixed point is reached. While several theoretical studies as well as applications have been conducted using FGCM-based models, convergence issues have been little analyzed [7,8].

In [7], the authors provided some sufficient conditions for the convergence of FGCMs to a unique fixed point, regardless of the initial values of neurons. These conditions are proven for the cases of the log-sigmoid and the hyperbolic tangent threshold functions and also generalized to arbitrary sigmoidal (S-shaped) threshold functions. The convergence conditions are expressed by the elements of the weight matrix and the maximal value of the derivative of the transfer function. Such conditions are applicable to a particular set of the FGCMs' universe, since a small number of neurons and low values for weights are required. Also, the estimation of the maximal value of the derivative could produce loose bounds.

Our previous research on the theoretical analysis of FCM-based models and their dynamics [4] motivated a similar approach to FGCMs. First, we propose novel simplified formulas to calculate the grey raw activation values in FGCMs. Such formulas are the basis to introduce several definitions, lemmas, theorems and corollaries, that allow studying the dynamic behavior of FGCMs equipped with F-functions. The proposed theorems and corollaries give sufficient conditions for the grey state vector to continuously shrink through the inference process and to converge to a so-called limit grey state. After that, we contrast our findings with the achievements depicted in [7] and the employment of FGCMs as predictors for FCMs is proposed.

The rest of this paper is organized as follows. Section 2 goes over the FGCMs' mathematical underpinnings. Section 3 introduces novel formulas for the grey raw activation values, while Sect. 4 enunciates two theorems and their respective corollaries giving conditions on the convergence of FGCM-based models. Section 5 compares our findings with those in the literature. In Sect. 6, we briefly summarize our achievements in this research.

2 Theoretical Background

As mentioned, FGCMs are an extension of traditional FCMs so that both weight and neurons' activation values are described with interval grey numbers.

2.1 Grey Theory

Let Ω be the universal set, then a grey set $\psi \subset \Omega$ is defined by two membership functions $\mu_\psi^- (\cdot) \in [0,1]$ and $\mu_\psi^+ (\cdot) \in [0,1]$ denoting the lower and upper membership functions, respectively, such that $\mu_\psi^- (\cdot) \leq \mu_\psi^+ (\cdot)$. Interval grey numbers with upper (g^+) and lower (g^-) limits are denoted as $g^\pm \in [g^-, g^+] \mid g^- \leq g^+$ [20]. The crisp value of a grey number is unknown, but the range in which the crisp value is located is known [9].

If $g^\pm$ only has an upper limit, it is denoted by $g^\pm \in (-\infty, g^+]$, but if $g^\pm$ only has a lower limit it is denoted by $g^\pm \in [g^-, +\infty)$. A black number with both unknown limits is denoted as $g^\pm \in (-\infty, +\infty)$ and it becomes a white number when both limits have the same value $g^- = g^+$. Although the length of a grey number with only one limit known $(g^\pm \in [g^-, +\infty)$ or $g^\pm \in (-\infty, g^+])$ is infinite, the grey number is not necessarily a black number because it is possible to know one of these limits.

Equations (1a)–(1d) show the grey arithmetic operations according to the common interval algebra:

$$g_a^\pm + g_b^\pm \in \left[g_a^- + g_b^-, g_a^+ + g_b^+ \right] \tag{1a}$$

$$g_a^\pm - g_b^\pm \in \left[g_a^- - g_b^+, g_a^+ - g_b^- \right] \tag{1b}$$

$$g_a^\pm \cdot g_b^\pm \in \left[\min \left\{ g_a^- \cdot g_b^-, g_a^+ \cdot g_b^+, g_a^- \cdot g_b^+, g_a^+ \cdot g_b^- \right\}, \right.$$
$$\left. \max \left\{ g_a^- \cdot g_b^-, g_a^+ \cdot g_b^+, g_a^- \cdot g_b^+, g_a^+ \cdot g_b^- \right\} \right] \tag{1c}$$

$$\frac{g_a^\pm}{g_b^\pm} \in \left[\min \left\{ \frac{g_a^-}{g_b^-}, \frac{g_a^+}{g_b^+}, \frac{g_a^-}{g_b^+}, \frac{g_a^+}{g_b^-} \right\}, \right.$$
$$\left. \max \left\{ \frac{g_a^-}{g_b^-}, \frac{g_a^+}{g_b^+}, \frac{g_a^-}{g_b^+}, \frac{g_a^+}{g_b^-} \right\} \right] \mid g_b^-, g_b^+ \neq 0. \tag{1d}$$

2.2 Fuzzy Grey Cognitive Maps

The mathematical formalism of grey numbers and the recurrent inference mechanism of traditional FCMs are the foundational underpinnings behind FGCM-based models. Roughly speaking, such models can be defined by means of the following 4-tuple:

$$\Theta = \langle C^\pm, A^\pm, W^\pm, f^\pm(\cdot), \ell(\psi) \rangle. \tag{2}$$

where $C^\pm = \{C_1^\pm, \ldots, C_i^\pm, \ldots, C_M^\pm\}$ is the set of M neurons with grey states $A^\pm = \{A_1^\pm, \ldots, A_i^\pm, \ldots, A_M^\pm\}$ such that $A_i^\pm \in [0,1]$, $f^\pm(\cdot)$ represents the grey transfer function and $\ell(\psi)$ is the range of the activation space. The grey weight connecting neurons $C_i^\pm$ and $C_j^\pm$ is denoted by $w_{ij}^\pm \in [-1,1]$ and gathered into the grey weight matrix $W^\pm$, which is defined as follows:

$$W^{\pm} = \begin{bmatrix} w_{11}^{\pm} & \cdots & w_{1i}^{\pm} & \cdots & w_{1M}^{\pm} \\ \vdots & \ddots & \ddots & \ddots & \vdots \\ w_{j1}^{\pm} & \ddots & w_{jj}^{\pm} & \ddots & w_{jM}^{\pm} \\ \vdots & \ddots & \ddots & \ddots & \vdots \\ w_{M1}^{\pm} & \cdots & w_{Mi}^{\pm} & \cdots & w_{MM}^{\pm} \end{bmatrix}.$$

The recurrent reasoning process of FGCMs is devoted to updating neurons' grey activation values given an initial stimulus [15]. Thus, in each iteration t the model produces a grey state vector $A^{\pm(t)} = [A_1^{\pm(t)}, \ldots, A_i^{\pm(t)}, \ldots, A_M^{\pm(t)}]$ containing the grey activation values of all neurons in the model. The first state vector $A^{\pm(0)}$ in this sequence is either provided by experts when performing WHAT-IF simulations or derived automatically from data. This iterative procedures is given below:

$$A_i^{\pm(t+1)} = f_i^{\pm} \left(\sum_{j=1}^{M} w_{ji}^{\pm} \cdot A_j^{\pm(t)} \right) \tag{3}$$

such that

$$f_i^{\pm}(\bar{A}_i^{\pm(t+1)}) = [f_i(\bar{A}_i^{-(t+1)}), f_i(\bar{A}_i^{+(t+1)})] \tag{4}$$

where $\bar{A}_i^{-(t+1)}$ and $\bar{A}_i^{+(t+1)}$ represents the neuron's lower and upper grey raw activation values, respectively, while $f_i(.)$ denotes the transfer function. Notice that self-feedback is allowed, given that the expert can avoid this by explicitly setting $w_{jj}^{\pm} = [0, 0]$.

The FGCMs' inference process finishes when the stability appears or a maximal number of iterations is reached. After the inference process, the FGCM either settles down to a fixed pattern of activation values (grey fixed-point attractor), keep cycling between several fixed grey states (limit grey cycle) or behaves chaotically (grey chaotic attractor) [16]. The last state occurs when, instead of stabilizing, the FGCM continues to produce different results for each iteration. Also, this state is only attainable with a continuous activation function.

The most widely used transfer functions are [3] the *sigmoid* function and the *hyperbolic tangent*. The *bivalent*, *trivalent* and *threshold* functions have also been employed. The former have continuous open intervals as their image set, while the latter are bounded into closed intervals instead (they also have discrete image set). Generally speaking, any bounded and monotonically increasing function over the set of real numbers is a candidate transfer function, since the image set of a bounded function belongs to an interval.

Let F be the set of all monotonically increasing functions bounded into non-negative intervals. Let $F^0 \subset F$ and $F' \subset F$ be the subsets bounded into open intervals and closed intervals respectively. Also, let $f_i \in F$ be the transfer function used in the activation process of neuron $C_i^{\pm}$ (i.e., every neuron has its own transfer function). This means that f_i is bounded into a non-negative interval (either open or closed). Observe that the hyperbolic tangent and the trivalent functions do not belong to F. In this paper, we refer to an F-function as any function belonging to F. It should be highlighted that Eq. (4) holds because f_i is an F-function and thus monotonically increasing. Given that $A_i^{\pm} \in [0,1]$, in this paper we assume that the image set of F-functions belongs to $[0,1]$.

3   Raw Activation Values of Grey Neurons

Equations (1a)–(1d) provide tools for operating with grey values and they involve finding the extreme values of sets. When applied to the inference mechanism of FGCMs, the latter equations can be used in simpler ways given the intrinsic properties of these networks. The grey raw activation value for the neuron $C_i^{\pm}$ at the $(t+1)$-th iteration is

$$\bar{A}_i^{\pm(t+1)} = \sum_{j=1}^{M} w_{ji}^{\pm} \cdot A_j^{\pm(t)}. \tag{5}$$

Equation (5) shows the summation of M terms, where each term is the product of two grey values. Equation (1c) is employed to solve such product, where we would have a set with four elements and we would select the minimum and the maximum among them. The above summation has M terms, so we would repeat the procedure in Eq. (1c) M times and then, we would add up the results using Eq. (1a). For a computer it is not a hard procedure, but using the restrictions of the grey numbers used in Eq. (5), we propose a formula which speeds up the computations while yielding a simplified representation for $\bar{A}_i^{\pm(t+1)}$. Also, this formula avoids explicitly finding minimum and maximum values of sets.

The raw activation value can also be represented by following grey dot product:

$$w_i^{\pm}.A^{\pm(t)} = \sum_{j=1}^{M} w_{ji}^{\pm} \cdot A_j^{\pm(t)}. \tag{6}$$

As derived from Eq. (3), the grey dot product between $w_i^{\pm}$ and $A^{\pm(t)}$ is considered for every neuron $C_i^{\pm}$ in order to compute its activation value. In this research, we assume that each neuron is influenced by, at least, another neural processing entity. In the case of input neurons, their activation values either remain unchanged or become inactive (depending on the FGCM implementation). Whichever the case, their values are easy to predict. Based on Eq. (1c), but using the fact that $A_i^{\pm} \subseteq [0,1]$ and $w_{ji}^{\pm} \subseteq [-1,1]$, we propose new formulas to calculate the bounds for the grey dot product between $w_i^{\pm}$ and $A^{\pm(t)}$. The latter product is the grey raw activation value used to calculate the next activation value for the neuron $C_i^{\pm}$.

It is known that $A_j^{\pm(t)} = [A_j^{-(t)}, A_j^{+(t)}]$ denotes the grey number associated to the j-th neuron at the t-th iteration. Then, the lower limit of $w_i^{\pm}.A^{\pm(t)}$, denoted by $\bar{A}_i^{-(t+1)}$, is:

$$\sum_{j=1}^{M} \frac{w_{ji}^- \left(A_j^{+(t)}(1 - sgn(w_{ji}^-)) + A_j^{-(t)}(1 + sgn(w_{ji}^-))\right)}{2} \tag{7}$$

and the upper limit, denoted by $\bar{A}_i^{+(t+1)}$, is

$$\sum_{j=1}^{M} \frac{w_{ji}^+ \left(A_j^{-(t)}(1 - sgn(w_{ji}^+)) + A_j^{+(t)}(1 + sgn(w_{ji}^+))\right)}{2} \tag{8}$$

where $sgn(.)$ is the sign function

Proof. Previous equations are a compact representation based on the following ideas. Analyzing Eq. (6) we realize that we need to focus on the grey product $w_{ji}^\pm \cdot A_j^{\pm(t)}$ $\forall j$ to find the lower limit depicted in (7). As we know, the grey limits for $A_j^{\pm(t)}$ are non-negative while the grey limits for $w_{ji}^\pm$ could be either positive, negative or zero. According to the sign of $w_{ji}^\pm$, two cases emerge in order to find the lower limit of $w_{ji}^\pm \cdot A_j^{\pm(t)}$:

- IF $w_{ji}^- < 0$ THEN $(w_{ji}^\pm \cdot A_j^{\pm(t)})^- = w_{ji}^- A_j^{+(t)}$
- IF $w_{ji}^- \geq 0$ THEN $(w_{ji}^\pm \cdot A_j^{\pm(t)})^- = w_{ji}^- A_j^{-(t)}$

By applying the aforementioned reasoning and using the sign function ($sgn(.)$) we assemble both cases within a single formula without ramifications. This results in the lower limit shown at (7).

Analogously, to derive Eq. (8) we analyze the upper limit of the grey product $w_{ji}^\pm \cdot A_j^{\pm(t)}$ $\forall j$. Two cases arise:

- IF $w_{ji}^+ < 0$ THEN $(w_{ji}^\pm \cdot A_j^{\pm(t)})^+ = w_{ji}^+ A_j^{-(t)}$
- IF $w_{ji}^+ \geq 0$ THEN $(w_{ji}^\pm \cdot A_j^{\pm(t)})^+ = w_{ji}^+ A_j^{+(t)}$

From this point, the upper limit shown at (8) is derived in the same way as the lower limit.   □

4   Studying Convergence in FGCMs

In this section, we provide sufficient conditions for the existence and uniqueness of grey fixed-point attractors of FGCMs equipped with transfer F-functions. To present the first theorem about the FGCMs' behavior, we need to define when a grey value $a^\pm$ contains another $b^\pm$.

Definition 4.1. *The grey value $a^\pm$ contains the grey value $b^\pm$ if $a^- \leq b^-$ and $a^+ \geq b^+$.*

In other words, the interval where the crisp value of $a^\pm$ lies, contains the interval where the crisp value of $b^\pm$ is located.

Definition 4.2. *The grey vector $A^\pm$ contains the grey vector $B^\pm$ if they have the same length and every grey value from $A^\pm$ contains the corresponding grey value in $B^\pm$.*

Now, we are in conditions to introduce the theorem. This theorem asserts that if some grey state vector contains the next one, then every successive state vector contains the following. Nevertheless, it is possible that $A^{\pm(t)} = A^{\pm(t+1)}$, which implies that $A^{\pm(t)} = A^{\pm(t+k)} \ \forall k \in \mathbb{N}$. So, the grey state vectors may not shrink forever.

Theorem 4.1 (Weak Shrinking Grey State Vector). *In an FGCM Θ, $A^{\pm(t)}$ contains $A^{\pm(t+1)}$ $\forall t > t_0 : t_0 \in \mathbb{N}$, if $A^{\pm(t_0)}$ contains $A^{\pm(t_0+1)}$ with $f_i \in F \ \forall i \in \{1, 2, \ldots, M\}$.*

Proof. Let $A^{\pm(t_0)} = \{A_1^{\pm(t_0)}, \ldots, A_M^{\pm(t_0)}\}$, $A^{\pm(t_0+1)} = \{A_1^{\pm(t_0+1)}, \ldots, A_M^{\pm(t_0+1)}\}$ and $A^{\pm(t_0+2)} = \{A_1^{\pm(t_0+2)}, \ldots, A_M^{\pm(t_0+2)}\}$.

We must demonstrate that $A_i^{\pm(t_0+1)}$ contains $A_i^{\pm(t_0+2)}$ for every $i = 1, 2, \ldots, M$, to prove that $A^{\pm(t_0+1)}$ contains $A^{\pm(t_0+2)}$. The fact that $A^{\pm(t_0)}$ contains $A^{\pm(t_0+1)}$ means that $A_i^{\pm(t_0)}$ contains $A_i^{\pm(t_0+1)}$ for every $i = 1, \ldots, M$. Based on this knowledge, we will prove that $A_i^{\pm(t_0+1)}$ contains $A_i^{\pm(t_0+2)}$ for every $i = 1, \ldots, M$.

According to Eqs. 3 and 4, we have that $A_i^{\pm(t_0+1)} = f_i^\pm(\bar{A}_i^{\pm(t_0+1)})$ and $A_i^{\pm(t_0+2)} = f_i^\pm(\bar{A}_i^{\pm(t_0+2)}) \ \forall i$. Given that f_i is monotonically increasing, it suffices to prove that $\bar{A}_i^{\pm(t_0+1)}$ contains $\bar{A}_i^{\pm(t_0+2)} \ \forall i$. Based on Definition 4.1 we need to prove two inequalities.

- **Inequality 1:** $\bar{A}_i^{-(t_0+1)} \leq \bar{A}_i^{-(t_0+2)}$ Formula (7) leads us to prove that:

$$\sum_{j=1}^{M} \frac{w_{ji}^-\left(A_j^{+(t_0)}(1 - sgn(w_{ji}^-)) + A_j^{-(t_0)}(1 + sgn(w_{ji}^-))\right)}{2}$$

$$\leq \sum_{j=1}^{M} \frac{w_{ji}^-\left(A_j^{+(t_0+1)}(1 - sgn(w_{ji}^-)) + A_j^{-(t_0+1)}(1 + sgn(w_{ji}^-))\right)}{2}$$

It is sufficient to prove that, for every j:

$$w_{ji}^-\left(A_j^{+(t_0)}(1 - sgn(w_{ji}^-)) + A_j^{-(t_0)}(1 + sgn(w_{ji}^-))\right)$$
$$\leq w_{ji}^-\left(A_j^{+(t_0+1)}(1 - sgn(w_{ji}^-)) + A_j^{-(t_0+1)}(1 + sgn(w_{ji}^-))\right)$$

There are three possible scenarios depending on the sign of w_{ji}^-:

- **Scenario 1.** If $sgn(w_{ji}^-) = -1$ then

$$A_j^{+(t_0+1)}(1 - sgn(w_{ji}^-)) \leq A_j^{+(t_0)}(1 - sgn(w_{ji}^-))$$

$$A_j^{+(t_0)} \leq A_j^{+(t_0+1)}$$

which is true because the condition in Theorem 4.1 saying that $A^{\pm(t_0)}$ contains $A^{\pm(t_0+1)}$.

- **Scenario 2.** If $sgn(w_{ji}^-) = 1$ then

$$A_j^{-(t_0)}(1 + sgn(w_{ji}^-)) \leq A_j^{-(t_0+1)}(1 + sgn(w_{ji}^-))$$

$$A_j^{-(t_0)} \leq A_j^{-(t_0+1)}$$

which is true because the aforementioned condition in Theorem 4.1.

- **Scenario 3.** If $sgn(w_{ji}^-) = 0$ then the inequality holds because both sides of the inequality are zero.

– **Inequality 2:** $\bar{A}_i^{+(t_0+1)} \geq \bar{A}_i^{+(t_0+2)}$

Formula (8) leads us to prove that:

$$\sum_{j=1}^{M} \frac{w_{ji}^+ \left(A_j^{-(t_0)}(1 - sgn(w_{ji}^+)) + A_j^{+(t_0)}(1 + sgn(w_{ji}^+)) \right)}{2}$$

$$\geq \sum_{j=1}^{M} \frac{w_{ji}^+ \left(A_j^{-(t_0+1)}(1 - sgn(w_{ji}^+)) + A_j^{+(t_0+1)}(1 + sgn(w_{ji}^+)) \right)}{2}$$

It is sufficient to prove that, for every j:

$$w_{ji}^+ \left(A_j^{-(t_0)}(1 - sgn(w_{ji}^+)) + A_j^{+(t_0)}(1 + sgn(w_{ji}^+)) \right)$$

$$\geq w_{ji}^+ \left(A_j^{-(t_0+1)}(1 - sgn(w_{ji}^+)) + A_j^{+(t_0+1)}(1 + sgn(w_{ji}^+)) \right)$$

Again, three possible scenarios arise, depending on the sign of w_{ji}^+:

- **Scenario 1.** If $sgn(w_{ji}^+) = -1$ then

$$A_j^{-(t_0+1)}(1 - sgn(w_{ji}^+)) \geq A_j^{-(t_0)}(1 - sgn(w_{ji}^+))$$

$$A_j^{-(t_0+1)} \geq A_j^{-(t_0)}$$

which is true because the condition in Theorem 4.1 saying that $A^{\pm(t_0)}$ contains $A^{\pm(t_0+1)}$.

- **Scenario 2.** If $sgn(w_{ji}^+) = 1$ then

$$A_j^{+(t_0)}(1 + sgn(w_{ji}^+)) \geq A_j^{+(t_0+1)}(1 + sgn(w_{ji}^+))$$

$$A_j^{+(t_0)} \geq A_j^{+(t_0+1)}$$

which is true because the aforementioned condition in Theorem 4.1.

- **Scenario 3.** If $sgn(w_{ji}^+) = 0$ then the inequality holds because both sides of the inequality are zero.

At this point, we have demonstrated that $A_i^{\pm(t_0+1)}$ contains $A_i^{\pm(t_0+2)}$, from $A_i^{\pm(t_0)}$ contains $A_i^{\pm(t_0+1)}$. Based on this and proceeding inductively, we can prove that $A_i^{\pm(t_0+2)}$ contains $A_i^{\pm(t_0+3)}$ and so on. Then we confirm that $A^{\pm(t)}$ contains $A^{\pm(t+1)}$, $\forall t > t_0, t_0 \in \mathbb{N}$ if $A^{\pm(t_0)}$ contains $A^{\pm(t_0+1)}$, having $f_i \in F \ \forall i \in \{1, 2, \ldots, M\}$. □

As presented in Sect. 2, F-functions are bounded into either open or closed intervals. Therefore, we can associate a grey value with this kind of intervals.

Definition 4.3. *The induced grey value of neuron $C_i^\pm$ is such that its lower and upper limits are the same as the lower and upper limits of the interval that the f_i transfer function is bounded to.*

Note: It should be highlighted that function f_i is associated to neuron $C_i^\pm$.
A consequence of this definition is the following lemma:

Lemma 4.1. *The induced grey value of neuron $C_i^\pm$ contains $A_i^{\pm(t)} \ \forall t \in \mathbb{N} : t > 0$, which is the grey activation value of neuron $C_i^\pm$.*

Proof. The f_i transfer function is bounded to an interval containing $A_i^{\pm t} \ \forall t \in \mathbb{N} : t > 0$. This happens because the activation values are generated by f_i. When $t = 0$, the experts might decide to assign values out of this interval, but when the inference is triggered, the following activation values will meet the interval restriction. The lemma holds based on Definition 4.3, since the induced grey value of neuron $C_i^\pm$ has the same limits as the interval mentioned in such definition. □

Definition 4.4. *The induced grey vector for the FGCM Θ is such that the i-th component is the induced grey value for the neuron $C_i^\pm$.*

The following lemma extends Lemma 4.1 to the grey vector space. This lemma is the basis of a new corollary that will be defined next.

Lemma 4.2. *The induced grey vector for the FGCM Θ contains $A^{\pm(t)} \ \forall t \in \mathbb{N} : t > 0$, which is the grey state vector of the FGCM at t-th iteration.*

Proof. The demonstration results by applying Lemma 4.1 for every activation value in $A^{\pm t}$ and also for every $t > 0$. □

The following corollary shows that the first state vector $A^{\pm(0)}$ plays a major role in the FGCM's behavior.

Corollary 4.1.1. *In an FGCM Θ, $A^{\pm(t)}$ contains $A^{\pm(t+1)} \ \forall t \in \mathbb{N}$, if $A^{\pm(0)}$ is the induced grey vector of the FGCM, with $f_i \in F \ \forall i \in \{1, 2, \ldots, M\}$.*

Proof. We only need to prove that $A^{\pm(0)}$ contains $A^{\pm(1)}$. According to Lemma 4.2, the induced grey vector contains $A^{\pm(1)}$. Having that $A^{\pm(0)}$ is the induced grey vector, the demonstration is completed.   □

In order to affirm that grey state vectors will shrink forever and $A^{\pm(t)} = A^{\pm(t+k)}$ is not possible, for any $t, k \in \mathbb{N}$, we need to define when a grey value $a^{\pm}$ strictly contains $b^{\pm}$.

Definition 4.5. *The grey value $a^{\pm}$ strictly contains the grey value $b^{\pm}$ if $a^- < b^-$ and $a^+ > b^+$.*

Definition 4.6. *The grey vector $A^{\pm}$ strictly contains the grey vector $B^{\pm}$ if they have the same length and every grey value from $A^{\pm}$ strictly contains the corresponding grey value in $B^{\pm}$.*

Similarly to Theorem 4.1, the next theorem declares that if some grey state vector strictly contains the next one, then every successive state vector strictly contains the following. Thus, it is not possible that $A^{\pm(t)} = A^{\pm(t+1)}$, which implies that $A^{\pm(t)} \neq A^{\pm(t+k)} \; \forall k \in \mathbb{N}$. Hence, grey state vectors will shrink forever.

Theorem 4.2 (Strong Shrinking Grey State Vector). *In an FGCM Θ, $A^{\pm(t)}$ strictly contains $A^{\pm(t+1)} \; \forall t > t_0 : t_0 \in \mathbb{N}$, if $A^{\pm(t_0)}$ strictly contains $A^{\pm(t_0+1)}$ with $f_i \in F \; \forall i \in \{1, 2, \ldots, M\}$.*

Proof. The proof is analogous to the weak version of the theorem, except that all inequalities are turned into strict ones. This means that every occurrence of the $\leq$ and $\geq$ symbols is replaced with the $<$ and $>$ symbols, respectively. Still, the scenario where $sgn(w_{ji}^-) = 0$ needs special attention. When this happens, both sides of inequality are equal to zero. If it were true for a fixed i and every j, then we would obtain that the grey activation value for neuron $C_i^{\pm}$ remains constant through iterations of the FGCM. This would imply that any grey state vector does not strictly contain the next one. It turns out that this situation is impossible given that, as we explained in Sect. 3, each neuron is influenced by at least another neural processing entity. Thus, for every neuron $C_i^{\pm}$, there must be at least a non-zero incoming connection. Therefore, the strong version of the theorem is true.   □

To define a corollary for Theorem 4.2 we must go over F-functions and specially its subset F^0, which is a subset of F where functions are bounded into open intervals. Also, Definitions 4.4, 4.5 and 4.6 will be useful.

Lemma 4.3. *The induced grey value of neuron $C_i^{\pm}$ strictly contains $A_i^{\pm t} \; \forall t \in \mathbb{N} : t > 0$ if $f_i \in F^0 \; \forall i \in \{1, 2, \ldots, M\}$.*

Proof. Transfer functions are bounded into open intervals, which means that grey activation values will always be contained into a grey value with lower and upper limits matching the interval's limits. Such grey value is the induced grey value and this concludes the proof. Being more explicit, let us take an example whit the sigmoid function $f(x) = 1/(1 + e^{-\lambda(x-h)})$, which belongs to F^0 because its image set is $(0, 1)$. In this case, the grey activation values of every neuron will take values between 0 and 1, but not inclusive. Also, according to Definition 4.3, the lower and upper limits of the induced grey value of any neuron are 0 and 1, respectively. Therefore, the lemma holds.   □

To define a corollary for Theorem 4.2, the following lemma extends Lemma 4.3 to the grey vector space.

Lemma 4.4. *The induced grey vector for the FGCM Θ strictly contains $A^{\pm(t)}$ $\forall t \in \mathbb{N} : t > 0$ if $f_i \in F^0$ $\forall i \in \{1, 2, \dots, M\}$.*

Proof. The demonstration results by applying Lemma 4.3 for every activation value in $A^{\pm(t)}$ and also for every $t > 0$.   □

Again, the first state vector $A^{\pm(0)}$ plays a major role in the FGCM's behavior, together with the fact that the transfer function belongs to F^0.

Corollary 4.2.1. *In an FGCM Θ, $A^{\pm(t)}$ strictly contains $A^{\pm(t+1)}$ $\forall t \in \mathbb{N}$, if $A^{\pm(0)}$ is the induced grey vector of the FGCM, with $f_i \in F^0$ $\forall i \in \{1, 2, \dots, M\}$.*

Proof. We only need to prove that $A^{\pm(0)}$ strictly contains $A^{\pm(1)}$. According to Lemma 4.4, the induced grey vector contains $A^{\pm(1)}$. Having that $A^{\pm(0)}$ is the induced grey vector, the demonstration is completed.   □

The above results lead to the question of whether the grey state vectors will shrink until they become a vector of white numbers or not. Becoming a white number would imply that, for every activation value, both grey limits have the same value. This would indicate that every FGCM converges to a white fixed-point attractor. This situation is false, as we know by other studies [4]. Such concerns serve as a motivation to define the limit grey state.

Let $\mathcal{G}$ be the set of all grey values with non-negative limits and let $\mathcal{S}^M$ be the set of all M-ary Cartesian products over the elements in $\mathcal{G}$. Formally, $\mathcal{S}^M = \{\mathcal{I}_1 \times \mathcal{I}_2 \times \dots \times \mathcal{I}_M : \mathcal{I}_i \in \mathcal{G}, \forall i = 1, 2, \dots, M\}$. Grey state vectors of an FGCM with M neurons, belong to $\mathcal{S}^M$.

Definition 4.7. $A^{\pm(\infty)} \in \mathcal{S}^M$ is the **limit grey state** of Θ, such that $A^{\pm(\infty)} = \lim_{t \to \infty} A^{\pm(t)}$.

Theorem 4.3. *The limit grey state $A^{\pm(\infty)}$ of FGCM Θ always exists if Θ fulfills the premises in Theorems 4.1 or 4.2.*

Proof. The shrinkage of the state vectors implies a contraction of the activation values. Also, from a mathematical point of view, iterative grey state vectors (e.g., $A^{\pm(0)}, A^{\pm(1)}, \dots, A^{\pm(t)}, \dots$) are a sequence of elements over $\mathcal{S}^M$ and iterative grey activation values of a neuron are a sequence of elements over $\mathcal{G}$. A sequence over $\mathcal{G}$ can be interpreted as two other sequences: the sequence of lower bounds and the sequence of upper bounds of the iterative grey activation values (both sequences are defined over the set of real numbers). We say that a sequence over $\mathcal{G}$ is *convergent*, if the sequences of lower and upper bounds are also convergent.

Theorems 4.1 and 4.2 imply that grey activation values associated with neurons become smaller from one iteration to the following, meaning that the lower bound becomes bigger and the upper bound becomes smaller. This suggests that the lower bounds sequence increases and the upper bounds sequence decreases. Besides, both sequences are bounded from each other, so the lower bounds sequence is a lower bound for the higher bound sequence and vice versa. Thus, both sequences are convergent because the *monotone convergence theorem* [1] and have a limit (in the extreme case, the limit is a closed interval with identical lower and upper bounds). Now, we have that the sequence of grey activation values is convergent, which implies the convergence of the sequence of grey state vectors. At this point, there is no doubt about the existence and unicity of a limit for iterative feasible state spaces. □

As long as the FGCM is equipped with transfer F-functions and any activation vector contains (or strictly contains) the next one, it will converge to a grey fixed-point attractor, the so-called limit grey state. Corollaries 4.1.1 and 4.2.1 are useful when the initial stimulus is unknown or the experts want to analyze the FGCM's behavior under full uncertainty to arrive at conclusions.

5   Comparison with Previous Convergence Results

In this section we discuss some results reported in the literature, specifically in [7]. Then we explain the reach and depth of our theoretical findings while we compare them with the latter results. Finally, we show the broad applicability of our results, particularly predicting the dynamic behavior of FCMs.

5.1   Literature Results

As mentioned earlier, in [7], conditions for the existence and uniqueness of fixed points of FGCMs are presented. The authors expanded and corrected the ideas presented by [2], who firstly addressed the convergence issues of FCMs from a mathematical perspective. Former conditions are expressed by matrix norms and applying the contraction mapping theorem [2] with suitable distance metrics. To do so, authors build on strict analytical bounds for the derivatives of the transfer functions. These functions may be the log-sigmoid threshold function, the hyperbolic tangent and, in general, the sigmoid-like (S-shaped) threshold function. Also, they assume that the human expert or the training process assigns the proper signs to the grey weights, so a weight is either non-positive or non-negative. This means that the type of relationship (direct or inverse) between the neurons is properly described by the FGCMs.

These sufficient conditions are summarized in Theorem 9 in [7]. This theorem is based upon some definitions that we are going to present briefly.

- W^* is the matrix where every element w_{ij}^* is the maximum between the absolute values of the lower and upper limits of the grey weight $w_{ij}^\pm$.
- The 1-norm of W^* is given by: $||W^*||_1 = \max_{1 \le j \le M} \left(\sum_{i=1}^{M} w_{ij}^* \right)$
- The ∞-norm of W^* is given by: $||W^*||_\infty = \max_{1 \le i \le M} \left(\sum_{j=1}^{M} w_{ij}^* \right)$
- The Frobenius norm of W^* is given by: $||W^*||_F = \left(\sum_{i=1}^{M} \sum_{j=1}^{M} (w_{ij}^*)^2 \right)^{\frac{1}{2}}$
- Let K be the maximal value of $f'(x)$, where f is an S-shaped transfer function. This real function is bounded, monotone increasing and continuously differentiable defined for all real values.

Essentially, Theorem 9 states that the FGCM has one and only one grey fixed point, regardless of the initial values of neurons, if at least one of the inequalities $||W^*||_1 < \frac{1}{K}$, $||W^*||_\infty < \frac{1}{K}$ or $||W^*||_F < \frac{1}{K}$ holds.

In the particular case of the log-sigmoid threshold function (the most widely used), taking into account its common set of parameters, the values of K are always higher than 1. So, it would be needed at least one norm (among the three) lower than 1. If we take a closer inspection on the norms' definitions, we notice that fulfilling this condition would need, we need FGCMs with a small number of neurons and with low values for the influences. In this case, if there exist only one grey weight whose upper limit is equal to 1, then no norm fulfills the inequality. In conclusion, the theorem conditions are quite restrictive to a particular set of the FGCMs' universe.

5.2   Our Findings

First, let us compare our definition of FGCM with the one reported in [7]. Two major differences appear:

- They assume that both limits of the grey weight connecting neurons are either non-positive or non-negative.
- Their transfer functions are S-shaped, bounded, monotone increasing and continuously differentiable.
- They use a single transfer function for the whole FGCM.

We do not restrict weights beyond the property $w_{ij}^\pm \in [-1, 1]$. A specific weight could normally be $w_{ij}^\pm = [-0.5, 0.5]$, which contradicts the assumption found in [7]. On the other hand, we allow transfer functions to be any bounded and monotonically increasing function, as long as its image set lies into $[0, 1]$ (intrinsic restriction in our FGCMs). We do not restrict to have a single transfer function in the model, since each neuron has one of these functions and the parameters may vary among them. Moreover, the estimation of K (the maximal value of $f'(x)$) may result hard working with more complex functions or loose bounds could be obtained.

By means of Theorems 4.1 and 4.2, we ensure the shrinkage of successive grey state vectors, where the only condition that must hold is that some state vector contains (or strictly contains) the next one. No restrictions are required for the weights' matrix or the number of transfer functions or the functions' parameter set. Beyond the shrinkage, we also prove the convergence to a unique grey fixed point, the so-called grey limit state.

Corollaries 4.1.1 and 4.2.1 lead us to a particular application for the FGCMs: the prediction of the state space of FCMs. When the first grey state vector is the induced grey vector, our results match the findings in [4]. In that paper we showed that, approximating the state space of an FCMs is useful to predict fixed-point attractors and to find hidden behavior even in unstable FCMs, without running the inference mechanism. Hence, FGCMs serve as predictors for FCMs that can notify human experts about the limitations of the FCM-model with no computational burden caused by the inference process.

6   Concluding Remarks

In this paper, motivated by our findings regarding the dynamic behavior of FCMs [4], we have introduced a set of mathematical entities (i.e., definitions, lemmas, theorems, corollaries) to uncover the behavior of FGCMs equipped with transfer F-functions. The research conducted in [7] is quite similar to ours, in the sense that our goals are the existence and uniqueness of grey fixed-point attractors. However, we use a completely different approach producing less restrictive rules for the existence and unicity of FGCMs. Furthermore, our definition of such models is wider when we refer to the influences among neurons, the usage of a transfer function per neuron and the parameters of these functions. Our research presents more flexible models, while brings forward less restrictive sufficient conditions in order to fulfill the convergence goal.

Theorems 4.1 and 4.2 ensure the shrinkage of the grey state vectors through the inference process, for FGCMs equipped with transfer F-functions. In case the conditions are met, the inference process leads to the grey limit state. This knowledge could be used by experts or maybe in the learning procedure, to ensure the convergence for this neural networks. Finally, we prove that FGCMs are predictors for FCMs' state spaces, disclosing hidden behavior with no need to trigger the inference mechanism.

References

1. Binmore, K.G.: Mathematical Analysis: A Straightforward Approach. Cambridge University Press, Cambridge (1977)
2. Boutalis, Y., Kottas, T.L., Christodoulou, M.C.: Adaptive estimation of fuzzy cognitive maps with proven stability and parameter convergence. IEEE Trans. Fuzzy Syst. **17**, 874–889 (2009)
3. Bueno, S., Salmeron, J.L.: Benchmarking main activation functions in fuzzy cognitive maps. Expert Syst. Appl. **36**(3), 5221–5229 (2009)
4. Concepción, L., Nápoles, G., Falcon, R., Bello, R., Vanhoof, K.: Unveiling the dynamic behavior of fuzzy cognitive maps. IEEE Trans. Fuzzy Syst. (2020). https://doi.org/10.1109/TFUZZ.2020.2973853

5. Felix, G., Nápoles, G., Falcon, R., Froelich, W., Vanhoof, K., Bello, R.: A review on methods and software for fuzzy cognitive maps. Artif. Intell. Rev. **52**(3), 1707–1737 (2017)
6. Froelich, W., Salmeron, J.L.: Evolutionary learning of fuzzy grey cognitive maps for the forecasting of multivariate, interval-valued time series. Int. J. Approx. Reason. **55**(5), 1319–1335 (2014)
7. Harmati, I.Á., Kóczy, L.T.: On the convergence of sigmoidal fuzzy grey cognitive maps. Int. J. Appl. Math. Comput. Sci. **29**, 453–466 (2019)
8. Harmati, I.Á., Kóczy, L.T.: On the convergence of fuzzy grey cognitive maps. In: Kulczycki, P., Kacprzyk, J., Kóczy, L.T., Mesiar, R., Wisniewski, R. (eds.) ITSRCP 2018. AISC, vol. 945, pp. 74–84. Springer, Cham (2020). https://doi.org/10.1007/978-3-030-18058-4_6
9. Nápoles, G., Salmeron, J.L., Vanhoof, K.: Construction and supervised learning of long-term grey cognitive networks. IEEE Trans. Cybern. 1–10 (2019)
10. Nápoles, G., Bello, R., Vanhoof, K.: How to improve the convergence on sigmoid fuzzy cognitive maps? Intell. Data Anal. **18**(6S), S77–S88 (2014)
11. Nápoles, G., Concepción, L., Falcon, R., Bello, R., Vanhoof, K.: On the accuracy-convergence tradeoff in sigmoid fuzzy cognitive maps. IEEE Trans. Fuzzy Syst. **26**(4), 2479–2484 (2018)
12. Nápoles, G., Papageorgiou, E., Bello, R., Vanhoof, K.: On the convergence of sigmoid fuzzy cognitive maps. Inf. Sci. **350**, 154–171 (2016)
13. Nápoles, G., Papageorgiou, E., Bello, R., Vanhoof, K.: Learning and convergence of fuzzy cognitive maps used in pattern recognition. Neural Process. Lett. **45**, 431–444 (2017)
14. Pedrycz, W., Jastrzebska, A., Homenda, W.: Design of fuzzy cognitive maps for modeling time series. IEEE Trans. Fuzzy Syst. **24**, 120–130 (2016)
15. Salmeron, J., Palos-Sanchez, P.: Uncertainty propagation in fuzzy grey cognitive maps with hebbian-like learning algorithms. IEEE Trans. Cybern. **49**, 211–220 (2019)
16. Salmeron, J.L.: Modelling grey uncertainty with fuzzy grey cognitive maps. Expert Syst. Appl. **37**(12), 7581–7588 (2010)
17. Salmeron, J.L.: An autonomous FGCM-based system for surveillance assets coordination. J. Grey Syst. **28**(1), 27–35 (2016)
18. Salmeron, J.L., Gutierrez, E.: Fuzzy grey cognitive maps in reliability engineering. Appl. Soft Comput. J. **12**(12), 3817–3823 (2012)
19. Salmeron, J.L., Papageorgiou, E.I.: A fuzzy grey cognitive maps-based decision support system for radiotherapy treatment planning. Knowl.-Based Syst. **30**(1), 151–160 (2012)
20. Yang, Y., John, R.: Grey sets and greyness. Inf. Sci. **185**(1), 249–264 (2012)

A Privacy-Preserving, Distributed and Cooperative FCM-Based Learning Approach for Cancer Research

Jose L. Salmeron[1,2(✉)] and Irina Arévalo[1]

[1] Universidad Pablo de Olavide, Km. 1 Utrera Road, 43013 Seville, Spain
salmeron@acm.org, iarebar@alu.upo.es
[2] Tessella, Altran World-Class Center for Analytics, c/ Campezo 1,
28022 Madrid, Spain

Abstract. Distributed Artificial Intelligence is attracting interest day by day. In this paper, the authors introduce an innovative methodology for distributed learning of Particle Swarm Optimization-based Fuzzy Cognitive Maps in a privacy-preserving way. The authors design a training scheme for collaborative FCM learning that offers data privacy compliant with the current regulation. This method is applied to a cancer detection problem, proving that the performance of the model is improved by the Federated Learning process, and obtaining similar results to the ones that can be found in the literature.

Keywords: Fuzzy Cognitive Maps · Federated Learning · Distributed Artificial Intelligence · Cancer diagnosis

1 Introduction

Distributed Artificial Intelligence is a subfield of Artificial Intelligence that studies the coordination among several semi-autonomous agents called participants. Such systems are able to solve more complex problems involving a large amount of data, but there are privacy concerns about sharing sensitive information.

Federated Learning is a novel approach to Distributed Artificial Intelligence that enables privacy-preserving communications by sharing the model (or gradients) instead of the data. A central server sends a model to be trained by the participants with their local data, who send the parameters of the model back to the server to be aggregated. After iterating this process, the output is a model that has been trained with the private information of all participants.

This method is especially useful when dealing with sensitive data, from domains such as finance or healthcare. In this paper, the authors propose a Federated Fuzzy Cognitive Map approach to help diagnose malignant breast tumor cells.

© Springer Nature Switzerland AG 2020
R. Bello et al. (Eds.): IJCRS 2020, LNAI 12179, pp. 477–487, 2020.
https://doi.org/10.1007/978-3-030-52705-1_35

The contributions of this paper can be summarized as follows:

- Distributed learning. The authors propose a PSO-based FCM learning in a distributed way.
- Privacy-preserving machine learning. The authors design a training scheme for collaborative FCM learning that offers data privacy. This proposal enables multiple participants to learn a FCM model on their own inputs, preserving the privacy of their own data and complying with data privacy regulations.
- Implementation. The authors evaluate the performance of the proposal with a well-known dataset of cancer diagnosis. The experimental results show that the proposal achieve a similar performance to other non-distributed methods and improves the performance of the non-collaborative approach.

The rest of this paper is organized as follows. We discuss existing fundamentals of FCM and the learning approach in Sect. 2. Distributed Artificial Intelligence is described in Sect. 3. Then, we present the methodological proposal in Sect. 3. Section 4 describes the details of the experimental approach and the results. Finally, we draw a conclusion in Sect. 5.

2   Fuzzy Cognitive Maps

2.1   Fundamentals

Fuzzy Cognitive Maps (FCMs) were initially proposed by Kosko [3]. FCMs represent concepts, variables or features as nodes, the relationships between them as arcs, and the strengths of those relations as weights. It means that a weight assesses how much node X causes node Y. The fuzzy weights for arcs are normalised on the range $\{[0, +1]|[-1, +1]\}$, depending if it includes negative values or not. The maximum negative influence is -1 and the maximum positive influence is $+1$. The value zero shows that there is no relationship between the concepts. For computational purposes, FCMs can be described via a weight matrix (connection or adjacency matrix) which contains all weight values of edges between the concepts.

The relationships between the nodes are expressed by their weights. That is, if there is a positive causality between two nodes, then $\varpi_{ij} > 0$. If there is a negative causality, then $\varpi_{ij} < 0$ and if there is no relationship between the two nodes, then $\varpi_{ij} = 0$. The state of the nodes together is shown in the state vector $c = [c_1, c_2, \ldots, c_N]$ that gives a snapshot of nodes at any point of the instant in the scenario.

From a formal point of view, it is possible to represent a FCM as a 4-tuple $\Phi = \langle c, \mathcal{W}, f, r \rangle$, where $c = \{c_i\}_{i=1}^n$ is the state of the nodes with n as the number of nodes, $\mathcal{W} = [\varpi_{ij}]_{n \times n}$ is the adjacency matrix representing the weights between the nodes, f is the activation function, and r is the nodes' range.

FCMs are dynamical systems involving feedback, where the effect of change in the state of a node may affect the state of other nodes, which in turn can affect the former node [7].

The dynamic starts with an initial vector state $c(0) = (c_1(0), \ldots, c_n(0))$, which represents the initial state (value) of each node. The new state of the nodes is computed as an iterative process. It includes an activation function [1] for mapping monotonically the node state into a normalized range $\{[0, +1]|[-1, +1]\}$. If the range is $[0, +1]$, the unipolar sigmoid is the most used activation function, but hyperbolic tangent is the most used when the range is $[-1, +1]$.

The component i of the vector state at time t, $c_i(t)$, can be computed as

$$c_i(t) = f\left(\sum_{j=1}^{n} \varpi_{ji} \cdot c_j(t-1)\right). \tag{1}$$

Some systems include nodes whose states should be steady because their states are not related with the dynamics of the system but their state has some influence on the state of the other nodes (i.e. sun radiation, wind speed and so on). In such cases, the state of the node is the same along the dynamics $c_i(t) = c(t-1) \mid c_i \in \mathcal{O}$, where $\mathcal{O}$ is the set of output concepts.

If the activation function f is unipolar sigmoid, then the component i of the vector state $c_i(t)$ at the instant t is computed as follows

$$c_i(t) = \left(1 + e^{-\lambda \cdot \sum_{j=1}^{n} \varpi_{ji} \cdot c_j(t-1)}\right)^{-1} \tag{2}$$

If the activation function f is hyperbolic tangent, then the component i of the vector state $c_i(t)$ at the instant t is computed as follows

$$c_i(t) = \frac{e^{\lambda \cdot \sum_{j=1}^{n} \varpi_{ji} \cdot c_j(t-1)} - e^{-\lambda \cdot \sum_{j=1}^{n} \varpi_{ji} \cdot c_j(t-1)}}{e^{\lambda \cdot \sum_{j=1}^{n} \varpi_{ji} \cdot c_j(t-1)} + e^{-\lambda \cdot \sum_{j=1}^{n} \varpi_{ji} \cdot c_j(t-1)}} \tag{3}$$

After the dynamics, the FCM reaches one of the three following states after a number of iterations: it settles down to either a fixed pattern of node values (the so-called hidden pattern), a limited cycle, or a fixed-point attractor.

2.2 Augmented FCMs

According to the FCM literature [4], an augmented adjacency matrix is built by aggregating the adjacency matrix of each FCM. The elements' aggregation depends on whether there are common nodes. If the adjacency matrices had no common nodes, the elements ϖ_{ij} in the augmented matrix $(\otimes_{i=1}^{N})$ are computed by adding the adjacency matrix of each FCM model $(\mathcal{W}_i)$.

The addition method when the adjacency matrices have not common nodes is known as direct sum of matrices, and the augmented matrix is denoted as $\otimes_{i=1}^{N} \varpi_i$. Given a couple of FCMs with no common nodes and even different number of nodes with adjacency matrices $[\varpi_{ij}]_{n \times n}$ and $[\varpi_{kl}]_{m \times m}$, the resulting augmented adjacency matrix is as follows

$$\otimes_{i=1}^{N} \varpi_i = \mathbf{diag}(\varpi_{jk}, \varpi_{lo})$$

$$= \begin{pmatrix} 0 & [\varpi_{jk}]_{r \times r} \\ [\varpi_{lo}]_{m \times m} & 0 \end{pmatrix} \tag{4}$$

where N is the number or adjacency matrices to add, zeroes are actually zero matrices and the dimension of $\otimes_{i=1}^{N}\varpi_i$ is $[\cdot]_{m+r\times m+r}$. In the case of common nodes, they would be computed as the average or weighted average of the states of the nodes in each adjacency matrix.

2.3   FCM for Classification

FCMs classification capabilities have been analysed by [8]. In general terms, the main goal of a conventional classifier is the mapping of an input to a specific output according to a pattern. In this proposal, the input concepts represent the features of the dataset, while the output concepts are the classes' labels where the patterns belong.

Figure 1 shows the typical topology of a FCM classifier where the state of the concepts c_1 and c_2 defines the class where the input vector state belongs. In that sense, if $c_1 > c_2$ the input vector state belongs to class 1 but if $c_1 < c_2$ the input vector state belongs to class 2. Note that $c_i \in \{[-1, +1], [0, +1]\}$, therefore if $c_1 = 0.03$ and $c_2 = 0.1$, then the input vector state belong to class 2.

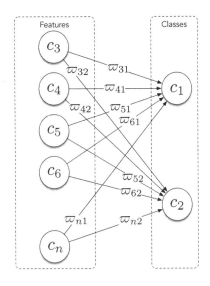

Fig. 1. Fuzzy Cognitive Maps classifier

2.4   PSO-Based FCM Learning

FCM learning endeavours are commonly focused on building the adjacency matrix based either on the available historical raw data or on expert knowledge. FCM learning approaches could be divided into three categories [10]: Hebbian, population-based, and hybrid, mixing the main aspects of Hebbian-based and population-based learning algorithm.

The goal of the Hebbian-based FCM learning approaches is to modify adjacency matrices leading the FCM model to either achieve a steady state or converge into an acceptable region for the target system.

Population-based approaches do not need the human intervention. They compute adjacency matrices from historical raw data that best fit the sequence of input state vectors (the instances of the dataset). The learning goal of FCM evolutionary learning is to generate optimal adjacency matrix for modeling systems behaviour.

In this sense, Salmeron et al. [11] proposed an advanced decision support tool based on consultations with a group of experienced medical professionals using FCMs trained with Particle Swarm Optimization (PSO). Also, Salmeron and Froelich [9] apply PSO for time series forecasting.

PSO is a bio-inspired, population-based and stochastic optimization algorithm. The PSO algorithm generates a swarm of particles moving in an n-dimensional search space which must include all potential candidate solutions.

In order to train the FCM adjacency matrices we take into account the kth particle's position (a candidate solution or adjacency matrix), denoted as $\varpi_k = (\varpi_{k_1}, \ldots, \varpi_{k_j})$ and its velocity, $v_k = (v_{k_1}, \ldots, v_{k_j})$. Note that each particle is a potential solution or FCM candidate and its position ϖ_k represents its adjacency matrix.

Each particle's velocity and position are updated at each time step. The position and the velocity of each particle is computed as follows

$$\varpi_k(t+1) = \varpi_k(t) + v_k(t) \tag{5a}$$

$$v_k(t+1) = v_k(t) + U(0, \phi_1) \otimes (\dot{\varpi}_k - \varpi_k(t)) + U(0, \phi_2) \otimes (\ddot{\varpi}_k - \varpi_k(t)) \tag{5b}$$

where $U(0, \phi_i)$ is a vector of random numbers generated from a uniform distribution within $[0, \phi_i]$, generated at each iteration and for each particle. Also, $\dot{\varpi}_k$ is the best position of particle k in all former iterations and $\ddot{\varpi}_k$ the best position of the whole population in all previous iterations and $\otimes$ is the component-wise multiplication.

The PSO algorithm's goal is to locate all the particles in the global optima to a multidimensional hyper-volume. The fitness function used in this research is the complement of the Jaccard similarity coefficient ($\overline{J} = (Y \times \hat{Y}) \setminus J$). The Jaccard score computes the average of Jaccard similarity coefficients between pairs of the sets of labels. The Jaccard similarity coefficient of the i-th samples, with a ground truth label set and a predicted label set. The complement operation is needed in terms of minimization of the fitness function. The Jaccard similarity coefficient's complement is computed as follows

$$\overline{J}(y_i, \hat{y}_i) = 1 - \frac{|y_i \cap \hat{y}_i|}{|y_i \cup \hat{y}_i|} \tag{6}$$

The fitness function is sampled after each particle position update and is the objective function used to compute how close a given particle is in order to be able to achieve the set aims.

3   Methodological Proposal

3.1   Fundamentals

Distributed Artificial Intelligence is a subset of Artificial Intelligence that allows the sharing of information among several agents or participants that interact by cooperation, by coexistence or by competition. Such system manages the distribution of tasks, being therefore more apt to solve complex problems, especially if they involve a large amount of data.

One of the methods available to construct a distributed artificial intelligence system is Federated Learning, proposed by McMahan et al. [5] and further developed in Konecny et al. [2] and McMahan and Ramage [6]. In such system, a central server constructs a model, usually a neural network, and sends it to the participants, who train the model in their private data. Their data never leaves their local devices, therefore ensuring privacy and security. The parameters of the participant's model are then averaged to obtain a global model. This process may be iterated till convergence.

Described in a formal way, a Federated Learning project is composed by a central server and the participants. The central server is responsible for managing the federated model and the communications with the participants. The participants own the datasets and train the partial models. The whole process is described in Fig. 2 and it is as follows:

1. The central server sends a federated model to each participant. If it is the initial iteration the federated model is proposed by the central server.
2. Each participant trains the received model with their own private dataset.
3. After the partial model is trained, each participant sends the parameters of the model or its gradients to the central server, encrypted to ensure privacy.
4. The central server aggregates the partial model and builds the federated model.
5. The central server checks the termination condition and if it is accomplished the federated model is finished, otherwise the process goes back to step 1.

When the researchers at Google first defined Federated Learning, their initial idea was to allow Android mobile phones to collaborative construct a prediction model without migrating the training data from the phone (see McMahan et al. [6] from the Google AI Blog). A first application they had was to use FL in Gboard on Android, the Google Keyboard, which predicts the most probable next phrase or word based on the user-generated preceding text. Recently, Federated Learning has improved this process, allowing the use of more accurate models with lower latency, ensuring privacy and less power consumption.

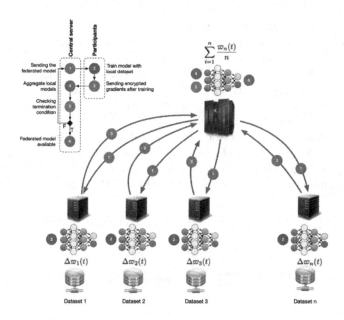

Fig. 2. Federated Learning process

One of the main advantages of Federated Learning is the promise of secure and private distributed machine learning, but there are risks associated with sharing data among several agents, such as the reconstruction of training examples from the neural network parameters, the uploading of private data from the agents to the central server, and the protection of the models as intellectual property of the companies. There is a large research interest in privacy-preserving methods applied to Federated Learning, such as the application of Differential Privacy, Secure Multi-Party Computation or Homomorphic encryption.

3.2 FCM Distributed Learning

The proposed methodology combines Federated Learning with learning FCMs using Particle Swarm Optimization. The process is shown in Fig. 3 and it is explained as follows.

1. Triggering the Federated Learning process. The central server triggers the process in the participants machines.
2. Training FCM in the local dataset. Each participant trains a local FCM with their own dataset. The authors apply PSO but this methodology is agnostic to the learning approach. The FCM dynamics is considered steady when the difference between two consecutive vector states is under $tol = 0.00001$
3. Sending the trained adjacency matrices and local accuracy for this stage to the central server. The local FCM is stored in the participant devices.

4. Weighting local FCMs using accuracy. The central server aggregates the local FCMs weighting by the accuracy. The aggregation method have been detailed as Sect. 2.2.
5. Aggregating Federated and Local FCMs. The participants aggregate the Federated FCM from the central server and their own local FCM.
6. Sending adjacency matrices and accuracy. Participants send again the local adjacency matrices and the new local accuracy.
7. Checking termination condition. The central server checks if the Federated process has been run 20 iterations as termination condition. If it is not accomplished then it goes back to the step 4.
8. If the termination condition is accomplished then a Federated FCM is achieved.

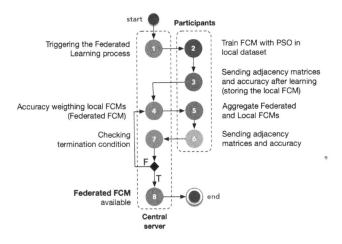

Fig. 3. Proposed methodology

The main contribution of this paper is the application of Federation Learning paradigm for privacy-preserving FCM distributed and coorperative learning.

4   Experimental Approach

4.1   Dataset

Breast cancer is one of the most common cancers among women, accounting for 25% of all cancer cases that affect women worldwide. According to the American Cancer Society, when breast cancer is detected early, and is in the localized stage, the 5-year relative survival rate is 99%, which makes the early diagnosis of breast cancer a main key in the prognosis and chance of survival of such types of cancer.

In recent years the use of Machine Learning algorithms in medicine has increased exponentially, with applications such as EEG analysis and Cancer detection. For example, automatized algorithms have been use to examine biological data such as DNA methylation and RNA sequencing to infer which genes can cause cancer and which genes can instead be able to suppress its expression.

In this paper the authors will use the Breast Cancer Wisconsin Dataset, created by Dr. William H. Wolberg, physician at the University Of Wisconsin Hospital at Madison, and made publicly available at the UC Irvine Machine Learning Repository. The dataset comprises data from digitized images of the fine-needle aspirate of a breast mass that describes features of the nucleus of the current image of 569 patients, of which 212 are malignant and 357 are benign cases.

The first two features correspond to the identifier number and the diagnosis status (our target). The remaining attributes are thirty real attributes that measure the mean, the standard error, and the worst radius, texture, perimeter, area, smoothness, compactness, concave points, concavity, symmetry, and fractal dimension of the nucleus of the solid breast mass (see Table 1). These data were

Table 1. Dataset details

Id	Description	Mean	Std
Diagnosis	Target	0.3726	0.4839
radius_mean	Mean of distances from center to points on the perimeter	14.1273	3.524
texture_mean	Standard deviation of gray-scale values	19.2896	4.301
perimeter_mean		91.969	24.299
area_mean		654.8891	351.9141
smoothness_mean	Local variation in radius lengths	0.0964	0.0141
compactness_mean	Perimeter^2 / area - 1	0.1043	0.0528
concavity_mean	Severity of concave portions of the contour	0.0888	0.0797
concave points_mean	Number of concave portions of the contour	0.0489	0.0388
symmetry_mean		0.1812	0.0274
fractal_dimension_mean	Coastline approx. - 1	0.0628	0.0071
radius_se	Mean of distances from center to points on the perimeter	0.4052	0.2773
texture_se	Standard deviation of gray-scale values	1.2169	0.5516
perimeter_se		2.8661	2.0219
area_se		40.3371	45.491
smoothness_se	Local variation in radius lengths	0.007	0.003
compactness_se	Perimeter^2 / area - 1	0.0255	0.0179
concavity_se	Severity of concave portions of the contour	0.0319	0.0302
concave points_se	Number of concave portions of the contour	0.0118	0.0062
symmetry_se		0.0205	0.0083
fractal_dimension_se	Coastline approx. - 1	0.0038	0.0026
radius_worst	Mean of distances from center to points on the perimeter	16.2692	4.8332
texture_worst	Standard deviation of gray-scale values	25.6772	6.1463
perimeter_worst		107.2612	33.6025
area_worst		880.5831	569.357
smoothness_worst	Local variation in radius lengths	0.1324	0.0228
compactness_worst	perimeter^2 / area - 1	0.2543	0.1573
concavity_worst	Severity of concave portions of the contour	0.2722	0.2086
concave points_worst	Number of concave portions of the contour	0.1146	0.0657
symmetry_worst		0.2901	0.0619
fractal_dimension_worst	Coastline approx. - 1	0.0839	0.0181

obtained using a graphical computer program called Xcyt, which is capable of perform the analysis of cytological features based on a digital scan. More details can be found in [12,13].

4.2 Results

After 20 iterations of the Federated Learning process, the Fuzzy Cognitive Map-based classifier is able to predict whether the tumor is malignant with an average accuracy of 0.9383 across all participants, improving the accuracy of a single Fuzzy Cognitive Map trained in the whole data, and the accuracy in each participant before the federation.

The goal of this paper is not the accuracy of the proposal but a distributed and privacy-preserving approach. Nevertheless, our results are similar to the ones found in literature [14] (Table 2).

Table 2. Results of the experiments

Participant	Accuracy pre-federated learning	Accuracy post-federated learning
1	0.7727	0.9091
2	0.9130	0.9130
3	0.8696	0.8696
4	0.9565	1.0000
5	1.0000	1.0000

5 Conclusions

This paper proposes an innovative methodology for learning Fuzzy Cognitive Maps with Federated Learning. It is a step forward for Distributed Artificial Intelligence and accomplishes the privacy-preserving requirements of the society.

In addition, the authors have developed a method for distributed Fuzzy Cognitive Maps that improves the accuracy of both the algorithm trained in the whole dataset in a local node and the participant's algorithms before the Federated Learning process.

This method was applied to a cancer detection problem, obtaining an accuracy of 0.9383. The participants in this process do not share their private data, therefore forming a privacy-preserving distributed system.

References

1. Bueno, S., Salmeron, J.L.: Benchmarking main activation functions in fuzzy cognitive maps. Expert Syst. Appl. **36**(3 Part 1), 258–268 (2009)
2. Konecný, J., McMahan, B., Ramage, D., Richtárik, P.: Federated optimization: distributed machine learning for on-device intelligence. ArXiv abs/1610.02527 (2016)
3. Kosko, B.: Fuzzy cognitive maps. Int. J. Man Mach. Stud. **24**(1), 65–75 (1986)
4. Lopez, C., Salmeron, J.L.: Modeling maintenance projects risk effects on erp performance. Comput. Stand. Interfaces **36**(3), 545–553 (2014)
5. McMahan, B., Moore, E., Ramage, D., y Arcas, B.A.: Federated learning of deep networks using model averaging. ArXiv abs/1602.05629 (2016)
6. McMahan, B., Ramage, D.: Google ai blog, April 2017. https://ai.googleblog.com/2017/04/federated-learning-collaborative.html
7. Nápoles, G., Jastrzebska, A., Mosquera, C., Vanhoof, K., Homenda, W.: Deterministic learning of hybrid fuzzy cognitive maps and network reduction approaches. Neural Networks **124**, 258–268 (2020)
8. Papakostas, G., Koulouriotis, D.: Classifying patterns using fuzzy cognitive maps. In: Glykas, M. (ed.) Fuzzy Cognitive Maps. Studies in Fuzziness and Soft Computing, vol. 247, pp. 291–306. Springer, Heidelberg (2010). https://doi.org/10.1007/978-3-642-03220-2_12
9. Salmeron, J.L., Froelich, W.: Dynamic optimization of fuzzy cognitive maps for time series forecasting. Knowl.-Based Syst. **105**, 29–37 (2016)
10. Salmeron, J.L., Mansouri, T., Moghadam, M.R.S., Mardani, A.: Learning fuzzy cognitive maps with modified asexual reproduction optimisation algorithm. Knowl.-Based Syst. **163**, 723–735 (2019)
11. Salmeron, J.L., Rahimi, S.A., Navali, A.M., Sadeghpour, A.: Medical diagnosis of rheumatoid arthritis using data driven pso-fcm with scarce datasets. Neurocomputing **232**, 65–75 (2017)
12. Street, W., Wolberg, W., Mangasarian, O.: Breast cancer diagnosis and prognosis via linear programming. Oper. Res. **43**(4), 570–577 (1995). https://doi.org/10.1287/opre.43.4.570
13. Street, W., Wolberg, W., Mangasarian, O.: Nuclear feature extraction for breast tumor diagnosis, vol. 1993, January 1999. https://doi.org/10.1117/12.148698
14. Wang, S., Wang, Y., Wang, D., Yin, Y., Wang, Y., Jin, Y.: An improved random forest-based rule extraction method for breast cancer diagnosis. Appl. Soft Comput. **86**, 105941 (2020)

Tutorials

fuzzy-rough-learn 0.1: A Python Library for Machine Learning with Fuzzy Rough Sets

Oliver Urs Lenz[1(✉)] , Daniel Peralta[1,2] , and Chris Cornelis[1]

[1] Department of Applied Mathematics, Computer Science and Statistics,
Ghent University, Ghent, Belgium
{oliver.lenz,chris.cornelis}@ugent.be
[2] Data Mining and Modelling for Biomedicine Group, VIB Center for Inflammation
Research, Ghent University, Ghent, Belgium
daniel.peralta@irc.vib-ugent.be
http://www.cwi.ugent.be, https://www.irc.ugent.be

Abstract. We present *fuzzy-rough-learn*, the first Python library of fuzzy rough set machine learning algorithms. It contains three algorithms previously implemented in R and Java, as well as two new algorithms from the recent literature. We briefly discuss the use cases of *fuzzy-rough-learn* and the design philosophy guiding its development, before providing an overview of the included algorithms and their parameters.

Keywords: Fuzzy rough sets · OWA operators · Machine learning · Python package · Open-source software

1 Background

Since its conception in 1990, fuzzy rough set theory [2] has been applied as part of a growing number of machine learning algorithms [17]. Simultaneously, the distribution and communication of machine learning algorithms has spread beyond academic literature to a multitude of publicly available software implementations [7,10,19]. And also during the same period, Python has grown from its first release in 1991 [13] to become one of the world's most popular high-level programming languages.

Python has become especially popular in the field of data science, in part due to the self-reinforcing growth of its package ecosystem. This includes *scikit-learn* [11], which is currently the go-to general purpose Python machine learning library, and which contains a large collection of algorithms.

Only a limited number of fuzzy rough set machine learning algorithms have received publicly available software implementations. Variants of Fuzzy Rough Nearest Neighbours (FRNN) [5], Fuzzy Rough Rule Induction [6], Fuzzy Rough Feature Selection (FRFS) [1] and Fuzzy Rough Prototype Selection (FRPS) [14,15] are included in the R package *RoughSets* [12], and have also been released for use with the Java machine learning software suite WEKA [3,4].

© Springer Nature Switzerland AG 2020
R. Bello et al. (Eds.): IJCRS 2020, LNAI 12179, pp. 491–499, 2020.
https://doi.org/10.1007/978-3-030-52705-1_36

So far, none of these algorithms seem to have been made available for Python in a systematic way. In this paper, we present an initial version of *fuzzy-rough-learn*, a Python library that fills this gap. At present, it includes FRNN, FRFS, FRPS, as well as FROVOCO [18] and FRONEC [16], two more recent algorithms designed for imbalanced and multilabel classification. These implementations all make use of a significant modification of classical fuzzy rough set theory: the incorporation of Ordered Weighted Averaging (OWA) operators in the calculation of upper and lower approximations for increased robustness [1].

We discuss the use cases and design philosophy of *fuzzy-rough-learn* in Sect. 2, and provide an overview of the included algorithms in Sect. 3.

2   Use Cases and Design Philosophy

The primary goal of *fuzzy-rough-learn* is to provide implementations of fuzzy rough set algorithms. The target audience is researchers with some programming skills, in particular those who are familiar with *scikit-learn*. We envision two principal use cases:

- The application of fuzzy rough set algorithms to solve concrete machine learning problems.
- The creation of new or modified fuzzy rough set algorithms to handle new types of data or to achieve better performance.

A third use case falls somewhat in between these two: reproducing or benchmarking against results from existing fuzzy rough set algorithms.

To facilitate the first use case, *fuzzy-rough-learn* is available from the two main Python package repositories, pipy and conda-forge, making it easy to install with both pip and conda. *fuzzy-rough-learn* has an integrated test suite to limit the opportunities for bugs to be introduced. API documentation is integrated in the code and automatically updated online[1] whenever a new version is released, and includes references to the literature.

We believe that it is important to make fuzzy rough set algorithms available not just for use, but also for adaptation, since it is impossible to predict or accommodate all requirements of future researchers. Therefore, the source code for *fuzzy-rough-learn* is hosted on GitHub[2] and freely available under the MIT license. We have attempted to write accessible code, by striving for consistency and modularity. The coding style of *fuzzy-rough-learn* is a compromise between object-oriented and functional programming. It makes use of classes to model the different components of the classification algorithms, but as a rule, functions and methods have no side-effects. Finally, subject to these design principles, *fuzzy-rough-learn* generally follows the conventions of *scikit-learn* and the terminology of the cited literature.

---

[1] https://fuzzy-rough-learn.readthedocs.io.

[2] https://github.com/oulenz/fuzzy-rough-learn.

3 Contents

fuzzy-rough-learn implements three of the fuzzy rough set algorithms mentioned in Sect. 1: FRFS, FRPS and FRNN, making them available in Python for the first time. In addition, we have included two recent, more specialised classifiers: the ensemble classifier FROVOCO, designed to handle imbalanced data, and the multi-label classifier FRONEC.

Together, these five algorithms form a representative cross-section of fuzzy rough set algorithms in the literature. In the future, we intend to build upon this basis by adding more algorithms (Table 1).

3.1 Fuzzy Rough Feature Selection (FRFS)

Fuzzy Rough Feature Selection (FRFS) [1] greedily selects features that induce the greatest increase in the size of the positive region, until it matches the size of the positive region with all features, or until the required number of features is selected.

The positive region is defined as the union of the lower approximations of the decision classes in X. Its size is the sum of its membership values.

The similarity relation R_B for a given subset of attributes B is obtained by aggregating with a t-norm the per-attribute similarities R_a associated with the attributes a in B. These are in turn defined, for any $x, y \in X$, as the complement of the difference between the attribute values x_a and y_a after rescaling by the sample standard deviation σ_a (1).

$$R_a(x, y) = \max(1 - \frac{|x_a - y_a|}{\sigma_a}, 0) \tag{1}$$

Table 1. Parameters of FRFS in *fuzzy-rough-learn*

Name	Default value	Description
n_features	None	Number of features to select. If None, will continue to add features until positive region size becomes maximal
owa_weights	deltaquadsigmoid (0.2, 1)	OWA weights to use for calculation of soft minimum in lower approximations
t_norm	'lukasiewicz'	T-norm used to aggregate the similarity relation R from per-attribute similarities

3.2   Fuzzy Rough Prototype Selection (FRPS)

Fuzzy Rough Prototype Selection (FRPS) [14,15] uses upper and/or lower approximation membership as a quality measure to select instances. It follows the following steps:

1. Calculate the quality of each training instance. The resulting values are the potential thresholds for selecting instances (Table 2).
2. For each potential threshold and corresponding candidate instance set, count the number of instances in the overall dataset that have the same decision class as their nearest neighbour within the candidate instance set (excluding itself).
3. Return the candidate instance set with the highest number of matches. In case of a tie, return the largest such set.

There are a number of differences between the implementations in [15] and [14]. In each case, the present implementation follows [14]:

– While [15] uses instances of all decision classes to calculate upper and lower approximations, [14] calculates the upper approximation membership of an instance using only instances of the same decision class, and its lower approximation membership using only instances of the other decision classes. This choice affects over what length the weight vector is 'stretched'.

Table 2. Parameters of FRPS in *fuzzy-rough-learn*

Name	Default value	Description
quality_measure	'lower'	Quality measure to use for calculating thresholds. Either the upper approximation of the decision class of each attribute, the lower approximation, or the mean value of both
aggr_R	np.mean	Function used to aggregate the similarity relation R from per-attribute similarities
owa_weights	invadd()	OWA weights to use for calculation of soft maximum and/or minimum in quality measure
nn_search	KDTree()	Nearest neighbour search algorithm to use

- In addition, [14] excludes each instance from the calculation of its own upper approximation membership, while [15] does not.
- [15] uses additive weights, while [14] uses inverse additive weights.
- [15] defines the similarity relation R by aggregating the per-attribute similarities R_a using the Łukasiewicz t-norm, whereas [14] recommends using the mean.
- In case of a tie between several best-scoring candidate prototype sets, [15] returns the set corresponding to the median of the corresponding thresholds, while [14] returns the largest set (corresponding to the smallest threshold).

In addition, there are two implementation issues not addressed in [15] or [14]:

- It is unclear what metric the nearest neighbour search should use. It seems reasonable that it should either correspond to the similarity relation R (and therefore incorporate the same aggregation strategy from per-attribute similarities), or that it should match whatever metric is used by nearest neighbour classification subsequent to FRPS. By default, the present implementation uses Manhattan distance on the scaled attribute values.
- When the largest quality measure value corresponds to a singleton candidate instance set, it cannot be evaluated (because the single instance in that set has no nearest neighbour). Since this is an edge case that would not score highly anyway, it is simply excluded from consideration.

3.3 Fuzzy Rough Nearest Neighbour (FRNN) Multiclass Classification

Fuzzy Rough Nearest Neighbours (FRNN) [5] provides a straightforward way to apply fuzzy rough sets for classification. Given a new instance y, we obtain class scores by calculating the membership degree of y in the upper and lower approximations of each decision class and taking the mean. This implementation uses OWA weights, but limits their application to the k nearest neighbours of each class, as suggested by [8] (Table 3).

3.4 Fuzzy Rough OVO Combination (FROVOCO) Multiclass Classification

Fuzzy Rough OVO COmbination (FROVOCO) [18] is an ensemble classifier specifically designed for, but not restricted to, imbalanced data, which adapts itself to the Imbalance Ratio (IR) between classes. It balances one-versus-one decomposition with two global class afinity measures (Table 4).

In a binary classification setting, the lower approximation of one class corresponds to the upper approximation of the other class, so when using OWA weights, the effective number of weight vectors to be chosen is 2. FROVOCO uses the IR-weighting scheme, which depends on the IR between the classes. If the IR is less than 9, both classes are approximated with exponential weights. If the IR is 9 or more, the smaller class is approximated with exponential weights,

Table 3. Parameters of FRNN in *fuzzy-rough-learn*

Name	Default value	Description
upper_weights	additive()	OWA weights to use in calculation of upper approximation of decision classes
upper_k	20	Effective length of upper weights vector (number of nearest neighbours to consider)
lower_weights	additive()	OWA weights to use in calculation of lower approximation of decision classes
lower_k	20	Effective length of lower weights vector (number of nearest neighbours to consider)
nn_search	KDTree()	Nearest neighbour search algorithm to use

while the larger class is approximated with a reduced additive weight vector of effective length k equal to 10% of the number of instances.

Provided with a training set X, and a new instance y, FROVOCO calculates the class score of y for a class C from the following components:

$V(C, y)$ **weighted vote** For each other class $C' \neq C$, calculate the upper approximation memberships of y in C and C', using the IR-weighting scheme. Rescale each pair of values so they sum to 1, then sum the resulting scores.

$mem(C, y)$ **positive affinity** Calculate the average of the membership degrees of y in the upper and lower approximations of C, using the IR-weighting scheme.

$mse_n(C, y)$ **negative affinity** For each class C', calculate the average positive affinity of the members of C in C'. Combine these average values to obtain the signature vector S_C. Calculate the mean squared error of the positive affinities of y for each class and S_C, and divide it by the sum of the mean squared errors for all classes.

Table 4. Parameters of FROVOCO in *fuzzy-rough-learn*

Name	Default value	Description
nn_search	KDTree()	Nearest neighbour search algorithm to use

The final class score is calculated from these components in (2).

$$AV(C, y) = \frac{V(C, y) + mem(C, y)}{2} - \frac{1}{m} mse_n(C, y). \qquad (2)$$

3.5 Fuzzy Rough Neighbourhood Consensus (FRONEC) Multilabel Classification

Fuzzy Rough Neighbourhood Consensus (FRONEC) [16] is a multilabel classifier. It combines the instance similarity R, based on the instance attributes, with label similarity R_d, based on the label sets of instances. It offers two possible definitions for R_d. The first, $R_d^{(1)}$, is simply Hamming similarity scaled to $[0, 1]$. The second label similarity, $R_d^{(2)}$, takes into account the prior probability p_l of a label l in the training set. Let L the set of possible labels, and L_1, L_2 two particular label sets. Then $R_d^{(2)}$ is defined as follows (Table 5):

$$a = \sum_{l \in L_1 \cap L_2} (1 - p_l)$$

$$b = \sum_{l \in L \setminus (L_1 \cup L_2)} p_l \qquad (3)$$

$$R_d^{(2)} = \frac{a + b}{a + b + \frac{1}{2} |L_1 \Delta L_2|}$$

Table 5. Parameters of FRONEC in *fuzzy-rough-learn*

Name	Default value	Description
Q_type	2	Quality measure to use for identifying most relevant instances: based on lower (1), upper (2) or both approximations (3)
R_d_type	1	Label similarity relation to use: Hamming similarity (1) or based on prior probabilities (2)
k	20	Number of neighbours to consider for neighbourhood consensus
weights	additive()	OWA weights to use for calculation of soft maximum and/or minimum
nn_search	KDTree()	Nearest neighbour search algorithm to use

Provided with a training set X, and a new instance y, FRONEC predicts the label set of y by identifying the training instance with the highest 'quality' in relation to y. There are three possible quality measures, based on the upper and lower approximations.

$$Q_1(y, x) = OWA_{w_l}(\{I(R(z, y), R_d(x, z))|z \in N(y)\})$$
$$Q_2(y, x) = OWA_{w_u}(\{T(R(z, y), R_d(x, z))|z \in N(y)\})$$
$$Q_3(y, x) = \frac{Q_1(y, x) + Q_2(y, x)}{2}$$

(4)

Where R_d is a choice of label similarity, T the Łukasiewicz t-norm, I the Łukasiewicz implication, and $N(y)$ the k nearest neighbours of y in X, for a choice of k.

For a choice of quality measure Q, FRONEC predicts the labels of the training instance with the highest quality. If there are several such training instances, it predicts all labels that appear with at least half.

3.6 OWA Operators and Nearest Neighbour Searches

Each of the algorithms in *fuzzy-rough-learn* uses OWA operators [20] to calculate upper and lower approximations. OWA operators take the weighted average of an ordered collection of real values. By choosing suitably skewed weight vectors, OWA operators can thus act as soft maxima and minima. The advantage of defining upper and lower approximations with soft rather than strict maxima and minima is that the result is more robust, since it no longer depends completely on a single value.

To allow experimentation with other weights, we have included a range of pre-defined weight types, as well as a general `OWAOperator` class that can be extended and instantiated by users and passed as a parameter to the various classes.

Similarly, users may customise the nearest neighbour search algorithm that is used in all classes except FRFS by defining their own subclass of `NNSearch`. For example, by choosing an approximative nearest neighbour search like Hierarchical Navigable Small World [9], we obtain Approximate FRNN [8].

Acknowledgement. The research reported in this paper was conducted with the financial support of the Odysseus programme of the Research Foundation – Flanders (FWO). D. Peralta is a Postdoctoral Fellow of the Research Foundation – Flanders (FWO, 170303/12X1619N).

References

1. Cornelis, C., Verbiest, N., Jensen, R.: Ordered weighted average based fuzzy rough sets. In: Yu, J., Greco, S., Lingras, P., Wang, G., Skowron, A. (eds.) RSKT 2010. LNCS (LNAI), vol. 6401, pp. 78–85. Springer, Heidelberg (2010). https://doi.org/10.1007/978-3-642-16248-0_16
2. Dubois, D., Prade, H.: Rough fuzzy sets and fuzzy rough sets. Int. J. General Syst. **17**(2–3), 191–209 (1990)

3. Hall, M., Frank, E., Holmes, G., Pfahringer, B., Reutemann, P., Witten, I.H.: The WEKA data mining software: an update. ACM SIGKDD Explor. Newslett. **11**(1), 10–18 (2009)
4. Jensen, R.: Fuzzy-rough data mining with Weka (2010). http://users.aber.ac.uk/rkj/Weka.pdf
5. Jensen, R., Cornelis, C.: A new approach to fuzzy-rough nearest neighbour classification. In: Chan, C.-C., Grzymala-Busse, J.W., Ziarko, W.P. (eds.) RSCTC 2008. LNCS (LNAI), vol. 5306, pp. 310–319. Springer, Heidelberg (2008). https://doi.org/10.1007/978-3-540-88425-5_32
6. Jensen, R., Cornelis, C., Shen, Q.: Hybrid fuzzy-rough rule induction and feature selection. In: Proceedings of the 2009 IEEE International Conference on Fuzzy Systems, pp. 1151–1156. IEEE (2009)
7. Jović, A., Brkić, K., Bogunović, N.: An overview of free software tools for general data mining. In: Proceedings of the 37th International Convention on Information and Communication Technology, Electronics and Microelectronics (MIPRO 2014), pp. 1112–1117. IEEE (2014)
8. Lenz, O.U., Peralta, D., Cornelis, C.: Scalable approximate FRNN-OWA classification. IEEE Transactions on Fuzzy Systems (to be published). https://doi.org/10.1109/TFUZZ.2019.2949769
9. Malkov, Y.A., Yashunin, D.A.: Efficient and robust approximate nearest neighbor search using hierarchical navigable small world graphs. IEEE Trans. Pattern Anal. Mach. Intell. **42**(4), 824–836 (2020)
10. Nguyen, G., et al.: Machine learning and deep learning frameworks and libraries for large-scale data mining: a survey. Artif. Intell. Rev. **52**(1), 77–124 (2019)
11. Pedregosa, F., et al.: Scikit-learn: machine learning in Python. J. Mach. Learn. Res. **12**(85), 2825–2830 (2011)
12. Riza, L.S., et al.: Implementing algorithms of rough set theory and fuzzy rough set theory in the R package "RoughSets". Inf. Sci. **287**, 68–89 (2014)
13. van Rossum, G., de Boer, J.: Interactively testing remote servers using the Python programming language. CWI Q. **4**(4), 283–303 (1991)
14. Verbiest, N.: Fuzzy rough and evolutionary approaches to instance selection. Ph.D. thesis, Ghent University (2014)
15. Verbiest, N., Cornelis, C., Herrera, F.: OWA-FRPS: a prototype selection method based on ordered weighted average fuzzy rough set theory. In: Ciucci, D., Inuiguchi, M., Yao, Y., Ślęzak, D., Wang, G. (eds.) RSFDGrC 2013. LNCS (LNAI), vol. 8170, pp. 180–190. Springer, Heidelberg (2013). https://doi.org/10.1007/978-3-642-41218-9_19
16. Vluymans, S., Cornelis, C., Herrera, F., Saeys, Y.: Multi-label classification using a fuzzy rough neighborhood consensus. Inf. Sci. **433**, 96–114 (2018)
17. Vluymans, S., D'eer, L., Saeys, Y., Cornelis, C.: Applications of fuzzy rough set theory in machine learning: a survey. Fundamenta Informaticae **142**(1–4), 53–86 (2015)
18. Vluymans, S., Fernández, A., Saeys, Y., Cornelis, C., Herrera, F.: Dynamic affinity-based classification of multi-class imbalanced data with one-versus-one decomposition: a fuzzy rough set approach. Knowl. Inf. Syst. **56**(1), 55–84 (2017). https://doi.org/10.1007/s10115-017-1126-1
19. Wang, Z., Liu, K., Li, J., Zhu, Y., Zhang, Y.: Various frameworks and libraries of machine learning and deep learning: a survey. Archives Comput. Methods Eng. 1–24 (2019). https://doi.org/10.1007/s11831-018-09312-w
20. Yager, R.R.: On ordered weighted averaging aggregation operators in multicriteria decisionmaking. IEEE Trans. Syst. Man Cybern. **18**(1), 183–190 (1988)

Comparative Approaches to Granularity in General Rough Sets

A. Mani$^{(\boxtimes)}$ ⓘ

HBCSE, Tata Institute of Fundamental Research, Indian Statistical Institute, Kolkata,
9/1B, Jatin Bagchi Road, Kolkata (Calcutta) 700029, India
a.mani.cms@gmail.com, mani@hbcse.tifr.res.in
https://www.logicamani.in

Abstract. A number of nonequivalent perspectives on granular computing are known in the literature, and many are in states of continuous development. Further related concepts of granules and granulations may be incompatible in many senses. This expository paper is intended to explain basic aspects of these from a critical perspective, their range of applications and provide directions relative to general rough sets and related formal approaches to vagueness. General granular principles related to knowledge are also mentioned.

Keywords: Granules · Primitive · Precision based and axiomatic granular computing · Rough objects · Mereology · High granular operator spaces · Contamination problem · Ontology

1 Introduction

In its simplest form, granules or information granules are building blocks of a reasoning or computational procedure in soft or hard contexts. Information granulation can be viewed as a human way of achieving complexity reduction (rather than data compression) that often plays a key role in divide-and-conquer strategies used in human problem-solving. Granulations are collections of granules that have been integrated by some processes that involve indistinguishability, similarity, proximity or functionality. Associated soft contexts typically involve vagueness, uncertainty, indecision or fuzziness and some level of indeterminacy. This has lead to many distinct mutually not-necessarily compatible approaches. For example, not all frameworks of granular computing used in general rough sets are compatible with those used in fuzzy sets.

A natural question is *do granulations come first or do granules come first?* If the goal is to perceive and classify objects irrespective of ontology or associated process, then the question is not particularly relevant. Some approaches to granularity as in the *classical granular computing approach* (CGCP) prefer to start from granulations and proceed to consider granules at multiple levels of precision. In the axiomatic approach (AGCP) [1], especially when ontology is important, it is more common to proceed from granules to granulations.

© Springer Nature Switzerland AG 2020
R. Bello et al. (Eds.): IJCRS 2020, LNAI 12179, pp. 500–517, 2020.
https://doi.org/10.1007/978-3-030-52705-1_37

But the converse approach is also relevant in AGCP. In adaptive systems, when granules are permitted to change relative to events or time or temporal instants, it makes sense to keep track of the changes through additional operators. This does suggest that a bottom up approach would be optimal in the scenario.

The number of distinct approaches to ideas of granularity depends on the perspective used. The level or qualitative description of granules involved may also be a key determiner of the perspective used. The major approaches are CGCP, AGCP, primitive granular computing paradigm and adaptive variants of the first two. Hierarchies within each of these types can also be formalized or specified.

1.1 Background

The concept of *information* can also be defined in many not necessarily equivalent ways. In the present author's view *anything that alters or has the potential to alter a given context in a significant positive way is information*. In the contexts of general rough sets, the concept of information must have the potential to alter supervenience relations in the contexts (A set of properties Q supervene on another set of properties T if there exist no two objects that differ on Q without differing on T), be formalizable and able to generate concepts of roughly similar collections of properties or objects. One of the popular abstractions is that of an information table.

Formally an *information table* $\mathfrak{I}$, is a tuple of the form

$$\mathfrak{I} = \langle \mathfrak{O}, \mathbb{A}, \{V_a : a \in \mathbb{A}\}, \{f_a : a \in \mathbb{A}\} \rangle$$

with $\mathfrak{O}$, $\mathbb{A}$ and V_a being respectively sets of *Objects*, *Attributes* and *Values* respectively. $f_a : \mathfrak{O} \longmapsto \wp(V_a)$ being the valuation map associated with attribute $a \in \mathbb{A}$. Values may also be denoted by the binary function $v : \mathbb{A} \times \mathfrak{O} \longmapsto \wp(V)$ defined by for any $a \in \mathbb{A}$ and $x \in \mathfrak{O}$, $v(a, x) = f_a(x)$.

Relations may be derived from information tables by way of conditions of the following form: For $x, w \in \mathfrak{O}$ and $B \subseteq \mathbb{A}$, $(x, w) \in \sigma$ if and only if $(\mathbf{Q}a, b \in B)\, \Phi(v(a, x), v(b, w),)$ for some quantifier $\mathbf{Q}$ and formula Φ. The relational system $S = \langle \underline{S}, \sigma \rangle$ (with $\underline{S} = \mathfrak{O}$) is said to be a *general approximation space* (S and $\underline{S}$ will be used interchangeably). In particular if σ is an equivalence relation then S is referred to as an *approximation space*. It should be noted that objects are assumed to be defined (to the extent possible) by attributes and associated valuations.

In classical rough sets, on the power set $\wp(S)$, lower and upper approximations of a subset $A \in \wp(S)$ operators, apart from the usual Boolean operations, are defined as per: $A^l = \bigcup_{[x] \subseteq A} [x]$, $A^u = \bigcup_{[x] \cap A \neq \varnothing} [x]$, with $[x]$ being the equivalence class generated by $x \in S$. If $A, B \in \wp(S)$, then A is said to be *roughly included* in B $(A \sqsubseteq B)$ if and only if $A^l \subseteq B^l$ and $A^u \subseteq B^u$. A is roughly equal to B $(A \approx B)$ if and only if $A \sqsubseteq B$ and $B \sqsubseteq A$. The positive, negative and boundary region determined by a subset A are respectively A^l, $(A^u)^c$ and $A^u \setminus A^l$ (c being the set complement).

In a general approximation space $S = \langle \underline{S}, R \rangle$, any subset $A \subseteq S$ will be said to be a R-*block* if and only if it is maximal with respect to the property $A^2 \subseteq R$. The set of all R-blocks of S will be denoted by $\mathcal{B}_R(S)$. If R is reflexive, then $\mathcal{B}_R(S)$ is a proper cover of S. These are examples of granules. Any map $n : H \longmapsto \wp(H)$ on a set H generates a set of granules called neighborhood granules [2] on H. These are called neighborhood maps if $x \in n(x)$ holds for all x. Specifically, the successor neighborhood generated by a point $x \in S$ is $[x] = \{a : Rax\}$ (Rax in infix form is aRx).

In any formal approach to vagueness, it is necessary to specify the environment or context of discourse, the main objects of interest, presumptions about how objects interact with the environment, and interpretation. Often people working in AI and ML refer to meta levels to partly specify this relative to what is known or assumed in the literature. This relative specification may not be always adequate (and requires elaboration) in a number of problems as indicated in [1,3]. Specific classes of domains that require different formalism are considered in [4,5].

In the context of general rough sets, various concepts of rough objects (including roughly equivalent objects) [1,3,6] with associated meta operations and rules correspond to semantic domains (or domains of discourse). In the context of relation based rough sets, the power set $\wp(S)$ (or generalizations thereof), lower and upper approximation operators, and other usual operations, generate a semantics. The associated semantic domain in the sense of a collection of restrictions on possible objects, predicates, constants, functions and low level operations on those is referred to as the classical semantic domain (meta-C) for general rough sets [3]. In contrast, the semantic domain associated with sets of rough objects is a rough semantic domain (meta-R). Many other domains, including hybrid semantic domains, can be generated [1]. In [7], the models refer to reasoning about the power set of the set of possible order-compatible partitions of the set of rough objects in the context, while in [8], the models refer to maximal sequences of mutually distinguishable objects.

The concept of *contamination* was introduced in [9] and explored in [1,3,8] by the present author. It is always relative to the application context and can be read as a realization of the meta principle *models should avoid making assumptions or simplifications that are not actualized in the application context in the contexts of human reasoning (or reasoning that involves causality as in human reasoning)*. A model is contaminated if and only if it does not satisfy the principle. Because of its focus on human reasoning (or reasoning that involves causality as in human reasoning), the problem of avoiding contamination may not always be important or may be solved in much weaker senses in specific application contexts of rough sets. For example, while computing attribute reducts of high dimensional noisy data, it may be more relevant to focus on quality of classification (especially when few preferences among attributes can be indicated or derived). On the other hand, while approximately designing the most tasty food for tigers under resource constraints, the addition of sodium glutamate and pepper to red meat (based on the experiences of non-vegetarian humans that possess far more sophisticated sense of taste) is not a good idea – in this scenario the approximations of *tasty food* are contaminated. Contamination may also be due to operations used in constructing approximations and rough objects [10].

Contamination avoidance is associated with a distinct minimalist approach that takes the semantic domains involved into account and has the potential to encompass the three principles of non-intrusive analysis. Some sources of contamination are those arising from assumptions about distribution of attributes, introduction of assumptions valid in one semantic domain into another by oversight [10], numeric functions used in rough sets (and soft computing in general) and fuzzy representation of linguistic hedges. It is essential for modeling relation between attributes [1,6,11,12]. A Bayesian approach to modeling causality between attributes is proposed in [13] – the approach tries to avoid contamination to an extent.

For basics of partial algebras, see [14]. A *partial algebra* P is a tuple of the form $\langle \underline{P}, f_1, f_2, \ldots, f_n, (r_1, \ldots, r_n) \rangle$ with $\underline{P}$ being a set, f_i's being partial function symbols of arity r_i. The interpretation of f_i on the set $\underline{P}$ should be denoted by $f_i^{\underline{P}}$, but the superscript will be dropped in this paper as the application contexts are simple enough. If predicate symbols enter into the signature, then P is termed a *partial algebraic system*.

Terms are defined in the following way:

- All variable symbols are term symbols;
- If $t_1, \ldots t_{r_i}$ are term symbols, then $f_i(t_1, \ldots t_{r_i})$ is also a term symbol;
- Nothing else is a term symbol.

When a term symbol t is interpreted on the partial algebra, then it is formally denoted by $t^{\underline{P}}$ and referred to as a term. The distinction between the two will be left to the context in this paper.

For two terms s, t, $s \stackrel{\omega}{=} t$ shall mean, if both sides are defined then the two terms are equal (the quantification is implicit). $\stackrel{\omega}{=}$ is the same as the existence equality (also written as $\stackrel{e}{=}$) in the present paper. $s \stackrel{\omega^*}{=} t$ shall mean if either side is defined, then the other is and the two sides are equal (the quantification is implicit). Note that the latter equality can be defined in terms of the former as

$$(s \stackrel{\omega}{=} s \longrightarrow s \stackrel{\omega}{=} t) \& (t \stackrel{\omega}{=} t \longrightarrow s \stackrel{\omega}{=} t)$$

2 Mereology

Mereology is a collective term for a number of philosophical and formal theoretical approaches to *parts and wholes, connectedness of objects,* and variants thereof. Many of these approaches are not mutually compatible and so the discipline should be regarded as a plural one that is united by the goal to study parts and wholes [15,16].

Five distinct phases in the development of mereology (based on significant methodological differences) are *ancient, medieval, universal parthood related, early twentieth century and modern mereologies*. The subject of mereology is common to most ancient cultures and philosophical debates associated concern questions related to the universality of parthood, the whole being a sum or fusion of its parts and concepts of emptiness. Many of these debates have had significant impact on subsequent developments. Gradation of wholes into *strong, weak and weaker wholes*, for example, can be related to debates about no component (like wheels, poles and axle) of a chariot having the property of being a chariot. A whole in which the parts exist relative to the whole and are mutually dependent on the same is said to be *strong*, while a *weak whole* is one in which parts are less united. The concept of emptiness or the empty is complicated in most mereologies and is of ancient origin.

Some important principles that may be accepted in a specific theory are the existence of mereological atoms (entities with no proper parts), *atomistic compositionality* (everything is ultimately composed of atoms), *extensionality* (no two composite wholes can have the same proper parts), and the *principle of unrestricted composition* (any group of objects composes a whole).

A major difference between mereology and set theory is that the latter is committed to the existence of abstract entities such as empty sets and classes. In the former, the whole can be as concrete as the part is. The idea of empty set is inadmissible in Lesniewski's mereology, and ideally it should be studied over categories or in a formal language. In most of this tutorial, parthood will be explored over a set-theoretic framework with its associated dualism. While the sum of certain things is unique whenever it exists, at least three concepts of mereological fusion are known. The third definition of fusion is that a *fusion* of b's is a sum of at least some bs. Thus a fusion of tomatoes may be the sum of all bright red ovaloid tomatoes. Variants of the third definition are used in this exposition. The *fusion axiom* is the principle that fusion is unrestricted. That is the principle that every plurality of objects has at least one fusion – this is not assumed.

For *ground mereology*, in a first order language enhanced with quantifiers, the binary parthood predicate **P** is assumed to be reflexive, antisymmetric and transitive. Theories that start from this mereology almost always assume a lot more. *In the axiomatic approach to granules, transitivity is not always assumed.* So associated mereology is quite distinct. From a basic parthood predicate **P** (irrespective of assumptions), the following derived predicates and partial operations $\oplus$, $\cdot$, $\ominus$ can be defined (some conditions are omitted below):

Overlap: $\mathbb{O}xa \leftrightarrow (\exists z)\,\mathbf{P}zx \wedge \mathbf{P}za$
Proper Part: $\mathbb{P}xa \leftrightarrow \mathbf{P}xa \wedge \neg\mathbf{P}ax,$
Overcross: $\mathbb{X}xa \leftrightarrow \mathbb{O}xa \wedge \neg\mathbb{P}xa$
Proper Overlap: $\mathbb{O}xa \leftrightarrow \mathbb{X}xa \wedge \mathbb{X}ax,$
wDifference1: $(\forall x, a, z)(x \ominus a = z \rightarrow (\forall w)(\mathbf{P}wz \leftrightarrow (\mathbf{P}wx \wedge \neg\mathbb{O}wa)))$
Sum1: $(\forall x, y, z)(x \oplus y = z \rightarrow (\forall w)(\mathbb{O}wz \leftrightarrow (\mathbb{O}wx \vee \mathbb{O}wy)))$
Product1: $(\forall x, y, z)(x \odot y = z \rightarrow (\forall w)(\mathbf{P}wz \leftrightarrow (\mathbf{P}wx \wedge \mathbf{P}wy)))$

3   General Rough Sets and Granularity

General rough sets can be studied for different purposes from the perspective of AGCP, CGCP or non-granular perspectives and in many different ways. Ideas of granularity used in fuzzy sets (see [17]) in particular are not always compatible with those used in general rough sets. It can however be said that granules (or information granules) are basically collections sharing some properties relating to indiscernibility, similarity or functionality at some levels of discourse.

3.1   Granules and Granulations

A granule may be vaguely defined as some concrete or abstract realization of relatively simpler objects through the use of which more complex problems may be solved. They exist relative to the problem being solved in question. In the present author's view at least some of the basic ideas of granular computing have been in use since the dawn of human evolution. In earlier papers [1,3,18], she has shown that the methods can be classified into the PGCP, CGCP and AGCP. Adding adaptive aspects and other time related constraints (especially for handling interactive or emergent systems [19,20]) leads to additional categories. Because they have been considered in the perspective of CGCP, they may be regarded as extensions of the same. In all theories or theoretical understandings of granularity, the term *granules* refer to parts or building blocks of the computational process and *granulations* to collections of such granules in the context.

3.2   Primitive Granular Computing

Even in the available information on earliest human habitations and dwellings, it is possible to identify a primitive granular computing process at work. This can for example be seen from the stone houses, dating to 3500 BCE, used in what is present-day Scotland. Related details can be found in [1].

The main features of primitive granular computing are that

- requirements associated with the problem are not rigidly specified;
- both vague and precise granules (more often the former) may be used;
- not much formalization is involved in the specifications (historically these become more complicated in mereological approaches) and that has never been part of the goals;
- scope for abstraction is relatively limited and
- the concept of granules used may be concrete or abstract (relative to all materialist and extended materialist viewpoints), but may be barely constrained by rules.

While the method may be of ancient origin, it is still used in a number of modern contexts. The diet of people living in regions close to the sea depends on seasonal fluctuations in the production of fish and other foods. These dynamics can be understood in the perspective of PGCP [1].

3.3   Classical Granular Computing Paradigm

In the context of commercial painting, different parts of navigation indicators can be painted with brushes of different sizes. The artist involved may be able to use many distinct subsets of brushes to paint the sign based on choice of style, the time required to complete the sign and quality. The entire thinking process associated with the execution of the job can be viewed from a granular computing paradigm based on approximate precision as opposed to exact precision (see [1]). One possible granular strategy in the situation is the following:

- draw outline of sign using stencils;
- identify sub-regions from the finest to the broadest;
- make an initial selection of brushes;
- paint and check the progress (and quality) of work produced, and finally
- stop or repeat steps using more appropriate brush sizes.

The strategy used in the example falls under the classical granular computing because painting brushes have fixed size. It differs from PGCP in that the form of the sign was preconceived and the tools including brushes do not have a role in determining the conception of the product.

Security personnel, while opening the gates of a building for incoming or outgoing vehicular traffic proceed to open gates from a granular perspective of approximation of the size or width of the vehicle involved in question. Granules of varying precision may be used in the process as opposed to the kind of precision supposed in the previous example. This also suggests a different axiomatic framework being employed in the rough computation. The extent to which gates have already been opened at a particular instant also has a role in influencing subsequent moves. If switching between levels of granularity is done, then it can also be argued that the solution used falls under CGCP and not PGCP. Because adaptivity is understood from a higher order perspective and in relation to features falling outside precision, this may be read from such a viewpoint as well.

In [3], the precision based granular computing paradigm was traced to [21] and named as the *classical granular computing paradigm* CGCP by the present author. More correctly, it is also an ancient method that has been identified as such in [1] and elsewhere by her. CGCP is often referred to as the granular computing paradigm and has since been used in soft, fuzzy and rough set theories in different ways [22–26]. Some of the paradigm fragments involved in applying CGCP are:

- PF-1: Granules can exist at different levels of precision.
- PF-2: Among the many precision levels, a precision level at which the problem at hand is solvable should be selected.
- PF-3: Granulations (granules at specific levels or processes) form a hierarchy (later development).
- PF-4: It is possible to easily switch between precision levels.
- PF-5: The problem under investigation may be represented by the hierarchy of multiple levels of granulations.

CGCP is Ancient. Many approximation methods used in mathematical practice essentially use CGCP for solving problems. Examples range from those relating algorithms for approximating π to finding square roots of numbers. An ancient procedure of computing square roots is the Babylonian method. It is at least 2500 years old and is essentially the following:

Babylonian Method

- Problem: To compute $\sqrt{a}$, $a \in R_+$ to some desired level of accuracy (specified in relative or absolute terms).
- Initialization: Select an arbitrary value a_o close to $\sqrt{a}$.
- Recursion Step: $a_{n+1} = 0.5(a_n + \frac{a}{a_n})$ for $n \in N$
- Repeat previous step
- stop if desired accuracy is attained

The algorithm is quadratically convergent and *good initialization* is necessary for fast convergence. In other words some idea about possible approximate solutions is also essential. It is a special case of many other methods including the Newton-Raphson method and the modern Householder's method. In fact, in mathematical contexts, it is possible to indicate concepts of precision in a number of ways:

- Fixed values of initialization correspond to bounds on the precision of the solution at different cycles of computation.
- If the precision of the solution desired is alone fixed, then wide variation in initialization would be admissible.
- If the time required for computation is alone fixed or specified by an interval, then again wide variation in precision of initialization would be admissible.

This suggests the following problem: *Can CGCP be classified or graded relative to possible ways in which the precision can be categorized?*

3.4   Axiomatic Granular Computing Paradigm

The axiomatic approach to granularity essentially consists in investigations relating to axioms satisfied by granules, the very definitions of granules and associated frameworks. Emphasis on axiomatic properties of granules can be traced to papers [7,9,27] in the year 2007. That is, if some covers used in constructing approximations are overlooked. Neighborhoods had been investigated by a number of authors (see references in [3,26,28–30]) with emphasis on point-wise approximations. A systematic axiomatic approach to granules and granulations has been due to the present author in [3,9]. Relatively more specific versions of this approach have rich algebraic semantics associated. Parts of the axiomatic approach developed by the present author for general rough sets have been known in some form in implicit terms. But these were not stressed in a proper way because of the partial dominance of the point-wise approach.

The axiomatic approach to granularity initiated in [9] has been developed by the present author in the direction of contamination reduction in [1,3,8,10,12]. The concept of admissible granules, mentioned earlier, was arrived in the latter paper. From the order-theoretic algebraic point of view, the deviation is in a new direction relative to the precision-based paradigm. The paradigm shift includes a new approach to measures.

In the present author's classification, a rough approximation operator may be *granular (in the axiomatic sense), co-granular, pointwise, abstract or empirical* [31]. Most of the point-wise approximations in cover or relation-based approaches are co-granular. In cover based rough sets, three kinds of approximations are mentioned in [28]. Of these the subsystem based approximations would fall under the axiomatic granular approach and are not non granular. This is because in the approach, granulations are necessarily set-theoretically derived from covers (while the approximations remain a simple union of granules). By *empirical approximations* is meant a set of approximations that have been specified in a concrete empirical context. These may not necessarily be based on known processes or definite attributes. Examples of such approximations have been discussed by the present author in rough contexts in [3,32].

4   High Granular Operator Spaces and Variants

Abstract frameworks for the axiomatic approach called *rough Y-systems* (RYS) were introduced and studied by the present author in [3] and other papers. Granular operator spaces (and variants), investigated by the present author in [1,33,34] in particular, are simplifications and higher order variants of RYS. They are meant for both abstract and concrete approximations that are granular in nature in the sense of the axiomatic approach, and are well suited for investigating semantic questions, representation, ontology, formulation of semantics and the inverse problem. Other abstract approaches to rough sets without any restrictions on granularity, but with additional assumptions about order structure and negations as in [35] are less related. For the connection of the present approach to the numeric function based rough mereological approach [36] the reader may refer to [1,3,37].

In a *high general granular operator space* (GGS), defined below, aggregation and co-aggregation operations ($\vee$, $\wedge$) are conceptually separated from the binary parthood (**P**), and a basic partial order relation ($\leqslant$). Parthood is assumed to be reflexive and antisymmetric. It may satisfy additional generalized transitivity conditions in many contexts. Real-life information processing often involves many non-evaluated instances of aggregations (fusions), commonalities (conjunctions) and implications because of laziness or supporting meta data or for other reasons – this justifies the use of partial operations. Specific versions of a GGS and granular operator spaces have been studied in [1] by the present author for handling a large spectrum of rough set contexts. GGS has the ability to handle adaptive situations as in [38] through special morphisms – this is again harder to express without partial operations.

The underlying set $\underline{S}$ may be a set of collections of attributes, objects with or without labels or anything else. In practice, the set of all attributes in a context need not be known exactly to the reasoning agent constructing the approximations. The element $\top$ may be omitted in these situations or the issue can be managed through restrictions on the granulation. Also, *it often happens that certain objects cannot be approximated in an acceptable way.* Therefore, it can be argued that the approximations operations used should be partial. Related abstractions (Pre-GGS) are not discussed in this tutorial.

Definition 1. *A High General Granular Operator Space (GGS) $\mathbb{S}$ shall be a partial algebraic system of the form $\mathbb{S} = \langle \underline{S}, \gamma, l, u, \mathbf{P}, \leqslant, \vee, \wedge, \bot, \top \rangle$ with $\underline{S}$ being a set, γ being a unary predicate that determines $\mathcal{G}$ (by the condition γx if and only if $x \in \mathcal{G}$) an* admissible granulation*(defined below) for $\mathbb{S}$ and l, u being operators : $\underline{S} \longmapsto \underline{S}$ satisfying the following ($\underline{S}$ is replaced with $\mathbb{S}$ if clear from the context. $\vee$ and $\wedge$ are idempotent partial operations and $\mathbf{P}$ is a binary predicate. Further γx will be replaced by $x \in \mathcal{G}$ for convenience.):*

$$(\forall x)\mathbf{P}xx \tag{PT1}$$

$$(\forall x, b)(\mathbf{P}xb \ \& \ \mathbf{P}bx \longrightarrow x = b) \tag{PT2}$$

$$(\forall a, b)a \vee b \stackrel{\omega}{=} b \vee a \ ; \ (\forall a, b)a \wedge b \stackrel{\omega}{=} b \wedge a \tag{G1}$$

$$(\forall a, b)(a \vee b) \wedge a \stackrel{\omega}{=} a \ ; \ (\forall a, b)(a \wedge b) \vee a \stackrel{\omega}{=} a \tag{G2}$$

$$(\forall a, b, c)(a \wedge b) \vee c \stackrel{\omega}{=} (a \vee c) \wedge (b \vee c) \tag{G3}$$

$$(\forall a, b, c)(a \vee b) \wedge c \stackrel{\omega}{=} (a \wedge c) \vee (b \wedge c) \tag{G4}$$

$$(\forall a, b)(a \leqslant b \leftrightarrow a \vee b = b \ \leftrightarrow \ a \wedge b = a) \tag{G5}$$

$$(\forall a \in S)\mathbf{P}a^l a \ \& \ a^{ll} = a^l \ \& \ \mathbf{P}a^u a^{uu} \tag{UL1}$$

$$(\forall a, b \in S)(\mathbf{P}ab \longrightarrow \mathbf{P}a^l b^l \ \& \ \mathbf{P}a^u b^u) \tag{UL2}$$

$$\bot^l = \bot \ \& \ \bot^u = \bot \ \& \ \mathbf{P}\top^l \top \ \& \ \mathbf{P}\top^u \top \tag{UL3}$$

$$(\forall a \in S)\mathbf{P}\bot a \ \& \ \mathbf{P}a\top \tag{TB}$$

Let $\mathbb{P}$ stand for proper parthood, defined via $\mathbb{P}ab$ if and only if $\mathbf{P}ab \ \& \ \neg\mathbf{P}ba$). A granulation is said to be admissible if there exists a term operation t formed from the weak lattice operations such that the following three conditions hold:

$$(\forall x \exists x_1, \dots x_r \in \mathcal{G}) \, t(x_1, x_2, \dots x_r) = x^l$$

$$\text{and } (\forall x) \, (\exists x_1, \dots x_r \in \mathcal{G}) \, t(x_1, x_2, \dots x_r) = x^u, \qquad \text{(Weak RA, WRA)}$$

$$(\forall a \in \mathcal{G})(\forall x \in \underline{S}) \, (\mathbf{P}ax \longrightarrow \mathbf{P}ax^l), \qquad \text{(Lower Stability, LS)}$$

$$(\forall x, a \in \mathcal{G})(\exists z \in \underline{S})) \, \mathbb{P}xz, \ \& \, \mathbb{P}az \ \& \ z^l = z^u = z, \qquad \text{(Full Underlap, FU)}$$

The conditions defining admissible granulations mean that every approximation is somehow representable by granules in a algebraic way, that every granule coincides with its lower approximation (granules are lower definite), and that all pairs of distinct granules are part of definite objects (those that coincide with their own lower and upper approximations). Special cases of the above are defined next.

Definition 2

- *In a* **GGS**, *if the parthood is defined by* **P**ab *if and only if* a ⩽ b *then the* **GGS** *is said to be a* high granular operator space **GS**.
- *A higher granular operator space (*HGOS*)* $\mathbb{S}$ *is a* **GS** *in which the lattice operations are total.*
- *In a higher granular operator space, if the lattice operations are set theoretic union and intersection, then the* **HGOS** *will be said to be a* set HGOS.

In [39], it is shown that the binary predicates can be replaced by partial two-place operations and γ is replaceable by a total unary operation. The results in a semantically equivalent partial algebra called a *high granular operator partial algebra* (GGSp).

Example 1. Suppose the problem at hand is to represent the knowledge of a specialist in automobile engineering and production lines in relation to a database of cars, car parts, calibrated motion videos of cars and performance statistics. The database is known to include a number of experimental car models and some sets of cars have model names, or engines or other crucial characteristics associated. Let $\mathbb{S}$ be the set of cars, some subsets of cars, sets of internal parts and components of many cars. $\mathcal{G}$ be the set of internal parts and components of many cars. Further let

- **P**ab express the relation that a is a possible component of b or that a belongs to the set of cars indicated by b or that
- a ⩽ b indicate that b is a better car than a relative to a certain fixed set of features,
- a^l indicate the closest standard car model whose features are all included in a or set of components that are included in a,
- a^u indicate the closest standard car model whose features are all included by a or fusion of set of components that include a
- ∨, ∧ can be defined as partial operations, while ⊥ and ⊤ can be specified in terms of attributes.

Under the conditions, $\mathbb{S} = \langle \underline{\mathbb{S}}, \mathcal{G}, l, u, \mathbf{P}, \leqslant, \vee, \wedge, \perp, \top \rangle$ forms a **GGS**. If the specialist has updated her knowledge over time, then this transformation can be expressed with the help of morphisms from a **GGS** to itself.

Granular operator spaces and variants (specifically high granular operator spaces) adhere to the weak definitions of granularity as per the axiomatic granular approach, do not assume a negation operation, their universe may be a collection of rough objects (in some sense), or a mix of rough and non rough objects or even a collection of all objects involved, the sense of parthood between objects is assumed to be distinct from other order relations, permit realistic partial aggregation and commonality operations, and numeric simplified measures are not assumed in general. These features are motivated by properties satisfied by models in real reasoning contexts, and help in avoiding contamination to a substantial extent.

4.1   Granularity Axioms

Even when additional lower and upper approximation operators are added to a GGS, the resulting framework will still be referred to as a GGS. In such a framework, granules definitely satisfy some of the following list of axioms (that are not assumed to be exhaustive). It is assumed that a finite number of lower ($\{l_i\}_{i=1}^n$) and upper ($\{u_i\}_{i=1}^n$) approximations are used. These have been grouped based on their role relative to approximations and ontology, and are known to have a central role in defining possible concepts of granules. For readability, the interpretations of the predicate γ are written out explicitly.

Representation Related Axioms. The central idea expressed by these axioms is that approximations are formed from granules through set theoretic or more general operations on granules that may be derived from the parthood relation used. In classical rough sets, every approximation is a union of equivalence classes (the granules). If $+$ is an aggregation operation (possibly related to the parthood used)

$$\forall i, \; (\forall x)(\exists a_1, \ldots a_r \in \mathcal{G}) \, a_1 + a_2 + \ldots + a_r = x^{l_i} \text{ and}$$
$$(\forall x)(\exists a_1, \ldots a_p \in \mathcal{G}) \, a_1 + a_2 + \ldots + a_p = x^{u_i} \quad \text{(Representability, RA)}$$

In the weaker versions below, approximations are assumed to be representable by derived terms instead of through aggregation of granules.

$$\forall i, \; (\forall x \exists a_1, \ldots a_r \in \mathcal{G}) \, t_i(a_1, a_2, \ldots a_r) = x^{l_i} \text{ and}$$
$$(\forall x)(\exists a_1, \ldots a_r \in \mathcal{G}) \, t_i(a_1, a_2, \ldots a_p) = x^{u_i} \quad \text{(Weak RA, WRA)}$$

The prefix *sub* as in Sub RA is used to indicate situations, where only a subset of approximations happen to be representable.

Crispness Axioms. As indicated before *an object is* crisp *in a sense if it is its own approximation in that sense.* This is quite different from claiming that *crisp objects are those that cannot be approximated by any other object.* While crispness of granules is not a given, they may possibly satisfy the following crispness axioms:

$$\text{For each } i, \; (\forall a \in \mathcal{G}) \, a^{l_i} = a^{u_i} = a \quad \text{(Absolute Crispness, ACG)}$$

Crispness Variants: By analogy, the crispness variants sub crispness (SCG), lower absolute crispness (LACG), upper absolute crispness (UACG), lower sub crisness (LSCG), and upper sub crispness (USCG) can be defined as for representation related axioms.

Mereological Axioms. The axioms for mereological properties of granules is presented next. The axiom of *mereological atomicity* says that no definite elements (relative to any permitted pair of lower and upper approximations) can be proper parts of granules.

$$\forall i, (\forall a \in \mathcal{G})(\forall x \in S)(\mathbf{P}xa, x^{l_i} = x^{u_i} = x \longrightarrow x = a)$$
$$\text{(Mereological Atomicity, MER)}$$

The axiom of *sub-mereological atomicity* says that no definite elements (relative to at least one specific pair of lower and upper approximations) can be proper parts of granules, while the axiom of *inward-mereological atomicity* says that no definite elements (relative to every permitted pair of lower and upper approximations) can be proper parts of granules.

$$(\forall a \in \mathcal{G})(\forall x \in S)(\mathbf{P}xa, \bigwedge_i (x^{l_i} = x^{u_i} = x) \longrightarrow x = a) \quad \text{(Inward MER, IMER)}$$

Stability Axioms. Stability of granules is that granules should preserve appropriate parthood relations relative to approximations. *Lower stability*, defined below, says that if a granule is part of an object, then the granule should still be part of the lower approximation of the object. In general, the same does not hold for all objects. Some stability axioms are

$$\forall i, (\forall a \in \mathcal{G})(\forall x \in S)(\mathbf{P}ax \longrightarrow \mathbf{P}(a)(x^{l_i})) \qquad \text{(Lower Stability, LS)}$$

$$\forall i, (\forall a \in \mathcal{G})(\forall x \in S)(\mathbf{O}ax \longrightarrow \mathbf{P}ax^{u_i}) \qquad \text{(Upper Stability, US)}$$

$$\text{LS \& US} \qquad \text{(Stability, ST)}$$

Overlap Axioms. The possible implications of the mereological overlap and underlap relations between granules is captured by these axioms. Some of these are

$$(\forall x, a \in \mathcal{G})\neg \mathbb{O}xa, \qquad \text{(No Overlap, NO)}$$

$$\forall i, (\forall x, a \in \mathcal{G})(\exists z \in S)\mathbb{P}xz, \mathbb{P}az, z^{l_i} = z^{u_i} = z \qquad \text{(Full Underlap, FU)}$$

$$\forall i, (\forall x, a \in \mathcal{G})(\exists z \in S)\mathbb{P}xz, \mathbb{P}az, z^{l_i} = z \qquad \text{(LU)}$$

Idempotence Axioms. Idempotence of approximation operators relative to granules are indicated by axioms such as

$$\forall i, (\forall x \in \mathcal{G}) x^{l_i} = x^{l_i l_i} \qquad \text{(l-Idempotence, LI)}$$

The *pre-similarity* axiom concerns the relation of commonalities between granules and parthood. It is redundant for classical rough sets with granules being the equivalence relations.

$$(\forall x, a \in \mathcal{G})(\exists z \in \mathcal{G}) \mathbf{P}(x \cdot a)(z) \qquad \text{(Pre-similarity, PS)}$$

Apparently the three axioms WRA, LS, LU hold in most of the known theories and with most choices of granules. This has been the main motivation for the definition of admissibility of a subset to be regarded as a granule in [3] and in the definition of GGS.

4.2   Specific Cases

Few examples that partially justify the formalism of the axioms are presented next. More details can be found in [1,3]. Let $S = \langle \underline{S}, R \rangle$ be a general approximation space, with granulation being $\mathcal{G}$ - the set of successor neighborhoods and

$$A^l = \cup\{g : g \subseteq A, g \in \mathcal{G}\}$$
$$A^u = \cup\{g : g \cap A \neq \emptyset \, g \in \mathcal{G}\}.$$

Theorem 1

- *If R is an equivalence, then all of* RA, ACG, MER, AS, FU, NO, PS, I, ST *hold, but* UU *does not hold in general.*
- *If R is a partial equivalence relation (symmetric, transitive and partially reflexive relation),* RA, MER, NO, UU, US *hold, but* ACG *may not.*
- *If R is a reflexive relation, then* RA, LFU *holds, but none of* MER, ACG, LI, UI, NO, FU *holds in general.*

Let $\langle S, (R_i)_i \in K \rangle$ be a multiple approximation space [40], then apart from the strong lower, weak lower, strong upper and weak upper approximations discussed in the paper a hierarchy of approximations can be defined and related properties can be studied [1].

In the perspective of the axiomatic approach, the next definition is natural:

Definition 3. *A specific mathematical approach to relation-based rough set is granular only if it can be rewritten in the form of a general granular operator space or a higher order granular operator space satisfying additional conditions.*

Some representation theorems that connect GGS with general approximation spaces are known and more are of natural interest [1].

5 Knowledge Representation and Granularity

From a theory of knowledge and application perspective, it is of much interest to study definitions, representation, ontology and relative consistency of knowledge among other things. Ontological correspondences between knowledge in different contexts, and problems of conflict representation and resolution are also of interest. The framework of high granular operator spaces (and partial algebras) can represent knowledge in a far more substantial way than is afforded by non granular extensions of the situation in classical rough sets. More so because it is easily extensible with ontology.

In classical rough sets, if $S = \langle \underline{S}, R \rangle$ is an approximation space, then approximations of subsets of $\underline{S}$ the form A^l and A^u represent clear and definite concepts [41]. Further every equivalence class interpreted as a granule is definite. R in this perspective encodes knowledge by way of the distribution of definite objects. If Q is another stronger equivalence ($Q \subseteq P$) on $\underline{S}$, then the state of the knowledge encoded by $\langle \underline{S}, Q \rangle$ is a *refinement* of that of $S = \langle \underline{S}, P \rangle$. Subsequent work on logics and semantics for comparing different types of knowledge and measures of relative consistency can be found in [42–44] and elsewhere.

This knowledge interpretation has been extended in a natural granular way to general approximation spaces by the present author in [9,11]. In [9], choice operations are used over granules in the context tolerances spaces for the construction of *definite* objects that correspond to clear concepts or beliefs with ontology. The upper approximation of an object may be a proper part of the upper approximation of the upper approximation of the same object in prototransitive rough sets considered in [11]. This itself has an impact on the granular axioms satisfied.

In general some axioms of interest are

K1 All Granules are atomic units of knowledge.
K2 Knowledge is characterized by granules.
K3 Maximal collections of granules subject to a concept of mutual independence are admissible concepts of knowledge.
K4 Parts common to subcollections of maximal collections of granules are also knowledge.
K5 Knowledge K_1 is *fully consistent* with another knowledge K_2 if and only if both generate the same granules.
K6 Knowledge K_1 is *fully inconsistent* with another knowledge K_2 if and only if no granule of one is included in a granule of the other.
K7 Some Granules are atomic units of knowledge.
K8 Every atomic unit of knowledge is a granule.
K9 Some collections of granules form a consistent unit of knowledge.

These axioms are not necessarily true in every context and stand to benefit much from additional ontologies that can specify rules of combination. This in turn makes the different semantic models that generalize high granular operator spaces (and partial algebras) all the more relevant [1,39,45].

References

1. Mani, A.: Algebraic methods for granular rough sets. In: Mani, A., Cattaneo, G., Düntsch, I. (eds.) Algebraic Methods in General Rough Sets. TM, pp. 157–335. Springer, Cham (2018). https://doi.org/10.1007/978-3-030-01162-8_3
2. Lin, T.Y.: Neighbourhood systems- applications to qualitative fuzzy and rough sets. In: Wang, P.P., et al. (eds.) Advances in Machine Intelligence and Soft Computing, Durham 1997, pp. 132–155. Duke University (1997)
3. Mani, A.: Dialectics of counting and the mathematics of vagueness. In: Peters, J.F., Skowron, A. (eds.) Transactions on Rough Sets XV. LNCS, vol. 7255, pp. 122–180. Springer, Heidelberg (2012). https://doi.org/10.1007/978-3-642-31903-7_4
4. Werner, K.: Enactment and construction of the cognitive niche: toward an ontology of the mind-world connection. Synthese **197**, 1313–1341 (2020)
5. Chandrasekharan, S., Stewart, T.C.: The origin of epistemic structures and proto-representations. Adapt. Behav. **15**(3), 329–353 (2007)
6. Mani, A.: Dialectical rough sets, parthood and figures of opposition-I. In: Peters, J.F., Skowron, A. (eds.) Transactions on Rough Sets XXI. LNCS, vol. 10810, pp. 96–141. Springer, Heidelberg (2019). https://doi.org/10.1007/978-3-662-58768-3_4
7. Mani, A.: Algebraic semantics of similarity-based bitten rough set theory. Fundam. Inf. **97**(1–2), 177–197 (2009)
8. Mani, A.: Knowledge and consequence in AC semantics for general rough sets. In: Wang, G., Skowron, A., Yao, Y., Ślęzak, D., Polkowski, L. (eds.) Thriving Rough Sets. SCI, vol. 708, pp. 237–268. Springer, Cham (2017). https://doi.org/10.1007/978-3-319-54966-8_12
9. Mani, A.: Choice inclusive general rough semantics. Inf. Sci. **181**(6), 1097–1115 (2011)
10. Mani, A.: Contamination-free measures and algebraic operations. In: 2013 IEEE International Conference on Fuzzy Systems (FUZZ), pp. 1–8. IEEE (2013)
11. Mani, A.: Algebraic semantics of proto-transitive rough sets. In: Peters, J.F., Skowron, A. (eds.) Transactions on Rough Sets XX. LNCS, vol. 10020, pp. 51–108. Springer, Heidelberg (2016). https://doi.org/10.1007/978-3-662-53611-7_3
12. Mani, A.: Ontology, rough Y-systems and dependence. Int. J. Comput. Sci. Appl. **11**(2), 114–136 (2014). Special Issue of IJCSA on Computational Intelligence
13. Yao, N., Miao, D., Pedrycz, W., Zhang, H., Zhang, Z.: Causality measures and analysis: a rough set framework. Expert Syst. Appl. **136**, 187–200 (2019)
14. Burmeister, P.: A Model-Theoretic Oriented Approach to Partial Algebras. Akademie-Verlag (1986, 2002)
15. Burkhardt, H., Seibt, J., Imaguire, G., Gerogiorgakis, S. (eds.): Handbook of Mereology. Philosophia Verlag, Munich (2017)
16. Gruszczyński, R., Varzi, A.: Mereology then and now. Log. Log. Philos. **24**, 409–427 (2015)
17. Zadeh, L.A.: Toward a theory of fuzzy information granulation and its centrality in human reasoning and fuzzy logic. Fuzzy Sets Syst. **90**(2), 111–127 (1997)
18. Mani, A.: Approximation dialectics of proto-transitive rough sets. In: Chakraborty, M.K., Skowron, A., Maiti, M., Kar, S. (eds.) Facets of Uncertainties and Applications. SPMS, vol. 125, pp. 99–109. Springer, New Delhi (2015). https://doi.org/10.1007/978-81-322-2301-6_8
19. Jankowski, A., Skowron, A., Swiniarski, R.: Interactive complex granules. Fundam. Inf. **133**(2–3), 181–196 (2014)
20. Skowron, A., Stepaniuk, J., Swiniarski, R.: Modeling rough granular computing based on approximation spaces. Inf. Sci. **184**, 20–43 (2012)

21. Moore, E.F., Shannon, C.E.: Reliable circuits using less reliable relays-I, II. Bell Syst. Tech. J. **191–208**, 281–297 (1956)
22. Zadeh, L.A.: Fuzzy sets and information granularity. In: Gupta, N., et al. (eds.) Advances in Fuzzy Set Theory and Applications, Amsterdam, North Holland, pp. 3–18 (1979)
23. Yao, Y.Y.: Information granulation and rough set approximation. Int. J. Intell. Syst. **16**, 87–104 (2001)
24. Yao, Y.: The art of granular computing. In: Kryszkiewicz, M., Peters, J.F., Rybinski, H., Skowron, A. (eds.) RSEISP 2007. LNCS (LNAI), vol. 4585, pp. 101–112. Springer, Heidelberg (2007). https://doi.org/10.1007/978-3-540-73451-2_12
25. Liu, G.: The axiomatization of the rough set upper approximation operations. Fundam. Inf. **69**(23), 331–342 (2006)
26. Lin, T.Y.: Granular computing-1: the concept of granulation and its formal model. Int. J. Granular Comput. Rough Sets Int. Syst. **1**(1), 21–42 (2009)
27. Wasilewski, P., Ślęzak, D.: Foundations of rough sets from vagueness perspective. In: Hassanien, A., et al. (eds.) Rough Computing: Theories, Technologies and Applications. Information Science Reference, pp. 1–37. IGI Global (2008)
28. Yao, Y.Y., Yao, B.X.: Covering based rough set approximations. Inf. Sci. **200**, 91–107 (2012)
29. D'eer, L., Restrepo, M., Cornelis, C., Gómez, J.: Neighborhood operators for covering-based rough sets. Inf. Sci. **336**, 21–44 (2016)
30. Samanta, P., Chakraborty, M.K.: Interface of rough set systems and modal logics: a survey. In: Peters, J.F., Skowron, A., Ślęzak, D., Nguyen, H.S., Bazan, J.G. (eds.) Transactions on Rough Sets XIX. LNCS, vol. 8988, pp. 114–137. Springer, Heidelberg (2015). https://doi.org/10.1007/978-3-662-47815-8_8
31. Mani, A.: Generalized ideals and co-granular rough sets. In: Polkowski, L., et al. (eds.) IJCRS 2017. LNCS (LNAI), vol. 10314, pp. 23–42. Springer, Cham (2017). https://doi.org/10.1007/978-3-319-60840-2_2
32. Mani, A.: Approximations from anywhere and general rough sets. In: Polkowski, L., et al. (eds.) IJCRS 2017. LNCS (LNAI), vol. 10314, pp. 3–22. Springer, Cham (2017). https://doi.org/10.1007/978-3-319-60840-2_1
33. Mani, A.: Antichain based semantics for rough sets. In: Ciucci, D., Wang, G., Mitra, S., Wu, W.-Z. (eds.) RSKT 2015. LNCS (LNAI), vol. 9436, pp. 335–346. Springer, Cham (2015). https://doi.org/10.1007/978-3-319-25754-9_30
34. Mani, A.: Probabilities, dependence and rough membership functions. Int. J. Comput. Appl. **39**(1), 17–35 (2016)
35. Cattaneo, G., Ciucci, D.: Algebraic methods for orthopairs and induced rough approximation spaces. In: Mani, A., Düntsch, I., Cattaneo, G. (eds.) Algebraic Methods in General Rough Sets, pp. 553–640. Birkhauser, Basel (2018)
36. Polkowski, L.: Approximate Reasoning by Parts. Springer, Heidelberg (2011). https://doi.org/10.1007/978-3-642-22279-5
37. Mani, A.: High granular operator spaces and less-contaminated general rough mereologies. Forthcoming 1–77 (2019)
38. Skowron, A., Jankowski, A., Dutta, S.: Interactive granular computing. Granular Comput. **1**(2), 95–113 (2016)
39. Mani, A.: Functional extensions of knowledge representation in general rough sets. In Bello, R., et al. (eds.) IJCRS 2020. LNAI, vol. 11499, pp. 19–34. Springer, Heidelberg (2020)
40. Khan, M.A., Banerjee, M.: Formal reasoning with rough sets in multiple-source approximation spaces. Int. J. Approx. Reason. **49**, 466–477 (2008)

41. Pawlak, Z.: Rough Sets: Theoretical Aspects of Reasoning About Data. Kluwer Academic Publishers, Dodrecht (1991)

42. Mani, A.: Granular foundations of the mathematics of vagueness, algebraic semantics and knowledge interpretation. University of Calcutta (2016)

43. Mani, A.: Towards logics of some rough perspectives of knowledge. In: Suraj, Z., Skowron, A. (eds.) Rough Sets and Intelligent Systems - Professor Zdzisław Pawlak in Memoriam. Intelligent Systems Reference Library, vol. 43, pp. 419–444. Springer, Heidelberg (2013). https://doi.org/10.1007/978-3-642-30341-8_22

44. Samanta, P., Chakraborty, M.K.: On extension of dependency and consistency degrees of two knowledges represented by covering. In: Peters, J.F., Skowron, A., Rybiński, H. (eds.) Transactions on Rough Sets IX. LNCS, vol. 5390, pp. 351–364. Springer, Heidelberg (2008). https://doi.org/10.1007/978-3-540-89876-4_19

45. Mani, A.: Towards student centric rough concept inventories. In: Bello, R., et al. (eds.) IJCRS 2020. LNAI, vol. 11499, pp. 251–266. Springer, Heidelberg (2020)

Author Index